NANOCOMPOSITES AND NANOMATERIALS IN BIOMEDICAL APPLICATIONS

NANOCOMPOSITES AND NANOMATERIALS IN BIOMEDICAL APPLICATIONS

Vivek Dave
Deependra Singh
Pallerla Naveen Reddy
Editors

APPLE ACADEMIC PRESS

First edition published 2026

Apple Academic Press Inc.
1265 Goldenrod Circle, NE,
Palm Bay, FL 32905 USA

760 Laurentian Drive, Unit 19,
Burlington, ON L7N 0A4, Canada

CRC Press
2385 NW Executive Center Drive,
Suite 320, Boca Raton FL 33431

4 Park Square, Milton Park,
Abingdon, Oxon, OX14 4RN, UK

Library and Archives Canada Cataloguing in Publication

CIP data on file with Canada Library and Archives

Library of Congress Cataloging-in-Publication Data

CIP data on file with US Library of Congress

ISBN: 978-1-779641-22-9 (hbk)
ISBN: 978-1-779641-23-6 (ebk)
DOI: 10.1201/9781779641236

About the Editors

Dr. Vivek Dave

Vivek Dave, PhD, is presently working as a Professor and Head of the Department of Pharmacy, and Dean of the School of Health Science at Central University South Bihar, Gaya, Bihar, India, where he played a crucial role in establishing the department as the founder. He has commendable experience, coupled with a sound theoretical background. He is also associated with various university committees like EC, AC, the university library purchase committee, the research innovation committee, etc. In addition, he acts as the chairperson of BOS, BOE, DRDC, DPC, FPAC, and so forth. He has more than 15 years of experience working in the field of novel drug delivery systems and nanotechnology and has already published several articles in journals of national/international repute. He has also recorded his expertise in the form of many chapters and books. He is the recipient of several awards at various conferences, seminars, and conventions, including a prestigious award of "Best Sustainability" for smart shoe sole at the International Conference on Future Healthcare and Economic Development by the Medical Device Innovation Center (MDIC), National Cheng Kung University, Taiwan, in 2019. His name was listed in the 2023 world's top 2% scientists published by Elsevier (Stanford). Dr. Dave is a regular reviewer for various national and international reputed journals of different publishing houses, namely Wiley, Springer, Elsevier, BMC, etc. His expertise in working with the formulation and development of novel drug delivery systems, like PEGylated liposomes, ethosomes, nanoparticles, and quantum dots, has led to high-impact research articles and book publications of international repute. Currently, he is working on nano-formulation for the therapeutic management of various diseases like Alzheimer's, inflammation, cancer, etc.

Dr. Deependra Singh

Deependra Singh, PhD is presently working as Professor at the University Institute of Pharmacy, Pt. Ravishankar Shukla University, Raipur, Chhattisgarh, India. He is a central council member and chairman of the education regulation committee, Pharmacy Council of India. He has received the highest acclaim as being featured among the top 2% scientists across the world in the field of pharmacy and pharmacology from 2020 to date by Stanford University, USA, Elsevier. He has around 20 years of teaching and research experience. His prime research area includes formulation of multiple bioactive delivery systems like multiphase hydrogels, nano¬systems, microspheres for wound management, advanced statistical and mathematical modeling methods, and designing novel systems for plant actives and extracts. He has more than 150 research and review papers, 37 book chapters published in national and international journals, and two books published (Elsevier). He has successfully completed projects funded by UGC, AICTE, DST, CCOST and several other industrial projects. He has citations of more than 9000, with an h-index of 37 and i10 index of 183 (Google Scholar) and Scopus h-index 33, with 4164 citations. He has two international patents, two Indian patents, and two filed. He is the recipient of the prestigious APTI CJ Shishoo Award for research in pharmaceutical sciences 2023, the Department of Health Research-ICMR Young Indian Biomedical Scientist Award 2019, UGC RAMAN International Award 2014, and UGC Research Award 2012 by the University Grants Commission, New Delhi, India. He has also won a number of prestigious awards at the state level for excellence in research, namely the Young Scientist Award of the state of Chhattisgarh 2006 in medical sciences. He is a state evaluator and trainee for the National Children Science Congress. He is a life member of various professional bodies such as Indian Pharmaceutical Association (IPA), Association of Pharmacy Teachers of India (APTI), Indian Society of Pharmacognosy (ISP) and serving as vice president, central zone APTI (elected); vice president, central zone IPGA (elected); secretary, Chhattisgarh state branch IPA. He was a visiting scientist at the National Center for Natural Products Research, School of Pharmacy, University of Mississippi, Mississippi, USA.

Mr. Pallerla Naveen Reddy

Naveen Reddy received his MPharm degree from Central University of South Bihar, Gaya, India, securing first position in all semesters and was also awarded a gold medal for academic excellence. Recently, he worked as a trainee research associate at Hetero Labs, India, on various ANDA, 505(b)(2), complex and simple ophthalmic formulations at the research and development and plant level. He also worked as a postgraduate research trainee at the Translational Health Science and Technology Institute, Faridabad, Haryana, India, on identifying the role of various inflammasomes and drug screening for non-alcoholic fatty liver disease and lung inflammation. He is skilled in handling different formulation-related equipment, molecular techniques, and cell culture methods. He carried out his dissertation work in the field of formulations and evaluation of diclofenac orodispersable tablets. He participated in laboratory animal techniques for the preclinical research workshop sponsored by SERB at CSIR-IIIM, and he had also actively participated in different national and international conferences. His expertise in formulations, molecular techniques, cell biology, and translation of products from research and development to production level proves his expertise in biomedical research. His current research interests are mainly focused on the 3D printing of medicines, nanomaterials, nanocomposites, and various drug delivery systems.

Contents

Contributors ... *xi*

Abbreviations ... *xvii*

Foreword by Prof. Giorgia Pastorin ... *xxiii*

Foreword Prof. Christopher R. McCurdy ... *xxv*

Preface ... *xxvii*

1. **Introduction and Mechanisms of Various Nanocomposites and Their Advantages and Disadvantages ... 1**
   Bhavani Boddeda, Akshay M. Baheti, Anil T. Pawar, Vani Setty, and Chandrasekar S. B.

2. **Different Types of Nanocomposites Used in Biomedical Applications ... 39**
   Abhishek Singh, Seema Yadav, Narhari N. Palei, Mouni Roy, Rabinarayan Parhi, Biswa Mohan Sahoo, and Bibhash Chandra Mohanta

3. **Basic and Modern Methods for the Preparation of Nanocomposites for Biomedical Applications ... 85**
   Reham S. Elezaby

4. **Drug Delivery Systems by Using Various Nanocomposites ... 111**
   Srijita Sen, Puja Kumari, Rashmi Sinha, Anjali Giri, and Om Prakash Ranjan

5. **Nanocomposites from Recycled Waste Materials and Their Utilization in Biomedical Fields ... 141**
   Rafwana Ibrahim, Jesil Mathew, and Shaila Angela Lewis

6. **Role and Applications of Biodegradable Nanocomposites vs. Nondegradable Nanocomposites in Biomedical Applications ... 165**
   Indhumathi Thirugnanasambandham, Parikshit Roychowdhury, Roshan P. Rao, Ashok Kumar Janakiraman, Muralikrishnan Dhanasekaran, Sachin Kumar Singh, and Gowthamarajan Kuppusamy

7. **Nanocomposites Applications in Pharmaceutical Medicines and Biotechnological Products ... 221**
   Lokesh Nama, Khushboo Choudhary, Krishna Murti, and Nitesh Kumar

8. Impact of Nanocomposites on Pollutants Prone to Ecosystem and Human Health.. 249
Preethi Addala, Pallerla Naveen Reddy, Jyothi Munjavarapu, and Deependra Singh

9. Application of Nanocomposites for Various Diseases.......................... 285
Kajal Kumari, Tripti Paliwal, Shruti Richa, Kanika Verma, and Swapnil Sharma

10. Applications of Nanocomposites for Cancer Therapy......................... 311
Ajay Singh, Bikram Singh, Ruhi Singh, Kuljit Singh, and Boobalan Gopu

11. Application of Nanocomposites in 3D and 4D Printing of Nanomaterials in Biomedical Research 341
Pallerla Naveen Reddy, Deependra Singh, Vivek Dave, and Ruchi Tandon

12. Biomedical Applications of Various Herbal Nanocomposites............ 373
Sumaiya Qasim and Surajpal Verma

13. Applications of Various Nanocomposites Used for Antimicrobial Activity, Tissue Engineering, Biosensors, and Dental......................... 397
Diksha Kumari, Parampreet Kour, Vishwani Jamwal, Boobalan Gopu, and Kuljit Singh

14. Future Aspects and Trends of Nanocomposites in Various Biomedical Applications.. 429
Priya Jha, Vivek Dave, Vaibhav Verma, and Deependra Singh

15. India, US, and Other Foreign Regulatory Guidelines for Nanocomposites.. 469
Mohammad Kashif Iqubal, Swaroop J. Pansare, Rizwan Shaikh, Bhanu P. Dongala, Sunil K. Thota, Naseem A Charoo, Mansoor A. Khan, and Ziyaur Rahman

Index*... *525

Contributors

Preethi Addala
School of Pharmaceutical Sciences and Technologies, Jawaharlal Nehru Technological University, Kakinada, Andhra Pradesh, India

Chandrasekar S.B.
Drugs Testing Laboratory, Drugs Control Department, Palace Road, Bangalore, Karnataka, India

Akshay M. Baheti
School of Health Sciences and Technology, Dr. Vishwanath Karad MIT World Peace University, Pune, Maharashtra, India

Bhavani Boddeda
Jawaharlal Nehru Technological University, Kakinada, Andhra Pradesh, India

Naseem A. Charoo
Adcan Pharma LLC, ICAD III, Mussafah, Abu Dhabi, UAE

Khushboo Choudhary
Department Of Pharmacology and Toxicology,
National Institute of Pharmaceutical Education and Research, Hajipur, Vaishali, Bihar, India

Vivek Dave
Department of Pharmacy, School of Health Sciences, Central University of South Bihar, Gaya, Bihar, India

Muralikrishnan Dhanasekaran
Department of Drug Discovery and Development, Harrison College of Pharmacy,
Auburn University Auburn, AL, USA

Bhanu P. Dongala
Department of Pharmaceutical Sciences, Irma Lerma Rangel School of Pharmacy,
Texas A&M Health Science Center, Texas A&M University, College Station, TX, USA

Reham S. Elezaby
Department of Pharmaceutics and Industrial Pharmacy, Faculty of Pharmacy, Ain Shams University, Cairo, Egypt

Anjali Giri
Department of Pharmaceutical Technology (Formulations),
National Institute of Pharmaceutical Education and Research (NIPER), Guwahati, Assam, India

Boobalan Gopu
Pharmacology Division, CSIR–Indian Institute of Integrative Medicine, Jammu,
Jammu and Kashmir, India
Academy of Scientific and Industrial Research (AcSIR), Ghaziabad, Uttar Pradesh, India

Rafwana Ibrahim
Department of Pharmaceutics, Manipal College of Pharmaceutical Sciences,
Manipal Academy of Higher Education, Manipal, Karnataka, India

Mohammad Kashif Iqubal
Department of Pharmaceutical Sciences, Irma Lerma Rangel School of Pharmacy,
Texas A&M Health Science Center, Texas A&M University, College Station, TX, USA

Vishwani Jamwal
Infectious Diseases Division, CSIR–Indian Institute of Integrative Medicine, Jammu,
Jammu and Kashmir, India
Academy of Scientific and Innovative Research (AcSIR), Ghaziabad, Uttar Pradesh, India

Ashok Kumar Janakiraman
Faculty of Pharmaceutical Sciences, UCSI University, Cheras, Kuala Lumpur, Malaysia

Priya Jha
Department of Pharmacy, School of Health Sciences, Central University of South Bihar, Gaya,
Bihar, India

Mansoor A. Khan
Department of Pharmaceutical Sciences, Irma Lerma Rangel School of Pharmacy,
Texas A&M Health Science Center, Texas A&M University, College Station, TX, USA

Parampreet Kour
Infectious Diseases Division, CSIR–Indian Institute of Integrative Medicine, Jammu,
Jammu and Kashmir, India

Nitesh Kumar
Department of Pharmacology and Toxicology,
National Institute of Pharmaceutical Education and Research, Hajipur, Vaishali, Bihar, India

Diksha Kumari
Infectious Diseases Division, CSIR–Indian Institute of Integrative Medicine, Jammu,
Jammu and Kashmir, India
Academy of Scientific and Innovative Research (AcSIR), Ghaziabad, Uttar Pradesh, India

Kajal Kumari
Department of Pharmacy, Banasthali Vidyapith, Banasthali, Rajasthan, India

Puja Kumari
Department of Pharmaceutical Technology (Formulations),
National Institute of Pharmaceutical Education and Research (NIPER), Guwahati, Assam, India

Gowthamarajan Kuppusamy
Department of Pharmaceutics, JSS College of Pharmacy,
JSS Academy of Higher Education and Research, Ooty, Nilgiris, Tamil Nadu, India

Shaila Angela Lewis
Department of Pharmaceutics, Manipal College of Pharmaceutical Sciences, Manipal Academy of
Higher Education, Manipal, India

Jesil Mathew
Department of Pharmaceutics, Manipal College of Pharmaceutical Sciences,
Manipal Academy of Higher Education, Manipal, Karnataka, India

Bibhash Chandra Mohanta
Department of Pharmacy, School of Health Science, Central University of South Bihar, Gaya, Bihar,
India

Jyothi Munjavarapu
School of Pharmaceutical Sciences and Technologies, Jawaharlal Nehru Technological University, Kakinada, Andhra Pradesh, India

Krishna Murti
Department Of Pharmacy Practice, National Institute of Pharmaceutical Education and Research, Hajipur, Vaishali, Bihar, India

Lokesh Nama
Department Of Pharmacology and Toxicology,
National Institute of Pharmaceutical Education and Research, Hajipur, Vaishali, Bihar, India

Narhari N. Palei
Amity Institute of Pharmacy, Amity University Uttar Pradesh, Lucknow Campus, Lucknow, Uttar Pradesh, India

Tripti Paliwal
Department of Pharmacy, Banasthali Vidyapith, Banasthali, Rajasthan, India

Swaroop J. Pansare
Department of Pharmaceutical Sciences, Irma Lerma Rangel School of Pharmacy,
Texas A&M Health Science Center, Texas A&M University, College Station, TX, USA

Rabinarayan Parhi
Department of Pharmacy, Susruta School of Medical and Paramedical Sciences,
Assam University (A Central University), Silchar, Assam, India

Anil T. Pawar
School of Health Sciences and Technology, Dr. Vishwanath Karad MIT World Peace University, Pune, Maharashtra, India

Sumaiya Qasim
School of Pharmaceutical Sciences, Delhi Pharmaceutical Sciences and
Research University (DPSRU), Mehrauli-Badarpur Road, Pushp Vihar, New Delhi, India

Ziyaur Rahman
Department of Pharmaceutical Sciences, Irma Lerma Rangel School of Pharmacy,
Texas A&M Health Science Center, Texas A&M University, College Station, TX, USA

Om Prakash Ranjan
Department of Pharmaceutical Technology (Formulations),
National Institute of Pharmaceutical Education and Research (NIPER), Guwahati, Assam, India

Roshan P. Rao
Department of Pharmaceutics, JSS College of Pharmacy,
JSS Academy of Higher Education and Research, Ooty, Nilgiris, Tamil Nadu, India

Pallerla Naveen Reddy
Department of Pharmacy, School of Health Sciences, Central University of South Bihar, Gaya, Bihar, India
Translational Health Science and Technology Institute, Faridabad, Haryana, India

Shruti Richa
Department of Pharmacy, Banasthali Vidyapith, Banasthali, Rajasthan, India

Mouni Roy
Department of Chemistry, Central University of South Bihar, Gaya, Bihar, India

Parikshit Roychowdhury
Department of Pharmaceutics, JSS College of Pharmacy,
JSS Academy of Higher Education and Research, Ooty, Nilgiris, Tamil Nadu, India

Biswa Mohan Sahoo
School of Pharmacy and Life Science, Centurion University of Technology and Management, Jatni, Khurda, Bhubaneswar, Odisha, India

Srijita Sen
Department of Pharmaceutical Technology (Formulations),
National Institute of Pharmaceutical Education and Research (NIPER), Guwahati, Assam, India

Vani Setty
Drugs Testing Laboratory, Drugs Control Department, Bangalore, Karnataka, India

Rizwan Shaikh
Department of Pharmaceutical Sciences, Irma Lerma Rangel School of Pharmacy,
Texas A&M Health Science Center, Texas A&M University, College Station, TX, USA

Swapnil Sharma
Department of Pharmacy, Banasthali Vidyapith, Banasthali, Rajasthan, India

Abhishek Singh
Amity Institute of Pharmacy, Amity University Uttar Pradesh, Lucknow Campus, Lucknow, Uttar Pradesh, India

Ajay Singh
Pharmacology Division, CSIR–Indian Institute of Integrative Medicine, Jammu,
Jammu and Kashmir, India
Academy of Scientific and Industrial Research (AcSIR), Ghaziabad, Uttar Pradesh, India

Bikram Singh
Pharmacology Division, CSIR–Indian Institute of Integrative Medicine, Jammu,
Jammu and Kashmir, India
Academy of Scientific and Industrial Research (AcSIR), Ghaziabad, Uttar Pradesh, India

Deependra Singh
University Institute of Pharmacy, Pt. Ravishankar Shukla University, Raipur, Chhattisgarh, India

Kuljit Singh
Infectious Diseases Division, CSIR–Indian Institute of Integrative Medicine, Jammu,
Jammu and Kashmir, India
Academy of Scientific and Industrial Research (AcSIR), Ghaziabad, Uttar Pradesh, India

Ruhi Singh
Pharmacology Division, CSIR–Indian Institute of Integrative Medicine, Jammu,
Jammu and Kashmir, India

Sachin Kumar Singh
School of Pharmaceutical Sciences, Lovely Professional University, Phagwara, Punjab, India
Faculty of Health, Australian Research Centre in Complementary and Integrative Medicine,
University of Technology Sydney, Ultimo, NSW, Australia

Rashmi Sinha
Department of Pharmaceutical Technology (Formulations),
National Institute of Pharmaceutical Education and Research (NIPER), Guwahati, Assam, India

Ruchi Tandon
Translational Health Science and Technology Institute, Faridabad, Haryana, India

Indhumathi Thirugnanasambandham
Department of Pharmaceutics, JSS College of Pharmacy,
JSS Academy of Higher Education and Research, Ooty, Nilgiris, Tamil Nadu, India

Sunil K. Thota
Department of Pharmaceutical Sciences, Irma Lerma Rangel School of Pharmacy,
Texas A&M Health Science Center, Texas A&M University, College Station, TX, USA

Kanika Verma
Department of Pharmacy, Banasthali Vidyapith, Banasthali, Rajasthan, India

Surajpal Verma
School of Pharmaceutical Sciences, Delhi Pharmaceutical Sciences and
Research University (DPSRU), Pusph Vihar, New Delhi, India

Vaibhav Verma
Department of Pharmacy, School of Health Sciences, Central University of South Bihar, Gaya,
Bihar, India

Seema Yadav
Amity Institute of Pharmacy, Amity University Uttar Pradesh, Lucknow Campus, Lucknow,
Uttar Pradesh, India

Abbreviations

3D	three dimensional
4D	four dimensional
AD	Alzheimer's disease
ADDL	amyloid β-derived diffusible ligand
ADSC	adipose-derived stem cells
AgNPs	silver nanoparticles
ANDAs	abbreviated new drug applications
APIs	active pharmaceutical ingredients
AUC	area under the plasma concentration curve
Aβ	amyloid-β
BCSFB	blood-cerebrospinal fluid barrier
BG	bioactive glass
BGTD	biologic and genetic therapies directorate
BLA	therapeutic biologics applications
BMSCs	bone marrow mesenchymal stem cells
BNC	bio-encapsulated chitosan-Ag nanocomplex
BTC	benzene tricarboxylic
C/Ppy	carbonaceous polypyrrole
CA	chitosan
CDE	Center for Drug Evaluation
CDs	carbon dots
CDSCO	Central Drugs Standard Control Organization
CdSe/ZnS	cadmium selenide/zinc sulphide
CF	chronic inflammatory
CFDI	Center for Food and Drug Inspection
cGMP	current good manufacturing practices
Ch-AgNPs	chitosan–silver nanocomplex
CI	confidence interval
CMAs	critical material attributes
CMAX	carboxy methyl arabinoxylan
Cmax	maximum plasma concentration
CMC	chemistry, manufacturing, and control
CMC	carboxymethyl cellulose
CMCs	ceramic matrix composites
CMNC	ceramic matrix nanocomposites
CNT-MMCs	carbon nanotube metal matrix composites

CNTs	carbon nanotubes
COPD	chronic obstructive pulmonary disease
CORES	collaborative opportunities for research excellence in science
CP	centralized procedures
CP	comparability protocol
CPPs	critical process parameters
CQAs	critical product quality attributes
CSO	chitosan oligosaccharide
CSP-Cr-NCs	chitosan–sodium alginate–polyethyleneglycol–crocin nanocomposites
CTD	common technical document
CuONPs	copper oxide nanoparticles
CUR	curcumin
CV	coefficient of variation
CVD	chemical vapor deposition
DBT	Department of Biotechnology
DC	direct current
DCA	decentralized procedures
DCGI	Drugs Controller General of India
DFO	deferoxamine
DFT	discrete Fourier transform
DG JRC	Director-General Joint Research Centre
DLS	dynamic light scattering
DMF	dimethyl formamide
DMFs	drug master files
DOE	design of experiments
DOX	doxorubicin
DST	Department of Science and Technology
EC	European Commission
ECA	European Chemical Agency
ECM	extracellular matrix
EDC	electronic data capture
EMA	European Medicines Agency
EMC	epoxy matrix composite
EMEA	European Medicines Evaluation Agency
EPR	enhanced permeability and retention
EPS	exopolysaccharides
ERAs	environmental risk assessments
EU	European Union
FDA	Food and Drug Administration
Fe_3O_4	magnetite
FFDCA	Federal Food, Drug, and Cosmetic Act

F-PEG-HMD	folate-polyethylene glycol-hydrophobically modified dextran
GBM	glioblastoma multiform
GC-NCs	gelatin–curcumin-coated nanocomposite
GLUT1	glucose transporter 1
GNFs	gelatin nanofibrils
GO	graphene oxide
GO-SnO_2-TiO_2	graphene oxide-tin dioxide-titanium dioxide
GPO	graphene oxide
HA	hyaluronic acid
HAP	hydroxyapatite
HDPE	high density polyethylene
HIV	human immunodeficiency virus
HPFB	health products and food branch
HRE	hypoxia response element
HRP	horseradish peroxidase
ICH	International Council for Harmonization
IND	investigational new drug
IONP	iron oxide nanoparticle
ISO	International Organization for Standardization
LBL	layer by layer
LDH	lactate dehydrogenase
LDHs	layered double hydroxides
LF	lactoferrin
LTCFs	long-term care facilities
MAA	marketing authorization application
MA-GG	sericin–chitosan-doped maleate gellan gum
MAPP	manual of policies and procedures
MBC	minimum bactericidal concentration
MBGs	mesoporous bioglasses
MDR	multi-drug resistant
MeNPs	metallic nanoparticles
MET	metformin
MHLW	ministry of health, labor, and welfare
MIC	minimum inhibitory concentration
MMNC	metal matrix nanocomposite
MMP	mitochondrial membrane potential
MNC	magnetic nanocomposite
MOF	metal organic framework
MRP	mutual recognition procedures
MRSA	methicillin-resistant Staphylococcus aureus
MSC	mesenchymal stem cell
MTT	3-[4,5-dimethylthiazol-2-yl]-2,5 diphenyl tetrazolium bromide

MWCNTs	multi-walled CNTs
NCE	new chemical entity
NCNST	National Center for Nanoscience and Technology
NCs	nanocomposites
NDAs	new drug applications
NEPA	National Environmental Policy Act
New-PPy	nanowire polypyrrole
NFs	nanofibers
NHS	National Health Service
NIH	National Institutes of Health
nm	nanometers
NMPA	National Medical Products Administration
NNI	National Nanotechnology Initiative
NP	national procedures
NPs	nanoparticles
NSTC	National Nanotechnology Standardization Technical Committee
NTF	nanotechnology task force
OD	orphan drug
OGP	osteogenic growth peptide
OTC	over the counter
PAI	pre-approval inspections
PAN	polyacrylonitrile
PANI	polyaniline
PAS	prior approval supplement
PASA	para-aminosalicylic acid
PAT	process analytical technology
PBAT	polybutylene adipate co-terephthalate
PBE	population bioequivalence
PBS	polybutylene succinate
PCL	polycaprolactone
PCS	photon correlation spectroscopy
PDAC	pancreatic ductal adenocarcinoma
PDGF	platelet-derived growth factor
PDI	polydispersity index
PDLA	poly (D-lactic acid)
PDLLA	poly(D,L-lactic acid)
PDUFA	Prescription Drug User Fee Act
PECs	predicted environmental concentrations
PED	pediatric
PEG	polyethylene glycol
PEI	polyethyleneimine

PEO	polyethylene oxide
PET	poly(ethylene phthalate)
PFOB	perfluorobromooctane
PGA	polyglycolic acid
PHA	polyhydroxyalkanoates
PHD	prolyl hydroxylase domain
PLA	poly lactic acid
PLGA	poly lactic-co-glycolic acid
PLLA	poly (L-lactic acid)
PMCs	post-marketing commitments
PMDA	Pharmaceuticals and Medical Devices Agency
PMNC	polymer matrix nanocomposite
PMRs	post-marketing requirements
PMS	PEG-modified 3-methacryloxypropyltrimethoxy-silane silica nanoparticles
PNP	polymeric nanocomposite particle
POSS	polyhedral oligomeric silsesquioxane
POSS-PCL	polyhedral oligomeric silsesquioxane–polycaprolactone
PP	polypropylene
PPgMA	PP grafted with maleic anhydride
PSD	particle size distribution
PSMA	prostate specific membrane antigen
PTX	paclitexal
PVA	polyvinyl alcohol
PVC	polyvinyl chloride
PVD	physical vapor deposition
PVP	polyvinylpyrolidone
QbD	quality by design
QTPP	quality target product profile
RDS	respiratory distress syndrome
REACH	registration, evaluation, authorization, and restriction of chemical substances
REMS	risk evaluation and mitigation strategies
rGO-FU	5-fluorouracil-loaded graphene oxide (GO)
RLD	reference-listed drugs
ROS	reactive oxygen species
RSP	rapid solidification process
RSRP	Regulatory Science Research Program
RV	resveratrol
SA	sodium ascorbate
SAC	Standardization Administration of China
ScNP	scintillating nanoparticles

SCPg-AI-Ncs	chitosan/polyethylene glycol/allyl isothiocyanate nanocomposites
SDT	sonodynamic therapy
SEM	scanning electron microscopy
SOR	sorafenib
SPION	supermagnetic iron oxide nanoparticles
SPOS	single-particle optical sizing
SQ	squaraine
STF	study tagging files
SWCNTs	single-walled CNTs
T2DM	type 2 diabetes mellitus
TB	tuberculosis
TDDS	targeted drug delivery systems
TEM	transmission electron microscopy
TERM	tissue engineering and regenerative medicine
TMP	ligustrazine hydrochloride
TPD	therapeutic product directorate
USP	United States Pharmacopoeia
UTIs	urinary tract infections
VEGF	vascular endothelial growth factor
VRE	vancomycin-resistant enterococcus
wFIONs	water-dispersible ferrimagnetic iron oxide nanocubes
ZLDH	zinc-layered double hydroxides
ZnNPs	zinc nanoparticles
ZrO_2	zirconium dioxide

Foreword

Prof. Giorgia Pastorin
Professor, Head of the Department,
Department of Pharmacy and Pharmaceutical Sciences,
National University of Singapore

This is an exciting book that spans over various disciplines in the field of Nanocomposites. The work starts with the description of key features that clearly define a nanocomposite, highlighting the pros and cons associated with different materials. It then expands to cover the most suitable biomedical applications: towards that purpose, it becomes of paramount importance to provide an in-depth description of the most used and innovative methods to produce nanocomposites with reproducible performance across various batches.

Of special note is the part on biodegradable materials and on nanocomposites derived from waste. This is because the topic of sustainability is relevant, now more than ever before, to advance research and technology alike towards more impactful and transformative solutions. Sustainability goes hand in hand with environmental impact, which is also nicely covered in terms of pollutants and human health. With regard to the latter, various approaches are proposed in this book for relevant biomedical applications such as cancer and infectious diseases. These conditions require the concerted efforts of medical doctors, basic scientists, and engineers, so that new opportunities, such as those provided by 3D-printing technologies, are created according to important regulatory considerations.

This is why this book is indicated to both experts, who can find updates on the latest trends, as well as to students that embark in the field of nanocomposites for the first time, as they will find inspiration in many of the described technologies.

Only when scientists, such as those reported in this book, come together to systemically compile the data reported in the literature and to provide detailed descriptions on fundamental concepts of nanocomposites, that a true advancement of knowledge and research happen.

G. Pastorin

Giorgia Pastorin

Foreword

Prof. Christopher R. McCurdy, FAAPS
Professor of Medicinal Chemistry
Frank A. Duckworth Eminent Scholar Chair in Drug Research and Development
Director, UF Translational Drug Development Core
2017-2018 President, AAPS College of Pharmacy,
University of Florida

It is with great pleasure that I introduce you to "Nanomaterials in Biomedical Research Drug Delivery System," a remarkable work by Vivek Dave, Deependra Singh, and Pallerla Naveen Reddy. I had the privilege of knowing editors for several years and have always been impressed by their dedication and expertise in the field of nanotechnology.

In today's rapidly advancing world, the intersection of nanotechnology and biomedical research holds immense promise. This book addresses the critical role of nanocomposites in enhancing drug delivery systems, a subject that is both timely and crucial. Authors contributed to this book brought a wealth of knowledge and a fresh perspective to this complex topic, making it accessible and engaging for readers.

Throughout this book, you will find a comprehensive exploration of the various types of nanocomposites used in biomedical applications, methods of preparation, and innovative drug delivery systems. The detailed discussions on biodegradable versus non-degradable nanocomposites, as well as the use of recycled materials, offer valuable insights into sustainable practices in biomedical research.

Their vision for harnessing nanocomposites to address pressing health challenges was both inspiring and thought-provoking. This book is a testament to that vision, meticulously researched and thoughtfully presented.

I wholeheartedly endorse this book for anyone interested in cutting-edge developments in biomedical research. Whether you are an academic professional or a curious learner, "Nanomaterials in Biomedical Research Drug Delivery System" will undoubtedly enrich your understanding and spark new ideas.

As you delve into these pages, I encourage you to embrace the innovative spirit that embodies. This book is not just a resource; it is an invitation to explore the future of medicine and contribute to the advancement of human health by understanding the potential of nanomaterials.

Christopher R. McCurdy

Preface

This book is the culmination of extensive research and collaboration aimed at providing a comprehensive overview of the latest developments and applications of nanocomposites in the biomedical field. Inspiration for this book stemmed from the rapid advancements in nanotechnology and its profound impact on biomedical research and drug delivery systems. Nanocomposites, with their unique properties and versatile applications, have the potential to revolutionize various aspects of healthcare, from diagnostics to treatment.

In the initial chapters, we delve into the introduction and mechanisms of various nanocomposites highlighting their advantages and disadvantages. We explore different types of nanocomposites used in biomedical applications and the various methods used in their preparation. Special emphasis is placed on drug delivery systems, showcasing how nanocomposites can enhance the efficacy and precision of therapeutic interventions. One of the key areas we address is the use of nanocomposites from recycled waste materials and their innovative applications in the biomedical field. This discussion extends to the role and applications of biodegradable versus nondegradable nanocomposites, providing insights into their environmental impact and sustainability. This book also covers the applications of nanocomposites in pharmaceutical medicines and biotechnological products, reflecting on their potential to improve patient outcomes. In light of growing environmental concerns, we examine the impact of nanocomposites on pollutants prone to the ecosystem and human health.

Furthermore, we explore the application of nanocomposites for various diseases and cancer treatment, illustrating their role in advancing medical therapies. This book also touches on the cutting-edge applications of nanocomposites for 3D and 4D printing of nanomaterials, which open new frontiers in personalized medicine and tissue engineering. In addition to these biomedical applications, we discuss the biomedical uses of various herbal nanocomposites and their potential for enhancing traditional medicine. The applications of nanocomposites in antimicrobial activity, tissue engineering, biosensors, and dental care are thoroughly examined, highlighting their multifaceted benefits.

We also consider the future aspects and trends of nanocomposites in biomedical applications, envisioning the innovations that lie on the horizon. Finally, the book concludes with an overview of the regulatory guidelines for nanocomposites in India, the United States, and other foreign countries, providing a global perspective on the regulatory landscape. This book combines knowledge from different fields, such as chemistry, materials science, biology, and medicine, to provide a comprehensive and interdisciplinary approach to the topics. Nanocomposites are an emerging technology with the potential to revolutionize drug delivery, and this book provides readers with the latest information on the subject.

This book would not have been possible without the support and encouragement of all the authors. We are deeply grateful to everyone who contributed their expertise and insights, enriching the content and ensuring its relevance and accuracy. As you embark on this journey through the world of nanocomposites, I hope the readers find the information both enlightening and inspiring. This book is a comprehensive resource for researchers, students, and healthcare professionals interested in exploring the transformative power of nanocomposites in biomedical research and drug delivery systems. May this book serve as a valuable resource in your research and professional endeavors. Thereby, the book's comprehensive coverage, expert contributors, interdisciplinary approach, practical applications, focus on emerging technology, and educational value make it unique, important, and appealing to potential purchasers.

—Vivek Dave
Deependra Singh
Pallerla Naveen Reddy

CHAPTER 1

Introduction and Mechanisms of Various Nanocomposites and Their Advantages and Disadvantages

BHAVANI BODDEDA[1], AKSHAY M. BAHETI[2], ANIL T. PAWAR[2], VANI SETTY[3], and CHANDRASEKAR S. B.[3,*]

[1]*Jawaharlal Nehru Technological University, Kakinada, Andhra Pradesh, India*

[2]*School of Health Sciences and Technology, Dr. Vishwanath Karad MIT World Peace University, Pune, Maharashtra, India*

[3]*Drugs Testing Laboratory, Drugs Control Department, Bangalore, Karnataka, India*

[*]Corresponding author.

ABSTRACT

Nanocomposites are multifunctional materials that combine nanoparticles with a matrix, enabling precise control over drug release kinetics, improved drug solubility, improved bioavailability, targeted delivery, and sustained release. Several mechanisms govern their performance which collectively influences drug release profiles. The stimuli-responsive behavior of nanocomposites elucidates the mechanism of drug release. The electric field, temperature, light, pH, and specific molecules are the most frequently occurring stimuli. Nanocomposites show unique mechanical, thermal, and electrical characteristics. They are used in several ways for energy storage and harvesting, defence aerospace and safety, environment remediation, electronics, food and agriculture, automobiles, textiles, and biomedical medicines.

Nanocomposites and Nanomaterials in Biomedical Applications. Vivek Dave, Deependra Singh, and Pallerla Naveen Reddy (Eds.)

DOI: 10.1201/9781779641236-1

1.1 INTRODUCTION

One of the most fascinating areas of study for researchers this century has been nanotechnology. The control and reformation of a subject at the nanoscale or atomic and molecular level in the size choice of roughly 1–100 nm is known as nanotechnology and is used to produce devices, materials, and systems that, due to their small structure, have fundamentally new properties and functions (Roco et al., 2011). Particles having a size range of 1–100 nm were classified as nanoparticles (Joudeh and Linke, 2022). Numerous fields benefit from this alteration in nanoparticle characteristics. Three groups of nanoparticles can be distinguished based on their sizes. The first classes of nanoparticles are spheres, which are three-dimensional objects at the nanoscale. The second type of tube is typically cylindrical and has two dimensions in the nanometer series and one measurement in the millimicron range. The class three type of nanoparticles consists of single-dimensional particles on the nanometer scale, referred to as nanoplates, which have two dimensions on the micrometer scale. Almost all technological fields recognize that nanotechnological research is one of the fastest-developing fields of study today. Current developments in the fields of material sciences and technologies include hydrogels, scaffolds, conductive polymers, biomaterials, nanoparticle drug delivery, nanofibers, membranes, nanocomposites and nanoceramic composites, nanotubes, nanofibers, nanorods, and nanofilms (Hasnain and Nayak, 2019). Composites are naturally occurring or synthetic solid materials created by combining two or multiple distinct constituent materials, each with important chemical or physical properties, to produce a new material that is superior to the original materials in a specific finished structure (Shin et al., 2012; Liu and Antonietti, 2014). When nanomaterials are incorporated in a continuous material phase to form a composite, retaining their own physical identities during and after production, one obtains a new class of materials called nanocomposites (Vinyas et al., 2019). Materials larger than one nanometer in size are called nanocomposites. When added in very small amounts to composites, these nanoparticles have the potential to significantly reduce weight, improve properties, and make composites appropriate for wider applications.

In order to create nanocomposites, a matrix of conventional material is mixed with nano-sized particles ranging from 0.5 to 5% by weight allowing the material's qualities to be improved. The improved properties of nanocomposite are strength, toughness, dimensional stability, modulus, electrical conductivity, the permeability of water and hydrocarbon, flame retardancy, thermal stability, chemical resistance, surface appearance, optical clarity, catalysis, separation, sorption, and fuel cells (Camargo et al., 2009). The properties of nanocomposites can be altered in comparison to their original composition because of their high surface-to-volume ratio. Since nanocomposites will be 1000 times more robust over bulk constituent material, they found applications in variable segments of industries, like food packaging, fuel cell, and thin-film capacitor industries in computer chips. There are four phases of nanocomposites: zero-dimensional (core-shell), one-dimensional (nanotubes and nanowires), two-dimensional (lamellar), and three-dimensional (matrix metal composites).

1.2 HISTORY OF NANOCOMPOSITES

Blumstein's initial documentation in 1961 introduced the concept of nanocomposites, while his subsequent research in 1965 showcased the enhanced thermal stability of poly methyl methacrylate (PMMA)-layered silicate nanocomposites (Gilman, 1999). Theng introduced the term nanocomposites in 1970 (Theng, 1970). Nanocomposites were first cited in the literature as early as 1950, with polyamide nanocomposites being reported as early as 1976. However, it was not until Toyota researchers initiated a comprehensive investigation of polymer/layered silicate clay mineral composites that nanocomposites gained broader attention in both academic and industrial research environments. The inception of polymer nanocomposites can be traced back to 1990. It was during this time that "Tayota" utilized clay/nylon-6 nanocomposites to manufacture timing belt covers for their cars (Akki et al., 2023).

The objective of nanocomposite development is to determine the most suitable raw materials and processes for producing specific nanomaterials by examining their applications, advantages, and disadvantages. The utilization of nanocomposites has a long-standing history. Egyptians combined straw with clay to create bricks. The Mongols utilized these composites in battles, and even in the 20th century, during World War II, composite-based materials were employed in military equipment. In contemporary times, a wide variety of composites are utilized in various industries. In 1981, the advent of the scanning tunneling microscope marked the true beginning of modern nanotechnology, enabling scientists and engineers to observe and control atoms at an individual level. Notably, recent progress in characterizing, producing, and manipulating materials at the nanometer scale has resulted in their growing utilization as fillers in innovative nanocomposites.

Nanocomposites exhibit a remarkable annual growth rate of 25% and are primarily sought after in the fields of engineering plastics and elastomers. Their potential is highly significant, making them valuable across a wide range of sectors, including packaging and biomedical applications (Akki et al., 2023).

1.3 TYPES OF NANOCOMPOSITES

Nanocomposite can be divided into two phases: the continuous and the discontinuous reinforcement phase. Nanocomposites were prepared from any blend of material with three basic building blocks, namely, ceramics, polymers, and metals. Nanocomposites can be made in different ways and are categorized into several kinds according to the types of reinforcing materials used. Nanocomposites can be categorized, based on the reinforcements or filler used, as ceramic matrix nanocomposites, metal oxide-based nanocomposites, polymer-based nanocomposites, noble-metal-based nanocomposites, and carbon-based nanocomposites. In a broad sense, nanocomposites can be categorized as polymers and nonpolymers.

1.3.1 POLYMER NANOCOMPOSITES

It can be categorized as organic-organic nanocomposites, inorganic nanocomposites; polymer-ceramic nanocomposites; inorganic-organic polymer nanocomposites; organic-inorganic hybrid nanocomposites; layered-polymer silicate nanocomposites; and polymer-polymer nanocomposites.

There are two parts to a polymer nanocomposite: the polymer and the fillers. The remarkable properties of polymers make them appealing smart materials and they are attaining more popularity in various fields and biomed applications. In this, polymeric matrices utilized are copolymers, homopolymers, polymer blends, and block copolymers (Oladele et al., 2020; Ali et al., 2015). Polymers comprise numerous monomer units that join to form polymeric chains. They can be synthetic or natural. The high-molecular-weight elastomers, low-molecular-weight resins, rubbers, fibers, plastics, thermoplastic, and thermosetting polymers can be used in nanocomposites (Wei et al., 2020). Nano-sized fillers, called nanofillers, act as reinforcing components and are an important part of the dispersed phase of polymer nanocomposites. Enhancing the magnetic, electrical, mechanical, flame-retardant, thermal, and gas-barrier qualities of polymers is the main objective of nanofillers. The nanofillers can impart structural and functional characteristics to polymer nanocomposites. Multiwalled carbon nanotubes, graphene quantum dots, metal oxide nanoparticles, graphene nanosheets, nano clay, nano-silica, and other nanofillers have all been recorded as being utilized in polymer nanocomposites until now (Al-Bataineh et al., 2021; Duc et al., 2021; Safari et al., 2021; Tai et al., 2003). Chemical bonds and intermolecular forces hold the polymer and nanofillers together in polymer nanocomposites (Saleh et al., 2020).

Depending upon the degree of partition between the nanoparticles, a polymer nanocomposite can have one of three morphologies. Microcomposites, intercalated nanocomposites, and exfoliated nanocomposites are these three possible types. Because the polymer cannot intercalate linking the filler layers in microcomposites, a composite with distinct phases is typically produced. In embedded polymer nanocomposites, single extended polymer chains are inserted between filler layers, consequential in an organized multilayer stratification with embedded polymer as well as filler layers. A layered nanocomposite is formed by uniform and complete dispersion of the filler layer in a continuous polymer matrix. There are several methods to prepare polymer nanocomposites, but the mainly used methods are melt extrusion, solution dispersion, and *in situ* polymerization. The method of preparation of the polymer nanocomposites has an impact on its final morphology.

1.3.2 NONPOLYMER-BASED NANOCOMPOSITES

They are categorized into metal-metal nanocomposites, metal-ceramic nanocomposites, and ceramic-ceramic nanocomposites. In the metal-metal nanocomposites category, a metal or an alloy serves as the matrix, and metal nanoparticles also function as the reinforcing phase. With their high ductility, toughness, strength, and modulus, these nanocomposites have qualities that are comparable to those of metals and ceramics and have widespread applications. In metal-ceramic nanocomposites, ceramic and metallic components are evenly distributed and one of the components must be in nano dimension. The most prevalent example is Ag-ceramic. Ceramic nanocomposites have high mechanical toughness and creep-resistance properties. SiC/Si_3N_4 composites show better performance under high temperature and oxidizing conditions.

1.4 METHOD OF PREPARATION OF NANOCOMPOSITES

Nanocomposites can be blended in various ways. A few methods are mentioned in the following sections.

1.4.1 *IN SITU METHOD*

In general, this method eliminates the need for solvents and produces a uniform circulation of the fillers inside the polymer matrix. It entails preparing a mixture with filler and monomer, which is subsequently polymerized using conventional polymerization processes like photo initiation (UV curing) or mechanical methods (ultrasonication, stirring, and so forth). It has been reported that using this method, nanocomposites having noncovalent or covalent bonds inert linking with constituents can be produced. This method can be applied for polymers that cannot be processed by melt blending and solution combination. This method can also be used to process insoluble and thermally unstable polymers (Ravichandran et al., 2018).

1.4.2 *MELT METHOD*

This method, also known as melt blending, melt compounding, or melt processing, is a simple and economical method that does away with the need for hazardous solvents. The filler is dissolved and mixed directly with the molten polymer. Techniques like extrusion and injection molding can be used to blend (Mittal et al., 2015).

1.4.3 *SOL–GEL METHOD*

The synthesis of nanomaterials by using the sol–gel method exhibits promising results owing to its mild reaction state and capacity to produce a wide range of materials and properties by building up the materials from molecular precursors. Precursors can be either organic or inorganic salts or metal alkoxides. Colloidal powder or films are the end products of the sol–gel process. Micro- and nanostructures can be created using the sol–gel technique. A simple wet chemical reaction based on hydrolysis and condensation is used in the process to create sol, which after aging yields information about an integrated network as a gel (Ravichandran et al., 2018).

1.4.4 CO-PRECIPITATION METHOD

In order to create metal-ceramic nanocomposites, co-precipitation is a commonly used technique. This method creates a uniform solution by dissolving raw materials (chloride or nitrate) in a solvent. Then, basic hydroxide (NaOH or NH_4OH) is added to the solution. This leads to the formation of precipitate, which is then washed. After the corresponding salts are washed, the heating treatment results in the synthesis of metal oxide nanocomposite (Adimule et al., 2017).

1.4.5 HYDROTHERMAL METHOD

This process uses wet-chemical techniques to crystallize the constituent materials into a nanostructure. Water is used as the solvent in this process, which is conducted in a sealed container with high vapor pressure and a temperature range of 100–250°C (Wu et al., 2008). The process of hydrothermal synthesis keeps porosity and composition under good control. The process is then known as the solvothermal method if other solvents are used than water (Dahiya and Manjeet, 2018).

1.4.6 MICROWAVE SYNTHESIS

Microwaves, which are electromagnetic radiation that fall between radio waves and the infrared region and have wavelengths ranging from 0.01 to 1 m, are important because they can cause material to nucleate or grow. Microwaves typically interact with polar molecules. This interaction causes heat to be produced inside the material and causes polar molecules to rotate and vibrate. The principle of microwave-assisted synthesis is to use the excitation caused by microwave electromagnetic radiation to align the material dipoles in an exterior field. A diminish in processing time and energy expenditure can be attributed to the generation of internal heat during the process of molecules aligning or orienting in reaction to an exterior electrical field. The substance is heated uniformly, which is the main benefit of this process. Quick microwave heating creates a homogenous reaction environment and allows for excellent product shape, size, and agglomeration control.

1.4.7 HIGH-ENERGY BALL MILLING

A simpler technique that typically uses a top-down approach to create powdered nanocomposite is ball milling. Powder materials with the appropriate composition and larger particle size (often measured in microns) are the first step in the process. This powder is placed inside a container with two or more heavy metals that are free to move around inside (a 2:1 ratio between the mass of the balls and the amount of material is typically preferred). The container is sealed tightly. The balls are typically composed of tungsten carbide or hardened steel. The material is then forced to be powdered into smaller particle sizes by the balls being made to rotate around a predetermined axis. The particle dimensions are adjusted by varying the ball rotation speed and the milling process duration. Mechanically alloyed nanocomposites' homogeneity, microstructural characteristics, and other attributes are influenced by the milling duration, rotation speed, energy, ball size, and content of reinforcing material (Joy et al., 2022).

1.4.8 SONOCHEMICAL METHOD

It has been reported that high-intensity ultrasound can produce high concentrations of H·and OH·radicals in water by exceeding the attraction force of molecules. Ultrasound is the sound that involves 20 kHz and 10 MHz and is undetectable to the human ear. This allows for the formation of an adequate number of cavitation bubbles. By absorbing vapor or gas from the solvent, these cavitation bubbles grow until they collapse at their maximum volume. Furthermore, the implosion of the bubble generates high pressure and a high temperature. By acting as a unique initiator, ultrasound promotes the breaking of chemical bonds and accelerates polymerization (Zhang et al., 2009; Xu et al., 2013).

1.4.9 SOLUTION DISPERSION METHOD

The nanofillers in this process swell in specific solvents such as toluene, water, and chloroform. When the nanofillers and polymer were combined in the solution, the solvents were positioned within the interlayer of nanofillers, causing the polymeric chains to intercalate. Polymer nano-composites are created upon removal of solvent, leaving the intercalated/exfoliated structure intact.

1.4.10 VAPOR-PHASE REACTION TECHNIQUES

This method evaporates a solid to form supersaturated vapor to produce nanomaterials with various particle sizes, shapes, and compositions. Arc discharge, laser ablation, electron beam evaporation, and Joule's heating can all produce supersaturated vapor. It is a process in which chemical reaction, mass transfer, nucleation, and solidification and condensation occur. This method eliminates the need for a calcination step. Powdered nanocomposite materials are produced commercially using vapour-phase techniques. The type of powder that is produced depends on the type of gas that is used. Metal oxide powders are typically created using gases that contain oxygen, while nonoxide powders are created using inert gases.

1.5 MECHANISM OF NANOCOMPOSITES

Currently, significant efforts are being undertaken to create safer and more effective delivery systems that transport bioactive molecules to certain tissues when and where they are needed. The two basic ways that nanocomposites are used to transport and release bioactive molecules are through stimuli-responsive delivery of bioactive molecules and the release of bioactive molecules during nanocomposite disintegration (Figure 1.1).

1.5.1 STIMULI-RESPONSIVE NANOCOMPOSITES

Bioactive molecules loaded with nanocomposites are stable at a steady state until exposed to stimulus namely temperature, pH, redox potential, enzymes, light, and so forth. Novel polymers known as stimuli-responsive polymers, sometimes cited as smart or intelligent polymers, modify their chemical and/or physical characteristics in response to external cues or signals. Temperature, pH, light, the electric field, and specific molecules are the most frequently occurring stimuli (McCarthy et al., 2021).

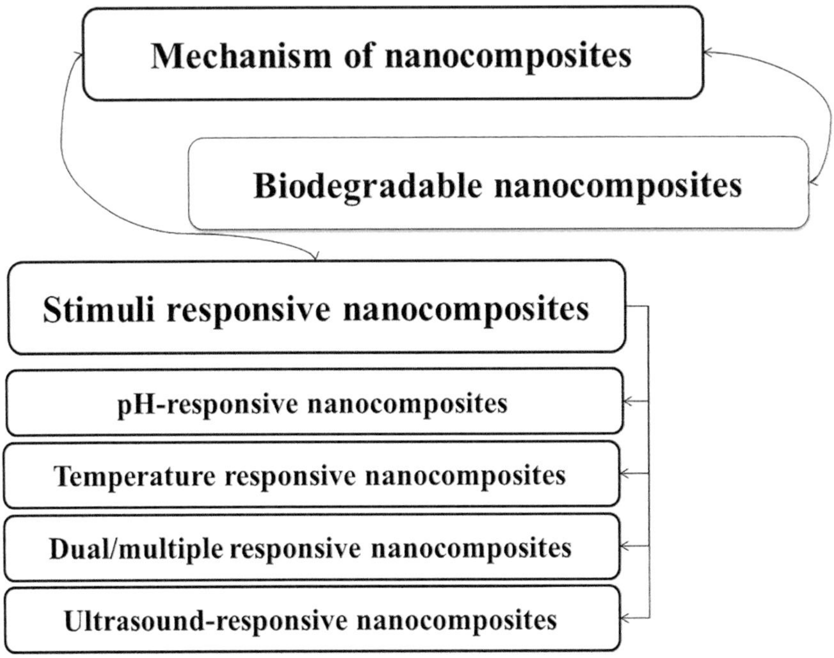

FIGURE 1.1 Mechanism of nanocomposites.

1.5.2 pH-RESPONSIVE NANOCOMPOSITES

A polymer is deemed pH sensitive if it contains pendant basic or acidic moieties that can donate or take in protons in reply to the pH changes of the surrounding environment. Changes in pH cause ionic interactions to begin and intensify, which in turn causes the polymer chain in an aqueous solution to expand or contract due to electrostatic repulsion. Weak polyacids and polybases are common examples of pH-responsive materials. When placed in basic conditions, polyacids like polymethacrylic acid (PMAAc) and poly acrylic acid (pKa of approximately 5) swell upon donating their protons. Whereas polybases like poly(N,N-diethylaminoethylmethacrylate) and poly(N,N-dim ethyl aminoethyl methacrylate) expand upon accepting protons due to Coulomb repulsion. Chitosan (CS), gelatin, albumin, poly (methacrylic acid-g-ethylene glycol) [P(MAA-g-EG)], poly(acrylic acid) (PAAc)/CS IPN, poly(ethylene imine) (PEI), poly(lysine) (PL), poly(L-histidine) (PLH), and poly (N,Ndialkylamino ethylmethacrylates) (PDAAEMA) are among the most-studied pH-responsive polymers (Deirram et al., 2019).

Graphene, cellulose nanowhiskers, and CS were cross-linked using synthetic dialdehyde and Schiff base chemistry to create pH-responsive injectable hydrogels. The anticancer medications curcumin and doxorubicin (DOX) were separately and jointly loaded into the resultant hydrogels. A pH-dependent release of the medications is revealed by *in vitro* release shown in phosphate buffer. When the pH was lowered from 7.4 to 5.4, independently loaded hydrogels released 28% more DOX and 30.1% more curcumin. When both medications are loaded into the same hydrogel, the pH is dropped from 7.4 to pH 5.4 and there is a 38.6% increase in curcumin release and an increase is 38% for DOX. It has been observed that as pH rises, the cross-linking of imine bonds weakens and the repulsive electrostatic interactions between the protonated amino groups intensify, increasing drug release (Omidi et al., 2020).

With chemically cross-linked polymers based on pH-responsive component N-acryloyl-N′-ethyl piperazine (AcrNEP) and temperature-responsive component N-isopropylacrylamide (NIPAM), hydrogel nanocomposites containing silver nanoparticles of size 15–21 nm were developed by diffusion and *in situ* chemical reduction. The study found that the maximum swelling resulting from ionization of the gel network causes the silver nanoparticles with pH 2.5 solution when exposed to external pH. In a pH 7 solution, no apparent release of nanoparticles was seen (Deen and Chua, 2015).

This study used an *in situ* cross-linking method catalyzed by horseradish peroxidase to generate a hydrogel for a hyaluronic acid (HA)/graphene oxide (GO) nanocomposite (NC). Incorporating GO into the HA matrix can improve the hydrogel's mechanical properties and extend the release of rhodamine B (RB), from the hydrogel which is the model drug, for an additional 33 h, when compared to hydrogels made entirely of HA. Furthermore, the NC hydrogels exhibited excellent pH-responsive discharge action mainly owing to numerous interactions involving GO and RB. The release of RB from the NC hydrogel was prolonged at low pH (pH 4.0) in the presence of GO, which could be attributed to the enhanced interactions between GO and HA as well as with RB (Song et al., 2015). Zhang et al. (2020a, b) developed a novel method with noninvasive magnetic resonance imaging (MRI) to monitor drug discharge from deliverance vehicles. The method uses S100 fibres and pH-responsive Eudragit L100 for encapsulating carmofur (used in the treatment of colon cancer) and super paramagnetic iron oxide nanoparticles (SPIONs). After being created by electrospinning, the fibres were found to be cylindrical and smooth. The nanofibers demonstrated pH-dependent dissolving characteristics. As a result of matrix swelling/dissolution in the physiological pH series, apparent pH-responsive proton relaxation velocity can be observed. These relaxivity variations show that SPION release is a consequence of pH-responsive dissolution. *In vitro* drug release demonstrated that the formulations gave instant release of carmofur at pH 6.5 and 7.4 (Zhang et al., 2020a, b).

A pH-dependent nanocomposite hydrogels were created by utilizing a gamma irradiation approach to graft gold nanoparticles and acrylamide monomer on CS (Cs-gPAAm/AuNPs) (Nasef et al., 2023). Gold nanoparticles enhance the capacity to kill elevated liver cancer cells, silver nanoparticles enhance the controlled discharge of anticancer drug fluorouracil along with its antimicrobial action and diminishing the cytotoxicity in nanocomposite hydrogels. The generated Cs-g-PAAm/Au-Ag-NPs nanocomposite hydrogel was shown to be extremely sensitive to pH changes during swelling tests at various pH levels. Strong antibacterial activity is demonstrated by bimetallic pH-responsive Cs-g-PAAm/Au-Ag-NPs nanocomposites. Fluorouracil drug loading in Cs-g-PAAm/Au-Ag-NPs was significant reaching 95 mg/g at pH 7.4 and a maximum drug release of 97% in 300 min (Nasef et al., 2023).

Xu et al. (2017) prepared an injectable, near-infrared (NIR)/pH-responsive poly (NIPAm-coMPCD)/gold nanorods (GNRs) nanocomposite hydrogel by copolymerizing nisopropylacrylamide (NIPAm) and methacrylated poly-β-cyclodextrin (MPCD) macromer. Through host–guest interaction, a prodrug of doxorubicin (AD-DOX) which is acid-labile and hydrophobic, is effectively loaded into the hydrogel. The gradual and constant release of DOX could be sustained for over a month by the nanocomposite hydrogel, which demonstrated a sustained drug release mechanism. Because of the breakdown of the acid-labile hydrazone bond linking the adamantyl group and DOX in acidic pH, the pH-responsive diffusion of DOX from nanocomposite hydrogel was seen. The diffusion of DOX starting from the networks was regulated by its hydrogel networks collapse due to the photothermal impact of GNRs and NIR irradiation could speed up this process. Because of the prolonged drug retaining in tumors and effective photothermal effect, the *in vivo* anticancer test exhibited the nanocomposite hydrogel's potential for chemophotothermal synergistic therapy with minimal side effects. Consequently, this injectable, pH- and NIR-responsive nanocomposite hydrogel showed significant promise as a long-term drug delivery platform for cancer therapy using chemophotothermal synergy (Xu et al., 2017).

Research has long focused on the transport of drugs to the intestinal tract, avoiding the harsh acidic atmosphere of the stomach. On the other hand, as many therapies activity and release rate depend on the system's pH, monitoring it during drug delivery is an essential diagnostic component. A unique nanocomposite hydrogel for near-neutral pH monitoring and intestinal drug delivery was created simultaneously. Carbon dots (CDs) were added to gelatin nanocomposites (GNCs) in different quantities using solvent-casting techniques. The CDs acted simultaneously as a chromophore and a cross-linker, exhibiting a dual role, thereby reducing the usage of toxic cross-linkers. In the pH region close to neutral, the prepared GNC hydrogel sample performed as both a pH sensor and could be valuable in quantitative pH assessments. For the *in vitro* drug release investigation, cefadroxil, a model antibacterial medication, was employed at 1.2 gastric pH and 7.4 intestinal pH. During the course of a 24-h investigation, a considerable and prolonged drug discharge pattern is seen in pH 7.4 as contrasting to the acidic environment. The drug discharge pattern showed the proportion of CDs cross-linked and even the rate of drug discharge was affected by the pH of the release environment. These findings suggest that GNC hydrogel may serve as a suitable pH-monitoring and oral medication delivery method in a pH range close to neutral (Bhattacharyya et al., 2020).

Quercetin (QC) has good anticancer properties, but because of poor solubility and short biological half-life, it is difficult to distribute the drug precisely over time. Therefore, it is important to fabricate a phase with stimulus-responsive possessions meant for targeted delivery, enhance encapsulation capacity, and extend the discharge pattern. For simultaneous sustained release of QC and improved loading efficiency, a unique technique is developed. This was accomplished by loading QC into an agarose (AG)-polyvinylpyrrolidone (PVP)-hydroxyapatite (HAp) hydrogel nanocomposite. The loading efficiency of the AG-PVP hydrogel was increased by up to 61% by adding HAp nanoparticles. Additionally, the interactions between the hydrogel polymers, drug, and nanoparticles made them pH-responsive in acidic environments and burst discharge in neutral environments. The drug release was further sustained by encapsulating the QC-loaded hydrogel in water to formulate oil-in-water nanoemulsions. Consequently, the discharge of QC was shown to be pH-responsive and to have a prolonged half-life of 96 h. To be more precise, the Korsmeyer-Peppas mathematical model states that at pH 7.4, the release mechanism was anomalous (diffusion-controlled), and at pH 5.4, anomalous transport (dissolution-controlled) (Samadi et al., 2021).

To improve combination therapy for multidrug-resistant cancer, for the co-delivery of cisplatin and DOX, metal-organic framework (MOF)-based polymer-coated hybrid nanoparticles (NPs) were fabricated. The MOF@polymer nanocarrier maximized the synergistic antitumor effect by combining the benefits of physiological stability, tumor microenvironment (TME) pH responsiveness, and high multidrug loading capacity. This allowed for the simultaneous transport of medicines into cancer cells. Interestingly, the hybrid nanocarrier preserves negative surface charge right through circulation for its stable and extended course of action *in vivo*. It then reveals an inner positive MOF to facilitate cellular internalization, multi-drug release, nuclear localization, and tumor infiltration following the external polymer degradation in the acidic TME. Studies using drug-resistant Michigan Cancer Foundation-7 (MCF-7)/ADR cancer *in vitro* and *in vivo* showed that the nanocarrier could enhance drug accumulation in solid tumors, produce amazing tumor elimination outcomes, and reduce side effects, all of which would improve the combination chemotherapy's safety and efficacy. MOF@polymer hybrid nanocarriers offer fresh perspectives on the creation of stimuli-responsive drug co-delivery systems.

A unique *de novo* drug delivery method for cancer chemotherapy was designed and developed by Abbasian et al. (2018). Phthalic anhydride was used to functionalize CS and then 4-cyano, 4-[(phenylcarbothioyl) sulfanyl] pentanoic acid was added considering it as chain transport mediator (CTA) to produce CS-CTA macroinitiator. Using the reversible addition-fragmentation chain transfer (RAFT) polymerization process, the synthesized CS–CTA macro initiators were subsequently copolymerized with methacrylic acid (MAA) monomers to generate CS–graft polymethacrylic acid (CS-g–PMAA) graft copolymer. Following that, physical interactions allowed GO nanosheets to be integrated into the copolymer that had been created. DOX hydrochloride, a universal anticancer medication, was incorporated into the CS-g-PMAA/GO nanocomposite. The CS-g-PMAA/GO nanocomposites that were developed demonstrated remarkable drug loading and entrapment efficiencies of 93.8 and 78.6%, respectively. The pH-dependent drug discharge profile of the DOX-loaded nanosystem was primarily caused by the presence of a PMAA segment, a pH-dependent biopolymer. With pH 4 being the greatest drug release value for the proposed nanosystem, minimal drug release behavior was seen in physiological conditions (pH 7.4 and 37°C) (Abbasian et al., 2018).

1.5.3 TEMPERATURE-RESPONSIVE NANOCOMPOSITE

The antibacterial action and biocompatibility of a temperature-responsive topical layer made of pullulan-g-Pt = NIPAM and impregnated with two distinct concentrations of silver nanoparticles (15 and 30 ppm) have been reported by Paneysar et al. (2022). The films were excellent materials for absorbing wound exudates because of their superior swelling capabilities. The films' antibacterial effectiveness against gram-negative *Escherichia coli* and gram-positive *Staphylococcus aureus* has been demonstrated at physiological temperature and compared to the commercial formulation. The temperature-responsive polymer-based nanocomposite films release Ag-NPs when the wound exudate temperature rises a little over the normal temperature (Paneysar et al., 2022).

Poly (N-isopropylacrylamide) (PNIPAM)-grafted multiwalled carbon nanotubes (MWNT-gPNIPAm) were fabricated onto a substrate modified with PNIPAm to create a temperature-sensitive polymer/carbon nanotube boundary with bio-electrocatalytic activity. These relatively wide (>6 μm) and extremely permeable nanocomposite films had strong conductivity and electrocatalytic action, as demonstrated by electron microscopy and electrochemical tests. The morphological changes in the polymers wrapping around the surface of the MWNT and the tethered PNIPAm chains on a substrate caused the permeability of the nanostructure to open, close, or change, and at the same time, an electron-transfer procedure occurred throughout the channels generating temperature within the nanostructure. Cyclic voltammetry and electrochemical impedance spectroscopy measurements have shown that 1,4-dihydro-β-nicotinamide adenine dinucleotide can be bioelectrocatalyzed with improved sensitivity when MWNTs with their huge surface area, superior biocompatibility, and good electron-transfer as well as electrochemical catalysis capabilities are combined with the responsive features of PNIPAm. The findings further indicated that the augmentation of the interfacial thermo-responsive behavior and enhanced surface wettability resulted from synergistic or cooperative interactions between PNIPAm and MWNTs. Potential uses for this bio-electrocatalytic "smart" system include the development of biosensors and biofuel cells with externally regulated activity (Zhao et al., 2012).

In order to create a solid gold nanoparticle core and a thermally responsive interpenetrating polymer network (IPN) shell that was surface functionalized or PEGylated with a covalently bound linear poly(ethylene glycol) chain layer, the goal of this study was to synthesize and characterize a polymer-metal nanocomposite system. Gold nanoparticles having of 50 nm diameter were created using the common reduction technique of gold chloride and citrate. After that, an *in situ* inverse emulsion polymerization was used to encapsulate these particles inside an IPN shell made of polyacrylamide (PAAm) and poly(acrylic acid) (PAA). Then, the primary amine groups of the PAAm network were covalently grafted with a linear methoxy-PEG N-hydroxysuccinimide (M.W. 3400) to PEGylate the surface of the nanocomposite system. When the temperature rises, IPN nanospheres show a positive abrupt sigmoidal enlargement reaction. As a result, these systems can be made into externally triggered intelligent therapeutic systems by heating the gold nanoparticle cores of these systems with a noninvasive laser. This allows for external control of the systems' swelling and consequently, their therapeutic release (Owens et al., 2007).

Using disulfide-based cross-linkers and the surfactant-free emulsion polymerization process, Ulasan et al. (2015) created biocompatible thermo responsive cross-linked poly (ethylene glycol) methyl ether methacrylate (PEGMA)-co-vinyl pyrrolidone nanoparticles (PEGMA NPs). A model of a small therapeutic molecule called rhodamine B was incorporated into PEGMA NPs, and these NPs' solution temperature was adjusted to 37°C, which is similar to body temperature. PEGMA NPs released their rhodamine B when reaching a temperature of 39°C (Ulasan et al., 2014).

Using gentamicin as the model drug, PNIPAM nanocomposites containing g-Fe_2O_3 and SBA-15 were created to create a temperature-induced release delivery system with magnetic targeting to certain tissues.

Despite the relatively low loading of gentamicin (28%), the nanocomposites managed to effectively produce temperature-dependent release. After an hour, the gentamicin release test conducted at 10°C attained a plateau of approximately 10%. However, a burst release occurred for 2 h in the discharge assay conducted at 40°C, which is nearer to the physiological human temperature. A plateau was reached until the completion of the assay, at which point the accumulated release of gentamicin reached almost 70% (Zhu et al., 2009).

1.5.4 DUAL/MULTIPLE-RESPONSIVE NANOCOMPOSITE

Kurd et al. (2016) developed a dual pH/thermal-response composite nanoparticle that serves as a carrier for cisplatin and assessed the discharge profile of the substance under various pH and temperature conditions that are pertinent to both the normal malignant environments during instances of hyperthermia. Poly (n-isopropyl acrylamide-methacrylic acid-hydroxy ethyl methacrylate) (P (NIPPAAM-MAA-HEM) was formulated by the emulsion polymerization method. Fe_3O_4 magnetic nanoparticles (MNPs) and cisplatin were incorporated into nanogel, by the enlargement and conjugation of cisplatin with the -COOH group of MAA. The profile of drug release at pH levels of 7.4 (physiological pH), 6.8 (tumor extracellular pH), and 5.3 (endosomal pH), as well as at physiological and hyperthermic temperatures of 37 and 43°C, were examined, in addition to the chemical and morphological characteristics of the drug-loaded composite nanoparticle. Enough heat is produced by the drug-loaded composite nanoparticles' magnetic saturation to be useful in cases of hyperthermia. The complex is stable enough to hold a drug loaded inside while keeping its size. Under high temperatures and a low pH environment related to the cancer media during hyperthermia, composite nanoparticles exhibit significant drug release. The dual pH/temperature-responsive nanocomposite demonstrated its effectiveness when used in conjunction with regulated drug administration to treat hyperthermia (Kurd et al., 2016).

For controlled release applications, nanocomposite double-network hydrogels (ncDN hydrogels) with determinable microstructures, characteristics, and multistimulus responsiveness are developed. Using electrostatic interactions, hydrogen bonding, amine-functionalized iron oxide nanoparticles ($IOPs\_NH_2$) having dynamic imine bonds, were employed as nanoparticle cross-linkers to concurrently attach the twin networks of polydextran aldehyde (PDA) and gelatin (Gel). Proteins are released quickly from temperature-responsive gel/PDA/IOP_NH_2 ncDN hydrogels and pH at high temperatures and acidic pH conditions. In addition, IOP_NH_2 gives ncDN hydrogels their magnetic responsiveness, which makes it possible to employ a magnetic field to produce heat and aid in the hydrogels' structural modification and uses (Lu and Yeh, 2021).

Using PNIPAM as a closely cross-linked first network, polyacrylic acid (PAA) as a freely crosslinked second network, and GO as an additive, pH- and temperature-responsive double network (DN) Hydrogels were formulated. GO sheets were first made by an oxidation reaction and they were subsequently silanized to dispersion in a NIPAM aqueous solution. NIPAM was free-radical polymerized at 20°C in a water bath. After being exposed to UV light, PNIPAM/AA/GO DN Hydrogels which were responsive to temperature and pH were formulated. The addition of the GO sheets causes the volume phase transition temperature (VPTT) of the innovative DN hydrogels to drop by 3°C, to around 29°C. The PNPAM/AA/GO DN hydrogels have high mechanical properties in addition to excellent swelling behavior. Furthermore, it was clear that the PNIPAM/AA/GO hydrogels were sensitive to pH below their VPTT (Li et al., 2013).

Efficient separation and enrichment procedures are necessary for the rapid identification of harmful microorganisms, especially in food samples. For the purpose of selectively enriching microorganisms, hydrophilic temperature-responsive boronate affinity magnetic nanocomposites have been developed. Using SIATRP polymerization, a new core-shell designed magnetic nanocomposite, Fe_3O_4@poly (NIPAm-co-AGE)@DFFPBA, was successfully created. It includes a very hydrophilic, pH- and thermo-responsive polymer layer. To achieve a high magnetic response, a solvothermal reaction was used to create the MNPs. Later, silica layers were added to ensure that the initiator would remain immobilized and that afterward, reactions would proceed. The water-soluble copolymer brushes were created by convention by attaching pendant epoxides to amino groups after the water-soluble co-monomers, allyl glycidyl ether, and N-isopropylacrylamide were synthesized on MNPs. Following the addition of 3, 5-difluoro-4-formylbenzeneboronic acid, the polymer network reinforced by MNPs demonstrated exceptional selectivity and binding capabilities for bacteria. By just changing the temperature and pH, it is also feasible to change the binding ability of the bacteria with affinity materials. It was also shown that the nanocomposite affinity material may be used to separate and enrich microorganisms in complex samples. The nanocomposite was used to separate bacteria from food samples by enriching *S. aureus* as well as Salmonella species. From the turbot extraction samples, 25% milk, and spiked tap water multiplex polymerase chain reaction (mPCR), which resulted in high bacteria enrichment and demonstrated great potential in the separation of bacteria from food samples (Zheng et al., 2021).

By using diffusion and *in situ* chemical reduction in chemically cross-linked polymers based on N-acryloyl-N′-ethyl piperazine (AcrNEP), and N-isopropylacrylamide (NIPAM), hydrogel nanocomposites containing silver nanoparticles of size 15–21 nm were created. The hydrogels of the nanocomposite were sensitive to variations in temperature and pH in the surrounding environment. The nanocomposite's silver nanoparticles provided extra physical crosslinking that affected media diffusion and penetration velocity. It was investigated that's how external pH variations influence the discharge of silver nanoparticles commencing the formulated hydrogel matrix. Because of the maximum swelling brought on by the ionization of the gel network, the rate at which silver nanoparticles were released was higher in a pH 2.5 solution. In a solution of pH 7, no apparent release of nanoparticles was seen. Bolu et al. (2020) created a dendritic micellar platform that is targeted towards trastuzumab to deliver docetaxel and optimize drug loading and release. Docetaxel release was sensitive to an acidic pH; during 30 h, 54% of drug release was recorded in acidic media, while only 30% was observed in neutral conditions (Bolu et al., 2020). Hosseinzadeh et al. (2019) proposed a unique approach to *in situ* nanoparticle creation by utilizing RAFT polymerization to produce a CS hydrogel composite containing magnetic Fe_3O_4 nanoparticles. The N-phthaloyl protective groups to the CS's amino groups were first added. The hydroxyl groups of CS were treated with a RAFT agent, and the resulting N-phthaloyl chitosan-RAFT was employed to esterify the hydroxyl side chains of MNPs. NIPAM and acrylic acid (AA) were used in a copolymerization process to create the magnetic nanocomposite. DOX was loaded into the nanocomposite. DOX was released from this nanogel over the course of 24 h under pH and temperature control. The drug is released at approximately 90 (pH 2) and 78% (pH 5.0) in buffered solutions at a temperature of 45°C. At pH 7.4, almost 40% was released at a temperature of 45°C. The observed values at 25°C were significantly lower than these values (Hosseinzadeh et al., 2019) Poly (N-(4-aminophenyl) methacrylamide))-carbon nano-onions (PAPMA-CNOs = fCNOs)/

diclofenac-complex integrated CS nanocomposite hydrogel nanoparticles (CNPs) were created by employing an ionic gelation technique. Drug release from CNPs *in vitro* was tested in gastrointestinal (GI) tract environments with varying pH levels and temperature ranges between 37 and 55°C. Dual (pH and thermo)-responsive regulated drug release was observed. In 15 days, CNPs showed the maximum drug release at 55°C when the pH was 7.4. The prolonged release of the encapsulated drug molecules was achieved by disassembling the structure of CNPs at 55°C, as revealed by the drug release data (Mamidi and Delgadillo, 2021).

To produce temperature and pH-responsive magnetite nanocomposite particles (MNCPs), ultrasonic-assisted emulsion copolymerization of NIPAM and 2-(N,N-diethylaminoethyl) methacrylate (DEAEMA) in the occurrence of Fe_3O_4 nanoparticles was used. Furthermore, MNCPs were loaded with the anticancer medication methotrexate (MTX) and the *in vitro* drug release of these MNCPs was examined under varied pH and temperature levels as well as in the occurrence of an alternating magnetic field. Specific conditions of pH 5.5 and 42°C showed the maximum rate of MTX release. Comparative research was done on the cell practicability of the treated MCF-7 human breast cancer cell line with free MTX-loaded MNCPs, MNCP, and MTXs or in conjunction using water-based magnetic hyperthermia (MHT). The anti-proliferative activity of MTX-loaded MNCPs with MHT was approximately 17% higher than that of free MTX (Najafipour et al., 2021). When an external alternating magnetic field with a frequency of 375 kHz and a field intensity of 4 Am^{-1} was applied *in vitro* to induce simultaneous hyperthermia and drug release, the discharge of MTX-loaded MNCPs was accelerated at relatively low doses.

Mesoporous silica, PNIPM-CS nanocomposite that responds to two stimuli were created and then loaded with DOX. Temperature and pH levels affected the release of DOX, with high temperatures and acidic environments favoring the drug's rapid release. When the temperature of the composites was higher than room temperature, DOX was released from them much more quickly. A significant factor in regulating the release of DOX is the coated stimuli-responsive polymer layers. The polymer chains in PNIPAM-CS can exhibit hydrophilicity and stretch significantly in water as temperature reduces beneath lower critical solution temperature (LCST). Hydrophobic interaction and hydrogen bonding could be used to load the DOX medicines onto the composites. The composites could grasp DOX and obstruct the drug discharge to some level. Therefore, a minute quantity of the drug could be released into the solution at room temperature. The drug molecules that are confined can escape as the temperature rises over the lower critical temperature (LCST) because the polymer chains start to contract (Huang et al., 2020).

Tetracycline hydrochloride (TH) was selected as a classical drug for exploration into double stimuli-responsive functionalized cellulose nanocrystal-poly [2-(dimethylamino)ethyl methacrylate] (CNC-g-PDMAEMA) strengthened poly(3-hydroxybutyrate-co-3-hydroxy valerate) (PHBV) electro spun composite membranes as drug release vehicles. Rigid CNC-gPDMAEMA nanoparticles were observed to improve the thermal, crystallization, and hydrophilic characteristics of PHBV. The pH of the release vehicle affected the amount of TH released, with pH = 5 PBS buffer exhibiting a greater degree of drug release than pH = 8 PBS buffer. More drug molecules were able to move freely in an acidic environment due to the charged tertiary amino group of the PDMAEMA polymer generating electrostatic repulsion linking polymer chains. The release pattern was promoted. Conversely, when the pH of the solution increased, the rate of drug discharge dropped. This might be explained by the amine group becoming hydrophobic and nonprotonated, which prevented the drug from being released at pH = 8. Furthermore, the drug's faster release in the acid solution would be advantageous for targeting tumor cells, since cancer cell surfaces have negatively charged components because, in the case of cancers, negative charge is translocated from the inner layer of the cell membrane to the cell surfaces. Therefore, the *in vitro* absorption of drug-loaded tumor tissues would be accelerated by the electrostatic interaction between the positive charge of CNC-g-PDMAEMA and the negative charge of the cancer cell surfaces (Chen et al., 2020).

The photothermal nanocomposite $MnFe_2O_4$@$mSiO_2$, loaded with the anticancer medication DOX hydrochloride, was synthesized with magnetic targeting and NIR/pH coupling sensitive mesoporous silica shell nanocarriers. The mesoporous SiO_2 has a greater positive surface zeta potential when the pH is lower. This will make it easier for DOX molecules to escape from mesoporous SiO_2 by reducing the electrostatic interaction between them and positive-charged DOX molecules. Furthermore, the thermal effect of MnSiP under 808 nm laser irradiation will raise the temperature, which will encourage the release of DOX from mesoporous SiO_2 (Chen et al., 2017).

Based on the tumor-specific mitochondria accumulation capacity and photothermal (PTT) therapy, a novel MoS_2/PDA-TPP nanocomposite encapsulated with the chemotherapeutic medication DOX was fabricated for TME dual response and synergistically improved antitumor therapy. The MoS_2/PDA-TPP nano platform that has been constructed can function as pH-dependent dissociation that allows rapid release of DOX in an acidic TME while also increasing PTT efficiency. Furthermore, the method demonstrates that via generating reactive oxygen species (ROS) and lowering mitochondrial membrane potential (MMP), MoS_2/PDA-TPP causes mitochondrial-dependent apoptosis. Most crucially, during the PTT treatment, tumor cell death can be efficiently induced by hyperthermia up to 50°C without severely inflaming nearby tissues (Zhang et al., 2022).

A stimuli-responsive paramagnetic nanocarrier was developed by Hosseini et al. (2018) to deliver MTX and DOX intracellularly to the MCF7 cell line. Combining the free radical polymerization techniques and ring opening, a new thermo/pH-sensitive amphiphilic paramagnetic nanocomposite made of hydrophilic biocompatible P(NIPAAm-co-HEMA-co-MAA-co-TMSPMA) and hydrophobic biodegradable polyl caprolactone (PCL) blocks are produced. Tumor niche-assisted release behavior was shown by the dual drug release pattern (higher drug release was seen at pH $\leq$ 5.4 and 41°C). The DAPI staining, MTT assay, real-time PCR analysis on MCF7 cell lines, and cell cycle demonstrated that the antitumor capacity of the DOX/MTX-loaded nanocomposite was much higher than that of free medicines (Hosseini et al., 2018). At a pH of 7.4 and temperature of 37°C, approximately 5 and 22% of the MTX and DOX were released, respectively. After 10 days, 90% of the medication for DOX and 45% for MTX was released at pH = 4 and 41°C. The amount of DOX emitted decreased from 95 to 45 to 15% at 41°C as the pH was raised from 4 to 5.4 to 7.4. At 37°C, on the other hand, the released quantities reached 75, 55, and 22%, respectively, with the same trend of pH shift (from 4 to 5.4 to 7.4). At 41°C, the amount of MTX released by raising the pH from 4 to 5.4 to 7.4 decreased from 45 to 30 to 5%, and at 37°C, the released amounts decreased from 40 to 35 to 5%, respectively. To simulate drug release, Kiaee et al. (2018) created pH-controlled and electric field-responsive CS-loaded nanoparticles encasing a model compound (fluorescein isothiocyanatedextran, or FTC). They suggest this as a possibly novel approach to managing compounds at the sites of wounds. The device consists of microelectrodes on a flexible substrate, an electronic driver, and a PEG-diacrylate/laponite hydrogel containing CS particles. In their work, applying an electric field causes the pH to rise quickly and under control. When 4 V is supplied in an on/off pattern lasting 2 s, the pH rises from 6 to 12 in 1 s and falls back to 6 in 2 s. It was possible to achieve 100% release of the FITC compound *in vitro* release tests conducted under these conditions. There was also good biocompatibility discovered in the pH and electro-responsive hydrogel (Kiaee et al., 2018).

By copolymerizing AA monomers (VCL-co-AA) and N-vinyl caprolactam (VCL) with a novel cross-linker, triethylene glycol dimethacrylate (TEGDMA), as a nontoxic and biocompatible component is developed. Aminoleslami et al. (2022) developed a pH and temperature-responsive polymeric nanogel. The chemotherapeutic agent DOX was loaded into the P(VCL-co-AA) with a notable loading of 83%. The drug release profile was examined under both physiological and malignant site simulation circumstances. P(VCL-co-AA) showed temperature- and pH-responsive behavior, and under the condition of the malignant site, it released the compound more quickly. When used as model malignant cells. Dox-loaded P(VCL-co-AA) had excellent toxic behavior on the MCF-7 cell line as model cancerous cells (Aminoleslami et al., 2022).

In order to achieve high biocompatibility, specificity, and excellent tumor-targeting properties, a polydopamine (PDA)-drug conjugate nanocomposite by targeted cancer photothermal chemotherapy was created via gradually copolymerizing pH-sensitive dopamine-derived prodrug (DA-DOX) and dopamine (DA) into the porous channels of zeolite imidazolate frameworks-8 (ZIF-8). This was abiding by PEGylation with amino-terminated folic acid polyethylene glycol. PDA was added, which improved ZIF-8's stability and dispersion while also bringing about the photothermal conversion effect. The acid-labile hydrazone linker between ZIF-8 and DA-DOX quickly broke down to release activated DOX in the acidic milieu of the tumor. Furthermore, the inflated photothermal conversion efficiency of the PDA constituent may cause hyperthermia, which could hasten drug release while concurrently thermally ablating tumor tissue to maximize DOX-induced CT and aid PTT in the elimination of tumor cells (Ma et al., 2022).

Glutathione (GSH) and magnetic-sensitive nanocomposites were synthesized by Yoon et al. (2021) as part of a dual-targeting approach.

Methoxy poly (ethylene glycol) was implanted into water-soluble CS in order to accomplish this goal. Following that, DOX was coupled via a disulfide bond to the CS backbone. To achieve magnetic sensitivity, iron oxide MNPs were further attached to the CS backbone. GSH addition enhanced the rate at which DOX was released from nanocomposites in a drug release study, suggesting that the nanocomposites were redox-responsive. In addition, in cell culture tests, an external magnetic stimulation concentrated the nanocomposites in the magnetic field, which facilitated the efficient internalization of the nanocomposites into cancer cells. Nanocomposites exhibited enhanced magnetic sensitivity and selectively targeted tumor tissues in the presence of external magnetic stimuli, as demonstrated in an animal investigation with mice carrying CT26 cells (Yoon et al., 2021).

In order to develop a unique tumor-targeting nanocomposite stage made of copper sulfide@polydopamine–folic acid/doxorubicin (CuS@PDA–FA/DOX) for photothermal and chemotherapeutic cancer treatment, Shang-Qing Zhang synthesized CuS@PDA nanoparticles. Compared to CuS/PDA alone, the nanocomposite platform exhibits higher photothermal conversion efficiency and ultrahigh loading levels.MCF-7 cells are far more folic acid receptors than A549 cells, which indicates that MCF-7 cells absorb CuS@PDA–FA/DOX nanocomposites considerably more readily than A549 cells do. When PTT and chemotherapy are used in combination, the CuS@PDA–FA/DOX system produces a more effective therapeutic result under NIR irradiation than when either treatment is used alone (Zhang et al., 2020a, b). MNPs were fabricated into the thermoresponsive Pluronic® P85 to facilitate the delivery of drugs. Beneath its transition temperature, Pluronic® P85, an impediment copolymer comprising PPO and PEO blocks, experiences a hydrophilic to hydrophobic transition. A critical micellar temperature (CMT) is the term used to describe this behavior. The authors covalently combined functionalized MNPs to Pluronic® P85 through amino functional groups to achieve MagPluronics. Curcumin was encapsulated in the nanocomposite arrangement, which was created at a temperature lesser than CMT. Under physiological settings, they demonstrated a continuous discharge of curcumin for almost 4 days. For stimuli-responsive drug delivery, the suggested approach shows a thermo-responsive system in which the MNPs function as a nucleating agent for the covalently attached thermoresponsive polymer to nanocarriers (Piao et al., 2016).

1.5.5 ULTRASOUND-RESPONSIVE NANOCOMPOSITES

Mesoporous silica nanoparticles (MSNs) coated and PDA were used by Li et al. (2018) to successfully establish ultrasound and pH-responsive dual-responsive drug release systems. Under alkaline conditions, oxidative self-polymerization produced the PDA shell on the MSN surface. Composite nanoparticles loaded with DOX hydrochloride were utilized to examine the reactive drug storage and release behavior. This type of hybrid material demonstrated a different pH response when DOX was released into an acidic environment. In addition, the ultrasonic cavitation effect of high-intensity ultrasound exposure enables the release of DOX loaded into the composite nanoparticles from the pores. By varying the on/off status, the DOX-releasing design might be optimized in a distinctive pulsatile manner. According to the results of the 3-(4,5-dimethylthiazol-2-yl)-2,5-diphenyltetrazolium bromide assay, blank nanoparticles were not toxic to HeLa cells, but DOX-loaded nanoparticles were able to stop tumor cells from growing. Additionally, these composite nanoparticles demonstrated reasonably high conversion efficiency (about 37%) together with an effective near-IR photothermal conversion capability (Li et al., 2018). In the fight against a variety of diseases, vaccines are one of the most vital tools in modern medicine. Vaccinations typically need to be administered repeatedly for the best possible benefits, which would be very uncomfortable for the patient. A remotely controlled tumor vaccine release and customized cancer immunotherapy are envisaged using an ultrasound-responsive, self-healing hydrogel structure loaded with nano vaccines. When treated with ultrasound, the gel may change into a sol state, releasing nano vaccines in a burst, and then self-healing to return to gel. The personalized nanovaccine system could prevent postoperative tumor metastases and recurrence for mice treated with a single subcutaneous injection of nanovaccine-loaded gel and multiple ultrasound treatments by eliciting antitumor immune responses that could effectively inhibit established tumors (Meng et al., 2021). Piezoelectric $FeWO_4$ nanorods, tannic acid (TA), carboxymethyl chitosan (CMCS), carbomer (CBM), and high degradability, expansibility, and adjustable rheological properties were combined to create adhesive piezoelectric antibacterial hydrogels by a simple assembly procedure. Utilizing ultrasound as an external mechanical wave, it is possible to initiate the piezoelectric effect of $FeWO_4$ and subsequently enhance the production of reactive oxygen kinds, demonstrating enhanced antibacterial efficacy and averting wound infection. Piezoelectric hydrogels have been shown to speed up full-thickness skin wound healing in bacteria-infected mice through a variety of mechanisms, including skin regeneration, inflammatory response inhibition, increased collagen deposition, and angiogenesis promotion (Xu et al., 2023).

1.5.6 BIODEGRADABLE NANOCOMPOSITES

Another way that bioactive chemicals trapped or encapsulated inside nanocomposites can be released is through the degradation of the nanocomposites, except in responsive systems. Bioactive compounds are enclosed within nanocomposite materials in delivery systems, and when the nanocomposites break down, the molecules are gradually released. Chemo-phototherapy drugs' penetration depth in the tumor location is limited by the intricate TME. Deep tumor chemo-phototherapy is intended to be achieved with a biodegradable amphiphilic gelatin (AG) enfolded nanocomposite (PRDCuS@AG) that is loaded with copper sulphide (CuS) and DOX. The arginine-conjugated polyamidoamine dendrimer is represented by PR in PRDCuS@AG. When exposed to NIR light, matrix metalloproteinases can quickly biodegrade PRDCuS@AG into PRDCuS. The resulting PRDCuS is capable of penetrating deep into the tumor tissue due to its dual cell-tissue penetration capabilities. With a 97% tumor inhibition rate, PRDCuS@AG demonstrates photoacoustic imaging-guided enhanced results of chemo-phototherapy. Additionally, PRDCuS@AG can break down even further into 3 nm ultra-small CuS, which the body can get rid of after therapy to prevent negative consequences (Pan et al., 2020). Using a combination of sugar particles as a porogen and Fe5Ag and Fe10Ag granules, biodegradable 60–75% porous scaffolds with pore sizes between 300 and 420 μm were created for load-bearing applications. Cold sintering and high-pressure consolidation were employed at room temperature to achieve this. To obtain the desired nanostructure, Fe-Ag powder was obtained by high energy milling followed by annealing in hydrogen flow at 550°C for 1 h, followed by cold sintering at 2.5 GPa pressure. The use of the product increases the size of pores by controlling pore volume. The presence of electric non-pairs in Fe5Ag and Fe10Ag increased the degradation rate compared to pure metal. After 28 days of degradation, there is a significant loss of flexural strength and a slight decrease in compressive strength. The compressive strength of both compositions remains within the trabecular bone strength range (Sharipova et al., 2019). Krishnan et al. (2020) demonstrated that methicillin-resistant Staphylococcus aureus-infected investigational osteomyelitis could be treated locally with vancomycin due to a biodegradable nanocomposite fibrous scaffold. Vancomycin may be added to pre-fabricated nanocomposites or integrated into the nanocomposites during scaffold production. Vancomycin was released constantly from both scaffold groups. The reason for this might be the release of vancomycin mainly depended on the porosity and chemical composition of the nanocomposites which in turn was associated with the degradation of one's (Krishnan et al., 2020).

1.6 APPLICATIONS OF NANOCOMPOSITES

It was generally known that nanocomposites had many uses in various industries as shown in Figure 1.2. Nanocomposites have applications in food and agriculture (bactericide, food packing, food analysis sensor, antimicrobial coating); electronics and information technology (ferrofluids, thermal, sound and electrical insulator, nano pigments, EMI shielding); environmental remediation (insect repellent blocking, heavy metal adsorption, wastewater, filtration, noise reduction); defence aerospace and safety (hyperspectral imaging, nuclear detection, deep space mission, military platforms, solid-state lasers); automobile and textile (anti-odor, moisture management, water and oil repellent, moisture management); infrastructure and civil structure (plastic and rubber, highway culverts, smokestacks, building and bridges); energy storage and harvesting (Pièze harvesters, H_2 storage, solar cells, fuel cells, battery electrodes) and biomedical applications.

1.6.1 BIOMEDICAL APPLICATIONS

Nanocomposite materials find extensive uses in medicine, including drug administration, imaging, disease detection and screening, gene therapy, DNA sequencing, tissue culturing, dental applications, wound healing, medical implants, biosensors, and cancer treatment. Foodborne disease inhibition and bacterial control or prevention can be greatly aided by nanocomposite materials that possess the resistant qualities of micro-organisms. The ideal nanocomposite systems with improved repairing capabilities can be shaped by incorporating metal nanoparticles into natural polymers. Processing of copolymer/clay nanocomposites into thin films, scaffolds/membranes, hydrogels, cell-filled 3D structures, and thin films has demonstrated its benefits in orthopedic implants, drug delivery, tissue regeneration, and microfluidic device fabrication. Polymer nanocomposites have been used in MRI, antibacterial activity, dental materials, drug delivery, bone engineering imaging, and so forth (Khan et al., 2023), making tumors more visible and easier to remove. In an effort to create a magnetic and fluorescent nanocomposite particle, researchers are combining magnetic and fluorescent nanoparticles. An MRI conducted before surgery reveals the tumor more clearly due to the magnetic characteristic of the nanocomposite particle. A physician may be able to see the tumor more clearly during surgery thanks to the nanocomposite particle's fluorescent quality. Studies have demonstrated that the implantation of a nanotube-polymer nanocomposite as a sort of framework to direct the formation of replacement bone accelerates the process of bone regrowth. After 10 days of use, a membrane containing graphene and blended CS over the polyvinyl surface led to full skin wound healing as reported by Lu et al. (2012). This effect was not observed in the control groups where graphene was absent (Lu et al., 2012) (Figure 1.2).

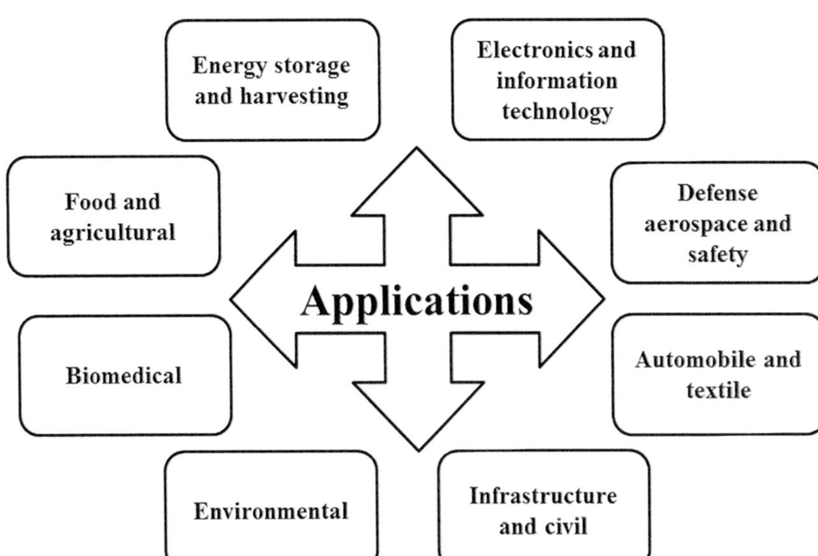

FIGURE 1.2 Applications of nanocomposites.

1.7 ADVANTAGES OF NANOCOMPOSITES

Many factors influence the various characteristics of nanocomposites, including surface energy, the type of nano-filler substrates, adhesion type at surfaces, matrix interface type, surface development, size and shape of the nano-filler, size, and shape of the nano-filler, distribution of nanoparticles in the matrix and specific surface area (Paszkiewicz et al., 2020). The significantly enhanced characteristics of nanocomposites comprise mechanical properties, namely, dimensional stability, modulus, and strength with reduced permeability toward hydrocarbons, water, and gases; thermal stability; flame resistance and heat distortion temperature (HDT); reduced emissions of smoke; resistance to chemicals; optical clarity; and electrical conductivity in comparison to conventionally filled polymers.

1.7.1 MAJOR ADVANTAGE OF NANOCOMPOSITES

When compared to single-component nanomaterials, nanocomposite materials have an advantage since their properties are influenced by their morphology and interface features in addition to their components (inorganic-inorganic, inorganic-organic, and organic). A nanocomposite is likely to exhibit superior electrical and ionic transport, superhydrophobicity/hydrophilicity, improved/enhanced electrical conductivity, better cell viability, targeted drug delivery, remarkable antibacterial activity, effective hyperthermia, and so forth, properties that a single component material typically is not able to exhibit.

1.7.2 MECHANICAL STRENGTH

Since they have a superior surface-to-volume ratio, nanoscale reinforcement, and surface chemistry than other classes of biomaterials, nanocomposites offer superior mechanical properties and cell-material interactions. Nanotechnology improves the properties of tensile modulus, tensile strength, and tensile HDT. Southern Clay Products' Cloisite® nylon nanocomposite, which has a 5% clay loading, is an example of this enhanced mechanical performance. The tensile strength of nylon is increased by

23% by nanoclays. Both the flexural and tensile moduli are increased by 56 and 69%, respectively. Furthermore, HDT has increased by 68%.

1.7.3 GAS BARRIERS

Nanocomposite improves gas barriers. According to studies, the "oxygen transmission rate (OTR) of nylon-6 nanocomposites can be nearly four times lower than that of unfilled nylon-6."

1.7.4 THERMAL STABILITY

The composition of the nanocomposite, the presence of fillers or nanoparticles, and the interfacial adhesion linking the fillers and polymer matrix are the elements that influence the thermal stability of nanocomposite films. It was discovered that the ferrite component in ferrite-polymer nanocomposites boosted the thermal stability of the composite. Thermal and thermal-oxidative stability of polyimide nanocomposites were improved by the inclusion of clay and graphene fillers. Similarly, the addition of $SrTiO_3$ nanoparticles enhanced the high-temperature energy storage properties of polyetherimide-based nanocomposite films. Metallic nanoparticles such as ZnO, SiO_2, TiO_2, Al_2O_3, and Fe_2O_3 co-assembly improved the thermal stability of cellulose nanocrystal films.

1.7.5 THERMAL/ELECTRIC CONDUCTIVITY

The thermal/electric conductivity of nanocomposite is due to its improved nanoparticle dispersion in the system depending on their size, shape, distribution; orientation, and interaction, nanoparticles can have a variety of effects on the matrix material's thermal conductivity. For example, phonon scattering can reduce thermal conductivity in materials by interfering with the movement of heat-carrying atoms; this impact is stronger for smaller and more distributed nanoparticles. Furthermore, a barrier to heat transfer may be created at the interface linking the nanoparticles and matrix, reducing thermal conductivity. Moreover, a continuous network of nanoparticles within the matrix material might create a conduit for heat transfer, improving thermal conductivity, when the volume percentage of nanoparticles reaches a crucial amount.

1.7.6 DRUG DELIVERY

Due to their nano size and large surface area, drugs delivered by means of nanoparticle-mediated delivery have the advantage of being able to freely circulate throughout the cell system, blood capillaries, and pores found at different surfaces and membranes. A lot of studies have been done on the use of metal-based nanocomposites, namely, titanium dioxide, nickel, silver, zinc oxide, iron oxide, gadolinium, and gold for drug delivery applications. The most important benefits of metal-based nanocomposites for drug administration are that (i) they allow for the inclusion of high drug dosages because of the huge surface area of the metallic nanoparticles; (ii) it is feasible to create pH-responsive delivery systems—this is so that pH-mediated delivery systems can use the metal ions to transport and release at the biological location; and (iii) at high temperatures, they remain stable (Díaz and Vivas-Mejia, 2013).

1.7.7 DRUG TARGETING

The field of nanocomposite science provides a versatile framework for the production of novel nanomaterials with distinct properties and functions, making them appropriate for a variety of up-and-coming applications, particularly in the field of drug targeting, particularly cancer targeting. The adoption of customized new drug delivery systems is necessary owed to the restrictions of conventional chemotherapy, such as regular dosing, more side effects, and deficiency in specificity of anticancer medications (Soni and Yadav, 2015). Because these medications are not cell-specific, tailored drug delivery is required. Effective multifunctional moieties are required in the design and construction of nanocomposites in order to target and regulate the distribution of the encapsulated anticancer medication.

1.8 DISADVANTAGES OF NANOCOMPOSITES

1.8.1 PARTICLE AGGREGATION

Because of the limited interfacial surface, the aggregation or agglomeration of nanoparticles diminishes the potential increase of mechanical properties in nanocomposites. Thus, achieving tiny nanoparticles and efficient

Oladele, I. O., Omotosho, T. F., & Adediran, A. A. (2020). Polymer-based composites: An indispensable material for present and future applications. *International Journal of Polymer Science, 4*, 47–52.

Omidi, S., Pirhayati, M., & Kakanejadifard, A. (2020). Co-Delivery of doxorubicin and curcumin by a ph-sensitive, injectable, and *in situ* hydrogel composed of chitosan, graphene, and cellulose nanowhisker. *Carbohydrate Polymers, 231*, 1145–1157

Owens, D. E., Eby, J. K., Jian, Y., & Peppas, N. A. (2007). Temperature-responsive polymer-gold nanocomposites as intelligent therapeutic systems. *Journal of Biomedical Materials Research Part A, 83*(3), 692–695.

Pan, X., Li, P., Bai, L., Ma, J., Li, S., Zhang, F., Liu, S., Wu, Q., Shen, H., & Liu, H. (2022). Biodegradable nanocomposite with dual cell-tissue penetration for deep tumor chemo-phototherapy. *Small, 16*(22), e2000809.

Paneysar, J. S., Barton, S., Ambre, P., & Coutinho, E. (2022). Novel temperature responsive films impregnated with silver nano particles (Ag-NPs) as potential dressings for wounds. *Journal of Pharmaceutical Science, 111*(3), 810–817.

Paszkiewicz, S., Pypeć, K., Irska, I., & Piesowicz, E. (2020). Functional polymer hybrid nanocom-posites based on polyolefins: A review. *Processes, 8*, 113.

Piao, S. H., Kwon, S. H., & Choi, H. J. (2016). Stimuli-responsive polymer-clay nanocomposites under electric fields. *Materials, 9*(1), 52.

Ravichandran, K., Praseetha, P. K., Arun, T., & Gobalakrishnan, S. (2018). Synthesis of nanocomposites. In S. M. Bhagyaraj, O. S. Oluwafemi, N. Kalarikkal & S. Thomas (Eds.), *Synthesis of inorganic nanomaterials*(pp. 141–168). Woodhead Publishing.

Roco, M. C., Mirkin, C. A., & Hersam, M. C. (2011). Nanotechnology research directions for societal needs in 2020: Summary of international study. *Journal of Nanoparticle Research, 13*, 897–919.

Safari, M., DeSousa, R. A., Salamat, M., Joudaki, J., Ghanbari, D., & Bakhtiari, A. (2021). Mechanical properties of green synthesized graphene nano-composite samples. *Applied Science, 11*, 46–48.

Saleh, T. A., Shetti, N. P., Shanbhag, M. M., Raghava, R. K., & Aminabhavi, T. M. (2020). Recent trends in functionalized nanoparticles loaded polymeric composites: An energy application. *Materials Science for Energy Technology, 3*, 515–525.

Samadi, A., Pourmadadi, M., Yazdian, F., Rashedi, H., Navaei, N. M., & Eufrasioda, S. T. (2021). Ameliorating quercetin constraints in cancer therapy with pH-responsive agarose-polyvinylpyrrolidone -hydroxyapatite nanocomposite encapsulated in double nanoemulsion. *International Journal of Biological Macromolecules, 182*, 11–25.

Sharipova, A., Gotman, I., Psakhie, S. G., & Gutmanas, E. Y. (2019). Biodegradable nanocomposite Fe-Ag load-bearing scaffolds for bone healing. *Journal of the Mechanical Behavior of Biomedical Materials, 98*, 246–254.

Shin, W. H., Jeong, H. M., Kim, B. G., Kang, J. K., & Choi, J. W. (2012). Nitrogen-doped multiwall carbon nanotubes for lithium storage with extremely high capacity. *Nano Letters, 12*, 283–288.

Song, F., Hu, W., Xiao, L., Cao, Z., Li, X., Zhang, C., Liao, L., & Liu, L. (2015). Enzymatically cross-linked hyaluronic acid/graphene oxide nanocomposite hydrogel with pH-responsive release. *Journal of Biomaterials Science, Polymer Edition, 26*(6), 339–352.

Soni, G., & Yadav, K. S. (2015). Applications of nanoparticles in treatment and diagnosis of leukemia. *Materials Science and Engineering C, 47*, 156–164.

Tai, W. P., Kim, Y. S., & Kim, J. G. (2003). Fabrication and magnetic properties of Al_2O_3/Co nanocomposites. *Materials Chemistry and Physics, 82*(2), 396–400.

Theng, B. K. G. (1970). Interactions of clay minerals with organic polymers. Some practical applications. *Clays and Clay Minerials, 18*, 357–362.

Ulasan, M., Yavuz, E., Bagriacik, E. U., Cengeloglu, Y., & Yavuz, M. S. (2015). Biocompatible thermoresponsive PEGMA nanoparticles crosslinked with cleavable disulfide-based crosslinker for dual drug release. *Journal of Biomedical Materials Research, 103*(1), 243–251.

Vinyas, M., Joseph, A., Harursampath, D., Loja, M., Trung, N. T. (2019). A comprehensive review on analysis of nanocomposites: From manufacturing to properties characterization. *Materials Research Express, 3*, 567–581.

Wei, H., Wang, H., Li, A., Cui, D., Zhao, Z., & Chu, L. (2020). Multi functions of polymer nanocomposites: Environmental remediation, electromagnetic interference shielding, and sensing applications. *ChemNanoMat, 6*, 174–184.

Wu, W., He, Q. G., & Jiang, C. Z. (2008). Magnetic iron oxide nanoparticles: Synthesis and surface functionalization strategies. *Nanoscale Research Letters, 3*, 397–415.

Xu, X., Huang, Z., Huang, Z., Zhang, X., He, S., Sun, X., Shen, Y., Yan, M., & Zhao, C. (2017). Injectable, NIR/pH-responsive nanocomposite hydrogel as long-Acting implant for chemophotothermal synergistic cancer therapy. *ACS Applied Materials and Interfaces, 9*(24), 20361–20375.

Xu, M., Wu, S., Ding, L., Lu, C., Qian, H., Qu, J., & Chen, Y. (2023). Engineering ultrasound-activated piezoelectric Hydrogels with antibacterial activity to promote wound healing. *Journal of Materials Chemistry B, 11*(19), 4318–4329.

Xu, H., Zeiger, B. W., & Suslick, K. S. (2013). Sonochemical synthesis of nanomaterials. *Chemical Society Reviews, 42*, 2555–2567.

Yoon, H. M., Kang, M. S., Choi, G. E., Kim, Y. J., Bae, C. H., Yu, Y. B., & Jeong, Y. I. (2021). Stimuli-responsive drug delivery of doxorubicin using magnetic nanoparticle conjugated poly(ethylene glycol)-*g*-chitosan copolymer. *International Journal of Molecular Science, 22*(23), 131–169

Zhang, W., Ding. M., Zhang, H., Shang, H., & Zhang, A. (2022). Tumor acidity and near-infrared light responsive drug delivery MoS2-based nanoparticles for chemo-photothermal therapy. *Photodiagnosis and Photodynamic Therapy, 38*, 102716.

Zhang, S. Q., Liu, X., Sun, Q. X., Johnson, O., Yang, T., Chen, M. L., Wang, J. H., & Chen, W. (2020a). CuS@PDA-FA nanocomposites: A dual stimuli-responsive DOX delivery vehicle with ultrahigh loading level for synergistic photothermal-chemotherapies on breast cancer. *Journal of Materials Chemistry B, 8*(7), 1396–1404.

Zhang, K., Park, B. J., Fang, F. F., & Choi, H. J. (2009). Sonochemical preparation of polymer nanocomposites. *Molecules, 14*(6), 2095–2110.

Zhang, Z., Wells, C. J. R., King, A. M., Bear, J. C., Davies, G. L., & Williams, G. R. (2020b). pH-responsive nanocomposite fibres allowing MRI monitoring of drug release. *Journal of Materials Chemistry B, 8*(32), 7264–7274.

Zhao, X., Liu, Y., Lu, J., Zhou, J., & Li J. (2012). Temperature-responsive polymer/carbon nanotube hybrids: smart conductive nanocomposite films for modulating the bio electro catalysis of NADH. *Chemistry, 18*(12), 3687–3694.

Zheng, H., Lin, H., Chen, X., Sui, J., Ullah, K. M., Ramesh, P. T., Han, X., & Cao, L. (2021). Tailor-made magnetic nanocomposite with pH and thermo-dual responsive copolymer brush for bacterial separation. *Food Chemistry, 358*, 129907.

Zhu, Y., Kaskel, S., Ikoma, T., & Hanagata, N. (2009). Magnetic SBA-15/poly (*N*-isopropylacrylamide) composite: Preparation, characterization and temperature-responsive drug release property. *Microporous and Mesoporous Materials, 123*(1–3), 107–112.

CHAPTER 2

Different Types of Nanocomposites Used in Biomedical Applications

ABHISHEK SINGH[1], SEEMA YADAV[1], NARHARI N. PALEI[1], MOUNI ROY[2], RABINARAYAN PARHI[3], BISWA MOHAN SAHOO[4], and BIBHASH CHANDRA MOHANTA[5,*]

[1]*Amity Institute of Pharmacy, Amity University Uttar Pradesh, Lucknow Campus, Lucknow, Uttar Pradesh, India*

[2]*Department of Chemistry, Central University of South Bihar, Gaya, Bihar, India*

[3]*Department of Pharmacy, Susruta School of Medical and Paramedical Sciences, Assam University (A Central University), Silchar, Assam, India*

[4]*School of Pharmacy and Life Science, Centurion University of Technology & Management, Jatni, Khurda, Bhubaneswar, Odisha, India*

[5]*Department of Pharmacy, School of Health Science, Central University of South Bihar, Gaya, Bihar, India*

*Corresponding author.

ABSTRACT

Nanocomposites, which consist of nanoscale materials incorporated into a matrix, have emerged as very promising alternatives for an array of biological applications, owing to their distinctive features. This chapter presents an inclusive summary of the various kinds of nanocomposites used in the biomedical field, emphasizing their compositions, functional characteristics, and possible uses. The selection of materials and their intended biological use determine the unique benefits and challenges

Nanocomposites and Nanomaterials in Biomedical Applications. Vivek Dave, Deependra Singh, and Pallerla Naveen Reddy (Eds.)

DOI: 10.1201/9781779641236-2

associated with each type of nanocomposite. In the field of biomedicine, the development and production of nanocomposites offer enormous prospective for improving the efficacy, accuracy, and safety of medical treatments. Additionally, these nanocomposites allow for the creation of new features that cannot be achieved with traditional materials. The choice of appropriate components for the synthesis of nanocomposites is determined by the unique needs of the intended biological use. Nanocomposites are utilized to fabricate advanced drug delivery systems, which would not only enhance drug efficacy, but also ensure the safety of medicines. Nanocomposites have numerous applications ranging from tissue engineering, cancer therapy, regenerative medicine, medical implants to biosensors. The development of advanced nanocomposites holds significant promise for enhancing patient care and advancing medical technologies.

2.1 INTRODUCTION

Since last few decades, nanotechnology is booming across all scientific fields and has brought numerous scientific developments. Healthcare systems are complex and challenging for all stakeholders, but the concept of nanotechnology has transformed the healthcare field tremendously and holds enormous potential not only to improve patient care, but also the quality of life. Rapid advancements in nanotechnology have revolutionized the healthcare system by integrating it into clinical practice (Sahoo et al., 2020; Mohanta et al., 2019; Palei et al., 2023). In this context, nanomaterials and nanocomposites have been introduced and applied across diverse fields, such as drug delivery, skin generation, ophthalmology, dentistry, surgery, bio-imaging, tissue engineering, regenerative medicine, food packaging, cosmetics, medical devices, and so forth.

Composites are materials which consist of a polymer, resin, a filler or organic/inorganic fiber, and an inorganic material. Normally, a composite comprises a filler dispersed into a polymer. A nanocomposite is a heterogeneous or hybrid solid material which is generally formed by joining two or more chemically and physically distinct phases of one, two, or three dimensions of nanometer range (usually less than 100 nm) (Hári and Pukánszky, 2011). Generally, the properties of the resulting products differ significantly from and are superior to those of the individual component materials. In its broadest definition, the term "nanocomposite" refers to several substances such as porous media, colloids, gels, and copolymers.

However, it is often referred to as a solid mixture consisting of a bulk matrix and one or more nano-dimensional phases that have distinct structural and chemical properties.

The concept of nanocomposite has been in existence since 1950s. In the mid-1950s, the nanoscale "organo-clays" have been employed to regulate the composition of gels or the flow of polymer solutions (as paint viscosifiers), for example, it is used as a thickening material in cosmetics, maintaining the preparations in homogenous form (Ruiz-Hitzky et al., 2016). Nanocomposite forms a mesmerizing interdisciplinary area which brings nanotechnology, biology, and material science together. The characteristics of the nanocomposites depend upon the individual parent compositions and their morphological as well as interfacial characteristic. These are polymeric systems with nanoscale fillers that enhance various properties. These materials are engineered to exhibit unique properties and functionalities that arise from the synergistic interactions between their components at the nanometer scale. The choice of appropriate components for nanocomposite fabrication is contingent upon the precise criteria of their intended biological use. Biodegradable and biocompatible polymers such as polyethylene glycol (PEG), poly lactic-co-glycolic acid (PLGA), and chitosan are often used as matrix materials due to their ability to break down naturally, compatibility with living organisms, minimal likelihood of causing an immune response, and adjustable mechanical properties. Conversely, inorganic nanoparticles (NPs) such as iron oxide, gold, silver, and silica (NPs) provide a wide range of functions, including improving imaging contrast for targeted drug delivery and using heat for cancer treatment (Silva et al., 2018). Polymeric nanocomposites, which incorporate both natural and synthetic biodegradable polymers along with organic or inorganic fillers, provide significant potential for emulating biological tissues, promoting cell proliferation, and facilitating effective drug delivery. Consequently, their application is found extensively in the field of tissue engineering. Polymer/metal nanocomposites offer durable and stable materials for conductive devices, sensors, and other biomedical products (Zare, 2016). Therefore, these materials are used in a wide range of industries, such as electronics, aerospace, automotive, energy, and biomedical sectors, due to their superior mechanical, thermal, electrical, or biological qualities compared to traditional composite materials (Tiwari and Panda, 2023). In this chapter, an overview of different nanocomposites, their significance, and various biomedical applications has been provided.

2.2 IMPORTANCE OF NANOCOMPOSITES

Nanocomposites are important because they can improve a variety of material properties such as better mechanical, thermal, electrical and optical properties, imaging capabilities, biocompatibility, stability, and durability as compared to the standard materials (Ge et al., 2022). Due to the above properties these are used in: (a) drug delivery system (controlled as well as targeted drug delivery);(b) functionalized nanocomposites enable precise targeting of disease tissues or cells, thus, improving diagnostic accuracy and treatment specificity (Pan et al., 2018) (c) augment the bioavailability and therapeutic efficacy of entrapped drugs (Bhakay et al., 2018), (d) tissue regeneration, imaging techniques, regenerative medicine (bone repair and wound healing), (e) cancer therapy, (f) exhibit superior mechanical strength, stiffness, and toughness compared to traditional materials making it apposite for load-bearing biomedical applications (e.g., bone implants and tissue scaffolds) (Rahmani et al., 2022), (g) modified biodegradable characteristic of nanocomposite allows controlled release of encapsulated drugs that helps in development of controlled release drug delivery systems (Zhou et al., 2018), (h) surface modification of nanocomposites provides exact control over bioactivity and interactions with biological systems, facilitating targeted drug administration, and molecular imaging (Wang et al., 2018), and (i) nanocomposites minimize the risk of immunogenic reactions and toxicity, ensuring safe and well-tolerated biomedical interventions (Falsafi et al., 2023). Various biomedical applications of nanocomposites are depicted in Figure 2.1.

2.3 LIMITATIONS OF NANOCOMPOSITES

Though nanocomposites offer numerous advantages for biomedical applications, they also exhibit certain disadvantages and face some challenges such as: (a) certain nanocomposites possess cytotoxicity and immunogenicity issues due to their accumulation in tissues as well as in organs (Rafieian et al., 2019), (b) undergo degradation over time affecting their long-term stability and performance (Liang and Wong, 2017), (c) complicated and time consuming regulatory approval process due to their unique features and associated uncertainties (Garcia et al., 2018), (d) difficulty in industrial-scale production of nanocomposites due to involvement of complex fabrication processes (Tan, 2020), (e) limited understanding about the long-term biological interactions and fate of nanocomposites in

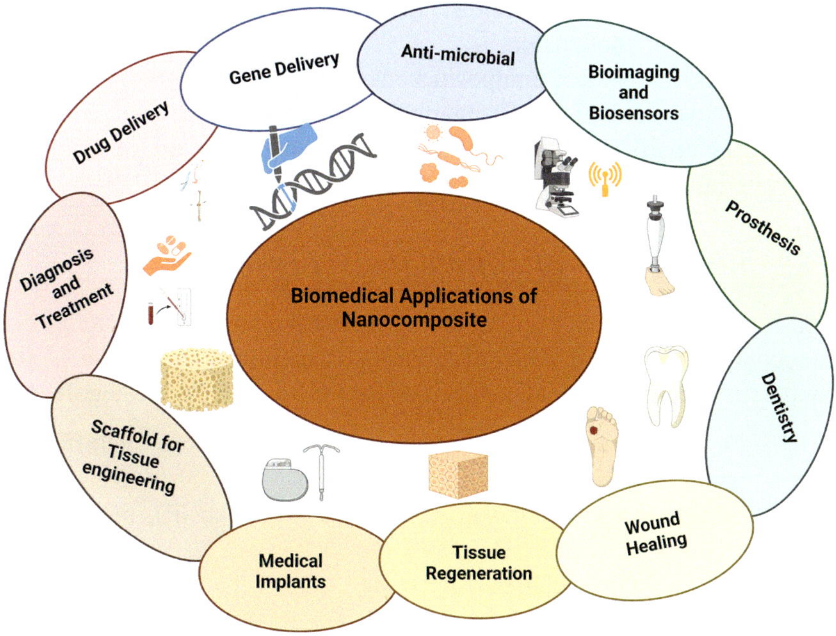

FIGURE 2.1 Different biomedical applications of nanocomposites.

the body (Omanović-Mikličanin et al., 2020), (f) amplify complexity and challenges in terms of design, characterization, and regulatory approval processes of nanocomposite-based biomedical devices (Ramezani and Ripin, 2023), and (g) causes environmental issues during its production, use and disposal (Sen, 2020).

2.4 CLASSIFICATION OF NANOCOMPOSITES

There are several ways to categorize the nanocomposites as described in the following sections.

2.4.1 ON THE BASIS OF MICRO-STRUCTURAL MAKEUP

The majority of developed nanocomposites that have been proven technologically significant are composed of two phases which are divided into three major categories:

1. Nanolayered composites: made up of alternating layers of nanoscale dimension,
2. Nanofilamentory composites: made up of a matrix with embedded nanoscale diameter filaments that are generally aligned; and
3. Nanoparticulate composites, which are made up of a matrix containing embedded nanoscale particles (Okpala, 2013).

2.4.2 ON THE BASIS OF COMPOUND INVOLVED

Considerable effort has been made to achieve the nanoscale structures of nanocomposite through innovative synthetic methods. On this basis, the nanocomposites are basically classified into following three categories:

1. Organic nanocomposite (e.g., copper NPs/Silver Schiff base nanocomposite, etc.),
2. Inorganic nanocomposite (e.g., gold NPs (AuNP), mesoporous silica NPs (MSNP), etc.); and
3. Organic/inorganic nanocomposite.

2.4.3 ON THE BASIS OF MATERIAL USED FOR ITS PREPARATION (HAFEEZ, 2022)

The nanocomposites are classified into three different categories:

1. Metal matrix-based nanocomposites (MMNCs) (e.g., iron oxide nanoparticle-based nanocomposites, silver polymeric nanocomposites, etc.),
2. Ceramic matrix-based nanocomposites (CMNCs); and
3. Polymer matrix-based nanocomposites (PMNCs).

2.4.4 BASED ON PRESENCE OR ABSENCE OF POLYMERIC MATERIAL IN NANOCOMPOSITE

The nanocomposites are classified on the basis of absence or presence of polymeric material into following categories:

1. Polymeric nanocomposite
2. Non-polymeric nanocomposite

2.4.4.1 POLYMERIC NANOCOMPOSITE

Polymeric nanocomposites are the ones in which polymers are involved in their preparation. Polymer nanocomposites (PNCs) are well known for being an effective way to provide materials extraordinary capabilities that are impossible to obtain with just one component. The polymeric NPs can be categorized into six subcategories: (a) polymer/PNC, (b) inorganic/organic polymeric nanocomposite, (c) inorganic-organic hybrid nanocomposite, (d) nanocomposite of polymer and layered silicate (clay), (e) polymer/ceramic nanocomposite, and (f) polymer/metal nanocomposite.

2.4.4.2 NON-POLYMERIC NANOCOMPOSITE

These are also known as inorganic nanocomposites. Non-polymer-based nanocomposites are those that don't include any polymers or components produced from polymers. The non-polymeric nanocomposites are further sub-classified into: (a) metal nanocomposite, (b) ceramic nanocomposite, and (c) ceramic-ceramic nanocomposite (Hule and Pochan, 2007). Examples of various nanocomposites are depicted in Figure 2.2.

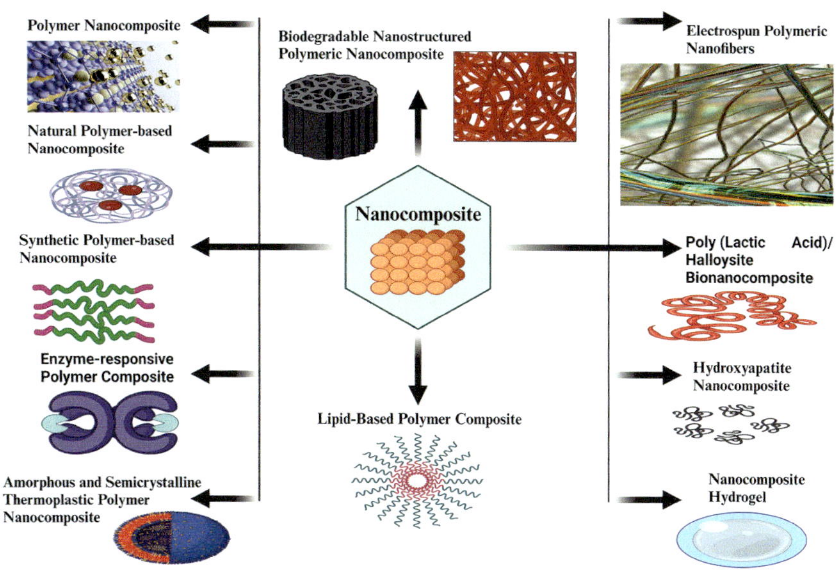

FIGURE 2.2 Examples of nanocomposites having significant biomedical importance.

In this chapter we have discussed some important nanocomposites having made significant contribution in biomedical fields.

1. Natural polymer-based nanocomposite
2. Synthetic polymer-based nanocomposite
3. Amorphous and semi-crystalline thermoplastic PNCs
4. Biodegradable nanostructured polymeric nanocomposite
5. Enzyme-responsive polymer composite
6. Hydroxyapatite nanocomposite
7. Nanocomposite hydrogel
8. Metallic nanocomposite
9. Silica nanocomposite
10. Hybrid nanocomposite
11. Ceramic nanocomposite
12. Quantum dot nanocomposite
13. Pullulan based nanocomposite

2.5 NATURAL POLYMER-BASED NANOCOMPOSITE

Natural polymer-based nanocomposites are composite materials composed of natural polymers as a matrix with nanoscale fillers for reinforcement. These materials combine the benefits of natural polymers, such as biocompatibility and biodegradability, with the additional qualities provided by nanofillers (Radoor et al., 2024). Natural polymer-based nanocomposites that are derived from biopolymers and conducting polymers (e.g., starch and protein-based bio-nanocomposites) have enormous scope in various biomedical fields owing to their improved properties (Bao et al., 2019).

2.5.1 DRUG DELIVERY

PNCs are designed to entrap the drug and deliver it to a specific location of the body where it is required. By modulating release kinetics and targeting mechanisms, these nanocomposites improve medication efficacy, reduce adverse effects, and allow for sustained or controlled drug release (Song and Jeong, 2014). Natural polymer-based nanocomposites, such as chitosan or alginate NPs encapsulated with pharmaceuticals or therapeutic agents, allow for targeted drug administration with regulated release rates. These nanocomposites improve medication stability, bioavailability, and tissue penetration, making them suitable for treating a variety of disorders such as cancer, infections, and inflammation (Tong et al., 2020).

2.5.2 SKIN REGENERATION AND WOUND HEALING

Nanocomposite scaffolds composed of biocompatible polymers and nanofillers can mimic the extracellular matrix of the skin. The natural polymer-based scaffolds promote cell adhesion, cell differentiation and cell proliferation, facilitating skin regeneration in wound healing, and tissue repairing process (Shokrani et al., 2022). Natural polymer-based nanocomposites placed into wound dressings or skin replacements aid in wound healing by creating an environment conducive to tissue regeneration. Natural polymer-based nanocomposites such as cellulose nano-crystals or nano fibrillated cellulose improve wound closure, reduce inflammation, and prevent microbiological infections (Renu et al., 2020).

2.5.3 DENTAL APPLICATIONS

Natural polymer-based nanocomposites find use as materials for various dental applications such as composites and cements. They improve mechanical properties, wear resistance, and esthetics of dental restorations, while also offering antimicrobial properties and fluoride release for caries prevention (Sonika et al., 2023).

2.5.4 OPHTHALMOLOGY

Natural polymer-based nanocomposite materials are being investigated for ocular drug delivery devices, contact lenses, and intraocular implants. They improve medication penetration, lengthen drug retention, and give long-term therapeutic effects, making them useful for treating ocular illnesses such as glaucoma, macular degeneration, and dry eye syndrome (Mozumder et al., 2017).

2.5.5 SURGERIES

Natural polymer-based nanocomposite-based surgical implants, sutures, and wound dressings have increased mechanical strength, biocompatibility, and antibacterial activity. These materials enhance tissue healing, reduce postoperative problems, and improve patient outcomes in a variety of surgical operations (Mozumder et al., 2017).

2.5.6 BIOIMAGING

Natural polymer-based nanocomposites are utilized as contrast agents employed in different biomedical imaging techniques (e.g., magnetic resonance imaging (MRI), fluorescence imaging, and computed tomography (CT). Functionalized NPs incorporated into polymers enable precise imaging of tissues, organs, and cellular structures for diagnostic and therapeutic purposes (Jia et al., 2016).

2.5.7 BONE TISSUE ENGINEERING

Nanocomposite scaffolds mimicking the structure and composition of bone extracellular matrix promote osteogenesis and bone regeneration. Incorporation of bioactive NPs enhances cell attachment, proliferation, and mineralization, making them a prospective candidate for bone tissue engineering and orthopaedic implants (Lowe et al., 2019).

2.5.8 GENE DELIVERY

PNCs act as gene therapy carriers, allowing nucleic acids like DNA and RNA to be delivered to target cells more efficiently. The surface modification and functionalization of NPs allow for more efficient cellular absorption, intracellular trafficking, and controlling gene expression that could be helpful in the treatment of various genetic diseases and cancer (Keles et al., 2016).

2.5.9 TISSUE ENGINEERING AND REGENERATIVE MEDICINE

Nanocomposite scaffolds made from natural polymers (e.g., collagen, gelatin, or hyaluronic acid) and supplemented with NPs like hydroxyapatite (HA) or graphene oxide replicate the extracellular matrix and offer structural support and biochemical cues for tissue regeneration of various organs and systems (e.g., cardiovascular, neurological, and musculoskeletal tissues). The mechanical, chemical, and biological properties of nanocomposites could be modified to facilitate cell adhesion, cell differentiation, cell proliferation, and tissue integration resulting in functional tissue substitutes for transplantation, repairing, and regeneration of damaged tissues and organs (Bharatiya et al., 2023).

2.5.10 AGRICULTURE AND ANALYTICAL RESEARCH

Moreover, due to their unique structure, functionality, conductivity, processability, responsiveness, biocompatibility, and biodegradability, these nanocomposites are appropriate for a wide range of applications, including agriculture and analytical research.

2.5.11 FOOD PACKAGING INDUSTRY

In food packaging industry, natural biopolymer-based nanocomposite films are preferred as they address environmental concerns and consumer demand for packaging of high-quality food products. These films exhibit improved properties, such as amplified modulus, mechanical strength, and water resistance with diminish gas permeability which in turn improves the quality of packaging as well as freshness of the food products (Rhim and Ng, 2007).

2.5.12 BIOMEDICAL SENSOR AND DEVICES

Using natural polymer-based nanocomposites in biomedical uses can be advantageous because of their biocompatibility, biodegradability, and tunable characteristics. Research has also investigated the possible uses of polymer/metal nanocomposites in biomedicine, including durable and strong materials, conductive devices, sensors, and biomedical objects (Zare et al., 2018).

2.6 SYNTHETIC POLYMER-BASED NANOCOMPOSITES

Synthetic polymer-based nanocomposites have emerged as interesting materials that can be used in a wide range of biomedical fields due to their adjustable characteristics, compatibility, and adaptability. Synthetic polymer-based nanocomposites are composite materials made from synthetic polymers as a matrix and nanoscale fillers for reinforcement. These materials combine the desirable features of synthetic polymers, such as tunable mechanical properties and chemical resistance, with the added functionality of nano-fillers (Gobi et al., 2021). Various biomedical applications of synthetic polymer-based nanocomposites are given below.

2.6.1 TISSUE ENGINEERING AND REGENERATIVE MEDICINE

Scaffolds made from synthetic polymer-based nanocomposites offer significant potential in the fields of tissue engineering and regenerative medicine. Nano-fillers, such as HA NPs, carbon nanotubes, or nanocellulose, are mixed with biodegradable polymers (e.g., poly (lactic acid), poly (caprolactone) to mimic the extracellular matrix and offer structural reinforcement for cell adhesion, cell growth, and tissue regeneration. The nanocomposite scaffolds aid in the repair and regeneration of injured tissues and organs, offering options for bone, cartilage, skin, and cardiovascular tissue engineering (Palit and Hussain, 2020).

2.6.2 IMPLANTS AND DEVICES

Synthetic polymer-based nanocomposites are used to create medical implants and devices with superior mechanical qualities, biocompatibility, and functionality. Nanofillers, such as titanium dioxide (TiO_2) NPs, carbon fibers, or shape-memory polymers are added to polymer matrices (e.g., polyether ether ketone, polyurethane) to augment the strength, durability, and bio-stability of implants such as orthopaedic implants, vascular stents, and dental prosthetics. Furthermore, nanocomposites enable the development of smart implants and devices that act in response to stimuli, allowing for controlled drug release, tissue regeneration, and physiological parameter monitoring (Velu et al., 2019).

2.6.3 BIOMEDICAL SENSOR AND DEVICES

Synthetic polymer-based nanocomposites are employed in the fabrication of biosensors and diagnostic devices for sensitive and selective detection of biomolecules and analytes. Nano-fillers like gold NPs, quantum dots (QD), or conducting polymers are integrated into polymer matrices (e.g., polyvinyl alcohol, and polyaniline) to enhance sensing capabilities, signal transduction, and biocompatibility. These nanocomposites enable the development of miniaturized, convenient, and cost-effective biosensing platforms for applications in disease diagnosis, point-of-care testing, and environmental monitoring (Miah et al., 2022).

2.6.4 BIOIMAGING CONTRAST AGENTS

Synthetic polymer-based nanocomposites are employed as contrast agents for biomedical imaging techniques (e.g., MRI, CT, and fluorescence imaging). Nano-fillers such as iron oxide NPs, gadolinium chelates, and QD are encapsulated or surface-functionalized with biocompatible polymers (e.g., PEG, polyvinylpyrrolidone) to improve imaging contrast, biocompatibility, and circulation time. These nanocomposites allow for non-invasive, high-resolution imaging of tissues, organs, and cellular structures for diagnostic and therapeutic applications (Bharatiya et al., 2023).

2.7 AMORPHOUS AND SEMICRYSTALLINE THERMOPLASTIC POLYMER NANOCOMPOSITES

Thermoplastic PNCs have received considerable attention in recent years owing to their unique amalgamated properties, which arise from the synergistic effects among the polymer matrix and nanofillers. Among various types of thermoplastic PNCs, those based on amorphous and semi-crystalline polymers, hold particular promise owing to their numerous applications in biomedical fields.

Amorphous thermoplastic polymers lack long-range molecular order and have a glassy appearance at normal temperature. Nanofillers can greatly improve these polymer's mechanical, thermal, and barrier characteristics. Polycarbonate (PC), polyethylene terephthalate (PET), and polymethyl methacrylate (PMMA) are among the most common amorphous polymers utilized in nanocomposites (Nazarychev and Lyulin, 2023).

On the other hand, semi-crystalline thermoplastic polymers possess both amorphous and crystalline regions within their molecular structure. These polymers exhibit unique mechanical properties due to the presence of crystalline domains, which contribute to rigidity, strength, and thermal stability (Zheng et al., 2023). Because of their unique combination of characteristics, amorphous and semi-crystalline thermoplastic PNCs are used in a wide range of biomedical fields which are as follows:

2.7.1 TISSUE ENGINEERING

Thermoplastic semi-crystalline PNCs are suitable components used as scaffolds in tissue engineering owing to their tunable characteristics and facile process as these affect cell attachment, cell differentiation, and cell proliferation. The human mesenchymal stromal cells are extra responsive to roughness of the surface that is provided by the nanocomposite in an osteogenic environment (Calore et al., 2021).

2.7.2 PACKAGING MATERIALS

Amorphous and semi-crystalline thermoplastic PNCs are used in the packaging industry to manufacture films, containers, and packaging trays for food products. Their barrier properties against moisture, gases, and odors help to extend the shelf life of packaged food products and simultaneously also maintain product freshness and quality (Díez-Pascual et al., 2014).

2.7.3 MEDICAL DEVICES

Nanocomposites play an imperative role in the production of medical devices and tools such as surgical instruments, implants, drug delivery systems, and diagnostic tools. Owing to some of their unique characteristics like biocompatibility, easy to sterilize, and improved mechanical strength make them apposite for a range of biomedical uses, such as implanted devices and tissue engineering (Ramezani and Ripin, 2023).

2.8 BIODEGRADABLE NANOSTRUCTURED POLYMERIC NANOCOMPOSITE

2.8.1 TISSUE ENGINEERING AND REGENERATIVE MEDICINES

Chitosan, a natural polysaccharide compound (linear bi-poly-aminosaccharide) when either blended with other polymers or incorporated into NPs forming nanocomposites, imparts improved biodegradability, controlled release of bioactive molecule, and mechanical strength. Owing to the above characteristics they are used in fabrication of next generation hydrogels and scaffolds for tissue engineering purpose (such as bone, cartilage, dental, hard, etc.) and regenerative medicines (Gupta et al., 2024).

2.8.2 GENE DELIVERY

Chitosan has amino and hydroxyl groups in their backbone and are non-allergenic as well as nontoxic. Due to their hydrophobicity, biocompatibility, and biodegradability, the chitosan incorporated into nanotube, bioglass, noble metals, and other moieties to form nanocomposite is used in gene delivery (Yaqoob et al., 2021).

2.8.3 FOOD PACKAGING

"Nanocellulose", is a term referring to nano-structured cellulose derived from plant sources, offers biodegradability, renewability, and high aspect ratio. Nanocellulose-based nanocomposites exhibit improved mechanical properties, biocompatibility, and barrier performance. Applications include food packaging, biomedical scaffolds, and reinforcement in polymer matrices (Wei et al., 2014).

2.8.4 DENTAL IMPLANTS

Nanodiamonds, consisting of SP^3 hybridized carbon (C) atoms arranged in a diamond lattice structure, offer extraordinary mechanical strength and surface functionalization. In nanocomposites, nanodiamonds improve mechanical properties, wear resistance, and thermal conductivity. Owing to the above properties the nanodiamonds are generally preferred to be utilized in dental implants (Zhang et al., 2018).

2.9 ENZYME-RESPONSIVE POLYMER-BASED NANOCOMPOSITE

Enzyme-responsive polymer composites are smart materials that undergo specific modifications in response to enzyme activity. These composites are created by integrating enzymes or enzyme-responsive moieties into polymer matrices, giving precise control over their characteristics and capabilities (Bhawani et al., 2021). Enzyme-responsive polymer composites are materials that can change their structure or functionality in response to the direct action of enzymes, even under mild conditions like aqueous environments. These composites are effective building blocks for various applications, particularly in drug delivery systems and detection of analytes (Chandrawati, 2016). The incorporation of enzymes in polymer matrices allows for stimuli-responsive behavior, enabling tailored responses to specific enzymatic triggers. The various applications of Enzyme-responsive polymer composite are given below:

2.9.1 DRUG DELIVERY SYSTEMS

Enzyme-responsive polymers can be employed to design drug carriers that release therapeutic agents in reply to specific enzymatic signals present in diseased tissues or organs. This enables drug targeting, controlled drug delivery, dropping off-target effects, and improving therapeutic efficacy (Hu et al., 2012).

2.9.2 BIOSENSORS

Enzyme-responsive polymer composites can be integrated into biosensor platforms for the detection of enzyme activity or the existence of specific enzymes in biological samples. Changes in the characteristics of the polymer matrix (e.g., conductivity, fluorescence, or mass) could be correlated with enzymatic activity, enabling sensitivity, and selectivity during detection of analytes (Wang et al., 2010).

2.9.3 BIOMEDICAL IMPLANTS

Enzyme-responsive polymers can be incorporated into implantable devices or scaffolds for tissue engineering applications. By responding to enzymatic cues present in the local microenvironment, these materials can modulate their properties to promote tissue regeneration, wound healing, or drug release at the implant site (Thornton et al., 2007).

2.9.4 ENVIRONMENTAL REMEDIATION

Enzyme-responsive polymer composites can be employed in environmental applications for the detection and degradation of pollutants. Enzyme-responsive polymers can selectively capture target pollutants from contaminated water or soil and trigger enzymatic reactions for their degradation or transformation into less toxic compounds (Mu et al., 2018).

2.10 HYDROXYAPATITE NANOCOMPOSITES

In recent years, biomaterials have made tremendous advances, particularly in the production of HA nanocomposites. HA nanocomposites are made by fusing HA NPs with a variety of polymer matrices, ceramics, or metals.

2.10.1 TISSUE ENGINEERING AND REGENERATIVE MEDICINE

The HA nanocomposites exhibit increased mechanical strength, osteoconductivity, and bioactivity, imitating bone's native mineral phase. Moreover, these are biocompatible and enable cell adhesion, proliferation, and mineralization. Surface modification approaches, such as biofunctionalization and the insertion of growth factors, seek to optimize cell-material interactions and promote tissue regeneration. Owing to the above characteristics, these nanocomposites hold considerable promise for employment in tissue engineering and in regenerative medicine (Aminzare et al., 2013).

2.10.2 ORTHOPAEDIC AND DENTISTRY SETTINGS

One of the primary benefits of HA nanocomposites is their potential to induce bone repair. These nanocomposites promote growth in tissue and regeneration by providing a scaffold that closely resembles the structure of natural bone. Furthermore, HA NPs in the composite matrix promote osteointegration, that lowers the likelihood of implant failure and enhances long-term clinical results. Besides bone tissue engineering, HA nanocomposites are used in dental implants to improve the stability and durability of the prostheses by better integrating with the surrounding tissues. Furthermore, HA nanocomposites are good carriers for regulated drug administration in orthopaedic and dentistry settings. By encapsulating therapeutic molecules inside the composite matrix, these nanocomposites allow for localized and sustained drug release, encouraging targeted treatment of bone-related disorders and injuries while minimizing systemic side effects (Grande et al., 2009). This adaptability in medication delivery allows for personalized therapy and specific treatment techniques in orthopaedics and dentistry.

2.11 NANOCOMPOSITE HYDROGEL

A nanocomposite hydrogel is a substance having 3-D network structure composed of microscopic NPs embedded in the framework of a base material and has a significant capacity for absorbing and retaining water. These hydrogels could assist in overcoming difficulties associated with the use of normal hydrogels in drug delivery systems. The primary polymers used in nanocomposite hydrogels are starch, chitosan, dextran, alginate, cellulose, and pullulan. Because of their potential uses in the biomedical, personal care, pharmaceutical, biosensor, and cosmetics industries, hydrogels have become increasingly popular in the recent decades (Merino et al., 2015).

2.11.1 STARCH-BASED NANOCOMPOSITE HYDROGELS

These hydrogels are made from plant polysaccharides comprising starch, which includes hydroxyl groups that facilitate hydrogel networking. The use of starch-based hydrogels (SHs) have numerous benefits in the biomedical field due to their unique characteristics with biological systems like biocompatibility, higher biodegradability, excellent mechanical characteristics, hydrophobic properties, permeability, and so forth. Starch based hydrogels are promising for the delivery of drugs, tissue engineering, wound repair, and other biological purposes (Qamruzzaman et al., 2022).

2.11.1.1 CONTROLLED RELEASE DRUG DELIVERY SYSTEM

Shalviri et al. (2010) investigated starch-xanthan gum hydrogels (SXH) as a drug carrier. The study implies that the developed SXHs could be employed as potential film-forming material for development of controlled release drug delivery system owing to their selective permeability (Shalviri et al., 2010).

2.11.1.2 BONE TISSUE ENGINEERING

Salgado et al. (2004) prepared innovative starch-based scaffolds with a specific interest in their application in tissue (orthopaedic) engineering. The prepared scaffolds were non-toxic, exhibited superior mechanical strength, did not impede cell growth, and promoted the growth of cells that resemble human osteoblasts (Salgado et al., 2004). Ngoenkam et al. (2010) developed thermo sensitive injectable hydrogel based nanocomposite by combining starch and chitosan with –glycerol phosphate. The developed nanocomposite was able to maintain chondrocyte phenotype and type II collagen (COL) mRNA and protein expression. It was also reported that the nanocomposite was able to form gels very quickly and confined at the site of injection. Thus, it could be employed as an injectable scaffold for regeneration of bone/cartilage tissue (Ngoenkam et al., 2010).

2.11.1.3 WOUND DRESSING

Batool et al. (2019) developed hydrogel based nanocomposite using green synthesized silver nanoparticle (AgNP), starch and polyvinyl alcohol (PVA). The tensile strength and the swelling tendency of nanocomposite were significantly higher. The antibacterial activity exhibited comparable results with Ciprofloxacin. Thus, it could be used as wound dressing applications helpful for rapid wound healing (Batool et al., 2019). Pal et al. (2006) developed hydrogel based nanocomposite using PVA, starch suspension, and glutaraldehyde as cross-linking agent. The tensile strength of the nanocomposite membrane was good enough as like an artificial skin. The developed nanocomposite was not only protecting the broken skin but was also able to maintain sufficient moisture content over the broken skin. Moreover, it maintains the various intrinsic factors required for tissue repair including growth factor and lysosomes in contact with the wound and absorbs exudate that ultimately helps in faster wound healing (Pal et al., 2006).

2.11.2 DEXTRAN BASED NANOCOMPOSITE HYDROGEL

Dextran nanocomposite hydrogels (DNHs) are the most advanced materials that combine NPs and dextran, a biocompatible polysaccharide, to produce adaptable structures with improved characteristics ideal for various types of biomedical applications. DNHs are enviable materials that not only are required for administration of drugs, cell therapy vectors but also for tissue regeneration devices owing to their biocompatibility and their capacity to protect and deliver the therapeutic agents (Nikpour et al., 2021).

2.11.2.1 CARDIAC TISSUE REPAIR AND REGENERATION

A thermosensitive hydrogel with chitosan, dextran, and β-Glycerol phosphate was produced to treat acute myocardial infarction. After being infused into the hydrogel, umbilical cord mesenchymal stem cells (UCMSCs) showed biocompatibility and myocardial development, suggesting the hydrogel's potential for heart repair (Ke et al., 2020).

2.11.2.2 GLUCOSE SENSOR

Liang et al. (2020) integrated CeO_2/MnO_2 hollow nanospheres into a self-healing hydrogel composed of oxidized dextran and quaternized chitosan for electrocatalysis. To improve the detection of glucose, glucose oxidase was added. A flexible chip containing hydrogel has been coated to avoid leaks, thereby enabling the fabrication of a glucose sensor. This sensor could self-heal for more than 30 days and had a large operating range, quick response time, and high sensitivity. It offers an effective means to handle the issues of leakage, mechanical damage, and sensitivity with continuous glucose monitoring (Liang et al., 2020). Gold NPs were altered with dextran for use as glucose sensors. Dextran/gold NPs were distributed with 16-mercaptohexadecanoic acid (MHA) and Tween 20 to form -COOH groups. Subsequently, the -OH functionality was introduced by adding 2-(2-amino-ethoxy) ethanol. By activating the -OH groups with epichlorohydrin, dextran-coated gold nanocomposite hydrogels were produced by coupling with amino groups (Sharma et al., 2018).

2.11.2.3 BONE REGENERATION

Ghaffari et al. (2020) developed a porous hybrid hydrogel containing dextran along with nanocrystalline β-tricalcium phosphate (β-TCP) for bone tissue creation. This combination enhanced porosity, swelling ability, and bioactivity, and, showed potential as scaffolds for bone regeneration (Ghaffari et al., 2020).

2.11.3 ALGINATE BASED NANOCOMPOSITE HYDROGEL

Alginate nanocomposite hydrogels (ANH) are novel materials composed of NPs with alginate, a naturally occurring hydrophilic biopolymer obtained from marine brown macroalgae, to form adaptable structures with improved characteristics ideal for a variety of biomedical uses.

2.11.3.1 EFFECTIVE DELIVERY OF VACCINE

Chitosan nanocomposites containing alginate were developed to deliver hepatitis B surface antigen (HBsAg) intra-nasally. It was reported that an alginate coat improves release of antigen from chitosan while simultaneously preventing enzymatic degradation (Zhou et al., 2015).

2.11.3.2 BONE-RELATED DISEASES TREATMENT

Nanoconjugate alginate hydrogel has proven to be extremely helpful in the construction of artificial bone in treating problems like trauma caused by congenital abnormalities, infections, and local damage (Liu and Ma, 2004).

2.11.3.3 NASAL DELIVERY OF DRUGS

Alginate lipid nanogels have been proposed as a promising technique for delivering corticosteroids via the nasal route. The experimental results demonstrated that these nanogels prolong the length of the drug's effect on the polyps beyond the nasal valve, with the most effective permeation happening at prolonged residence times (Costa et al., 2021). Alginate nanocomposite formulations are also effective in treating respiratory infections due to their high aerosolization efficiency.

2.11.3.4 ANTIDOT DELIVERY

ANH systems were developed to deliver vaccines against scorpion venoms. Using rabbit models, this study discovered that a drug enclosed in nanohydrogels showed strong immunoprotection against scorpion envenomation (Mohamed and Laraba-Djebari, 2016).

2.11.3.5 OCULAR DELIVERY OF ANTIBIOTICS

Alginate nanocomposites were used to investigate the ocular administration of the antibiotic daptomycin against bacterial endophthalmitis. Using epithelial cell models, it was shown that the mucoadhesive characteristics of these nanohydrogel systems revealed improved drug release permeability (Kopeček, 2007).

2.11.4 CHITOSAN BASED HYDROGELS

Chitosan nanocomposite hydrogels (CsNH) are innovative materials in which chitosan are combined with different NPs to form adaptable structures having improved characteristics suited for various applications in biomedical field.

2.11.4.1 BONE REGENERATION

Bone regeneration could be enhanced by a hydrogel nanocomposite consisting of bioactive glass: NPs, gelatin, and chitosan. The hydrogel's bioactivity was demonstrated *in vitro* through the formation of an apatite phase over the surface using simulated bodily fluid (SBF). Rat bone marrow mesenchymal stem cells (BMSCs) were used in *in vitro* experiments that showed greater cell density and 27% more cell proliferation than control. Biocompatibility was demonstrated *in vivo* in mice, with no obvious tissue changes or inflammation. Following the hydrogel treatment, 3D microcomputed tomography study confirmed increased tissue growth within bone cells (Moreira et al., 2019).

2.11.4.2 CANCER CHEMOTHERAPY

Abou Taleb et al. (2015) developed an oral DDS with sodium alginate/chitosan/hydroxyapatite (SA/CS/HAP) nanocomposite loaded with doxorubicin (DOX) with an aim of improving cancer treatment. The *in vitro* result showed that the nanocomposite releases DOX most efficiently (up to 95% for 10 h) at 37°C and pH 5, preserving the stability and efficacy of the drug (Abou Taleb et al., 2015). Similarly, modified magnetic Fe_3O_4 nanocomposite hydrogel made up through surface RAFT (reversible addition fragmentation chain transfer), co-polymer of acrylic acid and N-isopropyl acrylamide into chitosan could be employed for effective delivery of DOX in cancer chemotherapy (Hosseinzadeh et al., 2019).

2.11.4.3 ANTIBACTERIAL ACTIVITY

The LDH-Fe_2O_3 CsNH could be used for antibacterial activity against four bacterias: *S. aureus, E. coli* O157:H7, *Streptococcus pneumonia, and P. aeruginosa*. The hydrogel containing 64 μg of drug completely stopped the growth of any bacteria. *Candida glabrata* and *Candida albicans* were used in antifungal testing. Subsequently, the minimal fungicidal and inhibitory concentrations were established for the developed LDH@Fe_2O_3 nanocomposite (Gholami et al., 2022).

2.11.5 CELLULOSE NANOCOMPOSITE BASED HYDROGELS

Cellulose nanocomposite hydrogels (CNH) are advanced materials which integrate cellulose nanocrystals along with other polymeric components, for example, COL or chitosan, to form nanocellulose and bio nanocomposites. Nanocellulose and bio nanocomposites offer numerous ideal characteristics such as outstanding skeletal structure (hydrophilic functional groups), biodegradability, cyto/hemocompatibility, micro-nano size effect, and could maintain 3-D structure to a greater extent which offers enhanced mechanical strength ensuring moisture content. Thus, these serve as an important role in biomedical applications, in tissue engineering, antimicrobial therapy, drug delivery, as well as in disease diagnosis, 3-D organ printing, wound dressings, and biosensing (Wang et al., 2021).

2.11.5.1 PACKAGING INDUSTRY

AgNPs integrated into nano-fibrillated bacterial cellulose (AgNP/BC) use a photochemical reduction process involving UV radiation. Though the AgNP/BC is very much selective against growth of both gram-negative bacteria (*Escherichia coli*) and gram positive bacteria (*S. aureus*); it is more effective in inhibiting growth of *E. Coli* in comparison to *S. aureus.* Thus, AgNP/BC could be used in packaging and other biomedical fields (Audtarat et al., 2022).

2.11.5.2 BONE TISSUE ENGINEERING

Fang et al. (2009) created scaffolds composed of a nanocomposite material called bacterial cellulose/hydroxyapatite (BC-Hap). These scaffolds were then studied to understand their impact on the growth and development of human bone marrow-derived stromal cells (hBMSC) involved in bone formation. The human bone marrow-derived mesenchymal stem cells (hBMSC) exhibited enhanced biocompatibility, cell adhesion, and activity when cultured on the BC-HAp nanocomposites as compared to plain BC. This improvement can be attributed to the favorable pore size and enhanced inorganic elements present in the BC-HAp composites. Therefore, the (BC-Hap) nanocomposite has the potential to serve as a material for bone tissue engineering (Fang et al., 2009).

2.11.5.3 REDUCTION OF INTERNAL BLEEDING AND WOUND HEALING

Yuan et al. (2020) used an injured rat liver model to study the efficacy of an oxidized bacterial cellulose (OBC) and chitosan (CS) with collagen (COL) nanocomposite (OBC/COL/CS), to promote fast internal hemostasis. To improve hemostasis and wound healing, COL was used. It was reported that the nanocomposite had enough *in vivo* biodegradation, mechanical strength, antibacterial characteristics, and hemostatic activity. It also improved erythrocyte and platelet adherence and caused bleeding to stop rapidly, making it a promising treatment for reducing internal bleeding and boosting wound healing (Yuan et al., 2019).

2.12 METALLIC NANOCOMPOSITES

Metal/PNCs are composed of metal NPs as nanofiller and polymer as the matrix, frequently show a number of beneficial characteristics and have attracted considerable attention in various biomedical fields.

2.12.1 ANTI-ADHESIVE MEMBRANE AND WOUND DRESSING

Pseudomonas colonies were created in a membrane devoid of AgNPs, whereas the germs were eliminated in a polyamide-silver (Ag/PLA) membrane. According to this data, AgNPs are essential for reducing membrane biofouling, which could lengthen the membrane's lifespan. Additionally, the water fluxes as well as salt rejection capabilities of the membranes were slightly affected by the nanocomposite membranes. At different concentrations of silver (Ag), the ultrafine fibers made of silver/poly (lactic acid) (Ag/PLA) exhibited strong antibacterial action, resulting in a decline in the number of microorganisms, specifically against *S. aureus* and *E. coli*. The lifespan of this nanocomposite exceeds 20 days, but its antibacterial effectiveness reaches up to 94–58% within a mere 12-h timeframe. Consequently, these fibers are employed as membranes to prevent adhesion and as dressings for wounds (Xu et al., 2006).

2.12.2 TRIGGER AUTOPHAGIC CELL DEATH

Bai et al. (2017) investigated the possibility of zinc oxide (ZnO) NPs inducing autophagy using fluorescence microscopy to identify *LC3-II/I* expression using an LC3 antibody. Visualizing LC3 immunofluorescence in SKOV3 cells exposed to increasing concentrations of ZnO NPs revealed significant fluorescence and a crucial component of autophagosomes. Furthermore, ZnO NPs-treated SKOV3 cells increased the expression of *p53* and *LC3-I/II*, which further triggered autophagic cell death (Bai et al., 2017).

2.12.3 DRUG DELIVERY SYSTEM

The biocompatibility and the ability to release the drugs in a controlled manner are the unique characteristics of ZnO nanocomposites that make them a promising carrier in drug delivery system. Bai et al. (2017) investigated ZnO NPs composites as possible pharmaceutical/chemical carriers for increasing drug efficacy and reducing side effects. It was reported that the ZnO NPs composites having a crystal size of 20 nm could reduce ovarian cancer SKOV3 cell survival in a concentration-dependent manner (Bai et al., 2017). Zare-Akbari et al. (2016) studied that ZnO NPs and carboxymethyl cellulose together produce a pH-sensitive nanocomposite that results from the synergistic effects of both ZnO sensitivity and cellulose responsiveness (Zare-Akbari et al., 2016).

2.12.4 GENE DELIVERY

Gold NPs-siRNA, enclosed in a dual layer made from the cationic polymer PEI (acts as a proton sponge), gives stability against siRNAs degradation and is capable of efficiently delivering siRNAs into Vero cells and drastically inhibiting DENV-2 replication in the transfection model (Paul et al., 2014).

2.12.5 CANCER CHEMOTHERAPY

Qin et al. (2011) utilized TiO_2 NPs to administer the chemotherapy drug DOX) either through covalent bonding (TiO_2-DOX) or non-covalent binding (TiO_2/DOX). Due to the pH-sensitive nature of TiO_2/DOX, the generation of DOX in C6 Glioma cells was influenced. As a result, most of the TiO_2-DOX was found in the cytoplasm, with just a tiny amount that was able to enter the nuclei. Regarding the TiO_2/DOX case, a significant amount of DOX was observed within the nuclei. Due to enhanced cellular uptake, the combination of TiO_2/DOX demonstrated much higher toxicity against C6 Glioma tumour cells compared to free DOX (Qin et al., 2011).

2.12.6 BONE GRAFTING

Kunjalukkal Padmanabhan et al. (2013) investigated the application of nanocomposite scaffolds made of wollastonite and HA for bone grafting. They examined scaffolds with different percentages of wollastonite and reported that mechanical strength was greatly boosted by adding 50% wollastonite, the proportion of the (50/50) formulation exhibiting the maximum resistance. When immersed in SBF, the scaffolds formed a carbo-hydroxyapatite (CHP) layer over their surfaces, showing bioactivity. It was reported that an addition of 50% wollastonite, enhances the porous scaffold's strength as well as their bioactive and biodegradable properties that could be used for bone grafting (Kunjalukkal Padmanabhan et al., 2013).

2.12.7 ANTIMICROBIAL ACTIVITY

Anugrah et al. (2020) found that by inhibiting microbial activity, ZnO NPs and polysaccharides could extend the shelf life of food. They claimed that zinc NPs cause oxidative stress, which damages the cell walls of bacteria. When Zn_2+ is released in combination, it can pierce into bacterial cell membranes and pose a threat to their survival (Anugrah et al., 2020).

2.13 SILICA NANOCOMPOSITE

Silica nanocomposites (SNC) are those materials in which silica (SiO_2) NPs are mixed with polymers to form a new material with improved characteristics. The performance, advantages, and flexibility of SSN make them useful in a wide range of industries, including smart devices, coatings, biomedicine, and environmental technology (Zou et al., 2008).

2.13.1 CANCER CHEMOTHERAPY

Gao et al. (2018) employed folic acid as a coating for the silica layer to develop a specialized system capable of delivering magnetic targeted chemotherapy (MTCT). Despite the absence of a thermo-responsive nanogel, the release of DOX followed a consistent pattern regardless of whether an alternating magnetic field was present or not. Simultaneous application of MTCT demonstrated a synergistic apoptotic impact on human MCF-7 breast cancer cells *in vitro* (Gao et al., 2018).

Heggannavar et al. (2018) utilized transferrin conjugated on iron oxide magnetic silica nanoparticle (IOMSNs) surface to target U87 human glioblastoma cells. It was further reported that the developed nanocomposite had indicated that these nanocomposites are capable of crossing the blood-brain barrier (Heggannavar et al., 2018).

2.13.2 PHOTODYNAMIC THERAPY FOR CANCER

Liu et al. (2015) designed double silica-shelled up conversion nanoparticles (UCNPs) that can carry both photosensitizer (PS) molecules with a bioreductive pro-drug (tirapazamine, TPZ) for improved photodynamic treatment (PDT). The NPs, known as TPZ-UC/PS, make use of the idea that UC-PDT-induced hypoxia can bioreductively activate TPZ. When compared to UC-PDT alone, this novel method produces a synergistic therapeutic response that dramatically suppresses tumour growth. This study emphasizes the efficacy of silica-based nanocarriers for combination cancer therapy (Liu et al., 2015).

2.13.3 PACKAGING MATERIAL

Nanoclays are composed of layered silicate minerals in nanometer range which are generally derived from alkaline volcanic ash by hydro-thermal process. It exhibits improved mechanical, gas barrier properties, and flame retardancy in nanocomposites. Moreover, the high aspect ratio and surface area of nanoclays contribute to enhanced properties such as stiffness, toughness, and barrier performance. Thus, these are used as packaging films and flame-retardant coatings (Dillon et al., 2006).

2.14 HYBRID NANOCOMPOSITES

Hybrid nanocomposites are materials made by dispersing inorganic NPs in a macroscopic organic matrix. These nanocomposites integrate the distinct characteristics of the organic matrix, which is commonly a polymer, with those of the inorganic NPs, such as silica. The overall characteristics of the material are improved through the introduction of inorganic NPs into the organic matrix, resulting in improved mechanical, thermal, and electrical properties. Because of their adaptability and superior performance, hybrid nanocomposites have a wide range of applications in a wide range of fields, such as electronics, optical technology, magnetics, biomedicine, and so forth.

2.14.1 BIOSENSOR FOR DETECTION OF TOXIN

Listeriolysin (LLO), a pore-forming toxin is generated by listeria monocytogenes. Zhao et al. (2006) designed a liposome-doped SNC biosensor to detect the LLO. The developed hybrid nanocomposite provides a simple, cost-effective, and reliable substitute for the detection techniques used today by simulating whole-cell assays in LLO detection. Fluorescent dyes were trapped within porous silica by using alcohol free sol-gel manufacturing watermethod. The materials remained unchanged for a minimum of 5 months under normal conditions, and the immobilized liposomes detected LLO with kinetics that varied based on their concentration. Thus, the hybrid nanocomposites could successfully identify bacteria that produce hemolysin (Zhao et al., 2006).

2.14.2 CANCER DIAGNOSIS AND THERAPY

The MRI compatibility of uniform hybrid magnetite NPs, synthesized through the thermal decomposition of iron (III) acetylacetonate in an organic solvent and then modified with 2,3-dimercaptosuccinic acid, was examined. The study revealed a continuous decline in T 2-weighted MR signal strength as the particle size increased from 4 to 12 nm. This suggests an increase in saturation magnetization and results in darker MR pictures. By coupling the shell ligand together with the highly selective cancer-targeting antibody Herceptin, this material was also investigated for the diagnosis of cancer (Jun et al., 2005).

2.14.3 PROMOTE BONE AND CARTILAGE REGENERATION

Tanaka et al. (2005) reported that hybrid magnetic carriers, specifically magneto-liposomes containing TGF-β1, have the potential for disease treatment. TGF-β1, which is known to promote the production of bone and cartilage, was encapsulated within magneto-liposomes in an experiment performed on a rabbit model having articular cartilage abnormalities. This method reduces the noticeable negative effects associated with administering cytokines in high concentrations by facilitating targeted drug delivery. This study emphasizes the benefits and growing trends of employing hybrid magnetic carriers for targeted drug delivery in disease therapy (Tanaka et al., 2005).

2.14.4 THERAPEUTIC ENZYME IMMOBILIZATION

Yu et al. (2005) reported a simple procedure that uses mesoporous silicon spheres as substitute templates to encapsulate biomacromolecules in polyelectrolyte microcapsules. By using this method, it is easier to immobilize enzymes and build polyelectrolyte multilayer capsules, which produce microcapsules characterized by elevated enzyme loading and sustained enzyme activity. Enzymes, crucially, are contained in a "free" state that permits regulated release. Enzymes of different sizes and compositions can be used with this approach, which is adaptable and allows for the adjustment of release through changes in solution pH or salt content. This novel strategy has the potential for many different biomedical uses (Yu et al., 2005).

2.14.5 GENE DELIVERY

Patnaik et al. (2006) proposed a novel siRNA delivery approach using polyethyleneimine (PEI) NPs, alginate, alginic acid, and PEG. Alginate improves system compatibility, the cationic PEI NPs allow gene transfer across the plasma membrane, and the PEG shell improves blood circulation and helps the DDS evade immune cells. Their findings showed that the flexible, positively charged structure could readily interact with the negatively charged cell membrane (Patnaik et al., 2006).

2.14.6 NERVE SUBSTITUTE

Karami et al. (2018) successfully fabricated a nerve graft scaffold composed of chitosan and poly (hydroxybutyrate) nanostructures via electrospinning. It was reported that the biocompatible hydrophilic scaffold exhibits excellent mechanical, electrical, and morphological characteristics that are like those of a natural nerve (Karimi et al., 2018).

2.15 CERAMIC NANOCOMPOSITE

A ceramic nanocomposite is a ceramic matrix reinforced with nanophase materials such as NPs, nanofibers, and nanotubes. It results in improved thermal, mechanical, and functional properties by fusing the distinctive features of nanomaterials with the high strength and hardness of ceramics. Ceramic nanocomposites have numerous applications, including biomedicine, aerospace, and energy (Garmendia et al., 2013).

2.15.1 DENTAL PROSTHESIS

Philipp et al. (2010) investigated clinical efficacy of dental prosthesis frameworks composed of veneered ceria-stabilized tetragonal zirconia–alumina nanocomposites. After a year of operation, the developed nanocomposite was serving as an effective supporting material (Philipp et al., 2010).

2.15.2 TISSUE ENGINEERING

Several bioactive ceramics, such as biphasic calcium phosphate (BCP), tricalcium phosphate, HA, and bioactive glasses, have been used in bone tissue engineering during the past few decades. Wang et al. (2010) conducted a study to examine the *in vitro* reaction of scaffolds composed of porous nano-hydroxyapatite/polycaprolactone (nHAp/PCL). The bone marrow stromal cells (BMSCs) initially formed clusters or agglomerates on the PCL scaffold. However, after 7 days of culture, they fused together to create a large, flat layer of cells that covered the nHA/PCL scaffold. The hydrophobicity of the PCL scaffold, which could obstruct cell adhesion, is most likely the cause of its poor cell development. These findings imply that nHA/PCL scaffolds could have the potential for application in orthopaedic tissue engineering (Wang et al., 2010).

The nano-HAp and micro-hydroxyapatite (micro-HAp) composites exhibit improved mechanical characteristics and improved cellular responses in comparison to conventional polymeric scaffolds. Nejati et al. (2009) evaluated the cell affinity and cytocompatibility of mesenchymal stem cells (MSCs) of nano-HAp and micro-HAp in PCL composite scaffold. It was reported that the nanocomposite had better cell proliferation due to its increased surface area, indicating better cell viability and cytocompatibility (Nejati et al., 2009).

Bionanocomposites in tissue engineering and regenerative medicine (pp. 533–547). Woodhead Publishing Series in Biomaterials.

Aminzare, M., Eskandari, A., Baroonian, M. H., Berenov, A., Hesabi, Z. R., Taheri, M., et al. (2013). Hydroxyapatite nanocomposites: Synthesis, sintering and mechanical properties. *Ceramics International*, *39*, 2197–2206.

Anugrah, D. S. B., Alexander, H., Pramitasari, R., Hudiyanti, D., & Sagita, C. P. (2020). A review of polysaccharide-zinc oxide nanocomposites as safe coating for fruits preservation. *Coatings, 10*, 988.

Audtarat, S., Hongsachart, P., Dasri, T., Chio-Srichan, S., Soontaranon, S., Wongsinlatam, W., & Sompech, S. (2022). Green synthesis of silver nanoparticles loaded into bacterial cellulose for antimicrobial application. *Nanocomposites*, *8*(1), 34–46.

Bai, D.-P., Zhang, X.-F., Zhang, G.-L., Huang, Y.-F., & Gurunathan, S. (2017). Zinc oxide nanoparticles induce apoptosis and autophagy in human ovarian cancer cells. *International Journal of Nanomedicine, 12*, 6521–6535.

Bao, Z., Xian, C., Yuan, Q., Liu, G., & Wu, J. (2019). Natural polymer-based hydrogels with enhanced mechanical performances: Preparation, structure, and property. *Advanced Healthcare Materials, 8*, 1900670.

Batool, S., Hussain, Z., Niazi, M. B. K., Liaqat, U., & Afzal, M. (2019). Biogenic synthesis of silver nanoparticles and evaluation of physical and antimicrobial properties of Ag/PVA/starch nanocomposites hydrogel membranes for wound dressing application. *Journal of Drug Delivery Science and Technology, 52*, 403–414.

Bercea, M., Biliuta, G., Avadanei, M., Baron, R. I., Butnaru, M., & Coseri S. (2019). Self-healing hydrogels of oxidized pullulan and poly (vinyl alcohol). *Carbohydrate Polymers, 206*, 210–219.

Bhakay, A., Rahman, M., Dave, R. N., & Bilgili, E. (2018). Bioavailability enhancement of poorly water-soluble drugs via nanocomposites: Formulation–Processing aspects and challenges. *Pharmaceutics, 10*, 86.

Bharatiya, D., Patra, S., Parhi, B., & Swain, S. K. (2023). A materials science approach towards bioinspired polymeric nanocomposites: A comprehensive review. *International Journal of Polymeric Materials and Polymeric Biomaterials, 72*, 119–134.

Bhawani, S. A., Nisar, M., Tariq, A., Alotaibi, K. M., & Asaruddin, M. R. (2021). Enzyme-responsive polymer composites and their applications. In S. A. Bhawani, A. Khan & M. Jawaid (Eds.), *Smart polymer nanocomposites* (pp. 169–182). Woodhead Publishing Series in Composites Science and Engineering.

Calore, A. R., Srinivas, V., Anand, S., et al. (2021). Shaping and properties of thermoplastic scaffolds in tissue regeneration: The effect of thermal history on polymer crystallization, surface characteristics and cell fate. *Journal of Materials Research, 36*, 3914–3935.

Chandrawati, R. (2016). Enzyme-responsive polymer hydrogels for therapeutic delivery. *Experimental Biology and Medicine, 241*, 972–979.

Costa, C. P., Barreiro, S., Moreira, J. N., Silva, R., Almeida, H., Sousa Lobo, J. M., Silva, A. C. (2021). *In vitro* studies on nasal formulations of nanostructured lipid carriers (NLC) and solid lipid nanoparticles (SLN). *Pharmaceuticals (Basel), 14*(8), 711.

Díez-Pascual, A. M., Shuttleworth, P. S., Gónzalez-Castillo, E. I., Marco, C., Gómez-Fatou, M. A., & Ellis, G. (2014). Polymer blend nanocomposites: Effect of selective nanotube location on the properties of a semicrystalline thermoplastic-toughened epoxy thermoset. *Macromolecular Materials and Engineering, 299*, 1430–1444.

Dillon, D. R., Tenneti, K. K., Li, C. Y., Ko, F. K., Sics, I., & Hsiao, B. S. (2006). On the structure and morphology of polyvinylidene fluoride–nanoclay nanocomposites. *Polymer, 47*, 1678–1688.

Ebrahimian-Hosseinabadi, M., Ashrafizadeh, F., Etemadifar, M., & Venkatraman, S. S. (2011). Preparation and mechanical behavior of PLGA/nano-BCP composite scaffolds during *in vitro* degradation for bone tissue engineering. *Polymer Degradation and Stability, 96*, 1940–1946.

Falsafi, S. R., Topuz, F., Bajer, D., Mohebi, Z., Shafieiuon, M., Heydari, H., et al. (2023). Metal nanoparticles and carbohydrate polymers team up to improve biomedical outcomes. *Biomedicine & Pharmacotherapy, 168*, 115695.

Fang, B., Wan, Y.-Z., Tang, T.-T., Gao, C., & Dai, K.-R. (2009). Proliferation and osteoblastic differentiation of human bone marrow stromal cells on hydroxyapatite/bacterial cellulose nanocomposite scaffolds. *Tissue Engineering Part A, 15*, 1091–1098.

Gaharwar, A. K., Peppas, N. A., & Khademhosseini, A. (2014). Nanocomposite hydrogels for biomedical applications. *Biotechnology and Bioengineering, 111*, 441–453.

Garcia, C. V., Shin, G. H., & Kim, J. T. (2018). Metal oxide-based nanocomposites in food packaging: Applications, migration, and regulations. *Trends in Food Science & Technology, 82*, 21–31.

Gholami, M., Abbasi, N., Ghaneialvar, H., Karimi, E., Afzalinia, A., Zangeneh, M. M., & Yadollahi, M. (2022). Design and synthesis of chitosan magnetic nanocomposite hydrogel beads and investigation of their biological effects. *SSRN Electronic Journal,* 1–35.

Gupta, P., Sharma, S., Jabin, S., & Jadoun, S. (2024). Chitosan nanocomposite for tissue engineering and regenerative medicine: A review. *International Journal of Biological Macromolecules, 254*(1), 127660.

Hafeez, M. (2022). Recent progress and overview of nanocomposites. In A. Sharma (Ed.), *Nanocomposite materials for biomedical and energy storage applications* (pp. 1–30). Intech Open.

Hári, J., & Pukánszky, B. (2011). Nanocomposites: Preparation, structure, and properties. In M. Kutz (Ed.), *Applied plastics engineering handbook: Processing and materials* (pp. 109–142).William Andrew Publishing.

Hassan, A. H. A., & Cutter, C. N. (2020). Development and evaluation of pullulan-based composite antimicrobial films (CAF) incorporated with nisin, thymol and lauric arginate to reduce foodborne pathogens associated with muscle foods. *International Journal of Food Microbiology, 320*, 108519.

Hong, Z., Reis, R. L., & Mano, J. F. (2008). Preparation and *in vitro* characterization of scaffolds of poly (L-lactic acid) containing bioactive glass ceramic nanoparticles. *Acta Biomaterialia, 4*, 1297–1306.

Hu, J., Zhang, G., & Liu, S. (2012). Enzyme-responsive polymeric assemblies, nanoparticles and hydrogels. *Chemical Society Reviews, 41*, 5933–5949.

Hule, R. A., & Pochan, D. J. (2007). Polymer nanocomposites for biomedical applications. *MRS Bulletin, 32*, 354–358.

Gao, Q., Xie, W., Wang, Y., Wang, D., Guo, Z., Gao, F., et al. (2018). A theranostic nanocomposite system based on radial mesoporous silica hybridized with Fe_3O_4 nanoparticles for targeted magnetic field responsive chemotherapy of breast cancer. *RSC Advances, 8*, 4321–4328.

Garmendia, N., Olalde, B., & Obieta, I. (2013). Biomedical applications of ceramic nanocomposites. *Ceramic Nanocomposites,* 530–547. https://doi.org/10.1533/9780857093493.4.530.

Ge, X., Wong, R., Anisa, A., & Ma, S. (2022). Recent development of metal-organic framework nanocomposites for biomedical applications. *Biomaterials, 281*, 121322.

Ghaffari, R., Salimi-Kenari, H., Fahimipour, F., Rabiee, S. M., Adeli, H., Dashtimoghadam, E. (2020). Fabrication and characterization of dextran/nanocrystalline β-tricalcium phosphate nanocomposite hydrogel scaffolds. *International Journal of Biological Macromolecules, 148*, 434–448.

Gobi, R., Ravichandiran, P., Babu, R. S., & Yoo, D. J. (2021). Biopolymer and synthetic polymer-based nanocomposites in wound dressing applications: A review. *Polymers, 13*, 1962.

Grande, C. J., Torres, F. G., Gomez, C. M., & Bañó, M. C. (2009). Nanocomposites of bacterial cellulose/hydroxyapatite for biomedical applications. *Acta Biomaterialia, 5*, 1605–1615.

Heggannavar, G. B., Hiremath, C. G., Achari, D. D., Pangarkar, V. G., & Kariduraganavar, M. Y. (2018). Development of doxorubicin-loaded magnetic silica–pluronic F-127 nanocarriers conjugated with transferrin for treating glioblastoma across the blood–brain barrier using an *in vitro* model. *ACS Omega, 3*, 8017–8026.

Hosseinzadeh, S., Hosseinzadeh, H., Pashaei, S., & Khodaparast, Z. (2019). Synthesis of stimuli-responsive chitosan nanocomposites via RAFT copolymerization for doxorubicin delivery. *International Journal of Biological Macromolecules, 121,* 677–685.

Jia, L., Zhou, T., Xu, J., Xu, Z., Zhang, M., Wang, Y., et al. (2016). Visible light-induced lanthanide polymer nanocomposites based on clays for bioimaging applications. *Journal of Materials Science, 51*, 1324–1332.

Jun, Y., Huh, Y.-M., Choi, J., Lee, J.-H., Song, H.-T., Kim, S., et al. (2005). Nanoscale size effect of magnetic nanocrystals and their utilization for cancer diagnosis via magnetic resonance imaging. *Journal of the American Chemical Society, 127*, 5732–5733.

Karim & M. Jawaid (Eds.), *Polysaccharide-based nanocomposites for gene delivery and tissue engineering* (pp. 245–262). Woodhead Publishing.

Karimi, A., Karbasi, S., Razavi, S., & Zargar, E. N. (2018). Poly (hydroxybutyrate)/chitosan aligned electrospun scaffold as a novel substrate for nerve tissue engineering. *Advanced Biomedical Research, 7*, 44.

Ke, X., Li, M., Wang, X., Liang, J., Wang, X., Wu, S., et al. (2020). An injectable chitosan/dextran/β -glycerophosphate hydrogel as cell delivery carrier for therapy of myocardial infarction. *Carbohydrate Polymers, 229*, 115516.

Keles, E., Song, Y., Du, D., Dong, W.-J., & Lin, Y. (2016). Recent progress in nanomaterials for gene delivery applications. *Biomaterials Science, 4*, 1291–1309.

Khaliq, T., Sohail, M., Shah, S. A., Mahmood, A., Kousar, M., & Jabeen, N. (2022). Bioactive and multifunctional keratin-pullulan based hydrogel membranes facilitate re-epithelization in diabetic model. *International Journal of Biological Macromolecules, 209*, 1826–1836.

Kopeček, J. (2007). Hydrogel biomaterials: A smart future? *Biomaterials, 28*, 5185–5192.

Kunjalukkal Padmanabhan, S., Gervaso, F., Carrozzo, M., Scalera, F., Sannino, A., & Licciulli, A. (2013). Wollastonite/hydroxyapatite scaffolds with improved mechanical, bioactive and biodegradable properties for bone tissue engineering. *Ceramics International, 39*, 619–627.

Li, Y., Gan, Y., Li, C., Yang, Y. Y., Yuan, P., & Ding, X. (2020). Cell membrane-engineered hybrid soft nanocomposites for biomedical applications. *Journal of Materials Chemistry B, 8*, 5578–5596.

Liang, M., & Wong, K. L. (2017). Improving the long-term performance of composite insulators use nanocomposite: A review. *Energy Procedia, 110*, 168–173.

Liang, Z., Zhang, J., Wu, C., Hu, X., Lu, Y., Wang, G., et al. (2020). Flexible and self-healing electrochemical hydrogel sensor with high efficiency toward glucose monitoring. *Biosensors and Bioelectronics, 155*, 112105.

Liu, Y., Liu, Y., Bu, W., Cheng, C., Zuo, C., Xiao, Q., et al. (2015). Hypoxia induced by upconversion-based photodynamic therapy: Towards highly effective synergistic bioreductive therapy in tumors. *Angewandte Chemie, 127*, 8223–8227.

Liu, X., & Ma, P. X. (2004). Polymeric scaffolds for bone tissue engineering. *Annals of Biomedical Engineering, 32*, 477–486.

Lowe, B., Hardy, J. G., & Walsh, L. J. (2019). Optimizing nanohydroxyapatite nanocomposites for bone tissue engineering. *ACS Omega, 5*, 1–9.

Lu, Z. S., & Li, C. M. (2011). Quantum dot-based nanocomposites for biomedical applications. *Current Medicinal Chemistry, 18*, 3516–3528. https://doi.org/10.2174/092986711796642634.

Merino, S., Martín, C., Kostarelos, K., Prato, M., & Vázquez, E. (2015). Nanocomposite hydrogels: 3D polymer–nanoparticle synergies for on-demand drug delivery. *ACS Nano, 9*, 4686–4697.

Miah, M. R., Yang, M., Khandaker, S., Bashar, M. M., Alsukaibi, A. K. D., Hassan, H. M. A., et al. (2022). Polypyrrole-based sensors for volatile organic compounds (VOCs) sensing and capturing: A comprehensive review. *Sensors and Actuators A: Physical, 347*, 113933.

Mohamed, F. A. N., & Laraba-Djebari, F. (2016). Development and characterization of a new carrier for vaccine delivery based on calcium-alginate nanoparticles: Safe immunoprotective approach against scorpion envenoming. *Vaccine, 34*, 2692–2699.

Mohanta, B. C., Dinda, S. C., Palei, N. N., & Deb, J. (2019). Solid lipid based nano-particulate formulations in drug targeting. In R. K. Tyagi, N. Garg, R. Shukla & B. P. Singh (Eds.), *Role of novel drug delivery vehicles in nanobiomedicine* (pp. 1–30). IntechOpen Limited.

Moreira, C. D. F., Carvalho, S. M., Florentino, R. M., França, A., Okano, B. S., Rezende, C. M. F., et al. (2019). Injectable chitosan/gelatin/bioactive glass nanocomposite hydrogels for potential bone regeneration: *In vitro* and *in vivo* analyses. *International Journal of Biological Macromolecules, 132*, 811–821.

Mozumder, M. S., Mairpady, A., & Mourad, A. I. (2017). Polymeric nanobiocomposites for biomedical applications. *Journal of Biomedical Materials Research Part B: Applied Biomaterials, 105*, 1241–1259.

Mu, J., Lin, J., Huang, P., & Chen, X. (2018). Development of endogenous enzyme-responsive nanomaterials for theranostics. *Chemical Society Reviews, 47*, 5554–5573.

Mushtaq, A., Zhao, R., Luo, D., Dempsey, E., Wang, X., Iqbal, M. Z., et al. (2021). Magnetic hydroxyapatite nanocomposites: The advances from synthesis to biomedical applications. *Materials & Design, 197*, 109269.

Song, S. W., Jeong, Y., & Kwon, S. (2014). Photocurable polymer nanocomposites for magnetic, optical, and biological applications. *IEEE Journal of Selected Topics in Quantum Electronics, 21*, 324–335.

Tan, D. Q. (2020). Review of polymer-based nanodielectric exploration and film scale-up for advanced capacitors. *Advanced Functional Materials, 30*, 1808567.

Tanaka, H., Sugita, T., Yasunaga, Y., Shimose, S., Deie, M., Kubo, T., Murakami, T., & Ochi, M. (2005). Efficiency of magnetic liposomal transforming growth factor-beta 1 in the repair of articular cartilage defects in a rabbit model. *Journal of Biomedical Materials Research Part A, 73*(3), 255–263.

Thornton, P. D., Mart, R. J., & Ulijn, R. V. (2007). Enzyme-responsive polymer hydrogel particles for controlled release. *Advanced Materials, 19*, 1252–1256.

Tiwari, A., & Panda, S. K. (2023). Fracture energy of CNT/epoxy nanocomposites with progressive interphase debonding, cavitation, and plastic deformation of nanovoids. *Fatigue & Fracture of Engineering Materials & Structures, 46*, 1170–1189.

Tong, X., Pan, W., Su, T., Zhang, M., Dong, W., & Qi, X. (2020). Recent advances in natural polymer-based drug delivery systems. *Reactive and Functional Polymers, 148*, 104501.

Vaiani, L., Boccaccio, A., Uva, A. E., Palumbo, G., Piccininni, A., Guglielmi, P., Cantore, S., Santacroce, L., Charitos, I. A., & Ballini, A. (2023). Ceramic materials for biomedical applications: An overview on properties and fabrication processes. *Journal of Functional Biomaterials, 14*(3), 146.

Velu, R., Calais, T., Jayakumar, A., & Raspall, F. (2019). A comprehensive review on bio-nanomaterials for medical implants and feasibility studies on fabrication of such implants by additive manufacturing technique. *Materials, 13*, 92.

Wang, C., Bai, J., Tian, P., Xie, R., Duan, Z., Lv, Q., & Tao, Y. (2021). The application status of nanoscale cellulose-based hydrogels in tissue engineering and regenerative biomedicine. *Frontiers in Bioengineering and Biotechnology, 9*, 732513.

Wang, C., Chen, Q., Wang, Z., & Zhang, X. (2010). An enzyme-responsive polymeric superamphiphile. *Angewandte Chemie, 122*, 8794–8797.

Wang, Y., Liu, L., & Guo, S. (2010). Characterization of biodegradable and cytocompatible nano-hydroxyapatite/polycaprolactone porous scaffolds in degradation *in vitro*. *Polymer Degradation and Stability, 95*, 207–213.

Wang, Z., Zhao, S., Kang, H., Zhang, W., Zhang, S., & Li, J. (2018). Mussel byssus-inspired engineering of synergistic nanointerfacial interactions as sacrificial bonds into carbon nanotube-reinforced soy protein/nanofibrillated cellulose nanocomposites: Versatile mechanical enhancement. *Applied Surface Science, 434*, 1086–1100.

Wei, H., Rodriguez, K., Renneckar, S., & Vikesland, P. J. (2014). Environmental science and engineering applications of nanocellulose-based nanocomposites. *Environmental Science: Nano, 1*, 302–316.

Wu, C.-J., Gaharwar, A. K., Schexnailder, P. J., & Schmidt, G. (2010). Development of biomedical polymer-silicate nanocomposites: A materials science perspective. *Materials, 3*, 2986–3005.

Xu, H., Sha, M. Y., Wong, E. Y., Uphoff, J., Xu, Y., Treadway, J. A., et al. (2003). Multiplexed SNP genotyping using the Qbead™ system: A quantum dot-encoded microsphere-based assay. *Nucleic Acids Research, 31*, e43–e43.

Xu, X., Yang, Q., Wang, Y., Yu, H., Chen, X., & Jing, X. (2006). Biodegradable electrospun poly(l-lactide) fibers containing antibacterial silver nanoparticles. *European Polymer Journal, 42*, 2081–2087.

Yaqoob, A. A., Ahmad, A., Ibrahim, M. N. M., & Rashid, M. (2021). Chitosan-based nanocomposites for gene delivery: Application and future perspectives. In S. A. Bhawani, Z.

Yu, A., Wang, Y., Barlow, E., & Caruso, F. (2005). Mesoporous silica particles as templates for preparing enzyme-loaded biocompatible microcapsules. *Advanced Materials, 17*, 1737–1741.

Yuan, H., Chen, L., & Hong, F. F. (2019). A biodegradable antibacterial nanocomposite based on oxidized bacterial nanocellulose for rapid hemostasis and wound healing. *ACS Applied Materials & Interfaces, 12*, 3382–3392.

Zare, Y. (2016). Study of nanoparticles aggregation/agglomeration in polymer particulate nanocomposites by mechanical properties. *Composites Part A: Applied Science and Manufacturing, 84*, 158–164.

Zare, E. N., Motahari, A., & Sillanpää, M. (2018). Nanoadsorbents based on conducting polymer nanocomposites with main focus on polyaniline and its derivatives for removal of heavy metal ions/dyes: A review. *Environmental Research, 162*, 173–195.

Zare, Y., & Shabani, I. (2016). Polymer/metal nanocomposites for biomedical applications. *Materials Science and Engineering: C, 60*, 195–203.

Zare-Akbari, Z., Farhadnejad, H., Furughi-Nia, B., Abedin, S., Yadollahi, M., & Khorsand-Ghayeni, M. (2016). PH-sensitive bionanocomposite hydrogel beads based on carboxymethyl cellulose/ZnO nanoparticle as drug carrier. *International Journal of Biological Macromolecules, 93*, 1317–1327.

Zhang, L., & Chen, L. (2016). Fluorescence probe based on hybrid mesoporous silica/quantum dot/molecularly imprinted polymer for detection of tetracycline. *ACS Applied Materials & Interfaces, 8*, 16248–16256.

Zhang, Y., Rhee, K. Y., Hui, D., & Park, S.-J. (2018). A critical review of nanodiamond based nanocomposites: Synthesis, properties and applications. *Composites Part B: Engineering, 143*, 19–27.

Zhao, J., Jedlicka, S. S., Lannu, J. D., Bhunia, A. K., & Rickus, J. L. (2006). Liposome-doped nanocomposites as artificial-cell-based biosensors: Detection of listeriolysin O. *Biotechnology Progress, 22*, 32–37.

Zhao, L., Zhou, Y., Zhang, J., Liang, H., Chen, X., & Tan H. (2023). Natural polymer-based hydrogels: From polymer to biomedical applications. *Pharmaceutics, 15*, 2514.

Zheng, Z., Liu, L., Li, M., Sun, W., Liang, Y., Chen, J., et al. (2023). Molecular complexation of polymers and metal oxide clusters for semicrystalline, thermoplastic anhydrous proton exchange membranes in fuel cells. *The Journal of Physical Chemistry Letters, 14*, 5966–5974.

Zhou, Q. T., Leung, S. S. Y., Tang, P., Parumasivam, T., Loh, Z. H., & Chan, H.-K. (2015). Inhaled formulations and pulmonary drug delivery systems for respiratory infections. *Advanced Drug Delivery Reviews, 85*, 83–99.

Zhou, F., Yan, C., Liang, T., Sun, Q., & Wang, H. (2018). Photocatalytic degradation of Orange G using sepiolite-TiO2 nanocomposites: Optimization of physicochemical parameters and kinetics studies. *Chemical Engineering Science, 183*, 231–239.

Zou, H., Wu, S., & Shen, J. (2008). Polymer/silica nanocomposites: Preparation, characterization, properties, and applications. *Chemical Reviews, 108*, 3893–3957.

CHAPTER 3

Basic and Modern Methods for the Preparation of Nanocomposites for Biomedical Applications

REHAM S. ELEZABY

Department of Pharmaceutics and Industrial Pharmacy, Faculty of Pharmacy, Ain Shams University, Cairo, Egypt

ABSTRACT

A nanocomposite is a smart category of nanomaterials with superior mechanical, physical, thermal, optical, and barrier properties. It consists of multiphase solid material with distances in the nanoscale range between the different phases. Nanocomposites have huge applications in multidiscipline fields such as biomedical medicine and drugs, environment remediation, food and agriculture, and so forth. Polymer-based nanocomposites such as polymer/ceramic, inorganic/organic polymer, and inorganic/organic hybrid, and non-polymer-based nanocomposites such as metal and ceramic nanocomposites are available. Different methods are available to prepare nanocomposites such as *in situ* polymerization, melt mixing, solution mixing, precipitation, sol-gel process, template-assisted method, layer by layer, and electrospinning. Appropriate selection of the method of preparation of nanocomposites is crucial to suit the matrix type, nanoparticles type, and their general intended use. Consequently, in this chapter, we will discuss the different methods of preparation of nanocomposites, either basic or modern ones, their advantages and disadvantages, and how can the method of preparation affect the different properties of nanocomposites.

Nanocomposites and Nanomaterials in Biomedical Applications. Vivek Dave, Deependra Singh, and Pallerla Naveen Reddy (Eds.)

DOI: 10.1201/9781779641236-3

3.1 INTRODUCTION

Composite materials in which at least one of its components has nanometer-sized dimensions are called nanocomposites. They are very smart nanomaterials known for their excellent performance and superior properties (Mousumi, 2020; Muhammed Shameem et al., 2021). They are composed of two or more distinct components, each having unique chemical and physical characteristics, and they typically have distinct phases divided by interfaces. Because of this, the nanocomposite exhibits unique properties that cannot be obtained from any one of the constituent parts alone. The larger part of nanocomposites is known as a matrix. Adding components to the matrix material to improve the properties of nanocomposites is a process known as reinforcement. These reinforcement materials are mostly nanocarriers that act as nano-filler materials. Different kinds of nanoparticles (NPs) and matrix materials are used to form nanocomposite. There are three classes of nanocomposites based on matrix material which are metal matrix nanocomposites (MMNC), polymer matrix nanocomposites (PMNC), and ceramic matrix nanocomposites (CMNC) (Camargo et al., 2009).

Due to the presence of different types of nanocomposites, different methods of preparation (Table 3.1) have emerged. Some of these methods can be used for the preparation of more than one type of nanocomposites, while others are used for the preparation of a certain type of nanocomposites. For the preparation of PMNC, *in situ* polymerization, melt mixing, solution mixing, sol-gel process, and electrospinning methods can be used (Pandey, 2020; Thomas et al., 2007). Meanwhile, for the preparation of CMNC, powder process, polymer precursor process, vapor deposition methods (CVD and PVD), and sol-gel process can be used (Camargo et al., 2009; Rane et al., 2018). Furthermore, in the case of MMNC, spray pyrolysis, liquid infiltration, rapid solidification process (RSP), high-energy ball milling, vapor deposition methods (CVD and PVD), and chemical processes (sol-gel, colloidal) can be used as preparation methods (Camargo et al., 2009; Rane et al., 2018).

Recently, some modern techniques such as layer by layer (LBL) method, template method, and green synthesis began to be used in the preparation of nanocomposites. These methods seem to be more cost-effective, safe, biocompatible, and eco-friendly. Besides, they could overcome many of the limitations of the conventional methods.

Consequently, in this chapter, we will discuss the methods used for the preparation of different types of nanocomposites, either the basic or modern ones, together with the advantages and disadvantages of each method.

3.2 BASIC METHODS OF PREPARATION OF NANOCOMPOSITES

The methods used for the preparation of nanocomposites differ according to the type of nanocomposites either PMNC, CMNC, or MMNC.

3.2.1 METHODS OF PREPARATION OF POLYMER MATRIX NANOCOMPOSITES

PMNC (Figure 3.1) are nanocomposites that contain a certain inorganic nanomaterial integrated into a suitable polymer. Different inorganic materials can be used such as carbon nanotubes, iron oxyhydroxides, magnetite, montmorillonite, silica, graphene oxide, graphene oxide-magnetite, bentonite, iron–nickel oxide, gold and silver NPs, titanium dioxide, zinc selenide, zinc oxide, vanadyl phosphate, and many more (Camargo et al., 2009).

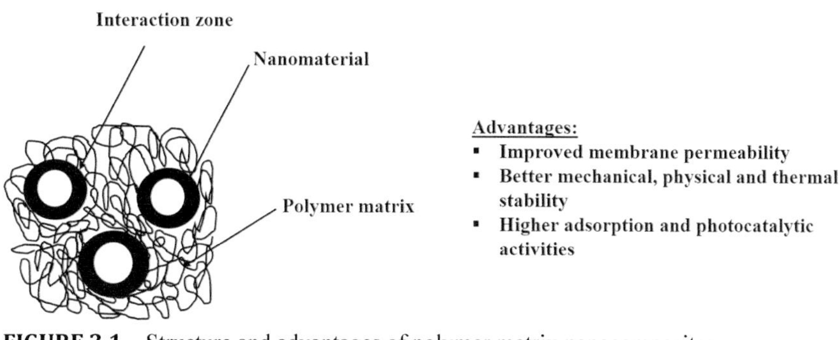

FIGURE 3.1 Structure and advantages of polymer matrix nanocomposites.

Because of their excellent mechanical properties, durability, and processing adaptability, thermoplastic polymers have been extensively utilized in the production of PMNCs. Polyethylene, polyvinylidiene fluoride, polypropylene, polystyrene, polyethylene terephthalate, poly (vinyl chloride), polyethersulfone, polyamide, polybutylene terephthalate, styrene acrylonitrile, polycarbonate, poly (p-phenylene oxide), and acrylonitrile butadiene are some examples of this special class. The inherent difficulties in using regular NPs are currently thought to be addressed in part by the use of PMNC. The properties of PMNC rely on the type of included nanoparticle, its size, concentration, shape, its interaction with the matrix material, and the method of preparation. Different methods are available for the preparation of PMNC which will be discussed in the upcoming section.

TABLE 3.1 Different Methods Used for the Preparation of Nanocomposites.

Categories of Nanocomposites	Polymer Matrix Nanocomposite	Ceramic Matrix Nanocomposite	Metal Matrix Nanocomposite
Examples	Polyester/TiO_2, polymer/CNT, polymer/layered double hydroxides.	Al_2O_3/SiO_2, SiO_2/Ni, Al_2O_3/TiO_2	Fe-Cr/Al_2O_3, Ni/Al_2O_3, Co/Cr, Fe/MgO
Methods of preparation	• *In situ* polymerization • Melt mixing • Solution mixing • Sol-gel process • Electrospinning	• Powder process • Polymer precursor process • Vapor deposition methods (CVD and PVD) • Sol-gel process	• Spray pyrolysis • Liquid infiltration • Rapid solidification process • High-energy ball milling • Vapor deposition methods (CVD and PVD) • Chemical processes (sol-gel, colloidal)

3.2.1.1 *IN SITU POLYMERIZATION*

In this method, the inorganic nanomaterial (Figure 3.2) is dispersed into a solution of the polymer matrix monomers. After the formation of a homogenous mixture, the polymerization process starts with the addition of a suitable activator/initiator/catalyst (Grothe et al., 2012; Paksoy et al., 2022). Ultrasounds may be used in this method to aid the dispersion. Moreover, the polymerization process may need exposure to heat or light. The polymer synthesized is called a thermoset.

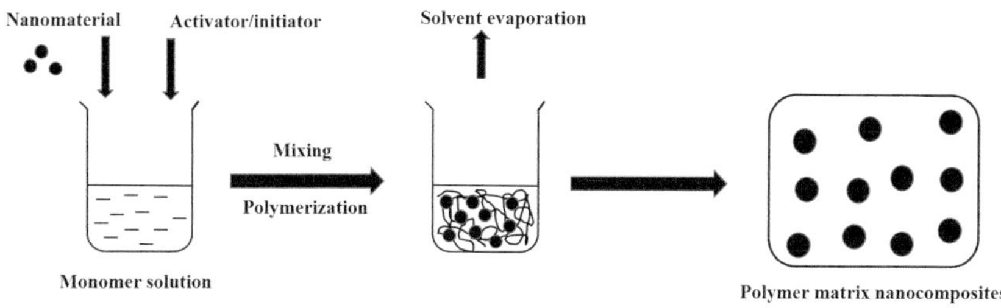

FIGURE 3.2 Preparation of polymer matrix nanocomposites by *in situ* polymerization method.

Different matrix polymers can be prepared by this method such as epoxy, cyanate, bismaliemide, and phenolic polymers. The technique of this method seems to be easy, and there is a thermodynamic compatibility at the reinforcement interface. Additionally, stronger bonds can be formed because there is no contamination at the reinforcement surfaces. Moreover, the method can be used in cases where melt and solutions methods cannot be used, for example, in the case of insoluble or thermally unstable polymer composites (Camargo et al., 2009; Tang et al., 2019). The method can also be used for the preparation of nanocomposites with nanotubes. However, the control of the polymerization process is not easy at all.

3.2.1.2 MELT METHOD

It is the most widely used and preferred method for the preparation of clay/polymer nanocomposites. In this method, the particles (i.e., intercalated clay or NPs) are dispersed in the polymer melt (Figure 3.3); then the nanocomposites are obtained by extrusion (Fawaz and Mittal, 2014; Verma et al., 2019).

Because no solvents are used, it has the benefit of being environmentally friendly. It is also thought to be compatible with industrial procedures like extrusion and injection molding, which increases its usability and lowers its cost. But still, the process‘s high temperatures can harm the surface modification of the filler. For instance, alkyl ammonium-modified organoclays typically break down over 140°C; nevertheless, the processing temperature of melt intercalation occurs between 190 and 220°C. Thus, obtaining optimal dispersion and exfoliation requires careful consideration of various factors, one of which is processing condition optimization. For example, degradation can be prevented by using more thermally stable modified materials or by operating at lower temperatures.

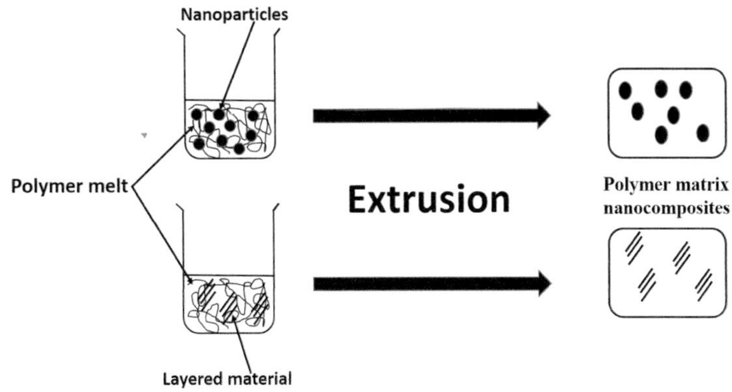

FIGURE 3.3 Melt method for the preparation of polymer matrix nanocomposites.

3.2.1.3 SOLUTION MIXING METHOD

This method is considered the simplest method used for the preparation of polymer nanocomposites. In this method, first, a solution of the polymer is mixed with a dispersion of the nanomaterial in the same solvent. This is followed by solvent evaporation and/or nanocomposite precipitation (Fawaz and Mittal, 2014). Chanakul et al. prepared polydiacetylene/zinc (II)/zinc oxide nanocomposites using this method. They simply mixed all the precursors in liquid phase (Figure 3.4) at room temperature followed by a short period of incubation at 4–70°C and then photopolymerization (Chanakul et al., 2021). They could obtain a small size of 60 nm. The method is suitable for large-scale production as it needs low energy and short processing time.

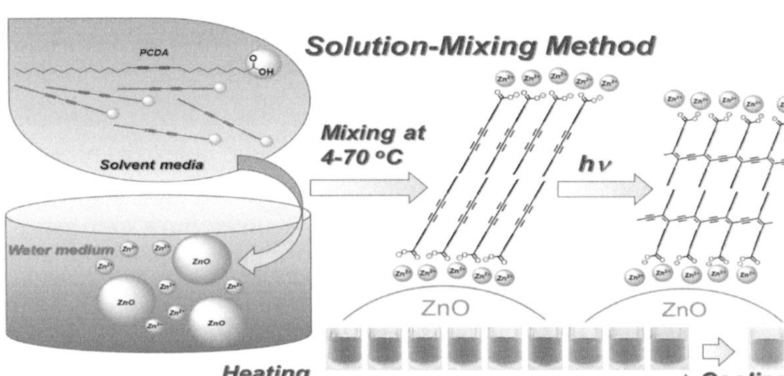

FIGURE 3.4 Preparing the reversible thermochromic PDA/Zn^{2+}/ZnO nanocomposites by the solution-mixing method.

Source: Reprinted from Chanakul et al. (2021) with permission from Elsevier.

Araby et al. prepared elastomer/graphene nanocomposites using the solution mixing method. They could use this method to overcome the challenge of graphene layer stacking (Araby et al., 2014). They proved that the solution mixing method was better than the melting method in enhancing the reinforcement effect of graphene platelets. This was explained on the basis that it keeps the high aspect ratio of graphene platelets, which results in a filler-filler network at a low volume fraction, and because it gives elastomer molecules more interlayer spacing for intercalating.

3.2.1.4 SOL-GEL METHOD

Generally, a more wet chemical method for preparing different nanostructures, especially metal oxide NPs, is the sol-gel process (Kianfar and Suksatan, 2021). It is one of the chemical methods used for the preparation of nanocomposite films. As it can be understood from the name of the method, it consists of two stages, first, the formation of sol, and second, the formation of gel. The term "sol" means a colloidal dispersion of solid particles in a liquid phase while the term "gel" indicates the formation of interconnected network between different phases.

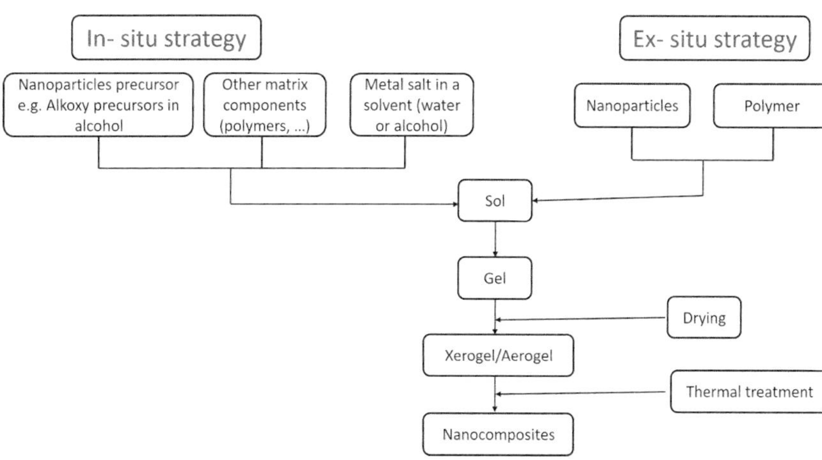

FIGURE 3.5 General scheme for the preparation of sol-gel nanocomposites by *in situ* and *ex situ* strategies.

Both *in situ* and *ex situ* strategies are used for the preparation of sol-gel nanocomposites (Figure 3.5). Through the *in situ* strategy, a sol comprising of sources for component materials is prepared (Bokov et al., 2021;

Guglielmi, 2014). Precursors for the matrix and the dispersed phase are prepared in a sol that typically begins with two different solutions. Although both might have undergone some sort of processing before mixing, the end product is a one-phase system. The sol usually contains metal alkoxides and metal chlorides, which are precursors for metal oxides. It usually also contains metallic stabilizers, catalysts, and other additives for porosity generation. However, in the *ex situ* strategy, separate synthesis of the NPs is performed first followed by mixing with the polymer. The sol subsequently goes through polycondensation and hydrolysis processes, which cause it to gradually change and eventually form a gel-like network. Significant shrinkage happens as the solvent liquid is eliminated during the drying process. The substance that is produced is called a xerogel. The network does not shrink when solvent removal takes place under hypercritical (supercritical) conditions; instead, an extremely porous, low-density substance known as aerogel is created. Viscos sintering, or the shrinkage of xerogel as a result of a tiny amount of viscous flow, is produced when xerogel is heat-treated at a high temperature, which successfully turns the porous gel into a nanocomposite. Different factors can affect the sol-gel method such as solvent type, timing, pH, precursors, water/metal ratio, and so forth. Controlling these different factors allows flexible control over the structural and chemical characteristics of the final product (Sahini et al., 2023).

The compatibility of this method with different polymers and polymerization processes, which permits the formation of NPs in the presence of organic molecules, is one of its most intriguing benefits (Camargo et al., 2009; Guglielmi, 2014). Moreover, the method is considered to be simple and does not require high processing temperatures. Additionally, the method is versatile and its products possess high purity. Furthermore, the method allows the formation of 3D polymers containing metal-oxygen bonds. Actually, the method is particularly useful for creating composite materials made of liquids or viscous fluids derived from alkoxides. However, compared with the mixing method, nanocomposites prepared by this method may suffer from more shrinkage and lower void spaces.

3.2.1.5 ELECTROSPINNING

A novel method for creating polymer nanofibers (NFs) with nanofillers using high voltage to create an electrostatic force that spins solutions into NFs, or polymer microfibers, is known as electrospinning (Kausar and

Ahmad, 2023; Quoc Pham et al., 2021). First, before electrospinning, the polymer is dissolved in a solvent (Figure 3.6), and the resulting homogenous solution is then collected into a syringe. The syringe's-controlled diameter metallic needle is attached at the tip. A syringe pump regulates the solution's flow rate. The distance between the needle tip and a metallic collector is adjusted to be between 10 and 30 cm. A high voltage (between 5 and 50 kV) will cause the polymer solution to be drawn from the needle to the collector by electromagnetic forces. This will result in the formation of charged solution jets, or Taylor cones, at the needle's tip.

The nanofiber fabrication process stability is determined by the Taylor cone formation. The polymer droplet will be ejected as a jet to the grounded target that serves as a collector when the voltage rises because the electric force will remove the droplet's surface tension. The polymer jet causes the creation of ultrathin fibers when it is subjected to the electromagnetic force. A nanofiber polymer mat is created on the collector surface during this process as the solvent evaporates.

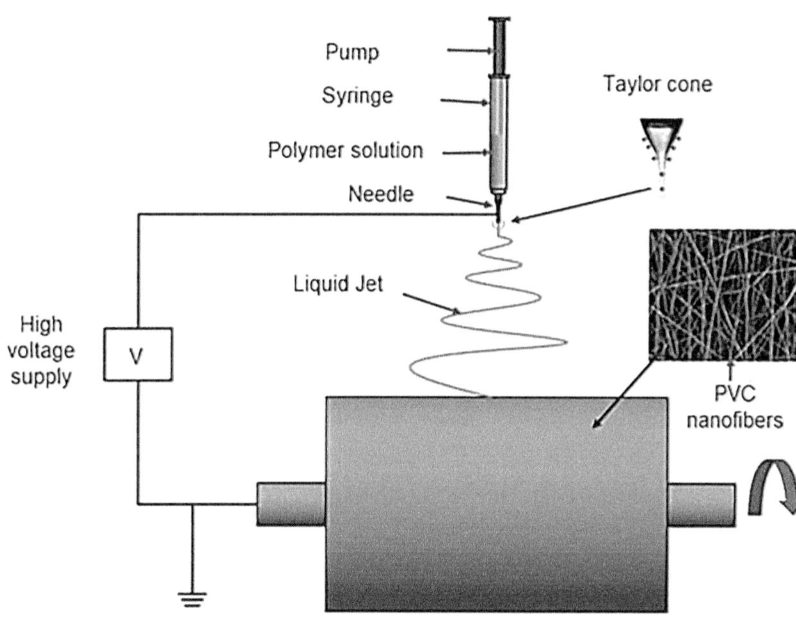

FIGURE 3.6 A schematic illustration of the electrospinning process setup.

Source: Reprinted from Quoc Pham et al. (2021), Copyright © 2021 by the authors. Licensee MDPI, Basel, Switzerland. https://creativecommons.org/licenses/by/4.0/

Many factors can affect the electrospinning process such as polymer solution parameters, process parameters, and environmental parameters (Jose Varghese et al., 2019; Refate et al., 2023). From the solution parameters, type, concentration and molecular weight of the polymer used together with the viscosity, surface tension, and electrical conductivity of the solution.

One of the primary variables affecting the formation of composites in electrospinning is viscosity. Because it dictates how the network structure and electrospun fiber are formed. Viscosity can be changed reasonably easily by: (1) adjusting the polymer and filler concentrations, (2) altering the component's feeding order while mixing, (3) adjusting the solution's pH, and (4) functionalizing the solution's component parts to customize how they interact with one another. Additionally, from the process parameters, nozzle diameter, applied voltage, flow rate, and collecting distance. Meanwhile, surrounding humidity and temperature can also affect the electrospinning process. All of these factors affect the properties of the prepared NFs such as particle size, morphology, and surface charge.

For the preparation of electrospun nanocomposites, NPs are first added to the polymer solution and then electrospun together. Emam et al. prepared cerium oxide (CeO_2)/polyacrylonitrile (PAN) nanocomposites. They optimized first different factors such as voltage, flow rate, distance, and polymer concentration in preparing plain PAN NFs (Emam et al., 2023). The optimum PAN NFs were loaded with CeO_2 NPs in accordance with the SEM micrographs (Figure 3.7) of the manufactured NFs. To create CeO_2 NP-loaded PAN NFs, CeO_2 NPs (0.25%, w/v) were first dispersed in dimethyl formamide (DMF). The dispersion was then gradually added to an 8% w/v PAN solution in DMF and stirred overnight at 50°C to create a homogenous mixture. The mixture was then electrospun into composite NFs at a feed rate of 0.5 mL/h, an applied voltage of 26 kV, and a needle-plate collector distance of 15 cm.

3.2.2 METHODS OF PREPARATION OF CERAMIC MATRIX AND POLYMER-CERAMIC NANOCOMPOSITES

A unique kind of nanocomposites is known as ceramic matrix nanocomposites (CMNC). They have both ceramic matrix material and ceramic reinforcement material. Sometimes, the two sections of the structure are

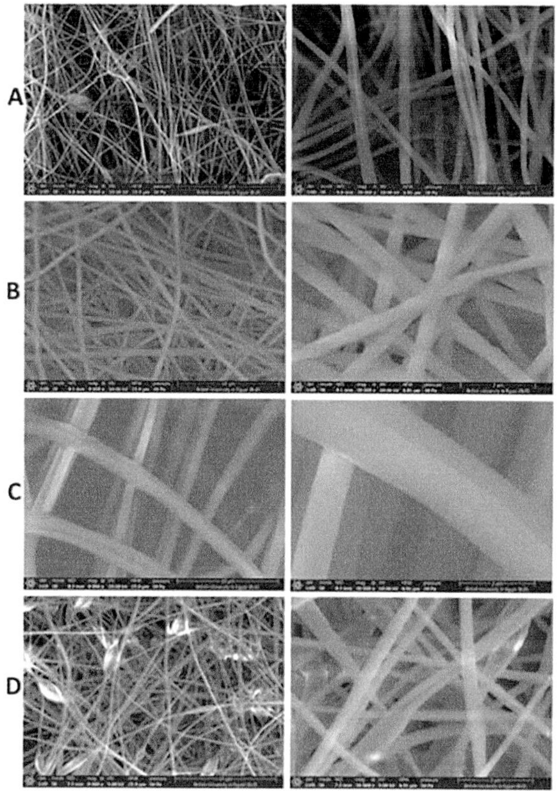

FIGURE 3.7 SEM images with magnification powers of 8000 × (left panel) and 30,000 × (right panel) of (A) blank PAN NFs (F1, 6%), (B) blank PAN NFs (F2, 8%), (C) blank PAN NFs (F3, 10%), and (D) CeO_2 NPs-PAN NFs (F4).

Source: Reprinted from Emam et al. (2023),

made of the same type of ceramic, and extra secondary fibers might be added (Bose et al., 2013). As a result, CMNC is regarded as a subset of ceramics and composite materials. The common reinforcement materials used in this type of nanocomposites are carbon (C), mullite or alumina silica (Al_2O_3-SiO_2), silicon carbide (SiC), and alumina (Al_2O_3). CMNCs are typically recognized via a fiber/matrix. Because of their exceptional qualities as being resistant to corrosion, high oxidation temperatures, and better wear resistance than metals in high-temperature settings, ceramic nanocomposites have found extensive industrial use (Kamalan et al., 2020). In addition, we have an intermediate class that is known as polymer-ceramic nanocomposites. The properties of polymer-ceramic nanocomposites are in the middle of those of organic polymers and inorganic ceramics. Hence, the main aim of preparing this type of composite is to have combinations of properties from both ingredients (Jayasuriya, 2017). In biomedical applications, polymer-ceramic nanocomposites present several potential benefits that surpass the capabilities of either polymer or ceramic material on their own. Hence, because inorganic materials are naturally stiffer and strong, nanocomposites with better mechanical properties can be produced by combining polymers with ceramic phases (Victor and Muthu, 2014).

For the preparation of ceramic matrix nanocomposites, numerous techniques have been reported. Conventional powder method, polymer precursor route, spray pyrolysis, vapor techniques (e.g., chemical vapor deposition (CVD) and physical vapor deposition (PVD)), and chemical methods such as the sol-gel, colloidal and precipitation processes, are the most often utilized techniques for nanocomposites preparation. The best methods for achieving good dispersion of ceramics into various types of polymeric matrices are solution-blending techniques, melt processing, and sol-gel techniques, according to some research studies (Ramdani, 2019).

3.2.2.1 POWDER METHOD

The process involves raw materials selection, which are primarily powders. The selected powders must have small and uniform particle sizes and high purity (Camargo et al., 2009). Next, the materials are mixed using wet ball milling or attrition milling techniques in organic or aqueous media. Then, the materials are dried using lamps, freeze-drying, or ovens. Finally, the solid material is consolidated using hot pressing, injection molding, gas pressure sintering, slip casting, or pressure filtration. A mechanical milling process can be used to produce nano-size particles in large quantities at a reasonable cost (Selvam et al., 2021). During the milling process, the kinetic energy generated by the ball is converted into powder. Numerous variables, including the types of mills, the powder used in the milling process, speed, ball size, wet or dry milling, temperature, and milling time, control the transfer of kinetic energy. Figure 3.8a shows the scheme for the preparation of Al_2O_3/SiC nanocomposite using the powder method. The method seems to be simple; however, it needs a very high temperature. Meanwhile, agglomeration, poor phase dispersion, and formation of secondary phases in the product can occur.

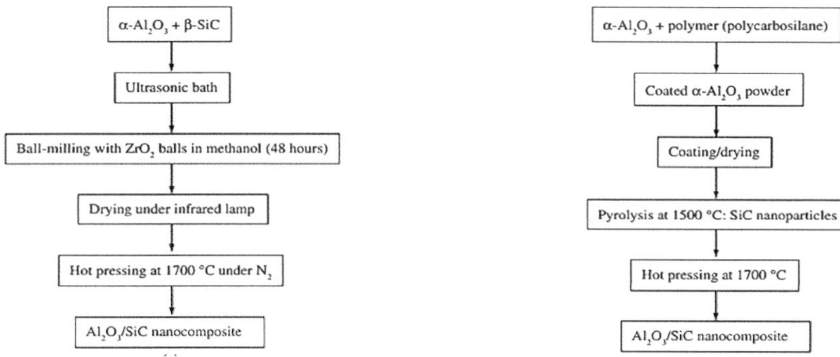

FIGURE 3.8 (a) Powder processing method, and (b) polymer precursor method.

Source: Reprinted from Camargo et al. (2009) with permission from Material Research.

3.2.2.2 POLYMER PRECURSOR METHOD

This method involves (Figure 3.8b) blending of the polymeric precursor and the matrix material together. Next, using a microwave oven, pyrolysis of the mixture is performed to produce the reinforcing particles (Camargo et al., 2009; Rane et al., 2018). Finer particles with better reinforcement dispersion can be prepared using this method. However, as a result of the agglomeration and dispersion of ultra-fine particles, phase-segregated and inhomogeneous materials can be formed. Meanwhile, this technique needs a very high temperature.

3.2.2.3 VAPOR TECHNIQUES

In PVD, a solid source is evaporated or otherwise atomized to deposit one or more depositing species into the coating chamber. In CVD, the depositing species are introduced into the coating chamber in a gaseous or vaporized state during the deposition process. A typical CVD system is depicted in Figure 3.9, where reactive gases, also known as precursor gases, are introduced into the chamber at the proper temperature. The hot substrate is in contact with the gases as they move through the reactor. Following their reaction with the substrate, a solid layer of a ceramic coating like Al_2O_3, SiC, or Cr_2O_3 is deposited on it. It should be mentioned that argon or another neutral gas is frequently used as a diluent in the CVD process. Finally, the produced volatile byproducts are eliminated by the reaction chamber's gas flow. The two limiting challenges in this process are typically the deposition temperature and pressure. The main advantage of this method is that the low temperature prevents autodoping and prevents impurities from being incorporated into the reactor walls. The primary drawback is the absence of extremely volatile, nontoxic, and nonpyrophoric precursors (Dahmen and Meyers, 2003).

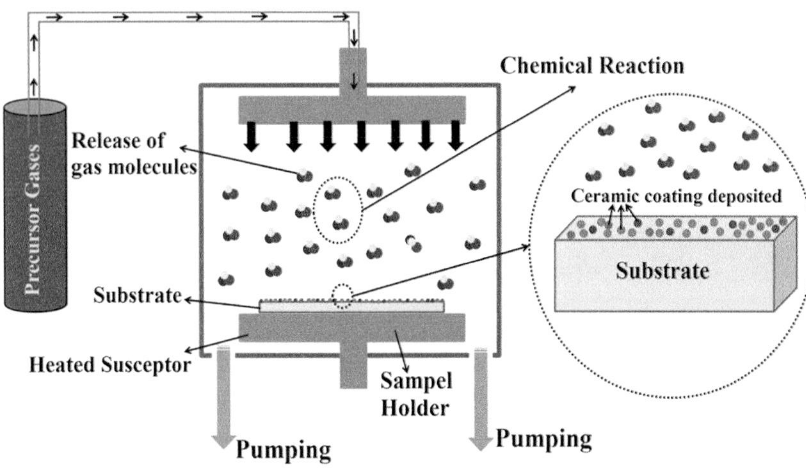

FIGURE 3.9 Deposition of ceramic nanocomposite coatings using a CVD system.

Source: Reprinted from Sabzi et al. (2023), Copyright © 2023 by the authors. Licensee MDPI, Basel, Switzerland. https://creativecommons.org/licenses/by/4.0/

On the other side, PVD is a thin-film deposition technique that involves evaporating a solid material under vacuum conditions. PVD is a widely used coating technique that creates nanocomposite coating by using various NPs to reinforce the ceramic matrix (Parvej et al., 2022). As shown in Figure 3.10, a vacuum chamber, a pump to maintain the vacuum atmosphere, a high-voltage DC power supply, a substrate (target electrode) connected to the power supply's negative voltage, and a nanoparticle source (sputtering target) connected to the power source's positive voltage make up the setup. A power source generates an extremely high DC voltage at the start of PVD. The cathode's metal or alloy is turned to vapor by the high voltage, and eventually, it enters the plasma state to mix with the process gas. At last, the atoms create a nanocomposite by condensing on the target electrode surface. Hence, at the target electrode, NPs from the sputtering target combine to form a nanocomposite. The three main benefits of PVD coatings are increased oxidation resistance, wear resistance, and hardness. The main drawbacks are the slow rate of deposition, high capital cost, difficulty coating undercuts, and similar surface features (Devasia et al., 2021; Ilangovan et al., 2021).

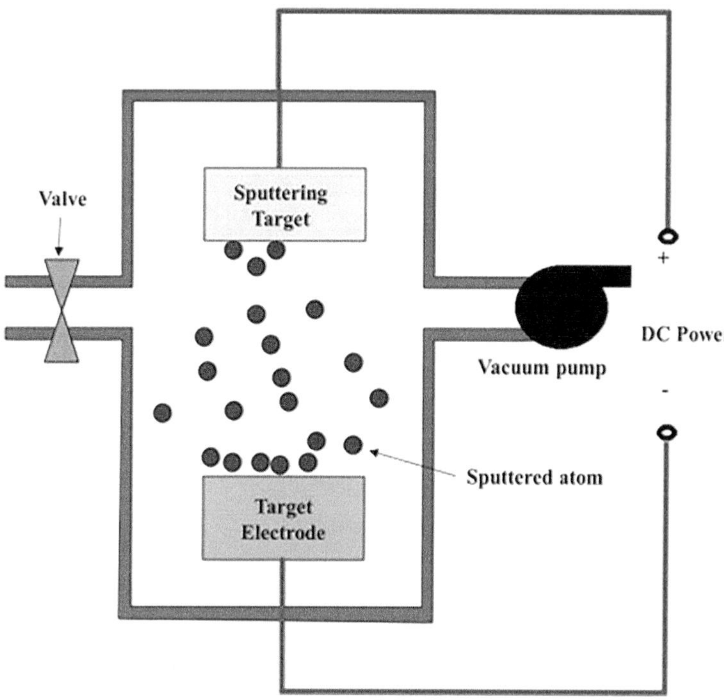

FIGURE 3.10 Schematic diagram of the PVD technique.
Source: Reprinted from Parvej et al. (2022) with permission from Elsevier.

3.2.3 METHODS OF PREPARATION METAL NANOCOMPOSITES

Nanocomposites in which the matrix is composed of metal or alloy and the reinforcement is mostly ceramic are known as metal matrix nanocomposites, or MMNCs. These nanocomposites combine the ductility and toughness of metal with the high modulus and strength of ceramics. They are appropriate for producing materials with high service temperature capabilities and high strength in shear and compression processes (Camargo et al., 2009; Rane et al., 2018).

3.2.3.1 SPRAY PYROLYSIS

Spray pyrolysis involves spraying a solution onto a heated surface, causing the constituents to react and form a chemical compound (Gavrilovic et al., 2018; Kumar et al., 2022). This results in the formation of a thin film. The choice of chemical reactants is made so that, at the temperature of deposition, any products that are not the intended compound are volatile. A suitable solvent is first used to dissolve the inorganic precursors (Figure 3.11) before being converted into a liquid mist using an ultrasonic atomizer. After that, the mist is transferred using a carrier gas to a heated chamber. In the next step, the droplets in the chamber are vaporized, trapped with a filter, and decomposed to yield oxide materials. The metal oxides are then reduced selectively to create the appropriate metallic materials. The primary advantages of this technique are reproductive size and quality, as well as, efficient production of ultra-fine, homogeneous, and spherical powders in multicomponent systems. However, the method is considered expensive if scaling up is desire.

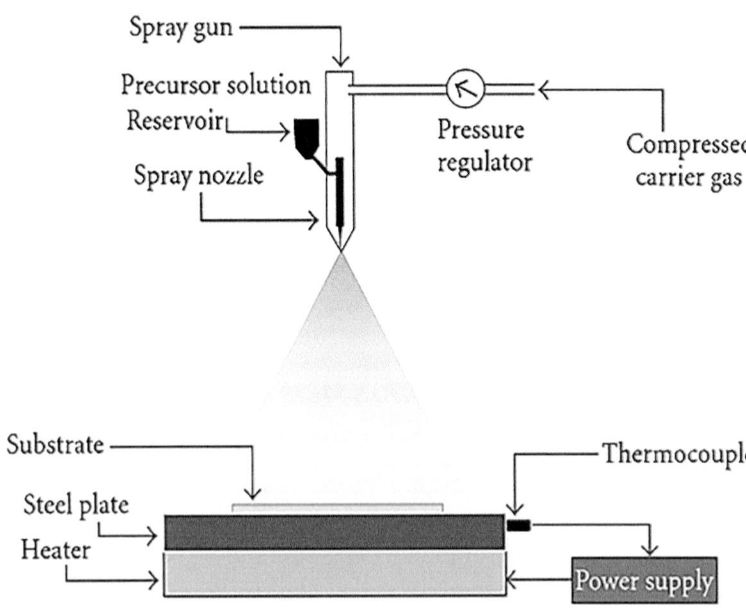

FIGURE 3.11 Scheme of spray pyrolysis.

Source: Reprinted from Patil et al. (2012), Copyright © 2012 by the authors. Licensee MDPI, Basel, Switzerland. https://creativecommons.org/licenses/by/4.0/

3.2.3.2 INFILTRATION

It is a liquid-state method for the preparation of metal nanocomposites. To prepare such nanocomposites, first, melting of the matrix metal is performed. After that, a prepared dispersed phase (ceramic particles, fibers) is soaked in it to fill the spaces between the dispersed phase inclusions (Alam et al., 2023). The driving force in a spontaneous infiltration process is usually the capillary force of the dispersed phase, while the driving force in a forced infiltration process is an external pressure that can be gaseous, mechanical, electromagnetic, centrifugal, or created by ultrasonic. The main limitation of this technique is that the processing techniques include using a high temperature, separating reinforcements, and creating unwanted products. However, on the other side, rapid solidification, quick contact times between the matrix and reinforcements, molding into various and nearly net shapes with varying stiffness, increased wear resistance and production on both an industrial and lab scale are the main advantages of this method. Infiltration works best when producing composites with relatively high-volume fractions of reinforcement and when producing composites with highly percolating reinforcement. In fact, infiltration is frequently the most effective and adaptable processing method for these different composites (Mortensen et al., 2000).

3.2.3.3 RAPID SOLIDIFICATION PROCESS (RSP)

As defined in scientific literature, rapid solidification refers to the process of rapidly converting a liquid state at a high temperature into a solid state at room temperature or ambient temperature by extracting thermal energy, including both superheat and latent heat (Lavernia and Srivatsan, 2009). Before solidification begins, there may be undercooling of up to 100°C due to the quick extraction of heat. This is directly compared to the few degrees obtained in the case of conventional casting, which is about 1°C or less per second. By imposing three things that are a high degree of undercooling before solidification, a high advance velocity during continuous solidification, and a high rate of cooling during solidification, it is possible to practically achieve rapid solidification. Large deviations from equilibrium are made possible by the rapid thermal energy extraction that takes place during RSP. The extension of solid solubility, improved compositional flexibility, reduction in the number and size of segregated phases, and preservation of a disordered crystalline structure in normally ordered materials and intermetallic compounds are just a few of the main advantages of these deviations. Meanwhile, the method is considered simple and effective. However, the method is suitable only for metal–metal nanocomposites with a chance of nonhomogeneous fine particle distribution and induced agglomeration (Camargo et al., 2009). In some instances, the matrix and reinforcements are mixed and made more wettable through the use of ultrasonics. RSP with ultrasonics offers excellent dispersion, even with small particles, free from any agglomerations (Li et al., 2004).

3.2.3.4 HIGH-ENERGY BALL MILLING

A common mechanical deformation method for creating powdered nanocrystalline metals or alloys is high-energy ball milling (Yang and Yang, 2015). Despite being a mechanical method, significant changes, either chemically or structurally, can occur to the material (Govindaraju et al., 2014). This method (Figure 3.12) involves filling a ball mill with a powder mixture and subjecting it to high-energy collisions from the balls (Rane et al., 2018). In high-energy ball milling, coarse-grained structures break apart as a result of strong cyclic deformation brought on by grinding stiff balls in a high-energy shaker mill. With this process, metals with minimum particle sizes of 4 to 26 nm can be produced successfully (Yang and Yang, 2015). High-energy ball milling makes it possible to change the circumstances in which chemical reactions usually take place by modifying the reactivity of the solids during the milling process or by inducing chemical reactions while the material is being milled. It is very possible to scale up the high-energy ball milling method to produce large amounts of material, and it is a simple process. Both homogenous distribution and uniform mixing are provided by the method.

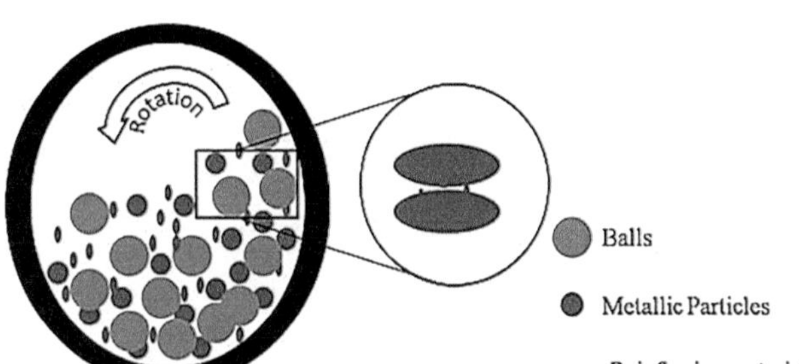

FIGURE 3.12 The preparation of metal nanocomposites using high-energy ball milling.

Source: Reprinted from Joy et al. (2022), Copyright © 2022 by the authors. Licensee MDPI, Basel, Switzerland. https://creativecommons.org/licenses/by/4.0/

3.2.3.5 COLLOIDAL METHOD

By using the colloidal method, which is a chemical process, metal particles are created by chemically reducing inorganic salts in a solution (Kamat et al., 2002). Then, the dry material is congealed, dried, and heated in a reducing atmosphere, like hydrogen, to produce the metal component and encourage selective oxide reduction. The method is considered simple with no need for high temperature and produces high purity products. Despite, poor bonding, high permeability, low wear resistance, and challenging porosity control are the main limitations. Another chemical procedure used to create metal matrix composites is the sol-gel process that was discussed before.

3.3 MODERN METHODS FOR THE PREPARATION OF NANOCOMPOSITES

3.3.1 LAYER-BY-LAYER METHODS

Multilayer film formation is primarily accomplished through the layer-by-layer (LPL) technique. It is applied to films using a range of nanomaterials to improve both the strength and durability of the final product. Hence, this method is applied to prepare films with exceptional properties for better performance in the field it is used in. Research has indicated that a wide variety of materials can be utilized to create multilayer structures (Rawtani and Agrawal, 2014). Olean-Oliveira et al. used cyclic electropolymerization to prepare the nanocomposite films, enabling the one-step assembly of thin films containing alternating azopolymers and reduced graphene oxide (rGO) layers (olean oliviera). Tian et al. prepared LPL nanocomposites that consisted of Co_3O_4 and reduced graphene (Tian et al., 2019).

As shown in Figure 3.13, the cobalt-containing precursor was first created through self-assembly assisted by a template. The structural guide, surfactant, was then added. After that, graphene was added, and the hydrothermal treatment was applied to the system. By annealing, the template was eliminated, yielding layer-by-layer porous nanocomposites. Graphene may play a role in the superior conduction of Co_3O_4 nanosheets. Zhao et al. reported a technique that uses shear flow to align two-dimensional nanosheets at an immiscible hydrogel/oil interface to produce nanocomposites with highly ordered layered structures (Zhao et al., 2020).

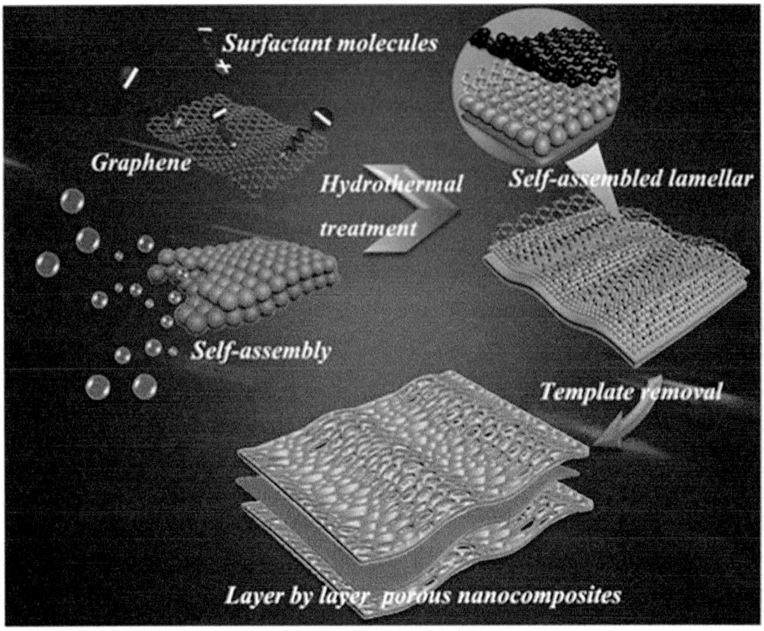

FIGURE 3.13 Preparation of layer-by-layer Co_3O_4/reduced graphene nanocomposites.
Source: Reprinted from Tian et al. (2019) with permission from Elsevier.

The observed improvement in mechanical properties was the consequence of a critical interphase formed by the well-aligned nanosheets. Zhao et al. believed that a wide range of structural composites could benefit from this approach, which is easily expandable to align various two-dimensional nanofillers and could result in the creation of high-performance composites.

3.3.2 TEMPLATE-ASSISTED METHODS

Growing focus on green chemistry means that creating an easy-to-use, low-impact way to synthesize NPs is becoming more crucial (Rane et al., 2018). One of the most promising techniques for creating monodispersed inorganic NPs is reportedly template synthesis, which uses the uniform void spaces of porous materials as guests to confine the synthesized NPs. For the synthesis, a template material must be used. The template serves two purposes. First, it serves as a skeleton to arrange the various functions of a device, the active components, and the various interfaces while enabling the optimal reproducibility of the structure. Throughout the particle preparation process, the technique can effectively control the morphology, particle size, and particle structure (Tijjani et al., 2022). By regulating crystal nucleation and growth during particle formation, the template method modifies the morphology of particles.

The nanoporous membranes that are usually used as templates are alumina or track-etched polymeric membranes, which are commonly used as ultrafiltration membranes. A recent development is the use of self-assembly pores in block copolymer structures to create metal nano-structures under high temperatures and electric fields (Ugo et al., 2007). Widely used techniques, such as precipitation or the sol-gel method, are used to prepare the templates. After the synthesis, the last step is to remove the template. Low-soluble molecules can be stabilized using surfactants as templates, which will stop particles from aggregating (Tijjani et al., 2022). The template method is simple to use and insensitive to preparation conditions. Meanwhile, utilizing the appropriate template is essential to this process. Capadonaa et al. reported the preparation of homogeneous polymer/nanofiber composites through the formation of a three-dimensional template (Capadona et al., 2007). They prepared the nanofiber template by the sol-gel process. When there was no polymer present in nanofiber dispersion, a nonsolvent was added. Next, nanofiber gel's ability to self-assemble was aided by solvent exchange. Finally, the polymer was infused into the gelled nanofiber scaffold by submersion in a polymer solution, prior to the drying and compacting of the nanocomposite.

3.3.3 BIOLOGICAL METHODS

Nanobiotechnology is a branch of nanotechnology that gained interest worldwide. Green nanotechnology is the best way to reduce the negative effects associated with producing NPs and its composites using alternative techniques (Pandit et al., 2022; Shafey, 2020). Green nanoparticle synthesis is a novel field of nanotechnology (Figure 3.14) that performs better than both physical and chemical approaches in terms of cost-effectiveness, biocompatibility, environmental friendliness, and scalability. Recently, as a substitute technique, metals and metal oxide NPs have been produced by plants, bacteria, fungi, and algae (Jain et al., 2020). Biotechnology and nanotechnology are connected through the green chemistry method of biological synthesis of NPs.

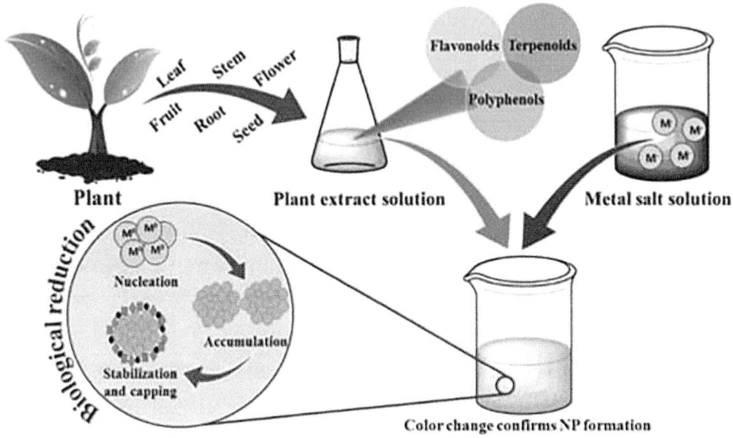

FIGURE 3.14 Schematic representation for plant-mediated biosynthesis of nanoparticles.

Source: Reprinted from Dikshit et al. (2021), Copyright © 2021 by the authors. Licensee MDPI, Basel, Switzerland. https://creativecommons.org/licenses/by/4.0/

It can be said that the biosynthetic approach is an effective way to create green NPs and nanocomposites. Bekru et al. used the aqueous extract of Verbascum sinaiticum Benth to prepare CuO-ZnO nanocomposites (Bekru et al., 2022). First, after gathering, drying by air, and washing in distilled water, the leaves of the *V. sinaiticum* Benth were ground into a powder using a grinder and then soaked in distilled water to obtain the extract by filtration. Using $Zn(CH_3COO)_2.2H_2O$, $Cu(CH_3COO)_2 \cdot H_2O$, and the leaves extract, CuO-ZnO NCs were synthesized by a microwave-assisted method.

3.4 CONCLUSION

Nanocomposites are a very smart class of nanomaterials. They can be adopted in different ways to obtain nanomaterials with superior properties which make them very attractive multifunctional materials for the future. Different methods are available for its preparation based on its type. Each method has its own advantages and disadvantages. Proper selection of the method of preparation affects greatly the properties of the prepared nanocomposites. Modern methods have appeared that are more cost-effective, biocompatible, safe, and eco-friendly. Hence, we need more research about these methods to be the principal methods of preparation of nanocomposites. Meanwhile, more modern techniques are needed to be developed for optimum preparation of nanocomposites.

KEYWORDS

- **ceramic nanocomposites**
- **layer by layer**
- **metal nanocomposites**
- **polymer nanocomposites**
- **template assisted**

REFERENCES

Alam, M. M., Imran, M., Abutaleb, A., Khan, M. E., Ali, W., Khan, M. E., Aslam, J., & Verma, C. (2023). 4 - Fabrication approaches of nanocomposites. In *Nanocomposites-advanced materials for energy and environmental aspects* (pp. 67–85). Woodhead Publishing.

Araby, S., Qingshi, M., Liqun, Z., Hailan, K., Peter, M., Youhong, T., & Jun, M. (2014). Electrically and thermally conductive elastomer/graphene nanocomposites by solution mixing. *Polymer*, *55*(1), 201–210.

Bekru, A. G., Tufa, L. T., Zelekew, O. A., Goddati, M., Lee, J., & Sabir, F. K. (2022). Green synthesis of a CuO/ZnO nanocomposite for efficient photodegradation of methylene blue and reduction of 4-nitrophenol. *ACS Omega 7*(35), 30908–30919.

Bokov, D., Turki Jalil, A., Chupradit, S., Suksatan, W., Javed Ansari, M., Shewael, I. H., Valiev, G. H., & Kianfar, E. (2021). Nanomaterial by sol-gel method: Synthesis and application. *Advances in Materials Science and Engineering*, *2021*, 5102014.

Bose, N. R., Banerjee, R., & Manna, I. (2013). 1 - Thermal shock resistant and flame retardant ceramic nanocomposites. In *Ceramic nanocomposites* (pp. 3–50). Woodhead Publishing.

Camargo, P., Satyanarayana, K. G., & Wypych, F. (2009). Nanocomposites: Synthesis, structure, properties and new application opportunities. *Materials Research*, *12*, 1–39.

Capadona, J., Berg, O., Capadona, L., Schroeter, M., Rowan, S., Tyler, D., & Weder, C. (2007). A versatile approach for the processing of polymer nanocomposites with self-assembled nanofibre templates. *Nature Nanotechnology*, *2*, 765–769.

Chanakul, A., Saymung, R., Seetha, S., Traiphol, R., & Traiphol, N. (2021). Solution-mixing method for large-scale production of reversible thermochromic and acid/base-colorimetric sensors. *Colloids and Surfaces A: Physicochemical and Engineering Aspects*, *615*, 126241.

Dahmen, K. H., & Meyers, R. A. (2003). Chemical vapor deposition. In *Encyclopedia of physical science and technology (Third Edition)* (pp. 787–808). Academic Press.

Devasia, R., Painuly, A., Devapal, D., Sreejith, K. J., Joseph, K., Oksman, K., George, G., Wilson, R., & Appukuttan, S. (2021). 22 - Continuous fiber reinforced ceramic matrix composites. In *Fiber reinforced composites* (pp. 669–751). Woodhead Publishing.

Dikshit, P. K., Kumar, J., Das, A. K., Sadhu, S., Sharma, S., Singh, S., Gupta, P. K., & Kim, B. S. (2021). Green synthesis of metallic nanoparticles: Applications and limitations. *Catalysts*, *11*(8), 902.

Emam, M. H., Elezaby, R. S., Swidan, S. A., Loutfy, S. A., & Hathout, R. M. (2023). Cerium oxide nanoparticles/polyacrylonitrile nanofibers as impervious barrier against viral infections. *Pharmaceutics*, *15*(5), 1494.

Fawaz, J., & Mittal, V. (2014). Chapter 1: Synthesis of polymer nanocomposites: Review of various techniques. In *Synthesis techniques for polymer nanocomposites* (pp. 1–30). Willey.

Gavrilovic, T. V., Jovanovic, D. J., Dramicanin, M. D., Bhanvase, B. A., Pawade, V. B., Dhoble, S. J., Sonawane, S. H., & Ashokkumar, M. (2018). Chapter 2 - synthesis of multifunctional inorganic materials: From micrometer to nanometer dimensions. In *Nanomaterials for green energy* (pp. 55–81). Elsevier.

Govindaraju, N., Singh, R. N., Schulz, M. J., Shanov, V. N., & Yin, Z. (2014). Chapter 8 - Synthesis and properties of boron nitride nanotubes. In *Nanotube superfiber materials* (pp. 243–265). William Andrew Publishing.

Grothe, J., Kaskel, S., Leuteritz, A., Matyjaszewski, K., & MÃller, M. (2012). 8.08 - Nanocomposites and hybrid materials. In *Polymer science: A comprehensive reference* (pp. 177–209). Elsevier.

Guglielmi, M. (2014). Chapter 3: Synthesis strategies for the preparation of sol-gel nanocomposites. In M. Guglielmi, G. Kickelbick, & A. Martucci (Eds.), *Sol-gel nanocomposites* (pp. 51–82). Springer Link.

Ilangovan, R., Subha, V., Ravindran, R. S. E., Kirubanandan, S., Renganathan, S., Thomas, S., & Balakrishnan, P. (2021). Chapter 2 - Nanomaterials: Synthesis, physicochemical characterization, and biopharmaceutical applications. In *Nanoscale processing* (pp. 33–70). Elsevier.

Jain, A., Banik, B. K., & Banik, B. K. (2020). 8 - Green synthesis of nanoparticles and nanocomposites: Medicinal aspects. In *Green approaches in medicinal chemistry for sustainable drug design* (pp. 231–242). Elsevier.

Jayasuriya, C. K. (2017). Interfacial bonding in polymer-ceramic nanocomposites. In *Reference module in materials science and materials engineering*. Elsevier.

Jose Varghese, R., Sakho, E. h. M., Parani, S., Thomas, S., Oluwafemi, O. S., Wu, J., Thomas, S., Sakho, E. H. M., Kalarikkal, N., Oluwafemi, S. O., & Wu, J. (2019). Chapter 3 - Introduction to nanomaterials: Synthesis and applications. In *Nanomaterials for solar cell applications* (pp. 75–95). Elsevier.

Joy, J., Krishnamoorthy, A., Tanna, A., Kamathe, V., Nagar, R., & Srinivasan, S. (2022). Recent developments on the synthesis of nanocomposite materials via ball milling approach for energy storage applications. *Applied Sciences*, *12*(18), 9312.

Kamalan, K. A. M., Kuppusami, P., Rajendran, S., Nguyen, T. A. N. H., Kakooei, S., Yeganeh, M., & Li, Y. (2020). Chapter 16 - Corrosion behavior of ceramic nanocomposite coatings at nanoscale. In *Corrosion protection at the nanoscale* (pp. 295–314). Elsevier.

Kamat, P. V., Flumiani, M., & Dawson, A. (2002). Metal-metal and metal-semiconductor composite nanoclusters. *Colloids and Surfaces A: Physicochemical and Engineering Aspects*, *202*(2), 269–279.

Kausar, A., & Ahmad, I. (2023). Electrospinning processing of polymer/nanocarbon nanocomposite nanofibers—Design, features, and technical compliances. *Journal of Composites Science*, *7*(7), 290.

Kianfar, E., & Suksatan, W. (2021). Nanomaterial by sol-gel method: Synthesis and application. *Advances in Materials Science and Engineering*, *2021*, 1–21.

Kumar, P., Mathpal, M. C., Inwati, G. K., Swart, H. C., Roos, W. D., Pawade, V. B., Dhoble, S. J., & Swart, H. C. (2022). 13 - Graphene oxide based semiconducting nanomaterial™ composites for environmental applications. In *Nanoscale compound semiconductors and their optoelectronics applications* (pp. 407–431). Woodhead Publishing.

Lavernia, E., & Srivatsan, T. (2009). The rapid solidification processing of materials: Science, principles, technology, advances, and applications. *Journal of Materials Science*, *45*, 287–325.

Li, X., Yang, Y., & Cheng, X. (2004). Ultrasonic-assisted fabrication of metal matrix Nanocomposites. *Journal of Materials Science*, *39*, 3211–3212.

Mortensen, A., Kelly, A., & Zweben, C. (2000). 3.20 - Melt infiltration of metal matrix composites. In *Comprehensive composite materials* (pp. 521–554). Pergamon.

Mousumi, S. (2020). Chapter 6: Nanocomposite materials. In *Nanotechnology and the environment*. IntechOpen

Muhammed Shameem, M., Sasikanth, S. M., Annamalai, R., & Ganapathi Raman, R. (2021). A brief review on polymer nanocomposites and its applications. *Materials Today: Proceedings*, *45*, 2536–2539.

Paksoy, H., Åžahan, N., Konuklu, Y., & Cabeza, L. F. (2022). Encapsulation of phase change materials. *Encyclopedia of Energy Storage*, *1*, 498–525.

Pandey, P. P. (2020). Preparation and characterization of polymer nanocomposites. *Soft Nanoscience Letters*, *10*, 1–15.

Pandit, C., Roy, A., Ghotekar, S., Khusro, A., Islam, M. N., Emran, T. B., Lam, S. E., Khandaker, M. U., & Bradley, D. A. (2022). Biological agents for synthesis of nanoparticles and their applications. *Journal of King Saud University – Science*, *34*(3), 101869.

Parvej, M. S., Khan, M. I., & Hossain, M. K. (2022). Preparation of nanoparticle-based polymer composites. In *Nanoparticle-based polymer composites* (pp. 55–94). Elsevier.

Patil, G. E., Kajale, D. D., Gaikwad, V. B., & Jain, G. H. (2012). Spray pyrolysis deposition of nanostructured tin oxide thin films. *ISRN Nanotechnology*, *2012*, 275872.

Quoc Pham, L., Uspenskaya, M. V., Olekhnovich, R. O., & Olvera Bernal, R. A. (2021). A review on electrospun PVC nanofibers: Fabrication, properties, and application. *Fibers*, *9*(2), 12.

Ramdani, N. (2019). Chapter 4: Processing methods of polymer/ceramic composites. In N. Ramdani (Ed.), *Polymer and ceramic composite materials*. CRC Press.

Rane, A. V., Kanny, K., Abitha, V. K., Thomas, S., Mohan Bhagyaraj, S., Oluwafemi, O. S., Kalarikkal, N., & Thomas, S. (2018). Chapter 5 - Methods for synthesis of nanoparticles and fabrication of nanocomposites. In *Synthesis of inorganic nanomaterials* (pp. 121–139). Woodhead Publishing.

Rawtani, D., & Agrawal, Y. K. (2014). Emerging strategies and applications of layer-by-layer self-assembly. *Nanobiomedicine (Rij)*, *1*, 8.

Refate, A., Mohamed, Y., Mohamed, M., Sobhy, M., Samhy, K., Khaled, O., Eidaroos, K., Batikh, H., El-Kashif, E., El-Khatib, S., & Mehanny, S. (2023). Influence of electrospinning parameters on biopolymers nanofibers, with emphasis on cellulose & chitosan. *Heliyon*, *9*(6), e17051.

Sabzi, M., Mousavi Anijdan, S. H., Shamsodin, M., Farzam, M., Hojjati-Najafabadi, A., Feng, P., Park, N., & Lee, U. (2023). A review on sustainable manufacturing of ceramic-based thin films by chemical vapor deposition (CVD): Reactions kinetics and the deposition mechanisms. *Coatings*, *13*(1), 188.

Sahini, M. G., Banyikwa, A. T., Nayak, A. K., Hasnain, M. S., Laha, B., & Maiti, S. (2023). Chapter 29 - Superparamagnetic iron oxide nanoparticles for drug delivery applications. In *Advanced and modern approaches for drug delivery* (pp. 817–850). Academic Press.

Selvam, J. D. R., Dinaharan, I., Rai, R. S., & Brabazon, D. (2021). Matrix and reinforcement materials for metal matrix composites. In *Encyclopedia of materials: Composites* (pp. 615–639). Elsevier.

Shafey, A. M. E. (2020). Green synthesis of metal and metal oxide nanoparticles from plant leaf extracts and their applications: A review. *International Journal of Molecular Sciences, 9*(1), 304–339.

Tang, L. C., Zhao, L., Qiang, F., Wu, Q., Gong, L. X., Peng, J. P., Yaragalla, S., Mishra, R. K., Thomas, S., Kalarikkal, N., & Maria, H. J. (2019). Chapter twelve - Mechanical properties of rubber nanocomposites containing carbon nanofillers. In *Carbon-based nanofillers and their rubber nanocomposites* (pp. 367–423). Elsevier.

Thomas, S., Stephen, R., Bandyopadhyay, S., & Thomas, S. (2007). Polymer nanocomposites: Preparation, properties and applications. *Rubber Fibers Plastics International*, *2*, 49–56.

Tian, M., Miao, J., Cheng, P., Mu, H., Tu, J., & Sun, J. (2019). Layer-by-layer nanocomposites consisting of Co_3O_4 and reduced graphene (rGO) nanosheets for high selectivity ethanol gas sensors. *Applied Surface Science*, *479*, 601–607.

Tijjani, H., Olatunde, A., Zangoma, M. H., Egbuna, C., Danyaro, A. M., Abdulkarim, H., Mahmoud, F. A., & Muhammad, M. (2022). Chapter 3 - Nanoformulation of antioxidant supplements. In C. Egbuna, M. A. Găman, & J. Jeevanandam (Eds.), *Drug discovery update, applications of nanotechnology in drug discovery and delivery* (pp. 45–70). Elsevier.

Ugo, P., Moretto, L. M., & Zoski, C. G. (2007). 16.2 - Template deposition of metals. In *Handbook of electrochemistry* (pp. 678–709). Elsevier.

Verma, D., & Goh, K. L. (2019). Chapter 11 - Functionalized graphene-based nanocomposites for energy applications. In M. Jawaid, R. Bouhfid, & A. E. Kacem Qaiss (Eds.), *Functionalized graphene nanocomposites and their derivatives* (pp. 219–243). Elsevier.

Victor, S., & Muthu, J. (2014). Polymer ceramic composite materials for orthopedic applications-relevance and need for mechanical match and bone regeneration. *Journal of Mechatronics*, *2*(4), 241–344.

Yang, L., & Yang, L. (2015). 2 - Nanotechnology-enhanced metals and alloys for orthopedic implants. In *Nanotechnology-enhanced orthopedic materials* (pp. 27–47). Woodhead Publishing.

Zhao, C., Zhang, P., Zhou, J., Qi, S., Yamauchi, Y., Shi, R., Fang, R., Ishida, Y., Wang, S., Tomsia, A. P., Liu, M., & Jiang, L. (2020). Layered nanocomposites by shear-flow-induced alignment of nanosheets. *Nature*, *580*(7802), 210–215.

CHAPTER 4

Drug Delivery Systems by Using Various Nanocomposites

SRIJITA SEN, PUJA KUMARI, RASHMI SINHA, ANJALI GIRI, and OM PRAKASH RANJAN*

Department of Pharmaceutical Technology (Formulations), National Institute of Pharmaceutical Education and Research (NIPER), Guwahati, Assam, India

*Corresponding author.

ABSTRACT

In the field of nanotechnology, nanocomposites have become a prominent area of current research and development. In the case of pharmaceuticals, the failures of single-drug delivery carriers have insisted on focusing on nanocomposite-based drug delivery where one phase necessarily should be in the nanoscale to get the desired benefits of the nanosystem. To get stronger or superior therapeutic effects the nanocomposites prepared are evaluated and compared for their action after administration through different routes like oral versus parenteral, parenteral versus inhalation, and so forth. Even one or more components (can be nanosystem and/or micro/macrosystem) are specifically designed or conjugated to selective agents (chemical groups/ligands/bioactive agents/polysaccharides/modified or unmodified amino acids and many more) to target the desired location of drug release. Even this selection depends on the route of administration and dose of the therapeutic nanocomposite. Indeed, nanocomposite-based drug delivery frameworks provide the opportunity to incorporate stimuli-sensitive components which is a golden opportunity for advanced therapy, and the drug delivery path can be monitored. Therefore, different

Nanocomposites and Nanomaterials in Biomedical Applications. Vivek Dave, Deependra Singh, and Pallerla Naveen Reddy (Eds.)

DOI: 10.1201/9781779641236-4

nanocomposites designed for different routes of administration, for example, oral, parenteral, inhalation and nasal, topical, ocular, vaginal, and so forth are the backbone of this chapter.

4.1 INTRODUCTION

Nanocomposites are materials comprising multiple phases where one of the phases is at least in the nanometer dimension. The phase interface between the nanomaterial and the comparatively larger macro/micro-scale phase creates a phase interphase that builds up special characteristics of the material. This specialty has been found as a beneficial alternative to conventional drug delivery and even superior to that of single-phase nanocarrier. Scientists have evaluated different kinds of nanocomposites in targeted and controlled delivery of various therapeutic agents including antimalarial, anticancer, antimicrobial, anti-inflammatory, cardiovascular, antiemetic agents, and so forth. Even layered structure has been introduced in the nanoarchitecture to incorporate different biological agents or functional materials like nucleic acids, polypeptides, polysaccharides, and so forth to get biostability and specific targeting (Omanović-Mikličanin et al., 2020).

A nanocomposite-based drug delivery system offers many benefits in the treatment of various diseased conditions by delivering therapeutic agents to the specific site. It has grown in popularity recently as a potential solution to the issues associated with gene and medicine delivery. To deliver drugs and genes, a great deal of research has been done on nanocomposites with various biological characteristics and compositions. Achieving effective drug delivery requires a thorough understanding of how nanomaterials interact with the biological environment, a focus on cell-surface receptors, and molecular mechanisms of cell signaling involved in the pathobiology of the disease under consideration. The development of many forms of nanocomposites in recent years has enhanced their use in biomedical and pharmaceutical sciences. The nanocomposites have been used in the biomedical field as drug delivery systems, wound healing, antimicrobial agents, bone tissue engineering, and biosensors. The current review provides an updated summary of recent developments in the field of drug delivery systems by using various nanocomposites through a thorough examination of the discovery and applications in improving the efficacy of both new and old drugs (such as natural products) and biomedical filed (Patra et al., 2018; Yusuf et al., 2023). Different nanocomposite-based drug delivery systems classified according to the route of administration have been discussed below.

4.2 ORAL NANOCOMPOSITE DRUG DELIVERY SYSTEMS

The majority (84%) of the top 50 pharmaceutical medications are consumed orally. Since it offers a variety of advantages, such as the ability for self-administration, dosage flexibility, improved patient compliance, and the removal of sterile manufacturing conditions, which lowers production costs. Despite the clear advantages of the oral route, the GI permeability, chemical/enzymatic stability, and solubility in GI fluids limit the therapeutic efficiency of several drugs and biologics. A wide range of biological barriers, including pH shifts, digestion processes, clearance, peristalsis, mucus, and tight epithelial layers, also prevent effective drug delivery. Recently, nanosystems for drug delivery were combined with macro- and micro-structures to overcome these barriers more effectively which is known as oral nanocomposites (Bhakay et al., 2018). Depending on the composition and method of manufacture, oral nanocomposites can be solid or semisolid in nature and come in various shapes and sizes. Hydrogels, aerogels, sponges, films, microneedles, tablets, and capsules are some examples of macrosystems made of nanocomposite materials. Microspheres, microgels, and microcapsules make up microparticles, which are classified as microsystems (Andretto et al., 2021). Some of these examples of such nanocomposites have been described below and in Table 4.1.

4.2.1 AMORPHOUS NANOCOMPOSITES

Telmisartan (TEL) has an extremely weak and pH-dependent solubility and is a BCS class II medication. Most commercially available formulations of TEL use a solid dispersion and a high concentration of a strong alkalizer (such as sodium hydroxide, potassium hydroxide, meglium). The pH of TEL was modulated by these potent alkalizers to obtain the desired dissolution under certain physiological conditions. However, using solid dispersion technology made it impossible to achieve an improved dissolving profile irrespective of physiological pH and to avoid the impact of food on the TEL pharmacokinetics. Researchers employed two alternative solubility enhancement approaches at a time that is ternary-cyclodextrin complexation and particle size reduction, as part of a particle engineering approach to solving this issue. Meglumine was used as a ternary component to create ternary cyclodextrin nanocomposites of TEL, and the final dosage form was obtained as a tablet. The nano amorphization has enabled the pH-independent dissolution of TEM in both fasted and fed conditions with more than 90% drug release (Sangwai and Vavia, 2013).

4.2.2 ORAL EXOSOME NANOCOMPOSITES

It is difficult to provide antibodies orally due to different barriers yet administering them parenterally can have major negative effects. For the oral administration of anti-TNF-α antibodies, researchers create a biomimetic nanocomposite employing an exosome generated from ginger (GE) with an inorganic framework. Large mesoporous silicon nanoparticles (LMSNs) were specially designed for the stable and effective loading of infliximab (INF/LMSN), and then GEs were separated from ginger for further coating onto INF/LMSN to create INF/LMSN@GE. The INF/LMSN@GE targeted the colon with a high payload of INF. This transports over the intestinal epithelium to release INF in the lamina propria for suppressing TNF-α produced by the nearby M1-type macrophages present in the colon. INF levels after GE coating reached 51.5 %, which is high enough for oral delivery. INF released from LMSN@GE was only 5.12% over the first 2 h in the SGF. After switching the release medium to SIF, within 10 h 85.33% of INF had been released gradually from the pores of the LMSN. However, only 16.12% of INF was released from LMSN@GE in SIF. It will ensure that the encapsulated cargos are structurally stable and do not release prematurely, leading to a high efficiency of oral antibodies (Mao et al., 2021).

4.2.3 ORAL NANOCOMPOSITE FILMS

Nanocomposite films were prepared for buccal delivery of enriched flavonoid fraction of *Cecropia glaziovii* (EFF-Cg). Drug-loaded PLGA nanospheres were created and encapsulated in chitosan films. The impact of various EFF-Cg PLGA nanospheres concentrations in the chitosan matrix was assessed for morphological, physical, chemical, and mechanical properties. The synthesized nanoparticle exhibited a size of 226 ± 3 nm. In terms of mechanical qualities, the EFF-Cg PLGA NP in chitosan films shows appropriate strength. The study finds that the novel chitosan nanocomposite films have the potential to resolve the low bioavailability issue associated with EFF-Cg. It is suggested delivery system for the treatment of labial herpes infection (Santos et al. 2017). Another long-acting oral nanocomposite *in situ* gelling film containing resveratrol (Rv) has been developed to treat colorectal cancer. The sodium alginate dry films were used to enclose the manufactured wet granules of Rv-Soy protein (Rv-Sp). The prepared film showed an encapsulation efficiency of about 97.87% and drug release of 14.45% ± 0.043 after 8 h. In comparison to free Rv, the Rv-nanocomposite composition displayed concentration-dependent cytotoxicity and increased apoptotic activity. This significantly increased Bcl-2-associated protein X (Bax) and decreased the expression of B-cell lymphoma 2, interleukin 1 β, IL-6, and tumor necrosis factor-alpha, compared to that of free Rv in HCT-116 cells. Overall results suggest that long-acting Rv nanocomposite gels could be a potential therapy option for colorectal cancer (Md et al., 2021).

4.2.4 NANOCOMPOSITES SPONGE

Nanocomposite sponges can improve intestine residence time after oral delivery. Muco-penetrating PEGylated nanoemulsions were created to be placed onto mucoadhesive chitosan sponges. Using the freeze-casting technique, researchers successfully embedded mucin solution in chitosan sponges without affecting their physicochemical characteristics. The nanoemulsions were shown to be cytocompatible at a variety of concentrations when tested on the HCT 116 and Caco-2 cell lines. Upon oral treatment of the sponge to mice demonstrated its efficacy in lengthening intestine residence time up to 6 h (Rosso et al., 2021).

4.2.5 HYDROGEL NANOCOMPOSITES

A pH-responsive nanocomposite hydrogel made of chitosan that has been grafted with acrylamide monomer and gold nanoparticles (Cs-g-PAAm/

AuNPs) was prepared. Silver nanoparticles (AgNPs) are then layer-coated on the top of these nanocomposites to enhance the drug's-controlled release as well as its antibacterial impact. The silver and gold particles have the power to kill microorganisms because they prevent bacteria from receiving oxygen. Additionally, it prevents the movement of enzymes across the cancer cell's surface. The inclusion of Au and Ag nanoparticles in the Cs-g-PAAm hydrogel has been verified by energy-dispersive X-ray and scanning electron microscopy. The mean diameters of the prepared AuNPs and AgNPs in the Cs-g-PAAm/AuNPs and Cs-g-PAAm/AgNPs nanocomposite hydrogels were 40 and 69 nm, respectively. Au-Ag NPs increased the releasing efficiency of fluorouracil anticancer drug. The outcome demonstrated better drug release performance from the nanocomposite hydrogel in a neutral pH environment of intestinal fluid (Nasef et al., 2023).

4.3 PARENTERAL NANOCOMPOSITE DRUG DELIVERY SYSTEMS

4.3.1 INTRAMUSCULAR NANOCOMPOSITE-BASED DRUG DELIVERY

Modern medical science is continuously seeking a safe vehicle to deliver genes/nucleic acids for the treatment of different genetic diseases. Here the modified cationic polymeric thermoreversible injectable nanocomposite system is the gold standard for aiding increased transfection without significant negative impacts. Specifically, DNA-polyplexes are prepared using modified chitosan, for example chitosan–arginine, PEGylated chitosan, chitosan-polyethyleneimine, and glycol–chitosan–spermine have been shown to aid gene release upto 30 days from the intramuscular depot formed after being injected into the muscle tissue (Zlotnikov et al., 2023). Topical application of nanocomposites containing curcumin in between the incised surgical gap of the internal and external abdominal obliques muscle has been shown to control the release of curcumin in a ceaseless manner with a continual sustained way from the nanolayers. This prolonged release significantly brings forth antibiofilm as well as antibacterial action against *Pseudomonas aeruginosa* and *Streptococcus mutans*. The slow and sustained release has been achieved by incorporating the drug into the nanolayer interspaces of zinc-aluminum layered double hydroxide sheets. The wound rapidly healed by forming reticular

TABLE 4.1 Oral Nanocomposite Drug Delivery.

S. No.	Nanocomposite Type/ Composition	Nanosystem	API	Significance	Therapeutic Application	References
1	Polymer nanocomposites (eudragit, sodium alginate, chitosan)	Nanoparticle	S-nitroso glutathione	Mucoadhesion	Sustained oral delivery	Wu et al., 2015
2	Polysaccharide-based nanocomposites (chitosan and bovine serum albumin)	Nanosuspension	Insulin	Protects insulin from the acidic environment of the digestive system and enzymatic degradation, also enhancing its uptake by the intestinal cells	Diabetes	Collado-González et al., 2020
3	Bioadhesive nanocomposite (alginate beads)	Nanoemulsion	Tofacitinib	Bio and mucoadhesion	Intestinal disorders	Andretto et al., 2023
4	Fullerene nanocomposites	Nanoparticle	Rhamnolipid	Colon targeting	Inflammatory bowel disorder	Xia et al., 2023
5	pH-responsive nanocomposites (aminoclay)	Nanoparticle	Liraglutide	Colonic delivery	Type 2 diabetes	Song et al., 2023
6	Graft copolymeric nanocomposites (soluplus-based nanocomposite)	Electro sprayed nanoparticle	*Angelica gigas* Nakai	Improved oral absorption	As analgesic and anticancer	Lee et al., 2016
7	Oral films (carmellose polymer)	Organovermiculite solution	Chlorhexidine	Mucoadhesive	Oral infections	Gajdziok et al., 2015
8	Polymeric (polysaccharide-based) nanocomposite	Nanoemulsion	Curcumin	Increased stability and oral bioavailability of curcumin	As an anti-inflammatory and antioxidant	Lee et al., 2021
9	Polysaccharide shell nanocomposites (chitosan shell)	Lipoid S75, tristearate hybrid lipid nanocore	Sulpiride	Mucoadhesive, increased intestinal permeation, reduced irritant effects of sulpiride	Depressive disorder therapy	Mohyeldin et al., 2021

4.6.1 TOPICAL NANOCOMPOSITES HYDROGEL

Recently, nanocomposite hydrogel has attracted the interest of researchers as a biomaterial and can be used as a promising candidate for different drug delivery applications. Sara *et al.* successfully prepared thyme essential oil (TEO)-loaded sodium caseinate nanocomposite hydrogel as a novel drug delivery platform for its antibacterial activity and wound-healing properties. In this study, it has been found that TEO-loaded Na-CAS/Gelatin nanocomposite hydrogels have a slower release pattern as compared to the TEO-loaded Na-CAS micelles. The total release of TEO from Na-CAS and hydrogel systems was 90 and 71% respectively after 72 h (Alsakhawy et al., 2022). Pierre and coworkers synthesized thermos-responsive polymeric–hydrogel nanocomposites for transdermal delivery of hydrocortisone for treatment against Psoriasis Vulgaris. In this study, hydrocortisone (HCT)-loaded sorbitan monostearate (SMS)-polycaprolactone (PCL) nanoparticles were prepared and then incorporated into carboxymethyl cellulose-based hydrogel. The *in-vitro* release profile showed that HCT release from SMS-PCL nanoparticles encapsulated with CMC hydrogel was slower, sustained, and controlled as compared to the HCT nanoparticles. The cytotoxicity study also showed that the viability of HCT-loaded hydrogel had more effect than HCT nanoparticles. At 24 h the viability of the HCT nanoparticles was 95%, whereas the HCT-loaded hydrogel showed a viability of 89%. After 48 h, the HCT-loaded nanoparticles and HCT-loaded hydrogel had high viability (116 and 124%) respectively (Kondiah et al., 2022).

4.6.2 NANOCOMPOSITES HYDROGEL FILM

Thiol pectin-based Polyethylene glycol and Montmorillonite (PEG/ MMT) nanocomposite films were successfully prepared for the controlled delivery of neomycin sulfate for wound healing. The higher concentration of MMT (5%) was shown to have higher entrapment efficiency than the lower concentration of MMT (1%) approximately (78.37 and 55.76%) respectively. *In vitro* wound-healing investigation suggested that the composite-to-film conversion process is responsible for the majority of the antioxidant activity because optimized nanocomposites film had somewhat stronger antioxidant activity than optimized nanocomposites (Raghavendra Naveen et al., 2023).

4.6.3 NANOCOMPOSITES MICRONEEDLE PATCH

A nanocomposite microneedle patch (ZCQ/MN) containing copper/zinc dual-doped mesoporous silica nanoparticles loaded with quercetin (ZCQ) was developed as a combination therapy for androgenic alopecia (AGA). After being applied to the skin, the biodegradable microneedle slowly dissolved and released the ZCQ nanoparticles. ZCQ nanoparticles subcutaneously release quercetin (Qu), copper (Cu^{2+}), and zinc ions (Zn^{2+}) to cooperatively enhance hair follicle regeneration. The synergistic effect in hair follicle growth has been found due to the combinational delivery of quercetin with these metal ions (Cu^{2+}), and (Zn^{2+}). The ZCQ/MN group even obtained 95.33% hair coverage after 2 weeks, demonstrating that the newly regenerated hairs entirely covered the alopecia area. Therefore, a multi-component-based nanocomposite delivery system is a novel therapy for androgenic alopecia (Zhang et al., 2023). Similarly, a nanomaterial-strengthened dissolving microneedle patch for transdermal delivery has been reported. In this study, layered double hydroxide (LDH) nanoparticles are combined with sodium carboxymethylcellulose (CMC). It seems that adding (5%) LDH nanoparticles to CMC can significantly improve its mechanical strength and make CMC-LDH nanocomposite far more robust than pure CMC. Small, densely packed CMC-LDH microneedles more effectively penetrate human and pig skin than pure CMC ones. Attractively, the nanocomposite-microneedles dissolve in the skin and release payload in just one minute. These nanocomposite-microneedle arrays are beneficial for vaccine delivery, and the results demonstrated that significantly greater antibody response may be elicited when compared with subcutaneous injection (Yan et al., 2014).

4.6.4 NANOCOMPOSITES NANOFIBER

A multifunctional nanocomposite for smart chronic wound healing was made with a curcumin-loaded carbon nanotube incorporated in nanofibers. In this study, curcumin (Cur)-loaded carboxylated multiwalled carbon nanotube (CMWCNT) hybrid was embedded in a polyamide 6/poly (vinyl alcohol) nanofiber matrix (PA6/PVA/CMWCNTCur) for controlled drug release. This hybrid nanofibrous composite greatly improved cell viability and proliferation when tested against the L929 cell line. The results of the

DPPH assay showed excellent antioxidation activity due to the release of Cur. It has contributed to the desirable antibacterial activity against both gram-positive and gram-negative bacteria. Hence, this nanocomposite is a promising candidate for chronic wound-healing applications (Shakiba et al., 2023). A nanofibrous core-sheath nanocomposite for dual drug delivery system based on poly (vinyl alcohol) (PVA)/chitosan/lidocaine hydrochloride loaded with erythromycin gelatin nanoparticles (GNP) was successfully prepared. The *in-vitro* release study showed that after incorporating the GNP into a nanofibrous mat, burst release was reduced to approximately 10-fold, and controlled release of the drug was achieved. The total amount of lidocaine released from the sheath part of the matrix was approximately 85% in 72 h and of erythromycin from the core part was approximately 75% in 72 h was reported. Therefore, this nanofiber composite framework enables dual therapy of pain relief and anti-infective treatment at the same time which is beneficial for injury healing in a more patient compliance way (Fathollahipour et al., 2015).

4.7 OCULAR NANOCOMPOSITE DRUG DELIVERY SYSTEMS

Delivery of a therapeutic agent in an efficient way to one of the most delicate and complex organs like an eye is a challenge. Different multifunctional nanocomposites are under research to deal with these challenges of static barriers (sclera, conjunctiva, choroid, blood–retinal barrier) and dynamic barriers (blood and lymphatic flow, tear, and so forth). In particular, nanocomposites have been found as a promising approach to retaining drugs in the posterior segment of the eye where a single nanocarrier failed to retain the drug at the required therapeutic level even after crossing multiple physiological barriers because of swift leakage at the anterior segment and lower deep tissue penetration ability. Ocular nanocomposites have been developed in different delivery forms that are discussed below.

4.7.1 NANOCOMPOSITE EYE DROP

Eye drop is always a convenient ophthalmic therapy. However, the facilitation of drug delivery into the posterior segment of the eyes without invasive procedures is an outstanding success in ophthalmic disease therapy. Layered double hydroxide (LDH)-based nanocomposite

TABLE 4.3 Different Types of Nanocomposites and Their Application in Topical Drug Delivery.

S. No.	Nanocomposites Form (Macrosystem/ Microsystem)	Macro/Micro-System Composition	Nanosystem Composition	Drug(s)	Application	References
1	Nanofiber	Polyurethane, montmorillonite clay	Montmorillonite nano clay loaded with chlorhexidine acetate	Chlorhexidine acetate	Wound healing	Saha et al., 2014
2	Hydrogel film	Carboxymethyl cellulose	Graphene quantum dot nanoparticle	Doxorubicin	Anticancer	Javanbakht and Namazi, 2018
3	Hydrogel	Arabinoxylan, chitosan	Nanodrug made with graphene oxide	5-fluorouracil	Melanoma cancer	Nazir et al., 2021
4	Fibrous meshes	No macro/ microsystem	Spinned nanofiber	Black forest honey and manuka honey	Antibacterial	Matharu et al., 2022
5	Magnetic nanocomposites	Fe_3O_4, SiO_2, chitosan	Chitosan-coated magnetic nanoparticles	Nisin	Antimicrobial	Nasaj et al., 2022
6	Microneedle patch	Polyvinylpyrrolidone, glucose	Glucose oxidase encapsulated AgNPs	Apramycin	Antimicrobial	Zhao et al., 2022
7	Metallic nanocomposites	Chitosan, polyvinyl alcohol	AgNPs	Amphotericin B	Subcutaneous sporotrichosis	Artunduaga Bonilla et al., 2022
8	Photocrosslinkable hydrogels	Polyvinylpyrrolidone	TiO_2, ZnO, and bentonite nanofillers	Nafcillin	Antimicrobial	Toader et al., 2023
9	Nanoemulsion	Alginate, pectin	Alginate-pectin microparticles	Curcumin	Wound healing	Amante et al., 2023
10	Hydrogel	Sialic acid, chitosan	Amphiphilic chitosan incorporated 18β-glycyrrhetinic acid nanocrystals	18β-glycyrrhetinic acid	Anti-photo aging effect	Quan et al., 2023

eye drops with a particle size of 100–200 nm range are under research investigations due to their ability to transport through the conjunctiva and redistribution into the choroid-retina from sclera. Notably, embedding or coating the LDH nanocomposite with carboxymethyl chitosan derivative CMCS-glutathione-glycyl sarcosine (CMCG-GS) enhances the bioadhesive property aiding in penetration enhancement into the sclera. It has been confirmed from the HConEpiC cell-based *in vitro* permeability study that the drug-LDH composite is less permeable through conjunctiva than its modified form with CMCG-GS. Even the permeation efficiency of CMCG-DXEP-LDH is less than CMCG-GS-DXEP-LDH. This is owing to the fact of specific recognition of this hybrid nanocomposite through the PepT-1 receptor by the GS part that facilitates transcellular transport as shown in Figure 4.2 (Wang et al., 2020). In another study, dexamethasone–cyclodextrin complex has been formulated as nanocomposite eye drop where this API complex has been intercalated into the LDH-GS interlayers as an anionic carrier that will aid the transport of dexamethasone to choroid/retina from conjunctiva. The positive nanocomplex (dexamethasone-CM-β-CD@LDH-GS) gets trapped in the negatively charged scleral pores, and the negative part dexamethasone-CM-β-CD gets dissociated. This dissociated part further facilitates the drug delivery into the inner portion (Xu et al., 2021). Thus target-specifically modified nanocomposite eye drops possess great commercial potential to deliver drug from ocular surface.

4.7.2 OCULAR NANOCOMPOSITE CONTACT LENS

Prolonged release of therapeutic agents in ocular sites can be achieved in a noninvasive and patient compliance way through medicated contact lenses. Researchers have found that hydrogel nanocomposite-based lenses enable drug delivery over a 6-fold longer time than the simple polymer contact lens. This has been achieved by incorporating albumin-gelatin-based nanoparticles into the p(HEMA-co-AEMA) hydrogel matrix. From the mesoporous structure of the nanoparticle the drug internally diffused into the swollen hydrogel reservoir matrix which further regulates the release pattern based on the polymer network condition. As the nanoparticles are covalently attached to the matrix, the burst release has been avoided and the nanoparticles are protected from degradation. It gradually slows down the release. This is advantageous to deliver proteins for different inflammatory conditions of the eye and in the treatment of dry eye syndrome (Zhang et al., 2013).

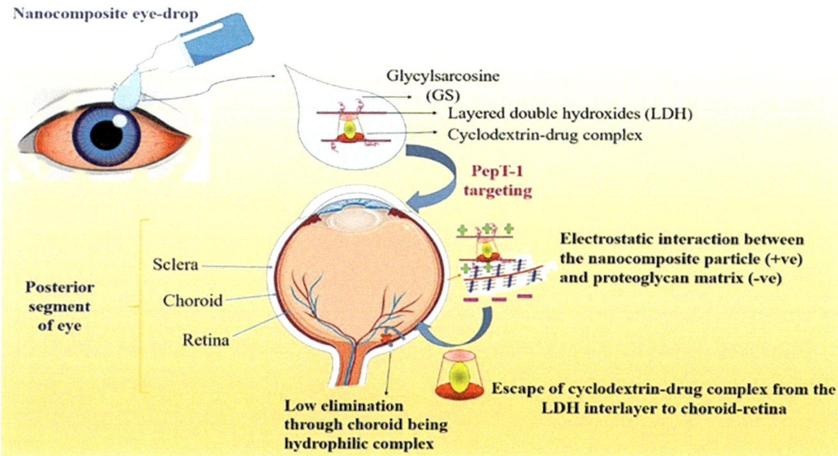

FIGURE 4.2 Painless delivery of drug to the posterior segment of the eye through conjunctiva-scleral pathway using dual-carrier nanocomposite-based eye drop.

4.7.3 NANOCOMPOSITE EYE DEPOT

Nanocomposite thermogels are excellent formulations that convert into a gel phase in response to the ocular fluid after injection and aid the sustained release of the drug molecules. The intraocular gel depot is the key to unlocking the prolonged release of the drug(s). Accounting for this principle, recently two kinds of *in situ* gelling nanocomposites integrating SLN and microemulsion for the loading of drugs have been prepared. Interestingly, three compartmental ocular flow cell models have been designed parallelly to mimic the mass transfer characteristics including the factor of anterior outflow and blood–retinal barrier. This has been aided in evaluating the slow release of the drug from a novel nanocarrier system in a simulated condition. The SLN-hydrogel depot has been shown to aid slow release of dodecyl-cefuroxime for up to one week. This slow-release nanocomposite will be a good choice to treat ocular problems in postoperative cases like after cataract surgery or in chronic eye diseases (Sapino et al., 2019).

4.7.4 INTRAOCULAR INJECTION

Few ocular disease conditions like macular degeneration need frequent intravitreal injections which is a risk factor for the patient to get infections. This medical condition urges for a formulation with multiple therapeutic benefits including anti-inflammatory, antioxidant, and antiangiogenesis. Favorably, the formulation should allow slow release of therapeutic agents over a period. Such kind of multifunctional nanoparticulate system has been designed using gold nanoparticles. The gold nanocore of 40–50 nm loaded with metformin is uniformly coated with polycatechin which is antioxidant in nature. This 20–30 nm thickened shell allows sustained release of metformin which is highly desirable. This nanocomposite system exhibits synergistic anti-inflammatory, antioxidant, and antiangiogenic properties. Indeed, this nanocomposite is immobilized with complement component protein C3 which can specifically bind to the complement receptor 1 localized in the drusen at the lesion site of the macula (Luo et al., 2021). This multifunctional nanocomposite can be extended to explore its efficacy in targeted delivery of drugs in other diseases.

4.8 VAGINAL NANOCOMPOSITES DRUG DELIVERY SYSTEMS

Controlled delivery of drugs locally or systematically through the vaginal route has a significant impact on better therapeutic effects for female reproductive diseases. A novel approach has been shown that crosslinked micellar nanocomposite hydrogel-based film made from gelatin and sodium alginate allows higher entrapment of drug within the nano micellar domain when SDS is added as an anionic surfactant. This allows the controlled release of ciprofloxacin as a model drug at both normal pH and infected pH conditions of the vagina rather than the burst release of the drug from the film. This pH-sensitive foldable film is a good biodegradable and economical candidate for bacterial vaginosis. Even this nanosystem should be evaluated for its application in the treatment of sexually transmitted diseases because of the virucidal activity of SDS (Adnan et al., 2020). Apart from the film, hydrogel rings are a good alternative for vaginal drug delivery due to their ease of application. Researchers have incorporated polyacrylamide-sodium carboxymethyl cellulose-montmorillonite nanoparticles within the hydrogel ring (5.5 cm diameter) which will allow prolonged release of the pharmacologically active substances depending upon the swelling behavior of the hydrogel. It has been observed that the incorporation of montmorillonite nanoparticles in the composite hydrogel increases the mucoadhesion property which is beneficial for the retention of NC in the vagina (Sharifzadeh et al., 2020; Güler et al., 2015). Chitosan-albumin-based nanocomposite system loaded with fluconazole has been shown to exert promising antifungal and biofilm-reducing activity against *Candida albicans* (Hatamiazar et al., 2023).

4.9 CONCLUSION

All the delivery routes that have been explored for nanocomposite delivery have a remarkable turning point to achieve better output in terms of therapeutic efficacy. The features associated with these innovative structures offer unique magnetic, electrical, and thermal conductivity, optical, antimicrobial, and other biological properties. Based on the requirement of localized and systemic effects these properties have been wisely utilized. In some cases, drug delivery has been improved by combining the imaging system or bimodal imaging with higher resolution and accuracy which is a great revolution in patient-based therapy. Recently, one carbon-based nanocomposite has completed clinical trial evaluation for its application in short- and long-term neurological disorders after being administered through the intracorporeal route [ClinicalTrials.gov ID: NCT04380194]. However, there is a need for more research in this area and a need to evaluate the patient-convenient routes for drug delivery.

KEYWORDS

- **hydrogel**
- **injection**
- **nanoparticle**
- **oral**
- **release**

REFERENCES

Abd Elwakil, M. M., Mabrouk, M. T., Helmy, M. W., Abdelfattah, E. Z. A., Khiste, S. K., Elkhodairy, K. A., & Elzoghby, A. O. (2018). Inhalable lactoferrin-chondroitin nanocomposites for combined delivery of doxorubicin and ellagic acid to lung carcinoma. *Nanomedicine*, *13*(16), 2015–2035. https://doi.org/10.2217/nnm-2018-0039.

Abdelaziz, H. M., Elzoghby, A. O., Helmy, M. W., Abdelfattah, E. A., Fang, J. Y., Samaha, M. W., & Freag, M. S. (2020). Inhalable lactoferrin/chondroitin-functionalized monoolein nanocomposites for localized lung cancer targeting. *ACS Biomaterials Science and Engineering*, *6*(2), 1030–1042. https://doi.org/10.1021/acsbiomaterials.9b01639.

Adnan, M., Santhosh Kumar, K., & Sreejith, L. (2020). Micellar nanocomposites hydrogels films for pH sensitive controlled drug delivery. *Materials Letters*, *277*, 128286. https://doi.org/10.1016/j.matlet.2020.128286.

Alsakhawy, S. A., Baghdadi, H. H., El-Shenawy, M. A., Sabra, S. A., & El-Hosseiny, L. S. (2022). Encapsulation of thymus vulgaris essential oil in caseinate/gelatin nanocomposite hydrogel: *In vitro* antibacterial activity and *in vivo* wound healing potential. *International Journal of Pharmaceutics*, *628*, 122280. https://doi.org/10.1016/j.ijpharm.2022.122280.

Amante, C., Andretto, V., Rosso, A., Augusti, G., Marzocco, S., Lollo, G., & Del Gaudio P. (2023). Alginate-pectin microparticles loaded with nanoemulsions as nanocomposites for wound healing. *Drug Delivery and Translational Research*, *13*(5), 1343–1357. https://doi.org/10.1007/s13346-022-01257-9.

Andretto, V., Rosso, A., Briançon, S., & Lollo G. (2021). Nanocomposite systems for precise oral delivery of drugs and biologics. *Drug Delivery and Translational Research*, *11*(2), 445–470. https://doi.org/10.1007/s13346-021-00905-w.

Andretto, V., Taurino, G., Guerriero, G., Guérin, H., Lainé, E., Bianchi, M. G., Agusti, G., & et al. (2023). Nanoemulsions embedded in alginate beads as bioadhesive nanocomposites for intestinal delivery of the anti-inflammatory drug tofacitinib. *Biomacromolecules*, *24*(6), 2892–2907. https://doi.org/10.1021/acs.biomac.3c00260.

Artunduaga Bonilla, J. J., Honorato, L., Guimarães, A. J., Miranda, K., & Nimrichter, L. (2022). Silver chitosan nanocomposites are effective to combat sporotrichosis. *Frontiers in Nanotechnology*, *4*. https://doi.org/10.3389/fnano.2022.857681.

Barakat, S. S., Nasr M, Ahmed, R. F., Badawy, S. S., & Mansour, S. (2017). Intranasally administered in situ gelling nanocomposite system of dimenhydrinate: Preparation, characterization and pharmacodynamic applicability in chemotherapy induced emesis model. *Scientific Reports*, *7*(1), 9910. https://doi.org/10.1038/s41598-017-10032-7.

Bayat Tork, M., Hemmati Nejad, N., Ghalehbagh, S., Bashari, A., Shakeri-Zadeh, A., & Kamrava, S. K. (2016). In situ green synthesis of silver nanoparticles/chitosan/poly vinyl alcohol/poly ethylene glycol hydrogel nanocomposite for novel finishing of nasal tampons. *Journal of Industrial Textiles*, *45*(6), 1399–1416. https://doi.org/10.1177/1528083714560255.

Bhakay, A., Rahman, M., Dave, R. N., & Bilgili, E. (2018). Bioavailability enhancement of poorly water-soluble drugs via nanocomposites: Formulation-processing aspects and challenges. *Pharmaceutics*, *10*(3), 86. https://doi.org/10.3390/pharmaceutics10030086.

Carling, C. J., & Brülls, M. (2021). Milling of poorly soluble crystalline drug compounds to generate appropriate particle sizes for inhaled sustained drug delivery. *International Journal of Pharmaceutics*, *593*, 120116. https://doi.org/10.1016/j.ijpharm.2020.120116.

Ceschan, N. E., Scioli-Montoto, S., Sbaraglini, M. L., Ruiz, M. E., Smyth, H. D. C., Bucalá, V., & Ramírez-Rigo, M. V. (2022). Nebulization of a polyelectrolyte-drug system for systemic hypertension treatment. *European Journal of Pharmaceutical Sciences*, *170*, 106108. https://doi.org/10.1016/j.ejps.2021.106108.

Collado-González, M., Ferreri, M. C., Freitas, A. R., Santos, A. C., Ferreira, N. R., Carissimi, G., Sequeira, J. A. D., & et al. (2020). Complex polysaccharide-based nanocomposites for oral insulin delivery. *Marine Drugs*, *18*(1), 55. https://doi.org/10.3390/md18010055.

Douafer, H., Andrieu, V., & Brunel, J. M. (2020). Scope and limitations on aerosol drug delivery for the treatment of infectious respiratory diseases. *Journal of Controlled Release*, *325*, 276–292. https://doi.org/10.1016/j.jconrel.2020.07.002.

El-Mekawy, R. E., & Jassas, R. S. (2017). Recent trends in smart and flexible three-dimensional cross-linked polymers: Synthesis of chitosan-ZnO nanocomposite hydrogels for insulin drug delivery. *MedChemComm*, *8*(5), 897–906. https://doi.org/10.1039/C6MD00580B.

Fathollahipour, S., Mehrizi, A. A., Ghaee, A., & Koosha, M. (2015). Electrospinning of PVA/chitosan nanocomposite nanofibers containing gelatin nanoparticles as a dual drug delivery system. *Journal of Biomedical Materials Research Part A*, *103*(12), 3852–3862. https://doi.org/10.1002/jbm.a.35529.

Gajdziok, J., Holešová, S., Štembírek, J., Pazdziora, E., Landová, H., Doležel, P., & Vetchý, D. (2015). Carmellose mucoadhesive oral films containing vermiculite/chlorhexidine nanocomposites as innovative biomaterials for treatment of oral infections. *BioMed Research International*, 1–15. https://doi.org/10.1155/2015/580146.

Ghasemian, E., Vatanara, A., Rouini, M. R., Najafabadi, A. R., Gilani, K., Lavasani, H., & Mohajel, N. (2016). Inhaled sildenafil nanocomposites: Lung accumulation and pulmonary pharmacokinetics. *Pharmaceutical Development and Technology*, *21*(8), 961–971. https://doi.org/10.3109/10837450.2015.1086369.

Güler, M. A., Gök, M. K., Figen, A. K., & Özgümüş, S. (2015). Swelling, mechanical and mucoadhesion properties of Mt/starch-g-PMAA nanocomposite hydrogels. *Applied Clay Science*, *112–113*, 44–52. https://doi.org/10.1016/j.clay.2015.04.019.

Guo, Y., Wang, Y., Li, S., Niu, L., Wei, D., & Zhang, S. (2017). DNA-spheres decorated with magnetic nanocomposites based on terminal transfer reactions for versatile target detection and cellular targeted drug delivery. *Chemical Communications*, *53*(35), 4826–4829. https://doi.org/10.1039/C7CC00310B.

Hatamiazar, M., Mohammadnejad, J., & Khaleghi, S. (2023). Chitosan-albumin nanocomposite as a promising nanocarrier for efficient delivery of fluconazole against vaginal candidiasis. *Applied Biochemistry and Biotechnology*, *196*, 701–716. https://doi.org/10.1007/s12010-023-04492-z.

Hickey, A. J. (2020). Emerging trends in inhaled drug delivery. *Advanced Drug Delivery Reviews*, *157*, 63–70. https://doi.org/10.1016/j.addr.2020.07.006.

Isabel, C. Z., Luis, A. D., Samuel, E. S., Elizabeth, P. S., Dea, H. R., & Sergio, A. A. (2022). Novel mucoadhesive PLGA-PVM/MA micro-nanocomposites loaded with felodipine intended for pulmonary administration by nebulization. *International Journal of Pharmaceutics*, *628*, 122295. https://doi.org/10.1016/j.ijpharm.2022.122295.

Javanbakht, S., & Namazi, H. (2018). Doxorubicin loaded carboxymethyl cellulose/graphene quantum dot nanocomposite hydrogel films as a potential anticancer drug delivery system. *Materials Science and Engineering: C*, *87*, 50–59. https://doi.org/10.1016/j.msec.2018.02.010.

Kabary, D. M., Helmy, M. W., Abdelfattah, E. Z. A., Fang, J. Y., Elkhodairy, K. A., & Elzoghby, A. O. (2018). Inhalable multi-compartmental phospholipid enveloped lipid core nanocomposites for localized MTOR inhibitor/herbal combined therapy of lung carcinoma. *European Journal of Pharmaceutics and Biopharmaceutics*, *130*, 152–164. https://doi.org/10.1016/j.ejpb.2018.06.027.

Kang, T., Cha, G. D., Park, O. K., Cho, H. R., Kim, M., Lee, J., Kim, D., & et al. (2023). Penetrative and sustained drug delivery using injectable hydrogel nanocomposites for postsurgical brain tumor treatment. *ACS Nano*, *17*(6), 5435–5447. https://doi.org/10.1021/acsnano.2c10094.

Kondiah, P. P. D., Rants'o, T. A., Mdanda, S., Mohlomi, L. M., & Choonara, Y. E. (2022). A poly (caprolactone)-cellulose nanocomposite hydrogel for transdermal delivery of hydrocortisone in treating psoriasis vulgaris. *Polymers*, *14*(13), 2633. https://doi.org/10.3390/polym14132633.

Lee, J. J., Nam, S., Park, J. H., Lee, S. Y., Jeong, J. Y., Lee, J. Y., Kang, W. S., Yoon, I., S., Kim, D. D., & Cho, H. J. (2016). Nanocomposites based on soluplus and angelica gigas Nakai extract fabricated by an electrohydrodynamic method for oral administration. *Journal of Colloid and Interface Science*, *484*, 146–54. https://doi.org/10.1016/j.jcis.2016.08.080.

Lee, J. Y., Lee, S., Choi, J. H., & Na, K. (2021). ι-carrageenan nanocomposites for enhanced stability and oral bioavailability of curcumin. *Biomaterials Research*, *25*(1), 32. https://doi.org/10.1186/s40824-021-00236-4.

Levina, A., Repkova, M., Kupryushkin, M., Seryapina, A., Shevelev, O., Pyshnyi, D., Zarytova, V., & Markel, A. (2022). In vivo hypotensive effect of aminosilanol-based nanocomposites bearing antisense oligonucleotides. *Journal of Drug Delivery Science and Technology*, *75*, 103612. https://doi.org/10.1016/j.jddst.2022.103612.

Lim, Y. H., Tiemann, K. M., Hunstad, D. A., Elsabahy, M., & Wooley, K. L. (2016). Polymeric nanoparticles in development for treatment of pulmonary infectious diseases. *WIREs Nanomedicine and Nanobiotechnology*, *8*(6), 842–871. https://doi.org/10.1002/wnan.1401.

Liu, Y., Zhu, M., Meng, M., Wang, Q., Wang, Y., Lei, Y., Zhang, Y., Weng, L., & Chen, X. (2023). A dual-responsive hyaluronic acid nanocomposite hydrogel drug delivery system for overcoming multiple drug resistance. *Chinese Chemical Letters*, *34*(1), 107583. https://doi.org/10.1016/j.cclet.2022.06.006.

Luo, L. J., Jian, H. J., Harroun, H. G., Lai, J. Y., Unnikrishnan, B., & Huang, C. C. (2021). Targeting nanocomposites with anti-oxidative/inflammatory/angiogenic activities for synergistically alleviating macular degeneration. *Applied Materials Today*, *24*, 101156. https://doi.org/10.1016/j.apmt.2021.101156.

Mahmoud, R., Safwat, N., Fathy, M., Mohamed, N. A., El-Dek, S., El-Banna, H. A., Farghali, A., & Abo El-Ela, F. I. (2022). Novel anti-inflammatory and wound healing controlled released LDH-curcumin nanocomposite via intramuscular implantation, in-vivo study. *Arabian Journal of Chemistry*, *15*(3), 103646. https://doi.org/10.1016/j.arabjc.2021.103646.

Mao, Y., Han, M., Chen, C., Wang, X., Han, J., Gao, Y., & Wang, S. (2021). A biomimetic nanocomposite made of a ginger-derived exosome and an inorganic framework for high-performance delivery of oral antibodies. *Nanoscale*, *13*(47), 20157–20169. https://doi.org/10.1039/D1NR06015E.

Matharu, R. K., Ahmed, J., Seo, J., Karu, K., Golshan, M. A., Edirisinghe, M., & Ciric, L. (2022). Antibacterial properties of honey nanocomposite fibrous meshes, *Polymers 14*(23), 5155. https://doi.org/10.3390/polym14235155.

Md, S., Abdullah, S., Alhakamy, N. A., Alharbi, W. S., Ahmad, J., Shaik, R. A., Ansari, M. J., Ibrahim, I. M., & Ali, J. (2021). Development, optimization, and in vitro evaluation of novel oral long-acting resveratrol nanocomposite in situ gelling film in the treatment of colorectal cancer. *Gels*, *7*(4), 276. https://doi.org/10.3390/gels7040276.

Mehanna, M. M., Mohyeldin, S. M., & Elgindy, N. A. (2019). Rifampicin-carbohydrate spray-dried nanocomposite: A futuristic multiparticulate platform for pulmonary delivery. *International Journal of Nanomedicine*, *14*, 9089–9112. https://doi.org/10.2147/IJN.S211182.

Misak, H., Zacharias, N., Song, Z., Hwang, S., Man, K. P., Asmatulu, R., & Yang, S. Y. (2013). Skin cancer treatment by albumin/5-Fu loaded magnetic nanocomposite spheres in a mouse model. *Journal of Biotechnology*, *164*(1), 130–136. https://doi.org/10.1016/j.jbiotec.2013.01.003.

Mohamed, A., Pekoz, A. Y., Ross, K., Hutcheon, G. A., & Saleem, I. Y. (2019). Pulmonary delivery of nanocomposite microparticles (NCMPs) incorporating MiR-146a for treatment of COPD. *International Journal of Pharmaceutics*, *569*, 118524. https://doi.org/10.1016/j.ijpharm.2019.118524.

Mohyeldin, S. M., Samy, W. M., Ragab, D., Abdelmonsif, D. A., Aly, R. G., & Elgindy, N. A. (2021). Hybrid lipid core chitosan-TPGS shell nanocomposites as a promising integrated nanoplatform for enhanced oral delivery of sulpiride in depressive disorder therapy. *International Journal of Biological Macromolecules*, *188*, 432–449. https://doi.org/10.1016/j.ijbiomac.2021.08.035.

Nasaj, M., Farmany, A., Shokoohizadeh, L., Jalilian, F. A., Mahjoub, R., Roshanaei, G., Nourian, A., Shayesteh, O. H., & Arabestani, M. R. (2022). Development of chitosan-assisted Fe_3O_4@SiO_2 magnetic nanostructures functionalized with nisin as a topical combating system against vancomycin-intermediate staphylococcus aureus (VISA) skin wound infection in mice. *Journal of Nanomaterials*, 1–17. https://doi.org/10.1155/2022/2914210.

Nasef, S. M., Khozemy, E. E., & Mahmoud, G. A. (2023). PH-responsive chitosan/acrylamide/gold/nanocomposite supported with silver nanoparticles for controlled release of anticancer drug. *Scientific Reports*, *13*(1), 7818. https://doi.org/10.1038/s41598-023-34870-w.

Nazir, S., Khan, M. U. A., Al-Arjan, W. S., Razak, S. I. A., Javed, A., & Kadir, M. R. A. (2021). Nanocomposite hydrogels for melanoma skin cancer care and treatment: In-vitro drug delivery, drug release kinetics and anti-cancer activities. *Arabian Journal of Chemistry, 14*(5), 103120. https://doi.org/10.1016/j.arabjc.2021.103120.

Omanović-Mikličanin, E., Badnjević, A., Kazlagić, A., & Hajlovac, M. (2020). Nanocomposites: A brief review. *Health and Technology*, *10*(1), 51–59. https://doi.org/10.1007/s12553-019-00380-x.

Patra, J. K., Das, G., Fraceto, L. F., & et al. (2018). Nano based drug delivery systems: Recent developments and future prospects. *Journal of Nanobiotechnology*, *16*(71). https://doi.org/10.1186/s12951-018-0392-8.

Quan, W., Kong, S., Li, S., Ouyang, Q., Lu, S., Guo, J., Wu, K., Zhao, W., & Luo, H. (2023). Anti-photoaging effects of nanocomposites of amphiphilic chitosan/18β-glycyrrhetinic acid. *Molecules*, *28*(11), 4362. https://doi.org/10.3390/molecules28114362.

Raghavendra Naveen, N., Posina Anitha, B. G., Goudanavar, P., & Fattepur, S. (2023). QbD assisted formulation design and optimization of thiol pectin based polyethyleneglycol and montmorillonite (PEG/MMT) nanocomposite films of neomycin sulphate for wound healing. *Journal of Drug Delivery Science and Technology*, *82*, 104348. https://doi.org/10.1016/j.jddst.2023.104348.

Rosso, A., Andretto, V., Chevalier, Y., Kryza, D., Sidi-Boumedine, J., Grenha, A, Guerreiro, F., & et al. (2021). Nanocomposite sponges for enhancing intestinal residence time following oral administration. *Journal of Controlled Release*, *333*, 579–592. https://doi.org/10.1016/j.jconrel.2021.04.004.

Vignesh, S., Sivashanmugam, A., Annapoorna, M., Janarthanan, R., Subramania, I., Nair Shantikumar, V., & Jayakumar, R. (2018). Injectable deferoxamine nanoparticles loaded chitosan-hyaluronic acid coacervate hydrogel for therapeutic angiogenesis. *Colloids and Surfaces B: Biointerfaces*, *161*, 129–138. https://doi.org/10.1016/j.colsurfb.2017.10.033.

Saha, K., Butola, B. S., & Joshi, M. (2014). Drug-loaded polyurethane/clay nanocomposite nanofibers for topical drug-delivery application. *Journal of Applied Polymer Science*, *131*(10). https://doi.org/10.1002/app.40230.

Sangwai, M., & Vavia, P. (2013). Amorphous ternary cyclodextrin nanocomposites of telmisartan for oral drug delivery: Improved solubility and reduced pharmacokinetic variability. *International Journal of Pharmaceutics*, *453*(2), 423–432. https://doi.org/10.1016/j.ijpharm.2012.08.034.

Santos, T. C., Rescignano, N., Boff, L., Reginatto, F. H., Simões, C. M. O., de Campos, A. M., & Mijangos, C. U. (2017). Manufacture and characterization of chitosan/PLGA nanoparticles nanocomposite buccal films. *Carbohydrate Polymers*, *173*, 638–644. https://doi.org/10.1016/j.carbpol.2017.06.014.

Sapino, S., Peira, E., Chirio, D., Chindamo, G., Guglielmo, S., Oliaro-Bosso, S., Barbero, R., & et al. (2019). Thermosensitive nanocomposite hydrogels for intravitreal delivery of cefuroxime. *Nanomaterials*, *9*(10), 1461. https://doi.org/10.3390/nano9101461.

Shakiba, M., Jahangiri, P., Rahmani, E., Hosseini, S. M., Bigham, A., Foroozandeh, A., Tajiki, A., et al. (2023). Drug-loaded carbon nanotube incorporated in nanofibers: A multifunctional nanocomposite for smart chronic wound healing. *ACS Applied Polymer Materials*, *5*(7), 5662–5675. https://doi.org/10.1021/acsapm.3c00965.

Sharifzadeh, G., Hezaveh, H., Muhamad, I. I., Hashim, S., & Khairuddin, N. (2020). Montmorillonite-based polyacrylamide hydrogel rings for controlled vaginal drug delivery. *Materials Science and Engineering: C*, *110*, 110609. https://doi.org/10.1016/j.msec.2019.110609.

Shi, X., Wang, Y., Sun, H., Chen, Y., Zhang, X., Xu, J., & Zhai, G. (2019). Heparin-reduced graphene oxide nanocomposites for curcumin delivery: *In vitro*, *in vivo* and molecular dynamics simulation study. *Biomaterials Science*, *7*(3), 1011–1027. https://doi.org/10.1039/C8BM00907D.

Silva, M., Silva, A., Fernandez-Lodeiro, J., Casimiro, T., Lodeiro, C., & Aguiar-Ricardo, A. (2017). Supercritical CO_2-assisted spray drying of strawberry-like gold-coated magnetite nanocomposites in chitosan powders for inhalation. *Materials*, *10*(1), 74. https://doi.org/10.3390/ma10010074.

Song, J. G., Kim, D. H., & Han, H. K. (2023). Fabrication and evaluation of a pH-responsive nanocomposite-based colonic delivery system for improving the oral efficacy of liraglutide. *International Journal of Nanomedicine*, *18*, 3937–3949. https://doi.org/10.2147/IJN.S413515.

Sou, T., & Bergström, C. A. S. (2021). Contemporary formulation development for inhaled pharmaceuticals. *Journal of Pharmaceutical Sciences*, *110*(1), 66–86. https://doi.org/10.1016/j.xphs.2020.09.006.

Taki, M., Tagami, T., Fukushige, K., & Ozeki, T. (2016). Fabrication of nanocomposite particles using a two-solution mixing-type spray nozzle for use in an inhaled curcumin formulation. *International Journal of Pharmaceutics*, *511*(1), 104–110. https://doi.org/10.1016/j.ijpharm.2016.06.134.

Teymouri Rad, R., Dadashzadeh, S., Vatanara, A., Alavi, S., Ghasemian, E., & Mortazavi, S. A. (2019). Tadalafil nanocomposites as a dry powder formulation for inhalation, a new strategy for pulmonary arterial hypertension treatment. *European Journal of Pharmaceutical Sciences*, *133*, 275–286. https://doi.org/10.1016/j.ejps.2019.04.001.

Tian, Y., Zheng, Z., Wang, X., Liu, S., Gu, L., Mu, J., Zheng, X., Li, Y., & Shen, S. (2022). Establishment and evaluation of glucose-modified nanocomposite liposomes for the treatment of cerebral malaria. *Journal of Nanobiotechnology*, *20*(1), 318. https://doi.org/10.1186/s12951-022-01493-8.

Toader, G., Podaru, I. A., Rusen, E., Diacon, A., Ginghina, R. E., Alexandru, M., Zorila, F. L., & et al. (2023). Nafcillin-loaded photocrosslinkable nanocomposite hydrogels for biomedical applications. *Pharmaceutics*, *15*(6), 1588. https://doi.org/10.3390/pharmaceutics15061588.

Wang, L., Xing, H., Zhang, S., Ren, Q., Pan, L., Zhang, K., Bu, W., & et al. (2013). A Gd-doped Mg-Al-LDH/Au nanocomposite for CT/MR bimodal imagings and simultaneous drug delivery. *Biomaterials*, *34*(13), 3390–3401. https://doi.org/10.1016/j.biomaterials.2013.01.070.

Wang, Y., Zhou, L., Fang, L., & Cao, F. (2020). Multifunctional carboxymethyl chitosan derivatives-layered double hydroxide hybrid nanocomposites for efficient drug delivery to the posterior segment of the eye. *Acta Biomaterialia*, *104*, 104–114. https://doi.org/10.1016/j.actbio.2020.01.008.

Wu, W., Gaucher, C., Fries, I., Hu, X. M., Maincent, P., & Sapin-Minet, A. (2015). Polymer nanocomposite particles of S-nitrosoglutathione: A suitable formulation for protection and sustained oral delivery. *International Journal of Pharmaceutics*, *495*(1), 354–361. https://doi.org/10.1016/j.ijpharm.2015.08.074.

Xia, Y., Hong, L., Zheng, J., Lu, Z., Zhang, Q., Chen, S., Pang, Z., & et al. (2023). Ulcerative colitis alleviation of colon-specific delivered rhamnolipid/fullerene nanocomposites *via* dual modulation in oxidative stress and intestinal microbiome. *Journal of Materials Chemistry B*, *11*(25), 5882–5897. https://doi.org/10.1039/D3TB00714F.

Xu, C., Lu, J., Zhou, L., Liang, J., Fang, L., & Cao, F. (2021). Multifunctional nanocomposite eye drops of cyclodextrin complex@layered double hydroxides for relay drug delivery to the posterior segment of the eye. *Carbohydrate Polymers*, *260*, 117800. https://doi.org/10.1016/j.carbpol.2021.117800.

Yan, L., Raphael, A. P., Zhu, X., Wang, B., Chen, W., Tang, T., Deng, Y., & et al. (2014). Nanocomposite-strengthened dissolving microneedles for improved transdermal delivery to human skin. *Advanced Healthcare Materials*, *3*(4), 555–564. https://doi.org/10.1002/adhm.201300312.

Yang, M., Yamamoto, H., Kurashima, H., Takeuchi, H., Yokoyama, T., Tsujimoto, H., & Kawashima, Y. (2012). Design and evaluation of poly (Dl-Lactic-Co-Glycolic Acid) nanocomposite particles containing salmon calcitonin for inhalation. *European Journal of Pharmaceutical Sciences*, *46*(5), 374–380. https://doi.org/10.1016/j.ejps.2012.02.024.

Yusuf, A., Almotairy, A. R. Z., Henidi, H., Alshehri, O. Y., & Aldughaim, M. S. (2023). Nanoparticles as drug delivery systems: A review of the implication of nanoparticles' physicochemical properties on responses in biological systems. *Polymers*, *15*, 1596. https://doi.org/10.3390/polym15071596.

Zhang, J., Bi, R., Hodge, W., Yin, P., & Tse, W. H. (2013). A nanocomposite contact lens for the delivery of hydrophilic protein drugs. *Journal of Materials Chemistry B*, *1*(35), 4388. https://doi.org/10.1039/c3tb20391c.

Zhang, Z., Li, W., Chang, D., Wei, Z., Wang, E., Yu, J., Xu, Y., & et al. (2023). A combination therapy for androgenic alopecia based on quercetin and zinc/copper dual-doped mesoporous silica nanocomposite microneedle patch. *Bioactive Materials*, *24*, 81–95. https://doi.org/10.1016/j.bioactmat.2022.12.007.

Zhao, M., Zhou, M., Gao, P., Zheng, X., Yu, W., Wang, Z., Li, J., & Zhang, J. (2022). AgNPs/NGOx/Apra nanocomposites for synergistic antimicrobial therapy and scarless skin recovery. *Journal of Materials Chemistry B*, *10*(9), 1393–1402. https://doi.org/10.1039/D1TB01991K.

Zhu, Y., Zhou, J., Hu, Y., Shi, H., Wu, Y., Pan, M., Chen, J., & et al. (2022). Single subcutaneous injection of the minocycline nanocomposite-loaded thermosensitive hydrogel for the effective attenuation of experimental autoimmune uveitis. *International Journal of Pharmaceutics*, *622*, 121836. https://doi.org/10.1016/j.ijpharm.2022.121836.

Zlotnikov, I. D., Malashkeevich, S. M., Belogurova, N. G., & Kudryashova, E. V. (2023). Thermoreversible gels based on chitosan copolymers as 'intelligent' drug delivery system with prolonged action for intramuscular injection. *Pharmaceutics*, *15*(5), 1478. https://doi.org/10.3390/pharmaceutics15051478.

CHAPTER 5

Nanocomposites from Recycled Waste Materials and Their Utilization in Biomedical Fields

RAFWANA IBRAHIM, JESIL MATHEW, and SHAILA ANGELA LEWIS*

Department of Pharmaceutics, Manipal College of Pharmaceutical Sciences, Manipal Academy of Higher Education, Manipal, India

*Corresponding author.

ABSTRACT

Recycled waste-derived nanocomposites are gaining a lot of interest in the biomedical industry and other industrial sectors as an environmentally acceptable and sustainable substitute for conventional materials. The growing need for environmentally friendly materials has made it necessary to investigate novel strategies, such as synthesizing nanocomposite materials from recycled waste components. These materials have enormous potential to solve environmental issues related to waste buildup and to provide novel approaches to biomedical problems. Recycled waste materials such as cellulose, starch, chitosan, and vegetable oils are commonly utilized in nanocomposites to improve their mechanical qualities and biocompatibility. The adaptability of nanocomposites made from recycled waste materials across a range of biomedical applications, including drug administration, gene therapy, tissue engineering, and antibacterial activity, is highlighted by the discussion of several fabrication techniques. The use of nanocomposites made from recycled waste materials in biomedical sectors to solve environmental issues, develop sustainable solutions, and improve human health and well-being

Nanocomposites and Nanomaterials in Biomedical Applications. Vivek Dave, Deependra Singh, and Pallerla Naveen Reddy (Eds.)

DOI: 10.1201/9781779641236-5

is highlighted in the chapter. This chapter covers an overview of the requirements for waste-derived nanocomposites, the kinds of recycled waste materials that are utilized, preparation techniques, and their biomedical applications.

5.1 INTRODUCTION

Nanocomposites are a growing category of materials that are gaining popularity due to their distinctive characteristics and applicability in a variety of fields (Hassan et al., 2021; Sharma et al., 2021). The increasing need for environmentally friendly and sustainable materials has spurred research into novel techniques for nanocomposites. The sustainable and environmentally beneficial characteristic of recovered waste materials makes their use in the synthesis of nanocomposites an interesting project. In place of conventional materials, recycled waste materials may be used to create eco-friendly and sustainable nanocomposites (Dong et al., 2015). These materials have the potential to revolutionize several biological applications, including tissue engineering, medication delivery, gene therapy, water purification, and medical imaging (Harish et al., 2022). Integrating recycled waste materials into nanocomposites allows researchers to address environmental issues while also developing creative answers to biological difficulties. Waste management and the demand for cutting-edge biomedical materials are two major issues that may be concurrently addressed in the biomedical disciplines by using nanocomposites made from recycled waste materials (Yu et al., 2021). In the realm of biomedicine, the use of nanocomposites made from recycled waste materials presents a more promising avenue for tackling environmental issues, developing long-term solutions, and improving human health and welfare (Jamp'ilek and Král'ová, 2020).

A viable and sustainable method is to use nanocomposites made from recycled waste materials in biomedical sectors. This tackles the environmental issues brought on by waste buildup and offers creative answers to problems in the biological field (Yu et al., 2021). This chapter explores nanocomposites made from waste materials that have been recycled and their processing techniques, with an emphasis on their potential uses in the biomedical field.

5.2 NEED FOR NANOCOMPOSITES FROM WASTE MATERIAL

The environmental challenges linked to petroleum-based polymers have become more evident in recent years. The substantial waste generated by these polymers poses significant problems, leading to environmental degradation and high recycling costs (Thakur and Kessler, 2014; Thakur and Singha, 2011b). Predictions for 2050 indicate a threefold increase in petroleum-based polymer waste, intensifying the ecological crisis (Thakur and Kessler, 2014). Despite their drawbacks, petroleum-based polymers offer a beneficial blend of structural characteristics, including flexibility, thermal performance, transparency, and strength, making them prevalent in various industrial applications due to their commendable chemical and physical properties.

However, current data on plastics recycling and recovery reveal that only a small margin of (39.5%) postconsumer plastic waste is suitable for reuse (Hestin et al., 2015). The remaining waste negatively impacts the environment by affecting air, soil, water, wildlife, habitats, and human health, thereby increasing the risk of illness and disease. Adopting recycled and eco-friendly materials is a crucial solution to address the adverse environmental effects (Thakur and Singha, 2011a; Wróblewska-Krepsztul et al., 2018).

Natural polymers present an attractive alternative with attributes such as biodegradability, biocompatibility, eco-friendliness, renewability, nontoxicity, and cost-effectiveness. However, their limited mechanical properties and water resistance constrain their widespread use. Nanocomposites from recycled waste resources have been developed to overcome these limitations using various methods to strengthen mechanical-features like tensile, flexibility, elasticity, and impact strength. Waste recycling sources such as β-cyclodextrin (β-CD), graphite, chitosan (CH), cellulose, and alginate have been extensively used to develop bio-nanocomposites.

In the present context, nanocomposites recycled from waste resources play a pivotal part and are expected to increase significanctly across diverse industrial field applications encompassing "drug delivery, fuel cells, tissue engineering, electronics, food packaging, environmental remediation, genetic engineering, and biomedical sciences" (Narancic et al., 2018). The necessity for recycled nanocomposites becomes apparent when addressing the ecological challenges of traditional petroleum-based polymers, fostering a more sustainable and eco-friendly approach to material utilization.

5.3 RECYCLED WASTE MATERIALS FOR NANOCOMPOSITES

Bio renewable-nanocomposites exhibit versatility through three distinct categories: those incorporating renewable additives, those comprising a bio renewable resource matrix, and those with a combination of both featuring both. Notable instances of these kinds of materials encompass chitosan, lignin, starch, chitin, hyaluronic acid, cellulose, vegetable oils, and various proteins. Within this framework, starch and cellulose play pivotal roles, with starch being widely utilized for its biodegradability, accessibility, low cost, and ease of manufacturing. Despite its advantages, starch-based composites may exhibit undesirable properties, necessitating structural modifications (Dubey et al., 2017; Hestin et al., 2015; Singha and Thakur, 2008; Thakur et al., 2010; Ulu et al., 2018; Wróblewska-Krepsztul et al., 2018).

Starch and cellulose play pivotal roles as components in plant-derived additives and matrices. The numerous advantages of using starch, a popular biopolymer, particularly in starch composites, make it a popular choice (Ulu et al., 2016c, 2016b, 2016a). These attributes make it valuable for enhancing biocompatibility in various diverse biomedical implementations and facilitating the attachment of principle drug to the composite. Mixtures holding over 85% starch find application in injection molding and also in foam production. However, there are certain disadvantages to starch-based composites, including increased hydrophilicity, noticeable biodegradability, low thermal strength, and a high decomposition temperature (Miculescu et al., 2017).

Consequently, occasional modification of the starch structure becomes necessary during the manufacturing process. Cellulose, a prevalent natural polymer found in diverse forms such as wood structures, microcrystalline cellulose, and cellulose nanocrystals, enhances nanocomposite strength as a natural fiber. Its widespread presence in the environment, sourced from wood, bacteria, tunicates, and algae, makes cellulose a preferred choice (Klemm et al., 2009). When integrated as an additive, it contributes to improved biocompatibility, biodegradability, and heightened mechanical properties. Additionally, chitosan, a linear amino polysaccharide, is prominently employed in composite production (Ates et al., 2018). The judicious selection and modification of these bio-renewable materials enable the tailoring of nanocomposite properties to meet specific requirements, particularly in medical applications where biocompatibility and substance adherence are crucial. Chitin, a biodegradable polysaccharide with notable mechanical stability, is characterized by nontoxic and physiologically inert property. It is a valuable part of polymer composite structures to enhance material mechanical properties.

Additionally, vegetable oils, cost-effective and widely accessible, contribute to composite structures through two approaches: synthesizing polymers and composites from unsaturated oils with double bonds or preparing polymers and composites through fatty acid extraction. The unsaturation levels of fats play a pivotal role in these methods, leading to the successful synthesis of polyester, polyurethane, or polyolefin copolymers from vegetable oils. These synthesized polymers find diverse applications in various fields (Thakur and Voicu, 2016; Ulu et al., 2020).

5.4 PREPARATION METHODS OF NANOCOMPOSITES FROM RECYCLED WASTE MATERIALS

Recycled nanocomposites can be conveniently manufactured through established methods used for standard composites. These include "solution-mixing, melt mixing, *in situ* polymerization, spray mixing melt intercalation, precipitation, extrusion, sol–gel processes," and so on. Additionally, hand lay-up technique employing fibers from natural sources allows the synthesis of layered composite structures. To create sustainable nanocomposites, innovative techniques like "electrospinning" and "nanofiber direct dispersion" are being employed more often. However, some stick primarily to the education research area and are not widely adopted in the manufacturing production sector. There is a preference for simpler and fewer processing steps in industrial production. Bio-renewable polymers and composites can be produced by modifying chitosan, starch, cellulose gluten, gelatine, etc. These modifications facilitate the transfer of fundamental characteristics of natural polymers to the resulting composites. Furthermore, a multiple-step method using agricultural wastes has been effectively used to produce polymers for biological purposes from renewable sources. Remarkably, techniques such as "melt intercalation, the extrusion process, and *in situ* polymerization," which are often employed to produce nanocomposites, have been effectively applied to the production of biological polymeric composites during the last ten years.

5.4.1 MELT MIXING METHOD

Melt mixing is commonly used in composite preparation. Above the melting point, by the application of heat, this method tends to transform the matrix material into a liquid. These composites are then shaped by carefully introducing an additive into this molten matrix. Fusible structures, such as gelatin, typically derived from recycled sources, serve as the cast, while various other recycled sources could be employed as additives. Particulate/layered additive composites are better made using structurally renewable fibers.

5.4.2 SOLUTION MIXING

Solution mixing is a technique commonly preferred for academic or institutional studies. In this case, an additive is mixed with the matrix material after it has been dissolved in a solvent. The process of "coating, moulding, or film casting" creates the final composite structure (Bledzki et al., 2007). However, this method has some drawbacks, including the need to manage the amount of solvent being used, concerns about toxicity, and the challenge related to solvent removal after the formation of the composite.

5.4.3 IN SITU POLYMERIZATION

It primarily involves polymerization of the monomer structure holding an additive. The additive is present during polymerization, allowing for its dispersion into the matrix. Renewable additives can undergo conversion into composites when incorporated into the polymerization of frequently employed organic monomers (Hiremath et al., 2021).

Another application of the *in situ* polymerization technique involves polymerizing the monomers obtained from renewable sources. It is noteworthy that terpenes and terpenoids may be combined with nanostructures like graphene, clay, and carbon nanotubes to create nanocomposites. Another interesting example is the generation of nanocomposites using common additives from "epoxidized monomers" derived from vegetable oils. This technique minimizes additive agglomeration, ensuring high homogeneity in the produced materials.

5.4.4 EXTRUSION

Extrusion is a key affordable and alternative approach for creating thermoplastic-based composites and nanocomposites. In order to mould the raw material into the necessary shape in the outlet block, this technique entails melting it and forcing it through a channel that is frequently fitted with one or more screws. During this procedure, the molten source material may have nanoscale modifications added to it to easily create nanocomposites. Few composites, reinforced with natural fibers, have been produced using the injection molding process. Extrusion technique is commonly employed for their practicality and suitability for mass production (Bledzki et al., 2007; Hiremath et al., 2021). It is crucial to remember that these approaches are particularly well-suited for creating thermoplastic composites, aiding in the creation of diverse technical items, and making it easier to use cutting-edge production procedures.

5.4.5 SOL–GEL PROCESS

The sol–gel process, which produces inorganic metal oxide additives to make nanocomposites from renewable resources, follows the principle of hydrolyzing metal alkoxide followed by subsequent condensation of the product. This technique uses pricey chemicals like moisture-resistant alkoxides, despite its promise. This method as well as hybrid composite preparation methods are primarily focused on overcoming the mechanical strength difficulty associated with biological composites. These approaches are restricted to academic settings and have not seen widespread industrial use (Zou et al., 2008). Adding an adaptable molecule to the matrix's surface, improving its mechanical characteristics, and eliminating the phase gap in the additive and the matrix material are all viable strategies.

5.4.6 HAND LAY-UP/SPRAY-MIXING PROCESS

These traditional methods are used to be employed in preparing renewable composites. The fibers of a prefabricated mould containing up to 30% bio additives are covered with matrix resin using hand tools like brushes. This straightforward production process is known as the hand lay-up technique. Until the required material thickness is reached, this procedure is repeated. This technique is commonly used for small-scale productions, especially with resins from renewable sources, such as vegetable oils. To produce a range of goods, several automated methods of hand lay-up and mix-coating procedures are used. The methods in question include spraying resins and natural fibers into or onto the mould at the same time (Bledzki et al., 2007). The additive phase, together with a resin and curing agent, is applied onto the container's surface in the form of fibers, nanoplates, or particles. The composite structure is formed when the outer layer is allowed to air out after spraying.

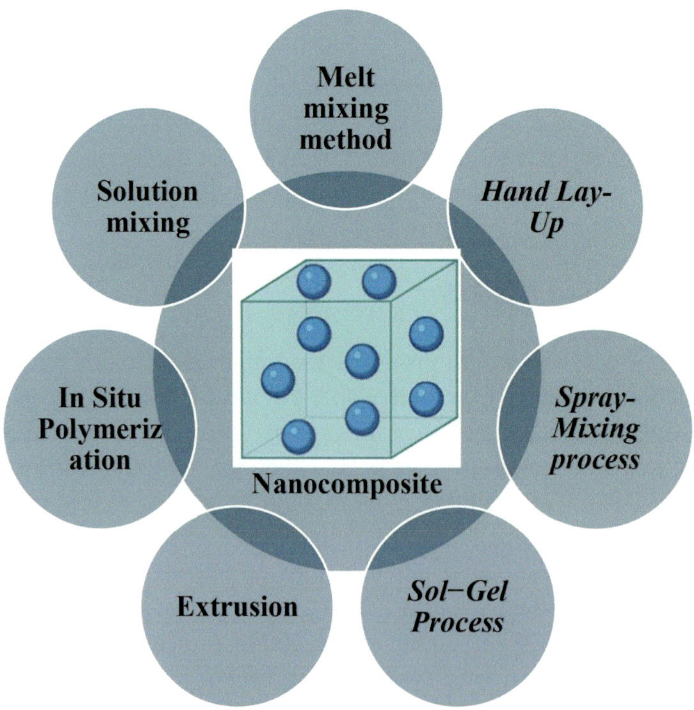

FIGURE 5.1 Preparation techniques of nanocomposites from recycled waste materials.

5.4.7 OTHER METHODS

Additional processes for creating sustainable nanocomposites include "pultrusion, centrifugal casting, electrospinning, and nanofiber direct dispersion technologies." Recognized for its ease of use, dependability, and affordability, the electrospinning process works especially well for creating nonbiological or biological polymer-based nanofibers with dimensions ranging from micrometers to nanometers. This process involves dissolving a solution of polymeric material with the necessary properties in a particular solvent, loading it into a syringe, and applying an electrical solution to the plate at a high potential—usually between 5 and 50 kV—to generate an electric field. The spherical polymeric droplet changes into a conical shape and finally develops into a fiber when the potential difference approaches a threshold. This fiber is collected by the use of a conducting collector. In recognition of the advantageous properties of nanofibrous polymer materials, electrospinning has been widely used to produce nanocomposites, particularly in various biomedical fields.

5.5 NANOCOMPOSITES FROM RECYCLED WASTE MATERIALS: BIOMEDICAL APPLICATIONS

The start of mankind's utilization of polymeric materials dates back to the use of natural sources such as wool, starch, cellulose, and proteins. Natural polymeric materials, which fall under the categories of proteins, polyester, mixed composites, and polysaccharides, nevertheless hold significant positions in a variety of sectors. Naturally occurring protein-based polymers with highly favored shapes like wool are also in great demand.

Additionally, several natural polymers sourced from plants contribute to renewable structures, including polyhydroxyalkonates, lignin, and so on. These polymers originate from either animals or plants and combine with hydrolytic or bacterial processes to form compounds that decompose in the soil. Biocompatibility, a crucial characteristic of medical materials, ensures that desired applications can be pursued without causing adverse events in the human body. While bio-polymers are often chosen for their biocompatible nature in clinical applications, there is a growing interest in nanocomposites derived from renewable sources.

Nanocomposites from biorenewable resources have garnered increased attention due to their versatility and seamless integration into various systems. This interest has surpassed that of synthetic materials, especially in biomedical applications. Notably, nanocomposites from renewable sources find significant applications in four major categories within the medical field: drug delivery, tissue engineering, gene therapy, and antimicrobial activity. The adaptability and diverse applications of these nanocomposites underscore their potential as promising materials in advancing biomedical technologies.

5.5.1 DRUG DELIVERY

In order to improve therapeutic efficacy and minimize adverse effects, formulations inside drug delivery systems are created to guarantee the dependable distribution of pharmacological substances to certain locations (Merino et al., 2015). It is imperative to achieve accurate medication targeting, which has led to the investigation of drug delivery systems with a variety of components. Notably, a great deal of research has been done on the potential of nanocomposites, especially those including biorenewable and biorecyled components, for drug delivery applications. These investigations take into account a variety of drug release mechanisms, including light-, pH-, and temperature-sensitive ones. Crucially, it is possible to effectively integrate the exteriors of the nanoparticles with different groups in order to enhance drug molecule adherence and enable the beneficial release of medications with a range of forms and functions.

For instance, hydroxyapatite (HA) and collagen alginate, a biological nanocomposite, are enhanced to serve as a bone filler and a drug delivery system. This nanocomposite has demonstrated promising results when combined with a growth factor known as a bone formation protein (Sotome et al., 2004). Additionally, fibroblast growth factor-2 has been added to collagen/HA scaffolds with porous topologies, and *in vivo* models have shown encouraging outcomes in terms of cartilage and bone regeneration (Maehara et al., 2010). Moreover, HA/CH nanocomposites have been employed as an administration matrix to provide exact vitamin release (T.-Y. Liu et al., 2006). The utilization of double-layered hydroxide nanocomposites, which act as carriers for the insertion of biopolymers in between the hydroxides, is another intriguing method for controlled drug release. Alginate, carrageenan, and pectin are examples of anionic polysaccharides that have been used to make N-hydroxyl-layered double hydroxide matrices. Furthermore, layered materials facilitate molecular transfer for biological applications, as demonstrated by layered composites of double hydroxide and polylactic acid functionalized with ibuprofen (IBU) (Dagnon et al., 2009).

The gradual release of the antihypertensive drug enalaprilat has been achieved by combining the xyloglucan with double-hydroxides (Ribeiro et al., 2009). Furthermore, Wu et al.'s multifunctional core-shell nanogels, which use hyaluronic acid chains as the targeting ligand, allow for concurrently measuring optical temperature and identifying specific tumors. Thermoresponsive nanogels with silver and gold in the bimetallic component nanoparticle core have been created for imaging applications based on poly(ethylene glycol). These nanogels release therapeutic medications in response to a localized rise in temperature in the targeted location with the application of near-infrared (NIR) irradiation (Wu et al., 2010). Lignin is a cheap, biorenewable, and environmentally benign material that is utilized in many medical applications. Researchers have created lignin nanoparticles (LNPs) with diameters ranging from 100 nm to micrometers using techniques such as antisolvent-precipitation. Diverse production methodologies have been employed in the synthesis of LNPs. These efforts highlight lignin's economic and ecologically positive characteristics, positioning it as a valuable candidate for encapsulating different nanoparticles in a range of pharmaceutical applications.

Nanoparticles play a pivotal role in modern drug delivery systems, offering substantial potential. LNPs exhibit nontoxicity toward various cells, including yeast and microalgae, making them valuable sources in drug administration applications as well as pharmaceutical products. Additionally, these can also serve to remove environmental pollution from being an adsorbent. For the synthesis of LNPs, the precipitation method was used by Richter et al. incorporating $Ag+$ and treating them with a coating solution of diallyl di-methyl-ammonium chloride. This biodegradable and environmentally friendly coating facilitated LNPs' adhesion to bacterial membranes, effectively eliminating various bacteria. The production of silver nanoparticles with enhanced antibacterial efficacy and biodegradable cores by Lievonen et al. reduced negative environmental impact.

Using a solvent exchange method, Lievonen et al. created spherically structured LNPs by dissolving lignin and tetrahydrofuran and with subsequent addition of water without chemically modifying lignin (Lievonen et al., 2016). The resulting LNPs, with diameters ranging from 200 to 500 nm, were gradually added to a poly diallyl di-methyl-ammonium chloride solution, leading to surface functionalization through the adsorption of poly diallyl di-methyl-ammonium chloride. Applications for these LNPs included medication delivery, antibacterial compounds, as well as nanocomposites. Lintinen et al. synthesized LNPs with iron, resulting in particles of different sizes by dissolving lignin in THF, mixing the solution with iron isopropoxide, and initiating condensation. Iron(III)-complexed LNPs exhibited varied magnetic characteristics, suggesting potential applications in making magnetic nanoparticles with antibacterial and biocompatible traits.

Figueiredo et al. refined three types of spherical LNPs through a "solvent exchange method" (Figueiredo et al., 2017). One of the LNPs that was combined with iron showed less cytotoxic effects on various cancer cells and super-paramagnetic behavior, which bodes well for the treatment of cancer. Consequently, pure LNPs were utilized for encapsulating water insoluble drugs, resulting in drug-administrating systems which could also be used in the field of cancer therapy.

Lignin nanocapsules, synthesized by Li et al., included variable sizes from 10 nm to 100s of nanometers and exhibited desirable properties (Li et al., 2016). Chen et al. developed of "pH-responsive lignin capsules" following the interfacial mini-emulsion method. Lignin capsules are particularly valuable in controlled-release applications. These capsules, with diameters ranging from 50–300 nm, exhibited pH-responsive properties due to β-thiopropionate cross-linkings.

Large specific surface areas of graphene nanocarriers stop drugs from releasing too soon before they reach target cells. Agglomeration in aqueous solutions, however, is a major problem. To solve this problem, Luo et al. synthesized nanocomposite drug carriers by homogeneously integrating graphene oxide within porous frameworks of bacterial cellulose. Ibuprofen was successfully loaded onto nanocomposites, indicating both a non-Fickian diffusion mechanism and effective drug release in drug release assays. When compared to bacterial cellulose, these nanocarriers showed increased cell viability (Luo et al., 2017).

5.5.2 GENE THERAPY

Biomaterials play a crucial role in various biomedical applications, with a particular emphasis on controlled gene delivery. Biodegradable cationic polymers have garnered significant attention for gene therapy applications due to their potential in addressing critical illnesses. The primary challenge lies in developing a secure, reliable, and effective vector for gene delivery while taking viral vector biosafety considerations into account (Keles et al., 2016; Mohajeri et al., 2019).

Nonviral gene delivery systems offer a promising solution to overcome these challenges (Mintzer and Simanek, 2009). Lipid-based biorenewable sources, with their ability for self-regulation through supramolecular interactions, form complex structures suitable for gene therapy applications. Specifically, composites derived from lipid/DNA structures have become essential in this context. The use of renewably sourced nanocomposite structures, particularly lipoplexes and polyplexes, has garnered considerable attention (Ates et al., 2020).

Different architectures are displayed by lipoplexes, such as condensed DNA covered in a lipid bilayer or big clumps around narrow fibers. After much investigation into the supramolecular architectures and thermodynamic stability of lipoplexes, the "sliding columnar phase" was postulated as a novel structure for the hexagonal packing of lipid-coated DNA. Because cationic polymers are known to connect to DNA, they have been used to create "polymer DNA complexes" that have surfaces being charged positively and measurable *in vitro* transfection activity (Ates et al., 2020). However, challenges such as elevated cytotoxicity and poor transfection efficiency limit the clinical applicability of certain cationic polymers.

Lipid nanoparticles have been employed for gene delivery, with research focusing on the preparation of lignin nanotubes using nanopore alumina membranes (Dai et al., 2021). Conclusively, the synthesis of lignin nanotubes using phosphoric acid at the conclusion of the reaction highlights the meticulous consideration of physicochemical characteristics. These characteristics, intricately linked to the composition, origin, and extraction method of lignin, play a pivotal role in altering various characteristics of lignin nanostructures. These lignin nanostructures are particularly well-suited for gene transfer applications due to their great biocompatibility and ability for DNA binding.

5.5.3 TISSUE ENGINEERING

The goal of tissue engineering is to promote tissue regeneration, the proliferation of healthy cells, and the generation of materials to replace damaged tissue (Bakhshandeh et al., 2017). Specifically designed compounds for this purpose need to exhibit morphological or structural characteristics suitable for their intended applications. Nanocomposites, known for their ability to mimic the natural morphology of the extracellular matrix surrounding cells, play a crucial role in promoting tissue structure repair.

In the broader field of tissue engineering, the most notable application of nanocomposites derived from biorecycled resources is observed in tissue engineering. Nanoscale proteoglycans, HA mineral nanocrystals, and collagen protein fibers are all involved in the hierarchical structure of natural bone. While HA nanocrystals help to give bones their stiffness and tensile strength, collagen supplies the structural matrix and flexibility with great strength. Replicating the natural bone structure entails synthesizing nanocomposites from polymer building blocks and HA (Hule and Pochan, 2007; Laurencin et al., 1990).

For use in bone regeneration in the clinical field, biodegradable as well as bioabsorbable nanocomposites that stimulate new bone production must be developed. In order to sustain cellular activity, nanocomposites should also have significant microporosity. For example, a composite material synthesized from bio-renewable resources at 37°C and pH 7 showcased the simultaneous production of collagen fibrils and HA nanoparticles, resulting in a porous structure. These scaffolds, exhibiting high elasticity, have been successfully implanted in animals for bone defect repair and reconstruction (Ates et al., 2020; Tampieri et al., 2008; Yokoyama et al., 2005).

Various approaches, such as coprecipitation methods and electrospinning followed by precipitation, have been employed to produce nanocomposites like CH–HA and HA–gelatin fibers, facilitating bone tissue regeneration. Additionally, three-component HA/CH–gelatin network composites produced through phase separation have demonstrated promise in inducing bone tissue generation. Another composite involving gelatin, montmorillonite (MMT), and CH exhibited adhesion and proliferation of mouse bone marrow stromal stem cells (Kim et al., 2005; Zhao et al., 2002; Zhuang et al., 2007).

Nanocomposites derived by combining alginate and polysaccharides of HA have been applied for the regeneration of bone, with alginate serving as a "polymeric sponge" for scaffolds. By using this technique, apatite crystals may be directly formed on spontaneous alginate strands, which promotes osteoblast growth (Zhao et al., 2002).

Furthermore, silk fibroin and hydroxyapatite nanocomposites prepared by the "coprecipitation" approach, as well as "porosity-structured silk fibroin networks" and "nano-hydroxyapatite composites," have demonstrated compatibility with osteoblasts and interactions with the extracellular matrix in three-dimensional scaffolds (Takeuchi et al., 2005).

Nanocomposites have been innovatively produced for tissue engineering applications employing biorenewable proteins and structured nanocrystalline HAs. These efforts collectively contribute to advancing the field of tissue engineering with versatile and biocompatible materials.

In a different study, a homogenous composite mix of "chitin and polybutylene succinate" containing chondroitin sulphate nanoparticles were created and subjected to its characterization (Deepthi et al., 2014). With the addition of CSNPs, the composite scaffold's porosity rose, resulting in ideal porosity and regulated swelling. The macroporous and biologically active characteristics of these nanocomposites led to enhanced adherence and proliferation of human dermal fibroblasts, suggesting potential applications as scaffolds in skin tissue engineering.

In orthopedic applications, the extensive usage of nanocomposites is especially noticeable in the area of bone restoration for certain types of injury. Bones benefit from nanocomposites that emulate their natural properties since they are nanocomposites made of a polymer matrix, usually collagen, and nanosized ceramics. Nanomaterials, such as HA and collagen-based nanocomposites with porous microstructures, have demonstrated significant potential in aiding the formation and growth of bone. Surface characteristics resembling natural bone are crucial for promoting cell and tissue interactions, contributing to the success of implants (Tulloch et al., 2002; Webster et al., 1999).

Nanosized materials, due to their high surface area and associated defects, influence surface properties relevant to cell adhesion, protein adsorption, and activity. Studies on nanocomposites, such as titania/poly(lactic-co-glycolic acid) (PLGA) and nano-HA/PLA scaffolds, have shown improved mineral deposition and osteoblast adherence (H. Liu et al., 2005; Wei and Ma, 2004; Yamasaki and Sakai, 1992).

In summary, nanocomposites with two to three components derived from biorenewable resources, mainly polysaccharides, and proteins, are prominently used in bone tissue engineering. The utilization of biorenewable-based nanocomposites holds significant promise for advancing tissue engineering applications in the future.

5.5.4 ANTIMICROBIAL ACTIVITY

Nanocomposites from recycled sources are increasingly being developed for antimicrobial applications, with a focus on utilizing polysaccharides and proteins as primary matrix components (Iroegbu and Ray, 2022). Utilizing antibacterial organic substances like essential oils together with inorganic metals or metal oxides has become commonplace in antimicrobial agents (Bhandari et al., 2022; Naseem and Durrani, 2021). Applications in food packaging and surface coatings have gained prominence, driven by the biodegradable nature of materials sourced from plants and animals.

Biopolymers such as casein, collagen, gelatin, soybean protein, and gluten from wheat are utilized in the production of films and coatings that provide selective barrier properties, effectively preventing moisture, gas, and solvents from permeating while also enhancing food safety by imparting antimicrobial characteristics. Nevertheless, obstacles such as expenses and mechanical durability have constrained the extensive adoption of these nanocomposites. To overcome these challenges, enhancements are made by incorporating metal oxide nanoparticles, nanoclays, metallic elements, carbon nanofibers, and nanotubes into the matrices to bolster their strength.

Biorenewable resources can be used to create antimicrobial food packaging, such as zein protein-containing nanocomposites found in large amounts in maize gluten meal. Plasticizers such as glycerol and PEG help zein films, especially those made from maize, show better qualities by reducing brittleness and improving processability (Taherimehr et al., 2021). Whey protein is another biopolymer explored for film applications, strengthened by nano-fillers such as nanoclay or SiO_2-coated TiO_2 particles (Gautam and Kumar, 2017). TiO_2/SiO_2 nano enhancers offer antibacterial properties and improve mechanical strength in biopolymer matrices, especially when exposed to UV radiation. Noteworthy examples involve nanocomposites with inhibitory effects on bacteria achieved by incorporating EDTA into films (Ates et al., 2020).

Chitosan, known for its antimicrobial, antiviral, and antitumoral activities, finds applications in preserving vegetables and fruits, extending shelf life by forming CH films with antimicrobial properties (Zhang et al., 2021). To address challenges like poor solubility and low miscibility with other polyelectrolytes, chitlac, a lactose-substituted chitosan, was developed. Chitlac nanoparticles exhibited antibacterial properties against various microorganisms when incorporated into hydrogels (Salomé Veiga and Schneider, 2013).

Cellulose-based nanocomposites, such as cellulose-silver nanocomposite fibers, show excellent antimicrobial properties. The reduction of $AgNO_3$ by the incorporation of gum acacia and guar gum results in a homogeneous distribution of AgNPs in the cellulose matrix, demonstrating antibacterial activity against *Escherichia coli* and potential applications in tissue scaffolds for wound healing.

Carrageenan, another polysaccharide, when combined with chitin nanofibrils, displays antimicrobial effects against specific bacteria. Similarly, nanocomposite films of CH-nanocellulose, combining matrix and reinforcements from biorenewable resources, exhibit increased thermal resistance and antimicrobial protection against various bacteria. Bacterial nanocellulose, though not inherently antimicrobial, can be functionalized with antibiotics, such as fusidic acid, to exhibit antibiotic activity against bacteria like *Staphylococcus aureus*. These bacterial nanocellulose composites show promise as wound-dressing materials.

5.6 FUTURE PERSPECTIVE

Promising opportunities exist for resolving environmental issues, creating sustainable solutions, and improving human health and well-being through the use of nanocomposites made from recycled waste materials in biomedical applications. The goal of ongoing research and development should be to better optimize these nanocomposites' mechanical strength, biocompatibility, and degradation kinetics to fulfill the unique needs of biomedical applications. Furthermore, extending the spectrum of accessible materials and improving the sustainability of nanocomposites may be accomplished by investigating new waste resources and creative synthesis methods. Moreover, for these nanocomposites to be widely used in biomedical contexts, efforts must be made to increase production scale and develop standardized procedures for their synthesis, characterization, and assessment. Overall, the future of nanocomposites created from recycled waste materials in biomedical applications looks promising, with prospects to solve important environmental issues while developing biomedical technology.

5.7 CONCLUSION

Furthermore, it should be noted that nanocomposites made of waste materials that have been recycled offer a viable and sustainable solution to environmental issues and the advancement of biomedical applications. Through the use of recycled waste materials, including cellulose, starch, chitosan, and vegetable oils, scientists may create nanocomposites that have improved mechanical characteristics and biocompatibility, making them suitable for a variety of biomedical uses. By combining cutting-edge manufacturing methods with cutting-edge waste resource discovery, nanocomposites made of recycled waste materials provide inventive answers to biomedical and environmental problems. Moving forward, additional research and development efforts are required to fully realize the promise of these nanocomposites in biomedical domains, eventually contributing to environmental sustainability while increasing human health and well-being.

KEYWORDS

- **biomedical**
- **drug delivery**
- **nanocomposite**
- **recycled waste materials**
- **renewable biowaste**

REFERENCES

Ates, B., Koytepe, S., Ulu, A., Gurses, C., & Thakur, V. K. (2020). Chemistry, structures, and advanced applications of nanocomposites from biorenewable resources. *Chemical Reviews, 120*(17), 9304–9362.

Ates, B., Ulu, A., Köytepe, S., Noma, S. A. A., Kolat, V. S., & Izgi, T. (2018). Magnetic-propelled $Fe_3\ O_4$—Chitosan carriers enhance L-asparaginase catalytic activity: A promising strategy for enzyme immobilization. *RSC Advances, 8*(63), 36063–36075.

Bakhshandeh, B., Zarrintaj, P., Oftadeh, M. O., Keramati, F., Fouladiha, H., Sohrabi-Jahromi, S., & Ziraksaz, Z. (2017). Tissue engineering; strategies, tissues, and biomaterials. *Biotechnology and Genetic Engineering Reviews, 33*(2), 144–172.

Bhandari, V., Jose, S., Badanayak, P., Sankaran, A., & Anandan, V. (2022). Antimicrobial finishing of metals, metal oxides, and metal composites on textiles: A systematic review. *Industrial & Engineering Chemistry Research, 61*(1), 86–101.

Bledzki, A. K., Mamun, A. A., & Faruk, O. (2007). Abaca fibre reinforced PP composites and comparison with jute and flax fibre PP composites. *Express Polymer Letters, 1*(11), 755–762.

Dagnon, K. L., Ambadapadi, S., Shaito, A., Ogbomo, S. M., DeLeon, V., Golden, T. D., Rahimi, M., Nguyen, K., Braterman, P. S., & D'souza, N. A. (2009). Poly (L-lactic acid) nanocomposites with layered double hydroxides functionalized with ibuprofen. *Journal of Applied Polymer Science, 113*(3), 1905–1915.

Dai, L., Liu, R., & Si, C.-L. (2021). Lignin-based materials for drug and gene delivery. In *Lignin-based materials for biomedical applications* (pp. 327–370). Elsevier.

Deepthi, S., Sidhy Viha, C. V, Thitirat, C., Furuike, T., Tamura, H., & Jayakumar, R. (2014). Fabrication of chitin/poly (butylene succinate)/chondroitin sulfate nanoparticles ternary composite hydrogel scaffold for skin tissue engineering. *Polymers, 6*(12), 2974–2984.

Dong, P., Prasanth, R., Xu, F., Wang, X., Li, B., & Shankar, R. (2015). *Eco-friendly polymer Nanocomposite—Properties and processing* (pp. 1–15). Springer.

Dubey, S. P., Thakur, V. K., Krishnaswamy, S., Abhyankar, H. A., Marchante, V., & Brighton, J. L. (2017). Progress in environmental-friendly polymer nanocomposite material from PLA: Synthesis, processing and applications. *Vacuum, 146*, 655–663.

Figueiredo, P., Lintinen, K., Kiriazis, A., Hynninen, V., Liu, Z., Bauleth-Ramos, T., Rahikkala, A., Correia, A., Kohout, T., Sarmento, B., et al. (2017). *In vitro* evaluation of biodegradable lignin-based nanoparticles for drug delivery and enhanced antiproliferation effect in cancer cells. *Biomaterials, 121*, 97–108.

Gautam, R. B., & Kumar, S. (2017). Development of protein based films with nanoparticle as strengthening material for biodegradable packaging—A review. *International Journal of Agriculture Innovations and Research, 5*(5), 790–805.

Harish, V., Tewari, D., Gaur, M., Yadav, A. B., Swaroop, S., Bechelany, M., & Barhoum, A. (2022). Review on nanoparticles and nanostructured materials: Bioimaging, biosensing, drug delivery, tissue engineering, antimicrobial, and agro-food applications. *Nanomaterials, 12*(3), 457.

Hassan, T., Salam, A., Khan, A., Khan, S. U., Khanzada, H., Wasim, M., Khan, M. Q., & Kim, I. S. (2021). Functional nanocomposites and their potential applications: A review. *Journal of Polymer Research, 28*(2), 36.

Hestin, M., Faninger, T., & Milios, L. (2015). Increased EU plastics recycling targets: Environmental, economic and social impact assessment. Retrieved, 17, 2018.

Hiremath, A., Murthy, A. A., Thipperudrappa, S., & KN, B. (2021). Nanoparticles filled polymer nanocomposites: A technological review. *Cogent Engineering, 8*(1), 1991229.

Hule, R. A., & Pochan, D. J. (2007). Polymer nanocomposites for biomedical applications. *MRS Bulletin, 32*(4), 354–358.

Iroegbu, A. O. C., & Ray, S. S. (2022). Recent developments and future perspectives of biorenewable nanocomposites for advanced applications. *Nanotechnology Reviews, 11*(1), 1696–1721.

Jamp'ilek, J., & Král'ová, K. (2020). Preparation of nanocomposites from agricultural waste and their versatile applications. In *Multifunctional hybrid nanomaterials for sustainable agri-food and ecosystems* (pp. 51–98). Elsevier.

Keles, E., Song, Y., Du, D., Dong, W.-J., & Lin, Y. (2016). Recent progress in nanomaterials for gene delivery applications. *Biomaterials Science, 4*(9), 1291–1309.

Kim, H.-W., Song, J.-H., & Kim, H.-E. (2005). Nanofiber generation of gelatin—Hydroxyapatite biomimetics for guided tissue regeneration. *Advanced Functional Materials, 15*(12), 1988–1994.

Klemm, D., Schumann, D., Kramer, F., Heßler, N., Koth, D., & Sultanova, B. (2009). Nanocellulose materials—Different cellulose, different functionality. *Macromolecular Symposia, 280*(1), 60–71.

Laurencin, C., Domb, A., Morris, C., Brown, V., Chasin, M., McConnell, R., Lange, N., & Langer, R. (1990). Poly (anhydride) administration in high doses *in vivo*: Studies of biocompatibility and toxicology. *Journal of Biomedical Materials Research, 24*(11), 1463–1481.

Li, H., Deng, Y., Liu, B., Ren, Y., Liang, J., Qian, Y., Qiu, X., Li, C., & Zheng, D. (2016). Preparation of nanocapsules via the self-assembly of Kraft lignin: A totally green process with renewable resources. *ACS Sustainable Chemistry \& Engineering, 4*(4), 1946–1953.

Lievonen, M., Valle-Delgado, J. J., Mattinen, M.-L., Hult, E.-L., Lintinen, K., Kostiainen, M. A., Paananen, A., Szilvay, G. R., Setälä, H., & Österberg, M. (2016). A simple process for lignin nanoparticle preparation. *Green Chemistry, 18*(5), 1416–1422.

Liu, H., Slamovich, E. B., & Webster, T. J. (2005). Increased osteoblast functions on nanophase titania dispersed in poly-lactic-co-glycolic acid composites. *Nanotechnology, 16*(7), S601.

Liu, T.-Y., Chen, S.-Y., Li, J.-H., & Liu, D.-M. (2006). Study on drug release behaviour of CDHA/chitosan nanocomposites—Effect of CDHA nanoparticles. *Journal of Controlled Release, 112*(1), 88–95.

Luo, H., Ao, H., Li, G., Li, W., Xiong, G., Zhu, Y., & Wan, Y. (2017). Bacterial cellulose/graphene oxide nanocomposite as a novel drug delivery system. *Current Applied Physics, 17*(2), 249–254.

Maehara, H., Sotome, S., Yoshii, T., Torigoe, I., Kawasaki, Y., Sugata, Y., Yuasa, M., Hirano, M., Mochizuki, N., Kikuchi, M., & others. (2010). Repair of large osteochondral defects in rabbits using porous hydroxyapatite/collagen (HAp/Col) and fibroblast growth factor-2 (FGF-2). *Journal of Orthopaedic Research, 28*(5), 677–686.

Merino, S., Mart'in, C., Kostarelos, K., Prato, M., & Vázquez, E. (2015). Nanocomposite hydrogels: 3D polymer—Nanoparticle synergies for on-demand drug delivery. *ACS Nano, 9*(5), 4686–4697.

Miculescu, F., Maidaniuc, A., Voicu, S. I., Thakur, V. K., Stan, G. E., & Ciocan, L. T. (2017). Progress in hydroxyapatite—Starch based sustainable biomaterials for biomedical bone substitution applications. *ACS Sustainable Chemistry \& Engineering, 5*(10), 8491–8512.

Mintzer, M. A., & Simanek, E. E. (2009). Nonviral vectors for gene delivery. *Chemical Reviews, 109*(2), 259–302.

Mohajeri, M., Behnam, B., & Sahebkar, A. (2019). Biomedical applications of carbon nanomaterials: Drug and gene delivery potentials. *Journal of Cellular Physiology, 234*(1), 298–319.

Narancic, T., Verstichel, S., Reddy Chaganti, S., Morales-Gamez, L., Kenny, S. T., De Wilde, B., Babu Padamati, R., & O'Connor, K. E. (2018). Biodegradable plastic blends create new possibilities for end-of-life management of plastics but they are not a panacea for plastic pollution. *Environmental Science \& Technology, 52*(18), 10441–10452.

Naseem, T., & Durrani, T. (2021). The role of some important metal oxide nanoparticles for wastewater and antibacterial applications: A review. *Environmental Chemistry and Ecotoxicology, 3,* 59–75.

Ribeiro, C., Arizaga, G. G. C., Wypych, F., & Sierakowski, M.-R. (2009). Nanocomposites coated with xyloglucan for drug delivery: *In vitro* studies. *International Journal of Pharmaceutics, 367*(1–2), 204–210.

Salomé Veiga, A., & Schneider, J. P. (2013). Antimicrobial hydrogels for the treatment of infection. *Peptide Science, 100*(6), 637–644.

Sharma, S., Sudhakara, P., Omran, A. A. B., Singh, J., & Ilyas, R. A. (2021). Recent trends and developments in conducting polymer nanocomposites for multifunctional applications. *Polymers, 13*(17), 2898.

Singha, A. S., & Thakur, V. K. (2008). Mechanical, morphological and thermal properties of pine needle-reinforced polymer composites. *International Journal of Polymeric Materials, 58*(1), 21–31.

Sotome, S., Uemura, T., Kikuchi, M., Chen, J., Itoh, S., Tanaka, J., Tateishi, T., & Shinomiya, K. (2004). Synthesis and *in vivo* evaluation of a novel hydroxyapatite/collagen—Alginate as a bone filler and a drug delivery carrier of bone morphogenetic protein. *Materials Science and Engineering: C, 24*(3), 341–347.

Taherimehr, M., YousefniaPasha, H., Tabatabaeekoloor, R., & Pesaranhajiabbas, E. (2021). Trends and challenges of biopolymer-based nanocomposites in food packaging. *Comprehensive Reviews in Food Science and Food Safety, 20*(6), 5321–5344.

Takeuchi, A., Ohtsuki, C., Miyazaki, T., Kamitakahara, M., Ogata, S., Yamazaki, M., Furutani, Y., Kinoshita, H., & Tanihara, M. (2005). Heterogeneous nucleation of hydroxyapatite on protein: Structural effect of silk sericin. *Journal of the Royal Society Interface, 2*(4), 373–378.

Tampieri, A., Sandri, M., Landi, E., Pressato, D., Francioli, S., Quarto, R., & Martin, I. (2008). Design of graded biomimetic osteochondral composite scaffolds. *Biomaterials, 29*(26), 3539–3546.

Thakur, V. K., & Kessler, M. R. (2014). Synthesis and characterization of AN-g-SOY for sustainable polymer composites. *ACS Sustainable Chemistry \& Engineering, 2*(10), 2454–2460.

Thakur, V. K., & Singha, A. S. (2011a). Physicochemical and mechanical behavior of cellulosic pine needle-based biocomposites. *International Journal of Polymer Analysis and Characterization, 16*(6), 390–398.

Thakur, V. K., & Singha, A. S. (2011b). Rapid synthesis, characterization, and physicochemical analysis of biopolymer-based graft copolymers. *International Journal of Polymer Analysis and Characterization, 16*(3), 153–164.

Thakur, V. K., Singha, A. S., Kaur, I., Nagarajarao, R. P., & Liping, Y. (2010). Silane functionalization of Saccaharum cilliare fibers: Thermal, morphological, and physicochemical study. *International Journal of Polymer Analysis and Characterization, 15*(7), 397–414.

Thakur, V. K., & Voicu, S. I. (2016). Recent advances in cellulose and chitosan based membranes for water purification: A concise review. *Carbohydrate Polymers, 146*, 148–165.

Tulloch, A. J. D., Bizios, R., & Siegel, R. W. (2002). Nanophase alumina/poly (L-lactic acid) composite scaffolds for biomedical applications. *MRS Online Proceedings Library (OPL), 740,* I6–3.

Ulu, A., Balcioglu, S., Birhanli, E., Sarimeseli, A., Keskin, R., Koytepe, S., & Ates, B. (2018). Poly (2-hydroxyethyl methacrylate)/boric acid composite hydrogel as soft contact lens material: Thermal, optical, rheological, and enhanced antibacterial properties. *Journal of Applied Polymer Science, 135*(35), 46575.

Ulu, A., Birhanli, E., Boran, F., Köytepe, S., Yesilada, O., & Ates, B. (2020). Laccase-conjugated thiolated chitosan-Fe3O4 hybrid composite for biocatalytic degradation of organic dyes. *International Journal of Biological Macromolecules, 150*, 871–884.

Ulu, A., Koytepe, S., & Ates, B. (2016a). Design of starch functionalized biodegradable P (MAA-co-MMA) as carrier matrix for l-asparaginase immobilization. *Carbohydrate Polymers, 153*, 559–572.

Ulu, A., Koytepe, S., & Ates, B. (2016b). Synthesis and characterization of biodegradable pHEMA-starch composites for immobilization of L-asparaginase. *Polymer Bulletin, 73*, 1891–1907.

Ulu, A., Koytepe, S., & Ates, B. (2016c). Synthesis and characterization of PMMA composites activated with starch for immobilization of L-asparaginase. *Journal of Applied Polymer Science, 133*(19).

Webster, T. J., Siegel, R. W., & Bizios, R. (1999). Osteoblast adhesion on nanophase ceramics. *Biomaterials, 20*(13), 1221–1227.

Wei, G., & Ma, P. X. (2004). Structure and properties of nano-hydroxyapatite/polymer composite scaffolds for bone tissue engineering. *Biomaterials, 25*(19), 4749–4757.

Wróblewska-Krepsztul, J., Rydzkowski, T., Borowski, G., Szczypiński, M., Klepka, T., & Thakur, V. K. (2018). Recent progress in biodegradable polymers and nanocomposite-based packaging materials for sustainable environment. *International Journal of Polymer Analysis and Characterization, 23*(4), 383–395.

Wu, W., Shen, J., Banerjee, P., & Zhou, S. (2010). Core-shell hybrid nanogels for integration of optical temperature-sensing, targeted tumor cell imaging, and combined chemo-photothermal treatment. *Biomaterials, 31*(29), 7555–7566.

Yamasaki, H., & Sakai, H. (1992). Osteogenic response to porous hydroxyapatite ceramics under the skin of dogs. *Biomaterials, 13*(5), 308–312.

Yokoyama, A., Gelinsky, M., Kawasaki, T., Kohgo, T., König, U., Pompe, W., & Watari, F. (2005). Biomimetic porous scaffolds with high elasticity made from mineralized collagen—An animal study. *Journal of Biomedical Materials Research Part B: Applied Biomaterials: An Official Journal of The Society for Biomaterials, The Japanese Society for Biomaterials, and The Australian Society for Biomaterials and the Korean Society for Biomaterials, 75*(2), 464–472.

Yu, S., Sun, J., Shi, Y., Wang, Q., Wu, J., & Liu, J. (2021). Nanocellulose from various biomass wastes: Its preparation and potential usages towards the high value-added products. *Environmental Science and Ecotechnology, 5*, 100077.

Zhang, X., Ismail, B. B., Cheng, H., Jin, T. Z., Qian, M., Arabi, S. A., Liu, D., & Guo, M. (2021). Emerging chitosan-essential oil films and coatings for food preservation—A review of advances and applications. *Carbohydrate Polymers, 273*, 118616.

Zhao, F., Yin, Y., Lu, W. W., Leong, J. C., Zhang, W., Zhang, J., Zhang, M., & Yao, K. (2002). Preparation and histological evaluation of biomimetic three-dimensional hydroxyapatite/chitosan-gelatin network composite scaffolds. *Biomaterials, 23*(15), 3227–3234.

Zhuang, H., Zheng, J. P., Gao, H., & De Yao, K. (2007). *In vitro* biodegradation and biocompatibility of gelatin/montmorillonite-chitosan intercalated nanocomposite. *Journal of Materials Science: Materials in Medicine, 18*, 951–957.

Zou, H., Wu, S., & Shen, J. (2008). Polymer/silica nanocomposites: Preparation, characterization, properties, and applications. *Chemical Reviews, 108*(9), 3893–3957.

CHAPTER 6

Role and Applications of Biodegradable Nanocomposites vs. Nondegradable Nanocomposites in Biomedical Applications

INDHUMATHI THIRUGNANASAMBANDHAM[1], PARIKSHIT ROYCHOWDHURY[1], ROSHAN P. RAO[1], ASHOK KUMAR JANAKIRAMAN[2], MURALIKRISHNAN DHANASEKARAN[3], SACHIN KUMAR SINGH[4,5], and GOWTHAMARAJAN KUPPUSAMY[1,*]

[1]*Department of Pharmaceutics, JSS College of Pharmacy, JSS Academy of Higher Education & Research, Ooty, Nilgiris, Tamil Nadu, India*

[2]*Faculty of Pharmaceutical Sciences, UCSI University, Cheras, Kuala Lumpur, Malaysia*

[3]*Department of Drug Discovery and Development, Harrison College of Pharmacy, Auburn University Auburn, AL, USA*

[4]*School of Pharmaceutical Sciences, Lovely Professional University, Phagwara, Punjab, India*

[5]*Faculty of Health, Australian Research Centre in Complementary and Integrative Medicine, University of Technology Sydney, Ultimo, NSW, Australia*

*Corresponding author.

Nanocomposites and Nanomaterials in Biomedical Applications. Vivek Dave, Deependra Singh, and Pallerla Naveen Reddy (Eds.)

DOI: 10.1201/9781779641236-6

ABSTRACT

Nanocomposites (NCs) have emerged as a promising field in the biomedical industry, wherein nanoscale components are integrated into a matrix to develop advanced materials with improved properties. This chapter provides a comprehensive overview of biodegradable and nondegradable NCs in biomedicine, encompassing their composition, properties, comparative advantages, and prospects, as well as the crucial kinetics of degradation and safety. Biodegradable NCs hold significant potential in biomedical applications, as they naturally degrade over time, minimizing long-term risks and environmental impacts. On the other hand, nondegradable NCs offer exceptional durability and stability, making them suitable for long-lasting biomedical devices. In the last decade, substantial research efforts have been dedicated to advancing both types of NCs, resulting in the development of novel materials with enhanced properties and functionalities for applications such as tissue engineering, drug delivery, medical implants, and diagnostics. Understanding and harnessing the properties of biodegradable and nondegradable NCs contribute to the development of advanced materials in the biomedical field. This chapter offers invaluable insights into the role and applications of biodegradable and nondegradable NCs, providing a comprehensive foundation for further research, innovation, a hybrid approach, and a synergistic effect within the dynamic field of biomedicine.

6.1 INTRODUCTION

Nanotechnology has emerged as a thriving field of research and development, permeating various technological disciplines. Within this domain, the exploration of nanocomposites (NCs) has played a pivotal role in advancing materials science. NCs, characterized by at least one phase exhibiting nanoscale dimensions, offer unique properties and versatile applications. This chapter delves into the role and applications of biodegradable and nondegradable NCs in the dynamic field of biomedicine. NCs are composite materials in which nanoscale components, such as nanoparticles, nanofibers, or nanotubes, are embedded within a matrix phase composed of metals, ceramics, or polymers (Khan et al., 2016). They harness the synergistic effects of the reinforcing phase's strength and the matrix phase's toughness, resulting in desirable properties unattainable in conventional materials alone (Xiong et al., 2018). The nanoscale nature of the reinforcement imparts high surface-to-volume ratios and increased ductility without sacrificing strength or scratch resistance (Jia and Kiick, 2009). In the context of biomedicine, the development of NCs has garnered significant attention. Biodegradable NCs hold promise for controlled degradation and compatibility with biological systems, making them suitable for drug delivery systems, tissue engineering scaffolds, and implantable devices (Sharma et al., 2021). Conversely, nondegradable NCs offer long-term stability, finding applications in areas requiring sustained structural integrity and functionality (Chung et al., 2011). This chapter provides a comprehensive review of both biodegradable and nondegradable NCs. It covers their composition, properties, relative benefits, and potential applications in the biomedical field. The discussion also extends to critical factors such as degradation kinetics and safety, emphasizing the pivotal role of kinetics within the dynamic landscape of biomedicine advancement. For a more profound understanding of NCs characteristics and performance, the selection of the optimal NCs becomes facilitated across diverse biomedical domains.

6.2 DISTINCTIVE MATERIALS USED IN BIODEGRADABLE AND NONDEGRADABLE NCs

In this section, a comprehensive distinction of materials has been undertaken, categorizing them into biodegradable and nondegradable constituents (shown in Figure 6.1) for nanocomposite fabrication.

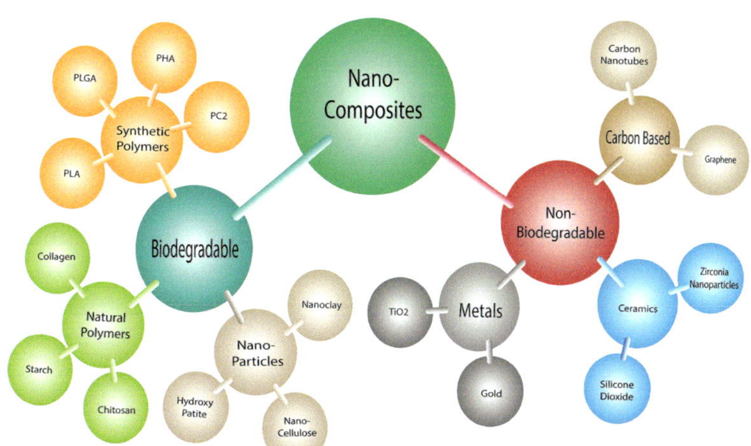

FIGURE 6.1 Charting biomedical nanocomposites: Varieties of biodegradable and nondegradable materials.

6.2.1 POLYLACTIC ACID (PLA)

Polylactic acid (PLA) is a widely used biodegradable polymer in the biomedical field, derived from renewable resources such as corn, sugar, potato, and other agricultural products (Sha et al., 2016). It offers several advantages, including producing water and carbon dioxide as harmless degradation products that can be easily eliminated from the human body, making PLA highly sought-after in the market (Balla et al., 2021). PLA exists in different forms, such as poly (D-lactic acid) (PDLA), poly (L-lactic acid) (PLLA), and the racemic blend D, L-PLA (PDLLA), each with its unique properties and applications (de França et al., 2022). While PLLA is preferred in applications requiring good mechanical behavior, PDLLA finds its use in drug delivery systems due to its monophasic structure (Capuana et al., 2022). However, pure PLA materials have limitations such as brittleness and uncontrolled degradation rates, prompting the development of PLA-based NCs that overcome these shortcomings by incorporating additives and fillers (Ranakoti et al., 2022). Among the PLA-based copolymers, poly (lactic-glycolic acid) (PLGA) has gained significant attention in the biomedical field and has been approved by the FDA for clinical use. PLGA is synthesized through ring-opening polymerization, using lactide monomers and polyglycolic acid (PGA). This efficient polymerization method allows the production of high-molecular-weight PLA and its copolymers with desired properties. PLGA offers advantages such as biodegradability, biocompatibility, and tunable degradation rates, making it suitable for various biomedical applications (Rocha et al., 2022). PLA-based NCs, including copolymers of PLA and nanofillers as annexing agents, have shown great potential in the biomedical field (Sha et al., 2016). These NCs incorporate various fillers such as clay minerals, carbon nanotubes (CNTs), graphene, graphene oxide (GO), organomontmorillonite, natural nanofibers, and calcium phosphate ceramics like HA (HA) and β-tricalcium phosphate. The addition of these fillers results in improved mechanical properties, biocompatibility, and bioactivity of PLA materials (Sha et al., 2016). The combination of different fillers often exhibits synergistic effects, further enhancing the overall performance of the NCs.

6.2.2 POLYHYDROXYALKANOATES (PHA)

Polyhydroxyalkanoates (PHA) are innovative and promising materials for NCs in biomedical applications. These biodegradable polymers are synthesized through microbial fermentation, where microorganisms convert sugar into fat for storage and energy production (Kalia et al., 2023). PHA contain carbon and hydrogen, with ester linkages between hydroxyl groups. The synthesis of PHAs involves the production of fatty acids by microbial enzymes known as synthases. These fatty acids are then polymerized to form PHA chains (Sehgal and Gupta, 2020). One notable advantage of PHAs is their ability to produce various types with different structures and properties. This versatility allows for the customization of PHA-based NCs to meet specific application requirements. PHA types can be classified based on their monomer composition and chain length. Short-chain-length PHAs (scl-PHAs) typically have 3–5 carbon atoms in their monomer units, while medium-chain-length PHAs (mcl-PHAs) have 6–14 carbon atoms. Long-chain-length PHAs (lcl-PHAs) have more than 14 carbon atoms. Among these types, mcl-PHAs are often favored for NC development in biomedical applications due to their balanced mechanical properties (Poltronieri and Kumar, 2017). In terms of degradation, PHAs are biodegradable and can be broken down by natural processes over time. This property is advantageous for biomedical applications, where the NC needs to gradually degrade as new tissue forms (Guo et al., 2022). The degradation rate of PHAs can be tailored by adjusting the chain length and monomer composition. For example, mcl-PHAs tend to degrade more slowly compared to scl-PHAs. Furthermore, the degradation of PHAs can occur through different mechanisms. Surface erosion involves the degradation of the outer layer of the NC, while bulk erosion involves the degradation of the entire material. The choice of degradation mechanism depends on the specific application and desired degradation profile (Dalton et al., 2022). When incorporated into NCs, PHAs offer several benefits. Their tunable chain length allows for the customization of mechanical properties, such as stiffness, strength, and elasticity. The addition of nanofillers, such as clay minerals, CNTs, or HA, can further enhance the mechanical properties and biocompatibility of PHA-based NCs (Xu et al., 2020).

6.2.3 STARCH

Starch, with its unique molecular structure and exceptional biocompatible characteristics, holds great promise for the development of innovative biocomposites. This natural polymer is composed of amylose and amylopectin, which contribute to its distinctive properties. Starch exhibits remarkable gelatinization behavior, making it highly processable and suitable for various applications (Faisal et al., 2022). To further enhance its properties, modified starches, such as hydroxypropyl starch or carboxymethyl starch, have been developed to improve stability, biodegradability, and mechanical strength (Amaraweera et al., 2021). One of the key advantages of starch-based materials is their excellent biocompatibility, making them ideal for biomedical applications. Starch-based products are well-tolerated by the human body, avoiding adverse reactions or toxicity concerns. This biocompatibility has led to their use in diverse biomedical fields, including tissue engineering and drug delivery systems (Marques et al., 2002). Starch can serve as both the matrix and reinforcement in NC materials, providing opportunities for creating biocomposites with enhanced mechanical properties and improved biocompatibility (Xie et al., 2014). In the realm of tissue engineering, starch-based biodegradable bone cement has demonstrated its ability to provide temporary structural support and gradually disappear, allowing for the ingrowth of new bone and complete healing. Additionally, starch-based polymers have been employed in bone tissue engineering scaffolds, where their suitable degradation rate promotes the formation of new tissue (Veeman et al., 2021). Starch-based materials have also shown promise in drug delivery systems. They offer advantages in terms of deliverability without the need for surgical removal. The biocompatibility of starch makes it a favorable material choice for drug delivery applications, ensuring the safe and controlled release of therapeutic agents (Sivamaruthi et al., 2022). Researchers have introduced the concept of starch-based NCs to overcome the limitations of biodegradable starch alone. By incorporating various nanofillers, such as particles, fibers, or layers, into the starch matrix, NCs with enhanced properties have been developed. Integrating nanofillers at the nanometer scale allows for forming strong molecular bridges within the starch matrix, resulting in improved mechanical properties, thermal stability, moisture resistance, oxygen barrier properties, and biodegradation rates (Xie et al., 2011). Furthermore, NCs offer advantages over conventional composites, as they require smaller amounts of fillers and can enhance the optical transparency of the composite (Castillo et al., 2017). Another exciting aspect is the combination of starch with HA, a widely used material in biomedical applications, particularly in contact with bone tissue. The incorporation of HA with starch addresses the brittleness issue associated with HA and enhances the final morphology of the composite material. Starch acts as a pore generator in membranes and porous ceramics, promoting mechanical strength, cell migration, bone oxygenation, and angiogenesis in bone tissue engineering applications. The use of nano-sized HA in conjunction with starch has also shown improved osteoblast adhesion compared to larger-sized ceramic materials (Turnbull et al., 2018).

6.2.4 CHITOSAN (CA)

Chitosan (CA), a unique polysaccharide used in the development of biodegradable and nondegradable NCs, exhibits fascinating properties at the molecular level. It is derived from chitin, a naturally abundant polymer found in the exoskeletons of crustaceans.

At the molecular level, CA is composed of repeating units of glucosamine and N-acetylglucosamine, which are joined together by β-1,4-glycosidic linkages. This molecular arrangement imparts specific characteristics to CA that make it highly versatile and adaptable for various applications (Iroegbu and Ray, 2023). One of the distinguishing features of CA is the presence of amino groups (-NH_2) along its polymer chain. These amino groups confer unique reactivity and functionality to CA, allowing for tailoring and copolymerization with other materials. The presence of amino groups also imparts CA with its inherent polycationic nature, making it an excellent candidate for applications involving interactions with negatively charged species, such as DNA or proteins (Sahoo et al., 2009). Crystallinity is another notable characteristic of CA at the molecular level. CA exhibits semicrystalline behavior, with regions of ordered molecular packing interspersed with less-ordered regions. This crystalline structure contributes to the mechanical strength and stability of CA-based NCs (Ioelovich, 2014). The isolation process of CA involves the deacetylation of chitin, where the acetyl groups are removed, resulting in varying degrees of deacetylation. This degree of deacetylation influences the physicochemical properties of CA, including its solubility, mechanical strength, and biocompatibility. By controlling the degree of deacetylation, the properties of CA can be tailored to meet specific application requirements (Pellis et al., 2022). CA also exhibits interesting biological properties at the molecular level. It is biocompatible, meaning it is well-tolerated by living organisms and does not elicit significant immune responses. Additionally, CA possesses antimicrobial properties, making it suitable for wound healing and tissue engineering applications (Dai et al., 2011). Furthermore, CA's mucoadhesive strength enables it to adhere to mucosal surfaces, making it ideal for drug delivery systems targeting mucosal tissues. Its solubility in acidic environments enhances its potential for drug delivery applications in the gastrointestinal tract (Ways et al., 2018). In the realm of NCs and nanomaterials, CA's molecular structure allows for the formation of nanoparticles and nanofibers. These nanostructures exhibit high surface area-to-volume ratios and can be functionalized with other materials to enhance properties such as drug loading capacity, mechanical strength, and controlled release capabilities (Sinha et al., 2022).

6.2.5 CELLULOSE

Cellulose, a remarkable natural biopolymer, holds great potential for biomedical applications, particularly when utilized in NCs. With its unique composition and polymerization characteristics, as well as its diverse forms, cellulose stands out as a remarkable material in the field of biomedicine. Cellulose exhibits a wide range of properties that make it highly desirable for biomedical applications. It possesses excellent mechanical strength, allowing it to endure rigorous conditions over time. Additionally, cellulose demonstrates exceptional stability, even in acidic environments, adding to its versatility. Its chirality, a fascinating characteristic, further enhances its appeal (Sinha et al., 2022). Beyond its strength and stability, cellulose offers more advantages. It is surprisingly lightweight, making it an ideal choice for reducing the overall weight of biomedical devices (Norizan et al., 2022). Moreover, cellulose is highly biodegradable, breaking down naturally without leaving harmful residues. This exceptional property ensures its compatibility with the human body, as it gradually degrades while new tissue forms (Chinta et al., 2021). At the molecular level, cellulose presents an abundance of hydroxyl functional groups on its surface. These groups enable chemical modification and enhance cellulose's wettability, enabling it to interact effectively with biological tissues. This unique feature facilitates cell adhesion and proliferation, promoting the growth of new tissue (Hu et al., 2014).

Cellulose's versatility extends to its various forms, each offering distinct advantages. Hydrogels, with their water-rich gel-like consistency, hold promise for controlled drug delivery, accelerated wound healing, and tissue engineering applications (Turunen et al., 2023). Cellulose-based membranes act as reliable shields, providing protection and precise filtering capabilities (Junter and Lebrun, 2017). Aerogels, on the other hand, possess highly porous structures, making them ideal for tissue engineering and advanced wound dressings (Abdul Khalil et al., 2022). Lastly, cellulose scaffolds provide a three-dimensional (3D) environment that supports cell growth and tissue regeneration (Mirtaghavi et al., 2020). Furthermore, cellulose's abundant hydroxyl (OH) groups and chemical adaptability make it suitable for surface modifications. Through grafting or functionalization, cellulose can be tailored to interact harmoniously with biological systems. This enables enhanced biocompatibility, improved cell adhesion, and the ability to perform specific biological functions (Trache et al., 2020). Such modifications pave the way for the development of advanced biomedical materials.

6.2.6 POLYCAPROLACTONE (PCL)

Polycaprolactone (PCL) is a biodegradable polyester that possesses a unique molecular structure characterized by a six-membered lactone ring. This ring structure provides PCL with high flexibility and crystallinity, making it an excellent choice for various biomedical applications. The repeating caprolactone units in PCL allow for the formation of long polymer chains, resulting in a relatively low melting point and the ability to be easily processed using conventional techniques (Dwivedi et al., 2020). Polybutylene succinate (PBS) is a biodegradable polyester composed of succinic acid and 1,4-butanediol units. At the molecular level, PBS exhibits a linear chain structure, with the butylene units providing flexibility and the succinate units contributing to the polymer's crystallinity. The specific arrangement of these units allows PBS to possess a unique combination of mechanical properties, such as good tensile strength, impact resistance, and thermal stability (Aliotta et al., 2022). Polybutylene adipate co-terephthalate (PBAT) is a biodegradable copolyester that consists of repeating units of 1,4-butanediol, adipic acid, and terephthalic acid. The incorporation of adipate and terephthalate units in PBAT gives it distinctive molecular properties. The butylene units contribute to the flexibility and elongation of PBAT, while the aromatic terephthalate units enhance their mechanical strength and thermal stability. PBAT's unique molecular

spectra, suggesting heightened interaction between the functionalized tubes, the polymer matrix, and physiological water solutions (Armentano et al., 2008). In the cross-linking/chain-scission mismatch, PLGA/SWCNTs-COOH systems displayed a predominant second mechanism. The presence of carboxylic groups in functionalized SWCNTs-COOH expedited the hydrolytic degradation of the PLGA matrix and the weight loss of the NCs. This behavior implies selective water interaction at the interface between the nanotubes and the polymeric matrix, akin to conventional composite fiber–matrix interfaces (Bian et al., 1991). The considerable interaction of functionalized carboxylate nanotubes promoted higher hydrolytic degradation in contrast to PLGA and PLGA/SWCNTs systems. Additionally, the enhanced dispersion of functionalized SWCNTs-COOH facilitated better surface interaction with biological environments than bundled SWCNTs, ultimately influencing the potential development and standardization of biodegradable biomaterials. Furthermore, the degradation rates of multiwalled CNTs (MWCNTs) composites in simulated body fluid (SBF) solution were inversely proportional to the nanotube percentage. These composites demonstrated negligible cytotoxic effects, especially with MWCNT loadings surpassing 1% wt. While previous reports focused on CNT functionalization, composite preparation, and property enhancement, the environmental durability of CNT NCs remains unexplored. CNT functionalization has expanded the scope of biocompatibility in carbon-based materials. Properly functionalized CNTs exhibit the potential to traverse cell membranes, thereby offering possibilities for the simultaneous utilization of targeting molecules, contrast agents, and drugs. While clinical utilization of CNTs remains premature, these innovative carriers warrant further investigation (Bianco et al., 2005). Allen et al. reported the *in vitro* catalytic biodegradation of CNTs using horseradish peroxidase (HRP) and low concentrations of hydrogen peroxide (Allen et al., 2008). The study explores potential biotechnological and natural methods, including plant H_2O_2 peroxidases, for degrading CNTs in environmental contexts. Notably, the research demonstrated that CNTs did not deactivate the enzyme HRP. Upon examination after 12 weeks, the majority of nanotubes had disappeared, with the emergence of globular material as the predominant species. These findings provide evidence for the HRP/H_2O_2-mediated biodegradation of CNTs over several weeks, suggesting a promising avenue for environmentally relevant degradation. This suggests that other peroxidases in plants and animals might also play a role in the oxidative degradation of CNTs, potentially mitigating cytotoxicity in applications like drug delivery, gene silencing, and tumor imaging. Further understanding of this biodegradation process could lead to more efficient drug delivery platforms, minimizing concerns about cytotoxic effects. Additional research is needed to determine the by-products of biodegradation and conduct practical cellular studies. Furthermore, studies should explore the mechanical properties, *in vivo* degradation, cell response, and toxicity aspects of these materials.

6.4 KINETICS OF STABILITY

In the realm of nondegradable NCs for biomedical applications, the kinetics of stability plays a crucial role in determining the long-term performance and durability of these advanced materials. Evaluating the stability of NCs involves a comprehensive assessment of their resistance to various environmental and biological factors over extended periods. Evaluating the stability of nondegradable NCs in the biomedical field demands thorough investigations into the intricate atomic-level interactions between nanoparticles and the matrix material. The choice of nanoparticles and their surface chemistry plays a pivotal role in establishing the composite's stability. For instance, consider the case of a poly (ethylene phthalate) (PET)/nano-TiO_2 fiber composite, where the tensile strength and elongation at break were found to be slightly reduced compared to the controlled PET fiber. This reduction was attributed to the addition of nano-TiO_2, which led to decreased interactions between the PET macromolecules. The outcome suggests that the introduction of nano-TiO_2 might have affected the overall intermolecular forces within the composite, thereby impacting its mechanical properties (Han and Yu, 2006). Similarly, when examining polystyrene NCs containing ZnO nanoparticles, both the tensile strength and elongation at break were observed to decrease. This decline implies that the interfacial adhesion between ZnO nanoparticles and the polystyrene matrix was not strong enough to withstand significant mechanical forces (Chae and Kim, 2005). One possible reason for this weaker adhesion could be the challenge of achieving a homogeneous dispersion of nanoparticles within the matrix. Nano-sized particles possess high surface energy, making them prone to aggregation, which can subsequently hinder their effective dispersion and interaction with the matrix material. The stability of NC materials is influenced by a myriad of factors, among which particle size and distribution, as well as the interface interaction between nanoparticles and the matrix, play a pivotal role. Uniform distribution and controlled particle size are crucial for enhancing overall stability, as they contribute to improved performance. Conversely, large particle agglomeration can lead to reduced stability and compromised mechanical properties. To illustrate this, a study by Bikiaris et al. examined polypropylene (PP)/SiO_2 NCs and attributed the less-improved mechanical properties of sample no. 4 to the presence of large agglomerates of SiO_2 nanoparticles in the PP matrix. To address this issue, the researchers suggested the addition of PP grafted with maleic anhydride (PPgMA) as a compatibilizer to promote good adhesion between the PP matrix and SiO_2 nanoparticles. The reactive groups of PP and PPgMA facilitated interfacial interactions, effectively preventing particle leaching or detachment and resulting in a reduction in the size of SiO_2 agglomerates as confirmed by morphological images. The introduction of PPgMA led to a further enhancement in mechanical properties, highlighting the pivotal roles of nanoparticle aggregation/agglomeration and interfacial properties in determining the mechanical performance of NCs (Bikiaris et al., 2005). In the current landscape of nondegradable NCs in the biomedical field, several materials have demonstrated exceptional stability and durability, offering promising solutions for various biomedical applications. The robustness of these NCs can be attributed to a careful consideration of the material properties of both the matrix and the nanoparticles, which play a vital role in influencing the overall stability and performance. Zirconia nanoparticles present a remarkable example of enhancing NC stability when integrated with a biocompatible ceramic matrix. Their bone-like crystalline structure provides exceptional mechanical strength and durability at the atomic level, closely resembling the properties of natural bone. Leveraging these advantageous characteristics, researchers have successfully incorporated ZrO_2 nanoparticles into ceramic matrices, resulting in biomimetic NCs with outstanding osseointegration and long-term stability. This unique combination makes such NCs highly suitable for load-bearing bone implants and dental prosthetics. For instance, a study conducted by Lohbauer et al. demonstrated the application of nanoscale ZrO_2 particles in dental adhesive systems. The researchers prepared ZrO_2 nanoparticles, ranging in size from 20 to 50 nm, using a laser vaporization method. Subsequently, these nanoparticles were seamlessly incorporated within the adhesive layer. The integration of ZrO_2 nanoparticles into the dental adhesive exhibited notable improvements in tensile strength. Furthermore, after implantation, the NC showed the ability to promote mineralization, which is a crucial aspect in dental applications. This remarkable outcome exemplifies the potential of ZrO_2-based NCs in dental restorations, where enhanced mechanical properties and biocompatibility are critical for long-term stability and successful clinical outcomes (Lohbauer et al., 2010). Long-term performance and durability assessments are imperative to gauge the reliability of nondegradable NCs in biomedical applications. Evaluating stability over extended periods involves subjecting the NCs to various stressors and environmental conditions representative of the intended application. Mechanical stress testing, exposure to simulated biological fluids, and temperature/humidity variations are some of the factors explored to determine the NCs' ability to maintain their structural integrity and functionality over time (Zare and Shabani, 2016). Techniques for studying stability and degradation resistance in nondegradable NCs provide critical insights into the behavior of these materials at the atomic level. Advanced imaging techniques such as Atomic Force Microscopy and Transmission Electron Microscopy allow researchers to visualize NC surfaces and study nanoparticle dispersion and interface interactions. X-ray diffraction analysis aids in monitoring changes in nanoparticle crystal structures over time, offering valuable insights into the stability of the composite (Venkateshaiah et al., 2020). Fourier transform infrared spectroscopy assists in identifying chemical bonds and functional groups, shedding light on the stability of surface modifications. Additionally, differential scanning calorimetry measures heat flow in NCs, uncovering changes in physical and chemical properties that may indicate degradation. Accelerated aging tests and biocompatibility assessments are also conducted to predict long-term stability and safety in relevant biomedical environments (Corcione and Frigione, 2012).

6.5 BIOMEDICAL APPLICATIONS

6.5.1 DRUG DELIVERY

In the realm of biomedical applications, NCs have emerged as a groundbreaking solution for drug delivery, offering tremendous potential to revolutionize the field. The rational design of NC-based drug delivery systems allows for precise control over drug release kinetics and therapeutic efficacy, making them highly desirable for targeted and personalized medicine. Biodegradable NCs are particularly advantageous for drug delivery applications that require sustained release and controlled degradation. These NCs are typically composed of natural or synthetic biodegradable polymers, such as CA, alginate, or PLA. The gradual degradation of the polymer matrix ensures a sustained and steady release of drugs over an extended period, making them suitable for delivering drugs with short half-lives or those requiring controlled release profiles. By tuning the degradation rate of the biodegradable polymer, researchers can achieve a desirable drug release profile, minimizing the need for frequent dosing and enhancing patient compliance (Marin et al., 2013). On the other hand, nondegradable NCs find their place in long-term drug delivery applications and implantable devices. These NCs are composed of polymers that do not undergo significant degradation, ensuring the structural integrity and mechanical stability of the drug delivery system over an extended period. Nondegradable NCs are often used for sustained-release drug delivery, providing a continuous and consistent drug release without the need for frequent interventions. The nondegradable nature of these NCs makes them suitable for implantable drug delivery devices, such as drug-eluting stents and intrauterine devices, which require long-term drug release and biocompatibility. Incorporating stimuli-responsive elements into NCs has revolutionized drug delivery capabilities, offering a new dimension of precision and control. Stimuli-responsive NCs possess the remarkable ability to release drugs in response to specific environmental triggers, such as changes in pH, temperature, or external stimuli like light, magnetic fields, or ultrasound. This unique feature allows for on-demand drug release, enabling precise and controlled delivery precisely when and where needed (Chung et al., 2011). A notable example of stimuli-responsive NCs comes from the work of Wu et al. In their study, they developed a light-responsive silica nanoparticle by covalently conjugating photoactive o-nitrobenzyl bromide molecules with amino groups on the particle surface. The researchers covalently attached drugs with carboxylic, phosphate, or hydroxyl groups to the nanoparticles and irradiated them at 310 nm. Upon irradiation, the photoactive groups underwent a transformation, leading to the irreversible cleavage of the drug–particle bond and inducing drug release (Wu et al., 2008). Another fascinating application of stimuli-responsive NCs is demonstrated by Hervault et al., who developed magnetic NCs (MNCs) composed of an iron oxide core and a pH- and thermo-responsive polymer shell. This dual-functionality NC acts as both a hyperthermic agent and a drug carrier. By conjugating the anticancer drug doxorubicin (DOX) to the pH- and thermo-responsive MNCs via an acid-cleavable imine linker, they achieved advanced features for targeted drug delivery. This conjugation allowed for the combination of magnetic targeting with dual pH- and thermo-responsive behavior, providing spatial and temporal control over the release of DOX (Hervault et al., 2016). Targeted drug delivery is another area where NCs shine. By functionalizing the surface of NCs with specific ligands or antibodies, researchers can achieve selective drug delivery to targeted tissues or cells. This targeted approach minimizes off-target effects and enhances therapeutic outcomes while reducing the required drug dosage and potential side effects. Targeted NCs hold great promise for precision medicine, as they can be tailored to address individual patient needs and unique disease characteristics (Batool et al., 2023). An exemplary illustration of the targeted drug delivery potential of NCs lies in their application for gliomas, a type of brain tumor. Researchers employed superparamagnetic Fe_3O_4 nanoparticles and loaded them onto the surface of GO through a chemical precipitation process, creating GO-Fe_3O_4 NCs. To achieve targeted delivery, the iron-transporting serum glycoprotein, lactoferrin (Lf), was chosen as the ligand due to its ability to attach to receptors overexpressed on the surface of glioma cells and vascular endothelial cells of the blood–brain barrier. Through EDC/NHS chemistry, the researchers developed the targeted drug delivery system Lf-GO-Fe_3O_4. The resulting nano-delivery structure displayed high drug-loading capability and demonstrated a pH-dependent drug release pattern. Comparing the targeted NC Lf-GO-Fe_3O_4-DOX with free DOX and DOX-GO-Fe_3O_4, the Lf-conjugated NC exhibited superior intracellular delivery efficiency and stronger cytotoxicity against C6 glioma cells. This study exemplifies the potential utility of Lf-conjugated GO-Fe_3O_4 NCs in therapeutic applications for the treatment of gliomas (Song et al., 2017). Controlled drug delivery from NCs ensures a steady and effective drug concentration, avoiding fluctuations that can occur with conventional drug formulations. Mathematical models play a crucial role in predicting drug release kinetics and optimizing NC formulations (Roy et al., 2021). These models consider various factors, such as nanoparticle size, surface charge, polymer matrix properties, drug loading, and environmental conditions, to accurately predict drug release profiles.

By understanding the interplay of these factors, researchers can tailor NC designs to achieve specific release profiles, ensuring optimal therapeutic outcomes for different drugs and medical conditions (Grassi and Grassi, 2005). An illustrative example of controlled drug delivery from NCs can be seen in the work of Jan et al. They developed a NC system using CA and Locust bean gum, cross-linked with glutaraldehyde. The researchers found that modifying the polymer composition influenced the size of the NCs and drug entrapment efficiency. The variations in either CS or Locust bean gum provided a means to control the rate of drug release from the NCs. Interestingly, the presence of both polymer components in the composites proved crucial in suppressing the burst release of the drug in the stomach and regulating the drug release profiles in the variable pH environments of the gastrointestinal tract. This was essential as CS particles alone were inadequate in resisting the burst effect at the acidic pH of the stomach. Consequently, this novel NC system showed great promise in reducing burst release, thereby minimizing gastrointestinal side effects associated with drug delivery. The controlled drug delivery application of this NC system holds the potential to significantly improve patient outcomes and safety (Jana and Sen, 2017).

6.5.2 TISSUE ENGINEERING AND REGENERATION

Tissue engineering, an innovative field in biomedical research, aims to create functional tissues and organs using biocompatible materials and living cells. In this context, NCs have emerged as a highly valuable tool due to their unique properties, versatility, and ability to mimic the natural ECM environment (Armentano et al., 2010). These NCs play multiple crucial roles in tissue engineering, including providing structural support, promoting cell adhesion and proliferation, delivering growth factors and therapeutic agents, and guiding tissue regeneration. By utilizing NCs in various forms such as hydrogels, nanofibers, and 3D scaffolds, researchers can engineer tissues for diverse applications (Verma et al., 2011). One primary challenge in tissue engineering lies in fabricating scaffolds that accurately mimic the architecture and mechanical properties of native tissues. NC scaffolds offer distinct advantages in this regard by incorporating nanoparticles into the scaffold matrix. This enhances the scaffold's mechanical strength, stiffness, and elasticity, making it more suitable for supporting and guiding tissue growth. The NC scaffold's porous structure further facilitates nutrient and oxygen diffusion, promoting cell survival and growth within the scaffold (Abdollahiyan et al., 2021). In the quest to encourage tissue regeneration, NCs can be tailored with cell-adhesive peptides or proteins, facilitating cell attachment and spreading. For instance, Li et al. conducted a study using laminin/CA/PLGA nerve conduit combined with Schwann cells and neural stem cells, which showed promising results in promoting nerve regeneration *in vivo* using Sprague Dawley rats as the model (Li et al., 2018). In another noteworthy research by Salehi et al., they fabricated a biodegradable nerve conduit comprising PLA, MWCNTs, and gelatin nanofibrils (GNFs), integrated with recombinant human erythropoietin-loaded CA nanoparticles (rhEpo-CNPs). The biomaterial showcased a highly porous structure (porosity above 85%) and demonstrated sustained release of rhEpo for up to 14 days. The conduit's mechanical strength surpassed that of the fresh transected adult rat sciatic nerve, as confirmed by mechanical tests. Additionally, MTT and LDH assays exhibited increased viability and proliferation of Schwann cells when exposed to PLA/MWCNTs/GNFs/rhEpo-CNPs, while *in vivo* studies demonstrated enhanced regeneration of the sciatic nerve in Wistar rats (Salehi et al., 2018). Moreover, the NC surface's nanoscale features effectively mimic the native ECM, providing cues that stimulate specific cell behavior and tissue differentiation (Dvir et al., 2020). This biomimetic approach fosters the formation of tissues with well-organized cell structures and improved functionality, further advancing the potential of NCs in tissue engineering applications.

The selection of biodegradable or nondegradable NCs depends on the tissue engineering application and scaffold characteristics. Biodegradable NCs are preferred for temporary tissue scaffolds that gradually degrade as new tissue forms. This eliminates the need for scaffold removal surgeries and minimizes long-term side effects (Iqbal et al., 2019). Conversely, nondegradable NCs are more suitable for long-term tissue replacements, such as heart valves or vascular grafts, where maintaining structural integrity is crucial (Xue et al., 2017). NCs can act as drug delivery carriers, enabling sustained and controlled release of growth factors or drugs to promote tissue repair. By modifying the NC surface to incorporate specific growth factor-binding molecules, targeted delivery to the desired tissue site becomes feasible. One such growth factor, the osteogenic growth peptide (OGP), naturally regulates the differentiation of bone marrow mesenchymal stem cells (BMSCs) into osteoblasts, making it essential for bone tissue engineering. Chen et al. (Chen et al., 2015) conducted a study using a mineral/OGP NC layer on titanium, demonstrating its positive impact on tissue regeneration. They observed a significant increase in the number of MSCs seeded on the mineral/OGP surface after 5 days, indicating enhanced osteogenic potential. Moreover, the release of alkaline phosphatase from MSCs was higher for the mineral/OGP layer, supporting the transition from osteogenic proliferation to differentiation. The combination of mineral/OGP facilitated osteogenic differentiation, as evident from the increasing alkaline phosphatase levels over time. Additionally, stimuli-responsive NCs can respond to environmental cues, such as changes in pH or temperature, triggering drug release at the appropriate time and location. NCs also enable the engineering of smart scaffolds with dynamic properties, responding to mechanical forces or biological signals. Ho et al. (2015) introduced a novel polymeric nanoparticle bilayer material featuring a unique core-shell structure termed "unibody core-shell." The unibody core-shell design offers multifunctionality, with a Gd_3+-based core for magnetic resonance imaging (MRI) contrast and a DOX-loaded shell for targeted cancer drug release. This innovative approach enables simultaneous imaging and therapeutic capabilities, making it a promising candidate for advanced cancer treatment strategies. This adaptability enhances the scaffold's ability to support tissue regeneration and integration effectively.

6.5.3 BIOIMAGING AND BIOSENSORS

In bioimaging, NC materials act as responsive components that can detect and respond to various *in vitro* conditions, such as magnetic resonance, photoacoustic, and fluorescence (Huang et al., 2022). These properties make NCs invaluable for noninvasive, real-time, and dynamic monitoring, particularly in crucial areas like anti-tumor therapies, tissue repair, and cell behavior observation. By incorporating NCs into hydrogels, the residence time of nanomaterials in target areas is prolonged, avoiding rapid elimination by the body and improving imaging outcomes. In the research conducted by Zhao et al., they successfully engineered a NC hydrogel comprising tumor-targeted nanoparticles (PL NPs) embedded within alginate hydrogels. This innovative combination allowed for prolonged luminescence in the near-infrared spectrum, enabling long-term imaging with minimized interference from *in situ* radiation (Zhao et al., 2020). In biosensors, NCs offer advantages in sensitivity, selectivity, and stability. Their high surface area-to-volume ratio facilitates the efficient immobilization of biological sensing elements, enhancing analyte binding and leading to highly specific and sensitive detection of target molecules (Pandey et al., 2008). Functionalizing NC surfaces with specific biomolecules enables precise and accurate analyte detection, making biosensors invaluable in medical diagnostics, environmental monitoring, and food safety testing (Minelli et al., 2010; Yu et al., 2021). Moreover, NC-based biosensors find applications in disease diagnosis and monitoring, providing rapid and sensitive detection of disease biomarkers, enabling early intervention, and improving patient outcomes. The stability and biocompatibility of NCs also make them suitable for implantable biosensors, offering continuous health monitoring for patients with chronic conditions. Buiculescu et al. presented a robust amperometric biosensor by covalently immobilizing acetylcholinesterase onto gold nanoparticles (Au NPs), followed by biosilica encapsulation. This NC design ensured enhanced enzyme stability and efficient electron transfer, resulting in prolonged biosensor activity over 4 months. The biosensor exhibited a high initial sensitivity of 27.58 nA mM^{-1} and maintained its activity, while a conventional acetylcholinesterase biosensor's activity dropped to 50% after only 42 days. This study exemplifies the potential of NCs in achieving long-term and stable biosensing applications (Chaniotakis and Buiculescu, 2012).

6.5.4 ANTIMICROBIAL ACTIVITY

NCs show great promise in fighting antimicrobial resistance, offering distinct advantages over traditional agents. These advanced materials combine different antimicrobial elements, enhancing their effectiveness against a broad range of microorganisms. NCs can be tailored for the controlled release of antimicrobial agents, maintaining optimal concentrations at infection sites while minimizing harm to healthy tissues (Xhazzy et al., 2023). An important advantage of NCs in antimicrobial applications is their potential to prevent the emergence of resistance. By targeting multiple pathways within microbial cells, NCs make it more difficult for resistance to develop, a significant concern with conventional treatments

(Makabenta et al., 2021). Moreover, NCs can be designed to specifically deliver antimicrobial agents to infection sites, reducing exposure to host cells and minimizing toxicity (Devnarain et al., 2021). NCs achieve antimicrobial effects through various mechanisms. Some disrupt microbial cell membranes, leading to their destruction, while others generate reactive oxygen species that damage microbial structures and biomolecules. Furthermore, certain NCs release antimicrobial agents directly into microbial cells, disrupting vital cellular processes and halting microbial growth (Wang et al., 2017). Different types of NCs are being explored for antimicrobial applications. Metal-based NCs containing nanoparticles of silver, copper, or zinc exhibit strong antimicrobial properties (Figure 6.3) (Xhazzy et al., 2023). Polymer-based NCs with materials like CA (Murugesan and Scheibel, 2021) or polyvinyl alcohol offer sustained release capabilities for long-lasting antimicrobial effects (Ruiz et al., 2019). Carbon-based NCs, such as CNT sand graphene, exert antimicrobial activity through physical interactions and reactive oxygen species generation (Azizi-Lalabadi et al., 2020). Simultaneously, nanoparticle-based drug delivery systems are proving valuable against intracellular microbial infections and various therapeutic challenges. Drugs loaded into nanoparticles through adsorption, encapsulation, or conjugation demonstrate improved efficacy and pharmacokinetics compared to free drugs. These nanoparticles extend systemic circulation, enhance drug solubility, and facilitate targeted delivery to specific cells and tissues. Moreover, they allow controlled and sustained drug release, reducing dosing frequency and enabling combination therapy with multiple agents (Singh and Lillard, 2009). When nanoparticles carrying therapeutic agents enter host cells through endocytosis, they release drugs intracellularly to combat infections caused by intracellular pathogens. This intracellular drug delivery approach is especially valuable for challenging intracellular pathogens. As a result, several nanoparticle-based drug delivery systems have gained clinical approval, effectively treating various diseases. Many other therapeutic nanoparticle formulations are currently in clinical trials, underscoring the vast potential of this technology in medical applications (McMillan et al., 2011). Polymeric nanoparticles, liposomes, dendrimers, and solid lipid nanoparticles are widely used as antimicrobial drug delivery systems, with some products already available in the pharmaceutical market (Muhamad et al., 2014). Amid these promising developments, addressing challenges related to NC biocompatibility and long-term safety is essential. Rigorous research and regulatory assessments are necessary to ensure their safety and effectiveness before widespread clinical use. The path of antimicrobial NCs holds promise for reshaping medical strategies and addressing the complex challenges posed by antimicrobial resistance.

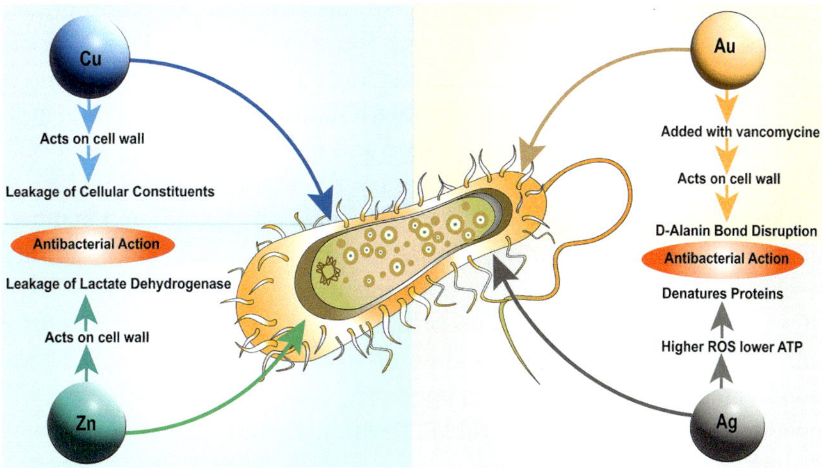

FIGURE 6.3 Mechanistic insights into metal-infused nanocomposites as antibacterial materials.

6.5.5 WOUND-HEALING APPLICATION

The occurrence of wounds, especially in chronic stages, represents a multifaceted process influenced by factors like underlying health conditions, impaired circulation, and compromised immune responses (Gould et al., 2015). This phenomenon often leads to delayed wound healing due to prolonged inflammation, disrupted cell migration, and compromised tissue regeneration. In this context, NCs have emerged as a promising strategy to address the complexities of chronic wound healing. These innovative materials synergize the strengths of both organic and inorganic components, yielding multifunctional advantages. Natural constituents such as polymers or biopolymers offer biocompatibility, while the incorporation of inorganic nanoparticles introduces unique properties like antimicrobial efficacy, controlled drug release, and augmented mechanical strength (Hamdan et al., 2017). The mechanistic underpinning of NC-assisted wound healing involves a multifaceted approach. Antimicrobial nanoparticles, like silver or zinc oxide nanoparticles, embedded within NCs, disrupt microbial cell membranes, curbing colonization and fostering a conducive environment for healing (Bui et al., 2017). Furthermore, the controlled release of growth factors and cytokines from NCs spurs cell migration, proliferation, and tissue regrowth (Nurkesh et al., 2020). Structural attributes of NCs are pivotal in orchestrating wound healing. Their elevated surface area facilitates efficient drug loading, enabling gradual and sustained release. The porous architecture mimics the ECM, serving as a scaffold for cell attachment, migration, and proliferation (Kalva et al., 2021). Notably, the biomimetic configuration of electrospun nanofibrous matrices enhances gas exchange and nutrient transport, promoting the essential cellular activities necessary for wound repair (Okur et al., 2020). NCs surmount the limitations of standalone natural and synthetic materials. For instance, while natural materials like collagen might lack mechanical resilience, synthetic polymers could be devoid of bioactivity. NCs circumvent these constraints by harnessing the synergistic benefits of both components. The incorporation of silver nanoparticles with natural biopolymer CA exemplifies this approach, endowing the composite with biocompatibility and antimicrobial potency, thereby expediting wound healing (Reddy et al., 2021). Research findings underscore the promise of NCs in chronic wound healing. The amalgamation of organic and inorganic constituents permits a tailored approach, addressing multiple facets of wound recovery. Integrating growth factors, enzymes, or stem cells within NCs amplifies therapeutic efficacy, promoting angiogenesis, collagen synthesis, and inflammation modulation—all integral to successful wound healing (Mihai et al., 2019). The mechanical properties of NCs can be tailored to mimic the native tissue environment, imparting structural support to the wound site. Moreover, the NC's topographical features guide cell adhesion, migration, and proliferation, facilitating tissue regeneration. Notably, electrospun nanofibrous mats offer a substantial surface-to-volume ratio that promotes cell attachment and gas exchange—crucial components of the healing process (Liu et al., 2021). Incorporating NCs transcends superficial applications. Scaffold-based strategies entail implanting NCs directly into the wound site, offering a 3D framework that encourages tissue growth, vascularization, and cell infiltration. These scaffolds enable controlled drug delivery, releasing growth factors or stem cells in a targeted manner, thereby fostering tissue regeneration from within (Jiang et al., 2023). Xiaocheng Wang et al. harnessed the photothermal conversion capacity of MXene nanosheets and the thermoresponsive contraction/expansion attributes of poly hydrogels. This innovative approach enabled the precise administration of vascular endothelial growth factor, facilitating the advancement of endothelial cell proliferation, migration, and proangiogenic effects (Wang et al., 2022). While NCs hold immense potential, their translation into clinical practice necessitates addressing challenges such as biocompatibility, degradation kinetics, and regulatory approval. Rigorous research is pivotal in ensuring the safety, effectiveness, and applicability of NC materials across various wound types. As the field of wound healing advances, NCs emerge as a promising avenue to revolutionize wound care, improve patient outcomes, and alleviate the burden of chronic wounds. Nonetheless, challenges persist in developing NCs for wound-healing applications, demanding meticulous considerations of biocompatibility, degradation kinetics, and long-term safety. Regulatory approvals for medical devices employing NCs entail comprehensive evaluations of their toxicity, stability, and efficacy. Table 6.1 highlights significant recent studies in the biomedical field, categorized by their applications in drug delivery, tissue engineering, bioimaging and biosensors, antimicrobial activity, and wound healing, as discussed in this chapter.

6.6 HYBRID APPROACHES AND SYNERGISTIC EFFECT

The strategic combination of biodegradable and nondegradable NCs in the biomedical field not only leverages the unique properties of each material but also addresses the inherent limitations that arise when using single-material systems. These limitations can hinder the full potential of a particular application and may include issues such as inadequate mechanical strength, limited drug release kinetics, and suboptimal degradation rates. In the case of using solely biodegradable materials, one major limitation is the rapid degradation that can lead to premature loss of structural integrity or uncontrolled drug release. Although biodegradable polymers provide excellent controlled release, their mechanical properties might not suffice for load-bearing applications or long-term implantable medical devices. Conversely, utilizing solely nondegradable materials may lead to biocompatibility concerns and long-term accumulation of NC remnants within the body. Additionally, the absence of controlled degradation mechanisms can impede tissue

TABLE 6.1 Recent Nanocomposite Types: A Summary of Features and Applications Over the Last Five Years.

S. No.	Nanocomposite Type	Features	Composition (Matrix + Nanomaterial)	Evaluation	Application
1	Tissue engineering (Prakash et al., 2020)	NC films with osteogenic potential for treating bone and bone-related disorders	Chitosan and PVA + graphene oxide/ hydroxyapatite/gold nanoparticles	Bioactivity and biodegradation	Bone tissue engineering
2	Drug delivery (Gunathilake et al., 2020)	Nanocomposite films which are activated by electrical signals to release poorly water-soluble drug	Polylactic acid/ CMC + zinc oxide nanocomposite films	Tensile strength, *in vitro* drug release (under DC current)	Drug delivery of poorly water-soluble drugs
3	Bioimaging (Nottelet et al., 2021)	Nanocomposites to be used as multifunctional scaffolds with magnetic and imaging properties	Super magnetic iron oxide nanoparticles (SPIONs) + polylactic acid nanofiber nanocomposites	Cytotoxicity, sensitivity of detection *in vivo* in rats	SPION/PLA core nanocomposites used as scaffolding for biomedical imaging and magnetic applications
4	Tissue engineering (Alam et al., 2020)	3D printed scaffolding which are porous and biodegradable with negligible loss in mass	PLA + carbon fibers and graphene nanoparticles PLA + ferromagnetic iron particles	Bioactivity and biodegradation	3D printed scaffolding for bone replacement and regeneration
5	Tissue engineering (Mukheem et al., 2022)	Novel scaffolding with high surface area, low cytotoxicity, and higher antimicrobial activity	Polyhydroxyalkanoates/ chitosan + tungsten disulphide nanocomposites	Bioactivity and biodegradation	Scaffolding with antibacterial and biological application
6	Drug delivery and bioimaging (Tanaya et al., 2023)	Fluorescent nanocomposites capable of acting as a drug delivery system and bioimaging agent for target sites	Quantum dots, gold and silver nanoparticles + polyhydroxyalkanoate	Drug release, fluorescence	Drug delivery to target sites and bioimaging of active sites

TABLE 6.1 *(Continued)*

S. No.	Nanocomposite Type	Features	Composition (Matrix + Nanomaterial)	Evaluation	Application
7	Antibacterial and biological (Mukheem et al., 2022)	Nanocomposites with the ability to act on multidrug-resistant microbes	Boron nitride doped polyhydroxyalkanoates/ chitosan nanocomposite	Host cytotoxicity, through a time-kill method against MDR microbes	Antibacterial and biological application
8	Drug delivery system (Mohammadi et al., 2021)	Single-step synthesis of nanocomposites which are loaded with drugs and administered orally	CMC/polyacrylic acid/ starch + modified iron oxide interpenetrating magnetic nanocomposite hydrogen	Drug release assay, drug-loading efficiency	Useful in the delivery of cancer drugs like 5-Fluorouracil orally without causing damage to the digestive system
9	Drug delivery (Nezami et al.,2020)	pH-Sensitive nanocomposites loaded with guaifenesin	Copolymerized itaconic acid onto starch + iron oxide nanocomposite	Drug release, pH-dependent release profile, drug-loading efficiency	pH-Dependent release of drug at site of the wound for wound healing
10	Drug delivery (Omrani et al., 2023)	Nanocomposite prepared to release macromolecules of curcumin in a pH-dependent atmosphere	Chitosan/starch/ molybdenum disulfide nanocomposites	Entrapment and drug-loading efficiency, cytotoxicity, and assay against MCF-7 (for anticancer activity)	Delivery of macromolecular curcumin to treat breast cancer
11	Tissue engineering (Gautam et al., 2021)	Novel electrospinning technique used to fabricate scaffolding for bone tissue regeneration	Gelatin-polycaprolactone + nanohydroxyapatite (electrospun) nanocomposite	Morphological study, degradation and cell culture of human osteoblasts	Scaffolding is used to provide support for bone and bone tissue healing

TABLE 6.1 *(Continued)*

S. No.	Nanocomposite Type	Features	Composition (Matrix + Nanomaterial)	Evaluation	Application
12	Tissue engineering (Mirmusavi et al., 2022)	Electrospun nanocomposite scaffolds with increased strength and cell response and facilitate natural healing mechanism	Polycaprolactone-chitosan/multiwalled carbon nanotube electrospun nanocomposite	Morphology, porosity, physicochemical structure, hydrophilicity, tensile strength, bioactivity, biodegradation, and cell response	Scaffolding for cartilage healing
13	Drug delivery (Gholamali et al., 2019)	Nanocomposite hydrogels formed *in situ* show swelling properties	Oxidised starch + copper oxide nanocomposite hydrogel	Swelling behavior, *in vitro* drug release	Helps in providing controlled and sustained release
14	Dental applications (Ansari et al., 2020)	Applied onto Ti6Al4V as *in situ* sol–gel for the purpose of dental fillings and adhesives	Polycaprolactone/fluoride substituted hydroxyapatite nanocomposites as an *in situ* sol–gel process	Conformation of *in situ* synthesis, surface morphology	As adhesives and fillings in dental procedures
15	Drug discovery (Chou et al., 2021)	Nanocomposites with excellent biocompatibility and can be used for 3D printing	Nano-hydroxyapatite + polycaprolactone nanocomposites	Tensile strength, fill density, biocompatibility	3D-printed medical implants which gradually release drug
16	Biomedical applications (Hada et al., 2022)	Shape memory polymers allow the nanocomposites to change from a programmed form to the original form	Polycaprolactone is a based shape memory polymeric nanocomposite	Toxicity, biocompatibility, tailorability, sterilization, and biocompatibility	Biocompatible shape polymer nanocomposites used for wound dressing, oesophageal stenosis, self-tightening sutures, bone tissue engineering, vascular graft, drug delivery, and bone defect repair

TABLE 6.1 *(Continued)*

S. No.	Nanocomposite Type	Features	Composition (Matrix + Nanomaterial)	Evaluation	Application
17	3D printing (Beesetty et al., 2020)	Inexpensive and mechanically strong used as a filament in 3D printers	Nanoclay/HDPE nanocomposites	Flowability, thermal, and mechanical properties	Used for 3D printing of thermally stable and significantly durable articles
18	Bioimaging (Mathew et al., 2019)	Amino-functioned carbon dots were synthesized and their luminescence properties were incorporated into the nanocomposites	Luminescent carbon dots + chitosan nanocomposites	External morphology, luminescence, atomic force microscopy, spherical morphology	Useful for bioimaging using the photoluminescent properly and superior biocompatibility
19	Bioimaging and herbaceutical detection (Pandey et al., 2019)	Hybrid nanomaterials with inherent physicochemical properties and compatibilities	Chitosanylated molybdenum trioxide-ruthenium nanocomposites	Antioxidant characteristics against reactive oxygen species and protein carbonyls and electrical transducing ability	Intracellular imaging, antioxidants, electrochemical sensor studies
20	Polymer design (Rajesh et al., 2019)	New nanocomposite polymer with enhanced tensile strength and reduced dielectric constant	Titanium dioxide-doped PVA/PVP nanocomposite	Mechanical study, dielectric studies, DC and AC conductivity studies	As a nanocomposite polymer for potential application in electronics with minimal loss in power
21	Tissue engineering (Ghosal et al., 2019)	Electrospun nanocomposite fibres are porous with a high surface area-to-volume ratio	PVP/polylactic acid/ polycaprolactone + titanium dioxide nanocomposite	Porosity, physicochemical structure, tensile strength, bioactivity, absorption of UV light	Biomedical application, specifically tissue engineering and wound healing
22	Drug delivery (Salahuddin et al., 2020)	Biodegradable, biocompatible, and nontoxic polymer-based nanocomposite capable of extended release of drug (norfloxacin)	Polylactic acid + titanium dioxide nanoparticles	Loading efficiency, release study, antibacterial activity, cytotoxicity study, antitumor evaluation	Drug release of norfloxacin for antibacterial and antitumor effect

TABLE 6.1 *(Continued)*

S. No.	Nanocomposite Type	Features	Composition (Matrix + Nanomaterial)	Evaluation	Application
23	Biomedical application (Taha et al., 2020)	Nanocomposites with remarkable enhancement of densification behavior, mechanical properties, and *in vitro* bioactivity of nanocomposites	Hydroxyapatite-based silicon dioxide nanocomposites	Mechanical property, *in vitro* bioactivity of nanocomposites	Biocompatible nanocomposites which can be used in various biomedical fields
24	Wound healing (Sun and Zheng, 2022)	Nanocomposites loaded with zinc with enhanced antibacterial and wound-healing properties	Zinc-silicon carbide nanocomposites	Morphology, zeta potential, biocompatibility, and *in vivo* healing efficacy	Useful as an *in vivo* wound-healing agent after a cesarean section
25	Drug delivery (Abdel-Bary et al., 2020)	Nanocomposites were synthesized which showed 84–87% with drug release 91%	Chitosan, magnetite, silicon dioxide, and graphene oxide nanocomposites	DFT calculations, drug-loading efficiency, drug-release potential	Drug delivery of cisplatin
26	Wound healing (Mohebali et al., 2020)	Nanocomposite-based wound dressing for the delivery of minocycline	Polyvinyl alcohol/halloysite nanotubes + silicon dioxide nanocomposite	Antibacterial study, water uptake, biodegradability, and *in vivo* activity	Burn wound healing and regeneration
27	Biomedical (Alturky et al., 2023)	Nanocomposites with distinct physical, tribo-mechanical, electrical, magnetic, and biological properties	Biologically extracted hydroxyapatite (from camel bone) hematite + silicon dioxide nanocomposites	Dielectric, physical, tribo-mechanical, magnetic, and antimicrobial properties	Bone replacement procedure
28	Drug delivery (Su et al., 2019)	Nanocomposites loaded with silver nanoparticles are useful for delivering antitumor drugs	Silver nanoclusters + zirconium metal-organic nanocomposite	Drug-loading efficiency, drug-release potential, antitumor properties on MCF-7 cells	Delivery of antitumor drugs in a biodegradable biocompatible carrier

TABLE 6.1 *(Continued)*

S. No.	Nanocomposite Type	Features	Composition (Matrix + Nanomaterial)	Evaluation	Application
29	Drug delivery (Gandara-Loe et al., 2020)	Metal-organic framework-based nanocomposites used to deliver drugs for ocular therapeutics	Polyurethane + zirconium-based MOF nanocomposite	Morphology, distribution within nanocomposite film, biocompatibility, and stability of loaded drug into nanocomposite	Delivery of drugs for ocular therapeutics such as lenses, lacrimal stoppers, and so on
30	Bone and dental (Al-Harbi et al., 2021)	Biopolymer nanocomposites	Cellulose acetate/ hydroxyapatite/ bioglass + zirconia nanocomposites	Mechanical performance, wettability, antibacterial activity, and toxicity	Dental applications
31	Theranostics (Choi et al., 2022)	Nanocomposites which help in utilizing Cherenkov radiation-induced cancer therapy	Radioisotope zirconia (89) + nanocomposites of titanium dioxide and manganese dioxide	Stability and tumor growth control in mice	Theranostic application in the treatment of cancer
32	Bioimaging (Murugesan and Scheibel, 2021)	Nanocomposites with stimuli-responsiveness, and can be used as biomedical sensors as well as antimicrobial agents	Bio enzymes, iron (III) oxide, polypyrrole + chitosan	Responsiveness to essential stimuli like the presence of glucose, cholesterol, etc.	As biosensors detect various chemical stimuli and initiate a programmed response
33	Bioimaging and biosensors (Dheyab et al., 2020)	Inorganic and organic coating on nanocomposites can be varied to specific requirements (as required for imaging, drug delivery, etc.)	Iron (III) oxide + gold nanocomposite	Surface morphology and characterization, uniformity of coating of organic or inorganic material onto nanocomposite	MRI, CT, drug delivery, biosensors, and hyperthermia
34	Drug delivery (Farani et al., 2020)	PEGylated and functionalized nanocomposites as a drug carrier for chemotherapy	PEGylated and functionalized GO nanocomposite	Drug release and drug loading, anticancer activity against MCF-7 cells	Localized photothermal ablation of cancer cells guided by the magnetic field

TABLE 6.1 *(Continued)*

S. No.	Nanocomposite Type	Features	Composition (Matrix + Nanomaterial)	Evaluation	Application
35	Drug delivery (Yu et al., 2019)	Rattle-type nanocomposites with movable cores, porous shells, and hollow interiors with high photothermal conversion efficiency	Gold nanorods/ porous silicon dioxide nanocomposites	Drug-loading capacity, photoconversion efficiency	Chemotherapy and photothermal therapy
36	Drug delivery (Doghish et al., 2022)	Nanocomposites with antibacterial, antifungal, and anticancer activities	Gold nanoparticles + carboxymethyl cellulose nanocomposites	Antibacterial activity, antifungal activity, anticancer activity	Against breast cancer, and certain bacterial and fungal infections
37	Wound healing (Deng et al., 2021)	Biomimetic injectable double network hydrogel embedded with metal-organic framework	Metal organic framework containing gold nanocomposite	Reactive oxygen species generation, antibacterial activity	Injectable double network hydrogel which could simultaneously provide antibacterial and healing capabilities
38	Drug delivery (Mdlovu et al., 2019)	Multifunctional functionalized nanocomposites loaded with doxorubicin for neuroblastoma treatment	Iron oxide nanoparticle core + pluronic f127 shells	Toxicity, *in vitro* drug release profile, pH sensitivitym and nontoxicity	Drug release of doxorubicin in the treatment of neuroblastoma
39	Wound healing (Zhang and Zhao, 2022)	Highly porous nanocomposite bandages were fabricated for burn injury	Gold-incorporated heparin-polyvinyl alcohol nanocomposites	Antimicrobial activity, biocompatibility, *in vivo* experiments	Bandages useful for perioperative care and in burn wound healing
40	Wound healing (El-Sayed et al., 2023)	Nanocomposites which are utilized as wet tissue adhesive	Dopamine-chitosan-iron oxide nanocomposites	*In vitro* scratch wound healing assay, cytotoxicity, antibacterial assessment, live and dead assay	Soft medical tissue adhesive for wound healing

TABLE 6.1 *(Continued)*

S. No.	Nanocomposite Type	Features	Composition (Matrix + Nanomaterial)	Evaluation	Application
41	Bioimaging and biosensors (Hejazi et al., 2023)	A nanocomposite-based bioimaging tool for the detection of cancer cells	Polyetherimide + hyaluronic acid functionalized gold nanoparticles	Flow cytometry and fluorescence imaging technique	Useful in the detection of MCF-7 cells
42	Bioimaging and biosensors (Cao et al., 2019)	Nanoscale metal-organic framework-based nanocomposites with good biocompatibility, excellent luminescence, and effective cellular uptake for bioimaging	Glutathione-stabilized gold nanoclusters	Limit of detection, fluorescence intensity, limit of detection	Biosensors for the detection of highly reactive oxygen species

integration and hinder the natural healing process. By combining biodegradable and nondegradable NCs, we can harmonize these disparities and create a superior platform that overcomes individual limitations. The controlled degradation of biodegradable components ensures precise drug release and gradual tissue regeneration, while the nondegradable nanoparticles impart mechanical strength, stability, and prolonged functionality to the overall composite. The complementary nature of these materials allows for precise tailoring of degradation rates, drug-release profiles, and mechanical properties to suit specific biomedical applications. Furthermore, the synergistic effect between biodegradable and nondegradable components in the NC fosters enhanced biocompatibility, reducing adverse reactions and promoting better tissue integration. The combination also opens up new avenues for the design of multifunctional NCs with diverse applications, such as smart drug delivery systems, implant coatings, and diagnostic agents. The ingenious merging of biodegradable and nondegradable NCs in biomedicine presents an innovative solution that overcomes the limitations of single-material systems. Illustrated by a study conducted by Petersson et al., involving biopolymer-based NCs, the positive impact of incorporating nano reinforcements into a PLA matrix is demonstrated. In this research, the addition of bentonite as a nano reinforcement remarkably improved tensile modulus and yield strength in the PLA/bentonite NC, owing to enhanced interaction and compatibility between the polymer matrix and reinforcing phase. Dynamic mechanical thermal analysis further validated improved temperature stability. Intriguingly, the addition of bentonite led to reduced oxygen permeability and decreased light transmission, promising enhanced packaging and biomedical applications. This study underscores the potential of NCs to engineer tailored materials with controlled properties, including degradation rates, drug-release behaviors, and mechanical strengths, ushering in a new era of precision in biomedicine (Petersson and Oksman, 2006). In another remarkable study, researchers utilized CA, a natural biopolymer, to modify the surface of electrospun PCL fibers. By incorporating CNTs into the composite, they successfully prepared negatively charged PCL fibers, leading to a significant enhancement in the mechanical strength of the electrospun PCL fibers. The CA-modified composite displayed exceptional antimicrobial properties, particularly against *Escherichia coli* and *Staphylococcus aureus*, showcasing its potential for combating bacterial infections. This antimicrobial performance was superior to that of both pure PCL fibers and PCL/CNT fibers. Additionally, the researchers conducted cell culture experiments to evaluate the biocompatibility of the composite materials. The results demonstrated that the CA-modified CS-PCL/CNT fiber mat facilitated enhanced cell attachment and distribution compared to the other fiber types. Living cells exhibited superior adhesion ability on the CS-PCL/CNT fiber mat further confirming the biocompatibility and potential utility of these composite materials in various biomedical applications. Notably, the incorporation of CNTs played a crucial role in improving the immobilization of CA onto the electrospun PCL fibers. The synergistic effects of CA and CNTs in the composite contributed to the overall enhancement in mechanical properties, antimicrobial performance, and cell attachment, making these materials highly promising for use as biomaterials. Lin et al. synthesized a multifunctional and biocompatible polyethylene glycol-CA-iron oxide NC incorporated with a near-infrared fluorescent cyanine dye (MTX-PEG-CS-IONPsCy5.5). This NC exhibited remarkable properties, including fluorescence and magnetic resonance imaging capabilities, making it suitable for imaging applications. Furthermore, its superparamagnetic and targeting characteristics enable self-targeted curative drug delivery. The incorporation of multiple functionalities in this hybrid NC underscores its potential as a versatile and effective therapeutic tool in cancer treatment (Lin et al., 2015).

6.7 EMERGING TRENDS AND FUTURE DIRECTIONS

In the future, the role and applications of biodegradable and nondegradable NCs in the biomedical field are likely to witness even greater advancements. The ongoing research and development in nanotechnology are expected to lead to breakthroughs in NC synthesis techniques, allowing for further optimization of their properties and functionalities. Novel materials with enhanced biocompatibility and tailored degradation profiles will continue to be explored, opening up new possibilities for tissue engineering and regenerative medicine. Additionally, the integration of NCs with emerging technologies, such as gene editing and nanorobotics, could pave the way for highly targeted and precise medical interventions. The development of smart NCs that respond to specific stimuli, such as changes in pH or temperature, will further enhance their therapeutic potential. These stimuli-responsive NCs could enable controlled drug release in response to specific physiological cues, improving treatment efficacy and minimizing side effects. In terms of clinical translation, efforts will be directed toward ensuring the safety and efficacy of NCs for human use. Robust preclinical testing and rigorous evaluation of long-term biocompatibility will be essential to gain regulatory approvals for their clinical applications. Furthermore, advancements in NC manufacturing techniques, such as continuous manufacturing and scalable production, will be crucial for facilitating large-scale production of these materials for widespread clinical use. Collaborations between academia, industry, and regulatory bodies will play a pivotal role in driving the future development of NCs in the biomedical field. These collaborations will foster knowledge exchange, address challenges, and expedite the translation of cutting-edge research into practical clinical applications. The future of biodegradable and nondegradable NCs in the biomedical field is highly promising. As research continues to push the boundaries of nanotechnology and materials science, NCs are expected to play a central role in shaping the future of medical treatments, diagnostics, and personalized therapies. Their potential to revolutionize healthcare, coupled with ongoing efforts to address challenges and ensure safety, positions NCs as a transformative technology in the field of biomedical engineering.

6.8 CONCLUSION

In conclusion, this chapter highlighted the diverse applications of biodegradable and nondegradable NCs in the field of biomedicine. It underscored their strengths and limitations, paving the way for the emergence of NCs as a promising solution to address these shortcomings. The blending of materials to create NCs presents a pathway to enhance the properties and functionalities of nanoparticles, thus opening new avenues for targeted therapies, drug delivery systems, and tissue engineering. Looking ahead, the integration of these advanced materials holds the potential to revolutionize medical treatments, offering tailored solutions to complex biomedical challenges and advancing the forefront of healthcare technologies.

KEYWORDS

- **biodegradable vs. nondegradable NCs**
- **biomedical applications**
- **hybrid NCs**
- **kinetics of degradation and stability**

REFERENCES

Abdollahiyan, P., Oroojalian, F., Hejazi, M., de la Guardia, M., & Mokhtarzadeh, A. (2021). Nanotechnology, and scaffold implantation for the effective repair of injured organs: An overview on hard tissue engineering. *Journal of Controlled Release, 333*, 391–417.

Abdul Khalil, H. P., Yahya, E. B., Tajarudin, H. A., Balakrishnan, V., & Nasution, H. (2022). Insights into the role of biopolymer-based xerogels in biomedical applications. *Gels, 8*(6), 334.

Abere, D. V., Ojo, S. A., Oyatogun, G. M., Paredes-Epinosa, M. B., Niluxsshun, M. C., & Hakami, A. (2022). Mechanical and morphological characterization of nano-hydroxyapatite (nHA) for bone regeneration: A mini-review. *Biomedical Engineering Advances,* 100056.

Abdel-Bary, A. S., Tolan, D. A., Nassar, M. Y., Taketsugu, T., & El-Nahas, A. M. (2020). Chitosan, magnetite, silicon dioxide, and graphene oxide nanocomposites: Synthesis, characterization, efficiency as cisplatin drug delivery, and DFT calculations. *International Journal of Biological Macromolecules, 154*, 621–633.

Alam, F., Varadarajan, K. M., & Kumar, S. (2020). 3D printed polylactic acid nanocomposite scaffolds for tissue engineering applications. *Polymer Testing, 81*, 106203.

Allen, B. L., Kichambare, P. D., Gou, P., Vlasova, I. I., Kapralov, A. A., Konduru, N., Kagan, V. E., & Star, A. (2008). Biodegradation of single-walled carbon nanotubes through enzymatic catalysis. *Nano Letters, 8*(11), 3899–3903.

Al-Harbi, N., Hussein, M. A., Al-Hadeethi, Y., & Umar, A. (2021). Cellulose acetate-hydroxyapatite-bioglass-zirconia nanocomposite particles as potential biomaterial: Synthesis, characterization, and biological properties for bone application. *Engineered Science, 17*, 70–82.

Aliotta L, Seggiani M, Lazzeri A, Gigante V, & Cinelli P. (2022). A brief review of poly (butylene succinate) (PBS) and its main copolymers: synthesis, blends, composites, biodegradability, and applications. *Polymers. 14*(4), 844.

Alturky, A., Alatawi, A., Abulyazied, D., Abomostaf, H., Elkomy, G., Taha, M., & Youness, R. (2023). Magnetic and dielectric properties of hybrid nanocomposites of biologically extracted hydroxyapatite/hematite/silicon dioxide for potential use in bone replacement applications. *ECS Journal of Solid State Science and Technology.*

Amaraweera, S. M., Gunathilake, C., Gunawardene, O. H., Fernando, N. M., Wanninayaka, D. B., Dassanayake, R. S., Rajapaksha, S. M., Manamperi, A., Fernando, C. A., Kulatunga, A. K., & Manipura, A. (2021). Development of starch-based materials using current modification techniques and their applications: A review. *Molecules, 26*(22), 6880.

Ansari, Z., Kalantar, M., Kharaziha, M., Ambrosio, L., & Raucci, M. G. (2020). Polycaprolactone/fluoride substituted-hydroxyapatite (PCL/FHA) nanocomposite coatings prepared by *in situ* sol-gel process for dental implant applications. *Progress in Organic Coatings, 147*, 105873.

Aqel, A., Abou El-Nour, K. M., Ammar, R. A., & Al-Warthan, A. (2012). Carbon nanotubes, science and technology part (I) structure, synthesis and characterisation. *Arabian Journal of Chemistry, 5*(1), 1–23.

Armentano, I., Dottori, M., Puglia, D., & Kenny, J. M. (2008). Effects of carbon nanotubes (CNTs) on the processing and in-vitro degradation of poly (DL-lactide-co-glycolide)/CNT films. *Journal of Materials Science: Materials in Medicine, 19*, 2377–2387.

Armentano, I., Dottori, M., Fortunati, E., Mattioli, S., & Kenny, J. M. (2010). Biodegradable polymer matrix nanocomposites for tissue engineering: A review. *Polymer Degradation and Stability, 95*(11), 2126–2146.

Ashammakhi, N., Ahadian, S., Xu, C., Montazerian, H., Ko, H., Nasiri, R., Barros, N., & Khademhosseini, A. (2019). Bioinks and bioprinting technologies to make heterogeneous and biomimetic tissue constructs. *Materials Today Bio, 1*, 100008.

Azizi-Lalabadi, M., Hashemi, H., Feng, J., & Jafari, S. M. (2020). Carbon nanomaterials against pathogens; the antimicrobial activity of carbon nanotubes, graphene/graphene oxide, fullerenes, and their nanocomposites. *Advances in Colloid and Interface Science, 284*, 102250.

Balla, E., Daniilidis, V., Karlioti, G., Kalamas, T., Stefanidou, M., Bikiaris, N. D., Vlachopoulos, A., Koumentakou, I., & Bikiaris, D. N. (2021). Poly (lactic Acid): A versatile biobased polymer for the future with multifunctional properties—From monomer synthesis, polymerization techniques and molecular weight increase to PLA applications. *Polymers, 13*(11), 1822.

Batool, S., Sohail, S., ud Din, F., Alamri, A. H., Alqahtani, A. S., Alshahrani, M. A., Alshehri, M. A., & Choi, H. G. (2023). A detailed insight of the tumor targeting using nanocarrier drug delivery system. *Drug Delivery, 30*(1), 2183815.

Beesetty, P., Patil, B., & Doddamani, M. (2020). Mechanical behavior of additively manufactured nanoclay/HDPE nanocomposites. *Composite Structures, 247*, 112442.

Bharadwaz, A., & Jayasuriya, A. C. (2020). Recent trends in the application of widely used natural and synthetic polymer nanocomposites in bone tissue regeneration. *Materials Science and Engineering: C, 110*, 110698.

Bikiaris, D. N., Chrissafis, K., Paraskevopoulos, K. M., Triantafyllidis, K. S., & Antonakou, E. V. (2007). Investigation of thermal degradation mechanism of an aliphatic polyester using pyrolysis–gas chromatography–mass spectrometry and a kinetic study of the effect of the amount of polymerisation catalyst. *Polymer Degradation and Stability, 92*(4), 525–536.

Bikiaris, D. N., Vassiliou, A., Pavlidou, E., & Karayannidis, G. P. (2005). Compatibilisation effect of PP-gMA copolymer on iPP/SiO_2 nanocomposites prepared by melt mixing. *European Polymer Journal, 41*(9), 1965–1978.

Bian, X. S., Ambrosio, L., Kenny, J. M., Nicolais, L., Occhiello, E., Morra, M., Garbassi, F., & Dibenedetto, A. T. (1991). The effects of surface treatments of fibers on the interfacial properties in single-fiber composites. *Journal of Adhesion Science and Technology, 5*(5), 377–388.

Bianco, A., Kostarelos, K., & Prato, M. (2005). Applications of carbon nanotubes in drug delivery. *Current Opinion in Chemical Biology, 9*(6), 674–679.

Bui, V. K., Park, D., & Lee, Y. C. (2017). Chitosan combined with ZnO, TiO_2 and Ag nanoparticles for antimicrobial wound healing applications: A mini review of the research trends. *Polymers, 9*(1), 21.

Cao, X., Cheng, S., You, Y., Zhang, S., & Xian, Y. (2019). Sensitive monitoring and bioimaging intracellular highly reactive oxygen species based on gold nanoclusters@ nanoscale metal-organic frameworks. *Analytica Chimica Acta, 1092*, 108–116.

Capuana, E., Lopresti, F., Ceraulo, M., & La Carrubba, V. (2022). Poly-L-lactic acid (PLLA)-based biomaterials for regenerative medicine: A review on processing and applications. *Polymers, 14*(6), 1153.

Castillo, L. A., López, O. V., Ninago, M. D., Versino, F., Barbosa, S. E., García, M. A., & Villar, M. A. (2017). Composites and nanocomposites based on starches: Effect of mineral and organic fillers on processing, structure, and final properties of starch. In *Starch-based materials in food packaging* (pp. 125–151). Academic Press.

Chae, D. W., & Kim, B. C. (2005). Characterization on polystyrene/zinc oxide nanocomposites prepared from solution mixing. *Polymers for Advanced Technologies 16*, 846–850. Doi: 10.1002/pat.673.

Chandra, P., Singh, J., Singh, A., Srivastava, A., Goyal, R. N., & Shim, Y. B. (2013). Gold nanoparticles and nanocomposites in clinical diagnostics using electrochemical methods. *Journal of Nanoparticles, 2013*.

Chaniotakis N, & Buiculescu R. (2012). Biosilica–nanocomposites–nanobiomaterials for biomedical engineering and sensing applications. *Biomedical Materials and Diagnostic Devices*. *24,* 323–338.

Chen, H., Bei, J., & Wang, S. (2000). Hydrolytic degradation of polyester–polyether block copolymer based on polycaprolactone/poly (ethylene glycol)/polylactide. *Polymers for Advanced Technologies, 11*(4), 180–184.

Chen, C., Kong, K., Zhang, S.-M., & Lee, I. S. (2015). Characterization and *in vitro* biological evaluation of mineral/osteogenic growth peptide nanocomposites synthesized biomimetically on titanium. *Applied Surface Science, 334*, 62–68.

Chinta, M. L., Velidandi, A., Pabbathi, N. P., Dahariya, S., & Parcha, S. R. (2021). Assessment of properties, applications and limitations of scaffolds based on cellulose and its derivatives for cartilage tissue engineering: A review. *International Journal of Biological Macromolecules, 175*, 495–515.

Choi, P. S., Lee, J. Y., Chae, J. H., Wadas, T., Cheng, Z., Hur, M. G., & Park, J. H. (2022). Theranostics through utilizing Cherenkov radiation of radioisotope Zr-89 with a nanocomposite combination of TiO_2 and MnO_2. *ACS Applied Materials & Interfaces, 15*(3), 3689–3698.

Chou, P. Y., Chou, Y. C., Lai, Y. H., Lin, Y. T., Lu, C. J., & Liu, S. J. (2021). Fabrication of drug-eluting nano-hydroxyapatite filled polycaprolactone nanocomposites using solution-extrusion 3D printing technique. *Polymers, 13*(3), 318.

Chung, J. H., Simmons, A., & Poole-Warren, L. A. (2021). Non-degradable polymer nanocomposites for drug delivery. *Expert Opinion on Drug Delivery, 8*(6), 765–778.

Corcione, C. E., & Frigione, M. (2012). Characterization of nanocomposites by thermal analysis. *Materials, 5*(12), 2960–2980.

Dai, T., Tanaka, M., Huang, Y. Y., & Hamblin, M. R. (2011). Chitosan preparations for wounds and burns: Antimicrobial and wound-healing effects. *Expert Review of Anti-Infective Therapy, 9*(7), 857–879.

Dalton, B., Bhagabati, P., De Micco, J., Padamati, R. B., & O'Connor, K. (2022). A review on biological synthesis of the biodegradable polymers polyhydroxyalkanoates and the development of multiple applications. *Catalysts, 12*(3), 319.

de França, J. O., da Silva Valadares, D., Paiva, M. F., Dias, S. C., & Dias, J. A. (2022). Polymers based on PLA from synthesis using D, L-lactic acid (or racemic lactide) and some biomedical applications: A short review. *Polymers, 14*(12), 2317.

Deng, Z., Li, M., Hu, Y., He, Y., Tao, B., Yuan, Z., Wang, R., Chen, M., Luo, Z., & Cai, K. (2021). Injectable biomimetic hydrogels encapsulating gold/metal–organic frameworks nanocomposites for enhanced antibacterial and wound healing activity under visible light actuation. *Chemical Engineering Journal, 420*, 129668.

Devnarain, N., Osman, N., Fasiku, V. O., Makhathini, S., Salih, M., Ibrahim, U. H., & Govender, T. (2021). Intrinsic stimuli-responsive nanocarriers for smart drug delivery of antibacterial agents—An in-depth review of the last two decades. *Wiley Interdisciplinary Reviews: Nanomedicine and Nanobiotechnology, 13*(1), e1664.

Dheyab M., Abdul Aziz A., Jameel M. S., Moradi Khaniabadi P. (2020). Recent advances in synthesis, medical applications and challenges for gold-coated iron oxide: Comprehensive study. *Nanomaterials. 11*(8), 2147.

Dinescu, S., Albu Kaya, M., Chitoiu, L., Ignat, S., Kaya, D. A., & Costache, M. (2019). Collagen-based hydrogels and their applications for tissue engineering and regenerative medicine. *Polymers and Polymeric Composites: A Reference Series, 10*, 978–973.

Djemai, A., Gladden, L. F., Booth, J., Kittlety, R. S., & Gellert, P. R. (2001). MRI investigation of hydration and heterogeneous degradation of aliphatic polyesters derived from lactic and glycolic acids: A controlled drug delivery device. *Magnetic Resonance Imaging, 19*(3–4), 521–523.

Doghish, A. S., Hashem, A. H., Shehabeldine, A. M., Sallam, A. A., El-Sayyad, G. S., & Salem, S. S. (2022). Nanocomposite based on gold nanoparticles and carboxymethyl cellulose: Synthesis, characterization, antimicrobial, and anticancer activities. *Journal of Drug Delivery Science and Technology, 77*, 103874.

Drury, J. L., & Mooncy, D. J. (2003). Hydrogels for tissue engineering: Scaffold design variables and applications. *Biomaterials, 24*(24), 4337–4351.

Dunn, A. S., Campbell, P. G., & Marra, K. G. (2001). The influence of polymer blend composition on the degradation of polymer/hydroxyapatite biomaterials. *Journal of Materials Science: Materials in Medicine, 12*, 673–677.

Dvir, T., Timko, B. P., Kohane, D. S., & Langer, R. (2020). Nanotechnological strategies for engineering complex tissues. *Nano-Enabled Medical Applications,* 351–382.

Dwivedi, R., Kumar, S., Pandey, R., Mahajan, A., Nandana, D., Katti, D. S., & Mehrotra, D. (2020). Polycaprolactone as biomaterial for bone scaffolds: Review of literature. *Journal of Oral Biology and Craniofacial Research, 10*(1), 381–388.

El-Sayed, N. M., El-Bakary, M. A., Ibrahim, M. A., Elgamal, M. A., ElZorkany, H. E., & Elshoky, H. A. (2023). Synthesis and characterization of mussel-inspired nanocomposites based on dopamine–chitosan–iron oxide for wound healing: *In vitro* study. *International Journal of Pharmaceutics, 632*, 122538.

Faisal, M., Kou, T., Zhong, Y., & Blennow, A. (2022). High amylose-based bio composites: Structures, functions and applications. *Polymers, 14*(6), 1235.

Farani, M. R., Khadiv-Parsi, P., Riazi, G. H., Ardestani, M. S., & Rad, H. S. (2020). PEGylation of graphene/iron oxide nanocomposite: Assessment of release of doxorubicin, magnetically targeted drug delivery and photothermal therapy. *Applied Nanoscience, 10*(4), 1205–1217.

Fiala, T., Barros, E. P., Ebert, M. O., Ruijsenaars, E., Riniker, S., & Wennemers, H. (2022). Frame shifts affect the stability of collagen triple helices. *Journal of the American Chemical Society, 144*(40), 18642–18649.

Fragogeorgi, E. A., Rouchota, M., Georgiou, M., Velez, M., Bouziotis, P., & Loudos, G. (2019). *In vivo* imaging techniques for bone tissue engineering. *Journal of Tissue Engineering, 10*, 2041731419854586.

Gandara-Loe, J., Souza, B. E., Missyul, A., Giraldo, G., Tan, J. C., & Silvestre-Albero, J. (2020). MOF-based polymeric nanocomposite films as potential materials for drug delivery devices in ocular therapeutics. *ACS Applied Materials & Interfaces, 12*(27), 30189–30197.

Gautam, S., Sharma, C., Purohit, S. D., Singh, H., Dinda, A. K., Potdar, P. D., Chou, C. F., & Mishra, N. C. (2021). Gelatin-polycaprolactone-nanohydroxyapatite electrospun nanocomposite scaffold for bone tissue engineering. *Materials Science and Engineering: C, 119*, 111588.

Gholamali, I., Hosseini, S. N., Alipour, E., & Yadollahi, M. (2019). Preparation and characterization of oxidized starch/CuO nanocomposite hydrogels applicable in a drug delivery system. *Starch-Stärke, 71*(3–4), 1800118.

Ghosal, K., Agatemor, C., Špitálsky, Z., Thomas, S., & Kny, E. (2019). Electrospinning tissue engineering and wound dressing scaffolds from polymer-titanium dioxide nanocomposites. *Chemical Engineering Journal, 358*, 1262–1278.

Gould, L., Abadir, P., Brem, H., Carter, M., Conner-Kerr, T., Davidson, J., DiPietro, L., Falanga, V., Fife, C., Gardner, S., & Grice, E. (2015). Chronic wound repair and healing in older adults: Current status and future research. *Wound Repair and Regeneration, 23*(1), 1–3.

Grassi, M., & Grassi, G. (2005). Mathematical modelling and controlled drug delivery: Matrix systems. *Current Drug Delivery, 2*(1), 97–116.

Guo, W., Yang, K., Qin, X., Luo, R., Wang, H., & Huang, R. (2022). Polyhydroxyalkanoates in tissue repair and regeneration. *Engineered Regeneration, 3*(1), 24–40.

Gunathilake, T. M., Ching, Y. C., Chuah, C. H., Hai, N. D., & Nai-Shang, L. (2020). Electro-stimulated release of poorly water-soluble drug from poly (lactic acid)/carboxymethyl cellulose/ZnO nanocomposite film. *Pharmaceutical Research, 37*, 1–20.

Hada, V., Hashmi, S. A., Mili, M., Gorhe, N., Sagiri, S. S., Pal, K., Chawdhary, R., Khan, M., Naik, A., Prashant, N., & Srivastava, A. K. (2022). Polycaprolactone-based shape memory polymeric nanocomposites for biomedical applications. *Food, Medical, and Environmental Applications of Nanomaterials,* 413–433.

Hamdan M. M., Siregar J. P., Rejab M. R., Bachtiar D., Jamiluddin J., Tezara C. (2017). Effect of maleated anhydride on mechanical properties of rice husk filler reinforced PLA matrix polymer composite. *International Journal of Precision Engineering and Manufacturing-Green Technology. 6,* 113–124.

Han, K., & Yu, M. (2006). Study of the preparation and properties of UV-blocking fabrics of a PET/TiO_2 nanocomposite prepared by *in situ* polycondensation. *Journal of Applied Polymer Science, 100*, 1588–1593. DOI: 10.1002/app.23312

Hejazi, M., Arshadi, S., Amini, M., Baradaran, B., Shahbazi-Derakhshi, P., Sameti, P., Soleymani, J., Mokhtarzadeh, A., & Tavangar, S. M. (2023). Hyaluronic acid-functionalized gold nanoparticles as a cancer diagnostic probe for targeted bioimaging applications. *Microchemical Journal, 193*, 108953.

Hervault, A., Dunn, A. E., Lim, M., Boyer, C., Mott, D., Maenosono, S., & Thanh, N. T. (2016). Doxorubicin loaded dual pH-and thermo-responsive magnetic nanocarrier for combined magnetic hyperthermia and targeted controlled drug delivery applications. *Nanoscale, 8*(24), 12152–12161.

Ho, L. C., Hsu, C. H., Ou, C. M., Wang, C. W., Liu, T. P., Hwang, L. P., Lin, Y. Y., & Chang, H. T. (2015). Unibody coreeshell smart polymer as a theranostic nanoparticle for drug delivery and MR imaging. *Biomaterials, 37*, 436–446. DOI: 10.1016/j.biomaterials.2014.10.006.

Hu, W., Chen, S., Yang, J., Li, Z., & Wang, H. (2014). Functionalized bacterial cellulose derivatives and nanocomposites. *Carbohydrate Polymers, 101*, 1043–1060.

Huang, S., Hong, X., Zhao, M., Liu, N., Liu, H., Zhao, J., Shao, L., Xue, W., Zhang, H., Zhu, P., & Guo, R. (2022). Nanocomposite hydrogels for biomedical applications. *Bioengineering & Translational Medicine, 7*(3), e10315.

Hurrell, S., & Cameron, R. E. (2001). Polyglycolide: Degradation and drug release. Part I: Changes in morphology during degradation. *Journal of Materials Science: Materials in Medicine, 12*, 811–816.

Iannace, S., Maffezzoli, A., Leo, G., & Nicolais, L. (2001). Influence of crystal and amorphous phase morphology on hydrolytic degradation of PLLA subjected to different processing conditions. *Polymer, 42*(8), 3799–3807.

Ioelovich, M. (2014). Crystallinity and hydrophility of chitin and chitosan. *Journal of Chemistry, 3*(3), 7–14.

Iroegbu, A. O., & Ray, S. S. (2023). Chitin nanomaterials as multifunctional systems in advanced applications–progress and challenges toward sustainability. *Macromolecular Materials and Engineering, 2023*, 2300053.

Iqbal, N., Khan, A. S., Asif, A., Yar, M., Haycock, J. W., & Rehman, I. U. (2019). Recent concepts in biodegradable polymers for tissue engineering paradigms: A critical review. *International Materials Reviews, 64*(2), 91–126.

Jafari, S., Mahyad, B., Hashemzadeh, H., Janfaza, S., Gholikhani, T., & Tayebi, L. (2020). Biomedical applications of TiO_2 nanostructures: Recent advances. *International Journal of Nanomedicine,* 3447–3470.

Jana, S., & Sen, K. K. (2017). Chitosan—Locust bean gum interpenetrating polymeric network nanocomposites for delivery of aceclofenac. *International Journal of Biological Macromolecules, 102*, 878–884.

Jia, X., & Kiick, K. L. (2009). Hybrid multicomponent hydrogels for tissue engineering. *Macromolecular Bioscience, 9*(2), 140–156.

Jian, J., Xiangbin, Z., & Xianbo, H. (2020). An overview on synthesis, properties and applications of poly (butylene-adipate-co-terephthalate)–PBAT. *Advanced Industrial and Engineering Polymer Research, 3*(1), 19–26.

Jiang, S., Wang, M., & He, J. (2021) A review of biomimetic scaffolds for bone regeneration: Toward a cell-free strategy. *Bioengineering & Translational Medicine, 6*(2), e10206.

Jiang, Z., Zheng, Z., Yu, S., Gao, Y., Ma, J., Huang, L., & Yang, L. (2023). Nanofiber scaffolds as drug delivery systems promoting wound healing. *Pharmaceutics, 15*(7), 1829.

Junter, G. A., & Lebrun, L. (2017). Cellulose-based virus-retentive filters: A review. *Reviews in Environmental Science and Bio/Technology, 16*, 455–489.

Kalia, V. C., Patel, S. K., & Lee, J. K. (2023). Exploiting polyhydroxyalkanoates for biomedical applications. *Polymers, 15*(8), 1937.

Kalva, S. N., Augustine, R., Al Mamun, A., Dalvi, Y. B., Vijay, N., & Hasan, A. (2021). Active agents loaded extracellular matrix mimetic electrospun membranes for wound healing applications. *Journal of Drug Delivery Science and Technology, 63*, 102500.

Khan, W. S., Hamadneh, N. N., & Khan, W. A. (2016). Polymer nanocomposites–synthesis techniques, classification and properties. *Science and applications of Tailored Nanostructures, 50*.

Kim, K., Yu, M., Zong, X., Chiu, J., Fang, D., Seo, Y. S., Hsiao, B. S., Chu, B., & Hadjiargyrou, M. (2003). Control of degradation rate and hydrophilicity in electrospun non-woven poly (D, L-lactide) nanofiber scaffolds for biomedical applications. *Biomaterials, 24*(27), 4ss977–4ss985.

Leal, D. A., Kuznetsova, A., Silva, G. M., Tedim, J., Wypych, F., & Marino, C. E. (2022). Layered materials as nanocontainers for active corrosion protection: A brief review. *Applied Clay Science, 225*, 106537.

León-López, A., Morales-Peñaloza, A., Martínez-Juárez, V. M., Vargas-Torres, A., Zeugolis, D. I., & Aguirre-Álvarez, G. (2019). Hydrolyzed collagen—Sources and applications. *Molecules, 24*(22), 4031.

Li, M., Komasa, S., Hontsu, S., Hashimoto, Y., & Okazaki, J. (2022). Structural characterization and osseointegrative properties of pulsed laser-deposited fluorinated hydroxyapatite films on nano-zirconia for implant applications. *International Journal of Molecular Sciences, 23*(5), 2416.

Li, S., & McCarthy, S. (1999). Further investigations on the hydrolytic degradation of poly (DL-lactide). *Biomaterials, 20*(1), 35–44.

Li, Y., Men, Y., Chen, X., & Wang, B. (2018). Laminin-chitosan-PLGA conduit co-transplanted with Schwann and neural stem cells to repair the injured recurrent laryngeal nerve. *Experimental and Therapeutic Medicine, 16*, 1250–1258.

Lin, J., Li, Y., Li, Y., Wu, H., Yu, F., Zhou, S., Xie, L., Luo, F., Lin, C., & Hou, Z. (2015). Drug/dye-loaded, multifunctional PEG–chitosan–iron oxide nanocomposites for methotraxate synergistically self-targeted cancer therapy and dual model imaging. *ACS Applied Materials & Interfaces, 7*(22), 11908–11920.

Liu, X., Xu, H., Zhang, M., & Yu, D. G. (2021). Electrospun medicated nanofibers for wound healing. *Membranes, 11*(10), 770.

Loher, S., Reboul, V., Brunner, T. J., Simonet, M., Dora, C., Neuenschwander, P., & Stark, W. J. (2006). Improved degradation and bioactivity of amorphous aerosol derived tricalcium phosphate nanoparticles in poly (lactide-co-glycolide). *Nanotechnology, 17*(8), 2054.

Lohbauer, U., Wagner, A., Belli, R., Stoetzel, C., Hilpert, A., Kurland, H. D., Grabow, J., & Müller, F. A. (2010). Zirconia nanoparticles prepared by laser vaporization as fillers for dental adhesives. *Acta Biomaterialia, 6*, 4539–4546.

Loo, S. C., Ooi, C. P., & Boey, Y. C. (2004). Radiation effects on poly (lactide-co-glycolide)(PLGA) and poly (L-lactide)(PLLA). *Polymer Degradation and Stability, 83*(2), 259–265.

Makabenta, J. M., Nabawy, A., Li, C. H., Schmidt-Malan, S., Patel, R., & Rotello, V. M. (2021). Nanomaterial-based therapeutics for antibiotic-resistant bacterial infections. *Nature Reviews Microbiology, 19*(1), 23–36.

Mathew, S. A., Praveena, P., Hubert, Y. S., Narayanan, V., & Stephen, A. (2019, Jul 11). Luminescent carbon dots/chitosan nanocomposite for bioimaging. *AIP Conference Proceedings, 2115*(1).

Marques, A. P., Reis, R. L., & Hunt, J. A. (2002). The biocompatibility of novel starch-based polymers and composites: *In vitro* studies. *Biomaterials, 23*(6), 1471–1478.

Marin, E., Briceño, M. I., & Caballero-George, C. (2013). Critical evaluation of biodegradable polymers used in nanodrugs. *International Journal of Nanomedicine,* 3071–3091.

Maquet, V., Boccaccini, A. R., Pravata, L., Notingher, I., & Jérôme, R. (2004). Porous poly (α-hydroxyacid)/Bioglass® composite scaffolds for bone tissue engineering. I: Preparation and *in vitro* characterisation. Biomaterials, *25*(18), 4185–4194.

Maynes, R., Ed. (2012). *Structure and function of collagen types*. Elsevier.

McMillan, J., Batrakova, E., & Gendelman, H. E. (2011). Cell delivery of therapeutic nanoparticles. *Progress in Molecular Biology and Translational Science, 104*, 563–601.

Mdlovu, N. V., Mavuso, F. A., Lin, K. S., Chang, T. W., Chen, Y., Wang, S. S., Wu, C. M., Mdlovu, N. B., & Lin, Y. S. (2019). Iron oxide-pluronic F127 polymer nanocomposites as carriers for a doxorubicin drug delivery system. *Colloids and Surfaces A: Physicochemical and Engineering Aspects, 562*, 361–369.

Mihai, M. M., Dima, M. B., Dima, B., Holban, A. M. (2019). Nanomaterials for wound healing and infection control. *Materials, 12*(13), 2176.

Minelli, C., Lowe, S. B., & Stevens, M. M. (2010). Engineering nanocomposite materials for cancer therapy. *Small, 6*(21), 2336–2357.

Mirmusavi, M. H., Ahmadian, M., & Karbasi, S. (2022). Polycaprolactone-chitosan/multi-walled carbon nanotube: A highly strengthened electrospun nanocomposite scaffold for cartilage tissue engineering. *International Journal of Biological Macromolecules, 209*, 1801–1814.

Mirtaghavi, A., Baldwin, A., Tanideh, N., Zarei, M., Muthuraj, R., Cao, Y., Zhao, G., Geng, J., Jin, H., & Luo, J. (2020). Crosslinked porous three-dimensional cellulose nanofibers-gelatine biocomposite scaffolds for tissue regeneration. *International Journal of Biological Macromolecules, 164*, 1949–1959.

Mohammadi, R., Saboury, A., Javanbakht, S., Foroutan, R., & Shaabani, A. (2021). Carboxymethylcellulose/polyacrylic acid/starch-modified Fe_3O_4 interpenetrating magnetic nanocomposite hydrogel beads as pH-sensitive carrier for oral anticancer drug delivery system. *European Polymer Journal, 153*, 110500.

Mohebali, A., Abdouss, M., & Taromi, F. A. (2020). Fabrication of biocompatible antibacterial nanowafers based on HNT/PVA nanocomposites loaded with minocycline for burn wound dressing. *Materials Science and Engineering:* C, *110*, 110685.

Muhamad II, Selvakumaran, S., Lazim, N. A. (2014). Designing polymeric nanoparticles for targeted drug delivery system. *Nanomedicine, 287*, 287.

Mukheem, A., Shahabuddin, S., Akbar, N., Ahmad, I., Sudesh, K., & Sridewi, N. (2022). Development of biocompatible polyhydroxyalkanoate/chitosan-tungsten disulphide nanocomposite for antibacterial and biological applications. *Polymers, 14*(11), 2224.

Munir, M. U., Salman, S., Ihsan, A., & Elsaman, T. (2022). Synthesis, characterization, functionalization and bio-applications of hydroxyapatite nanomaterials: An overview. *International Journal of Nanomedicine,* 1903–1925.

Murugesan, S., & Scheibel, T. (2020). Copolymer/clay nanocomposites for biomedical applications. *Advanced Functional Materials, 30*(17), 1908101.

Murugesan, S., & Scheibel, T. (2021). Chitosan-based nanocomposites for medical applications. *Journal of Polymer Science, 59*(15), 1610–1642.

Nezami, S., Sadeghi, M., & Mohajerani, H. (2020). A novel pH-sensitive and magnetic starch-based nanocomposite hydrogel as a controlled drug delivery system for wound healing. *Polymer Degradation and Stability, 179*, 109255.

Norizan, M. N., Shazleen, S. S., Alias, A. H., Sabaruddin, F. A., Asyraf, M. R., Zainudin, E. S., Abdullah, N., Samsudin, M. S., Kamarudin, S. H., & Norrrahim, M. N. (2022). Nanocellulose-based nanocomposites for sustainable applications: A review. *Nanomaterials, 12*(19), 3483.

Nottelet, B., Awada, H., Laurencin, D., Gilbert, R., Funnell, J., Franconi, F., Larionova, J., Félix, G., Lemaire, L., & Guari, Y. (2021). Core-shell PLA nanofibers@ SPIONs nanocomposites with enhanced properties for biomedical imaging and magnetic applications. In *31st Annual Conference of the European Society for Biomaterials.*

Nurkesh, A., Jaguparov, A., Jimi, S., & Saparov, A. (2020). Recent advances in the controlled release of growth factors and cytokines for improving cutaneous wound healing. *Frontiers in Cell and Developmental Biology, 8*, 638.

Okur, M. E., Karantas, I. D., Şenyiğit, Z., Okur, N. Ü., & Siafaka, P. I. (2020). Recent trends on wound management: New therapeutic choices based on polymeric carriers. *Asian Journal of Pharmaceutical Sciences, 15*(6), 661–684.

Omrani, Z., Pourmadadi, M., Yazdian, F., & Rashedi, H. (2023). Preparation and characterization of pH-sensitive chitosan/starch/MoS_2 nanocomposite for control release of curcumin macromolecules drug delivery; application in the breast cancer treatment. *International Journal of Biological Macromolecules,* 125897.

Pandey, G., Marimuthu, M., Kanagavalli, P., Ravichandiran, V., Balamurugan, K., & Veerapandian, M. (2019). Chitosanylated MoO_3–ruthenium (II) nanocomposite as biocompatible probe for bioimaging and herbaceutical detection. *ACS Biomaterials Science & Engineering, 5*(7), 3606–3617.

Pandey, J. K., Reddy, K. R., Kumar, A. P., & Singh, R. P. (2005). An overview on the degradability of polymer nanocomposites. *Polymer Degradation and Stability, 88*(2), 234–250.

Pandey, P., Datta, M., & Malhotra, B. D. (2008). Prospects of nanomaterials in biosensors. *Analytical Letters, 41*(2), 159–209.

Pellis, A., Guebitz, G. M., & Nyanhongo, G. S. (2022). Chitosan: Sources, processing and modification techniques. *Gels, 8*(7), 393.

Peña-Parás, L., Sánchez-Fernández, J. A., & Vidaltamayo, R. (2018). Nanoclays for biomedical applications. *Handbook of Ecomaterials,5*, 3453–3471.

Petersson, L., & Oksman, K. (2006). Biopolymer based nanocomposites: Comparing layered silicates and microcrystalline cellulose as nanoreinforcement. *Composites Science and Technology, 66*(13), 2187–2196.

Poltronieri, P., & Kumar, P. (2017). Polyhydroxyalkanoates (PHAs) in industrial applications. *Handbook of ecomaterials, 4*, 2843–2872.

Prakash, J., Prema, D., Venkataprasanna, K. S., Balagangadharan, K., Selvamurugan, N., & Venkatasubbu, G. D. (2020). Nanocomposite chitosan film containing graphene oxide/hydroxyapatite/gold for bone tissue engineering. *International Journal of Biological Macromolecules, 154*, 62–71.

Rajesh, K., Crasta, V., Rithin Kumar, N. B., Shetty, G., & Rekha, P. D. (2019). Structural, optical, mechanical and dielectric properties of titanium dioxide doped PVA/PVP nanocomposite. *Journal of Polymer Research, 26*(4), 99.

Ratner, B. D. (2008). Proteins controlled with precision at organic, polymeric, and biopolymer interfaces for tissue engineering and regenerative medicine. In *Principles of regenerative medicine* (pp. 734–742). Academic Press.

Ranakoti, L., Gangil, B., Mishra, S. K., Singh, T., Sharma, S., Ilyas, R. A., & El-Khatib, S. (2022). Critical review on polylactic acid: Properties, structure, processing, biocomposites, and nanocomposites. *Materials, 15*(12), 4312.

Reddy, M. S., Ponnamma, D., Choudhary, R., & Sadasivuni, K. K. (2021). A comparative review of natural and synthetic biopolymer composite scaffolds. *Polymers, 13*(7), 1105.

Rezvani Ghomi, E., Nourbakhsh, N., Akbari Kenari, M., Zare, M., & Ramakrishna, S. (2021). Collagen-based biomaterials for biomedical applications. *Journal of Biomedical Materials Research Part B: Applied Biomaterials, 109*(12), 1986–1999.

Rocha, C. V., Gonçalves, V., da Silva, M. C., Bañobre-López, M., & Gallo, J. (2022). PLGA-based composites for various biomedical applications. *International Journal of Molecular Sciences, 23*(4), 2034.

Roy, S. M., Garg, V., Barman, S., Ghosh, C., Maity, A. R., & Ghosh, S. K. (2021). Kinetics of nanomedicine in tumor spheroid as an *in vitro* model system for efficient tumor-targeted drug delivery with insights from mathematical models. *Frontiers in Bioengineering and Biotechnology, 9*, 785937.

Ruiz, S., Tamayo, J. A., Delgado Ospina, J., Navia Porras, D. P., Valencia Zapata, M. E., Mina Hernandez, J. H., Valencia, C. H., Zuluaga, F., & Grande Tovar, C. D. (2019). Antimicrobial films based on nanocomposites of chitosan/poly (vinyl alcohol)/graphene oxide for biomedical applications. *Biomolecules, 9*(3), 109.

Sahoo, D., Sahoo, S., Mohanty, P., Sasmal, S., & Nayak, P. L. (2009). Chitosan: A new versatile bio-polymer for various applications. *Designed Monomers and Polymers, 12*(5), 377–404.

Salahuddin, N., Abdelwahab, M., Gaber, M., & Elneanaey, S. (2020). Synthesis and Design of Norfloxacin drug delivery system based on PLA/TiO_2 nanocomposites: Antibacterial and antitumor activities. *Materials Science and Engineering: C, 108*, 110337.

Salehi, M., Naseri-Nosar, M., Ebrahimi-Barough, S., Nourani, M., Khojasteh, A., Hamidieh, A. A., Amani, A., Farzamfar, S., & Ai, J. (2018). Sciatic nerve regeneration by transplantation of Schwann cells via erythropoietin controlled-releasing polylactic acid/multiwalled carbon nanotubes/gelatin nanofibrils neural guidance conduit. *Journal of Biomedical Materials Research Part B: Applied Biomaterials, 106*, 1463–1476.

Sehgal, R., & Gupta, R. (2020). Polyhydroxyalkanoate and its efficient production: An eco-friendly approach towards development. *3 Biotech, 10*(12), 549.

Sha, L., Chen, Z., Chen, Z., Zhang, A., & Yang, Z. (2016). Polylactic acid based nanocomposites: Promising safe and biodegradable materials in biomedical field. *International Journal of Polymer Science, 2016*.

Sharma, S., Sudhakara, P., Singh, J., Ilyas, R. A., Asyraf, M. R., & Razman, M. R. (2021). Critical review of biodegradable and bioactive polymer composites for bone tissue engineering and drug delivery applications. *Polymers, 13*(16), 2623.

Shoulders, M. D., & Raines, R. T. (2009). Collagen structure and stability. *Annual Review of Biochemistry, 78*, 929–958.

Siddiqui, H. A., Pickering, K. L., & Mucalo, M. R. (2018). A review on the use of hydroxyapatite-carbonaceous structure composites in bone replacement materials for strengthening purposes. *Materials, 11*(10), 1813.

Singh, R., & Lillard Jr, J. W. (2009). Nanoparticle-based targeted drug delivery. *Experimental and Molecular Pathology, 86*(3), 215–223.

Sinha, A., Simnani, F. Z., Singh, D., Nandi, A., Choudhury, A., Patel, P., Jha, E., Kaushik, N. K., Mishra, Y. K., Panda, P. K., & Suar, M. (2022). The translational paradigm of nanobiomaterials: Biological chemistry to modern applications. *Materials Today Bio,* 100463.

Sivamaruthi, B. S., kumar Nallasamy, P., Suganthy, N., Kesika, P., & Chaiyasut, C. (2022). Pharmaceutical and biomedical applications of starch-based drug delivery system: A review. *Journal of Drug Delivery Science and Technology,* 103890.

Song, M. M., Xu, H. L., Liang, J. X., Xiang, H. H., Liu, R., & Shen, Y. X. (2017). Lactoferrin modified graphene oxide iron oxide nanocomposite for glioma-targeted drug delivery. *Materials Science and Engineering: C, 77*, 904–911

Slepičková Kasálková, N., Slepička, P., & Švorčík, V. (2021). Carbon nanostructures, nanolayers, and their composites. *Nanomaterials, 11*(9), 2368.

Smart, S. K., Cassady, A. I., Lu, G. Q., & Martin, D. J. (2006). The biocompatibility of carbon nanotubes. *Carbon, 44*(6), 1034–1047.

Sun, J., & Zheng, X. (2022). Fabrication of zinc-loaded silicon carbide nanocomposite for *in vitro* cell viability and *in vivo* wound dressing care. *Journal of Microencapsulation, 39*(4), 341–351.

Su, F., Jia, Q., Li, Z., Wang, M., He, L., Peng, D., Song, Y., Zhang, Z., & Fang, S. (2019). Aptamer-templated silver nanoclusters embedded in zirconium metal–organic framework for targeted antitumor drug delivery. *Microporous and Mesoporous Materials, 275*, 152–162.

Taha, M. A., Youness, R. A., & Ibrahim, M. (2020). Biocompatibility, physico-chemical and mechanical properties of hydroxyapatite-based silicon dioxide nanocomposites for biomedical applications. *Ceramics International, 46*(15), 23599–23610.

Tanaya, K., Mohapatra, S., Maity, S., & Samantaray, D. P. (2023). Fluorescent-pigmented microbial polymer nanoparticles: An approach for replacement of bioimaging agent in the drug delivery system. *Applied Biochemistry and Microbiology, 59*(3), 378–383.

Trache D, Tarchoun AF, Derradji M, Hamidon TS, Masruchin N, Brosse N, & Hussin M. H. (2020). Nanocellulose: From fundamentals to advanced applications. *Frontiers in Chemistry*. *8,* 392.

Turon, P., Del Valle, L. J., Alemán, C., & Puiggalí, J. (2017). Biodegradable and biocompatible systems based on hydroxyapatite nanoparticles. *Applied Sciences, 7*(1), 60.

Turnbull, G., Clarke, J., Picard, F., Riches, P., Jia, L., Han, F., Li, B., & Shu, W. (2018). 3D bioactive composite scaffolds for bone tissue engineering. Bioactive materials*, 3*(3), 278–314.

Turunen, S., Kalpio, T., Lindahl, C., Shanthinathan, C. J., Akhter, T., Concaro, S., & Simonsson, S. (2023). Future solutions for osteoarthritis using 3D bioprinting of articular cartilage. In *Handbook of surgical planning and 3D printing* (pp. 335–369). Academic Press.

Veeman, D., Sai, M. S., Sureshkumar, P., Jagadeesha, T., Natrayan, L., Ravichandran, M., & Mammo, W. D. (2021). Additive manufacturing of biopolymers for tissue engineering and regenerative medicine: An overview, potential applications, advancements, and trends. *International Journal of Polymer Science, 2021*, 1–20.

Venkateshaiah, A., Padil, V. V., Nagalakshmaiah, M., Waclawek, S., Černík, M., & Varma, R. S. (2020). Microscopic techniques for the analysis of micro and nanostructures of biopolymers and their derivatives. *Polymers, 12*(3), 512.

Verma, S., Domb, A. J., & Kumar, N. (2011). Nanomaterials for regenerative medicine. *Nanomedicine, 6*(1), 157–181.

Wang, C., Huang, W., Zhou, Y., He, L., He, Z., Chen, Z., He, X., Tian, S., Liao, J., Lu, B., & Wei, Y. (2020). 3D printing of bone tissue engineering scaffolds. *Bioactive Materials, 5*(1), 82–91.

Wang, L., Hu, C., & Shao, L. (2017). The antimicrobial activity of nanoparticles: Present situation and prospects for the future. *International Journal of Nanomedicine,* 1227–1249.

Wang, X., Yu, Y., Yang, C., Shang, L., Zhao, Y., & Shen, X. (2022). Dynamically responsive scaffolds from microfluidic 3D printing for skin flap regeneration. *Advanced Science, 9*(22), 2201155.

Ways, T. M., Lau, W. M., & Khutoryanskiy, V. V. (2018). Chitosan and its derivatives for application in mucoadhesive drug delivery systems. *Polymers, 10*(3), 267.

Wen, X., & Tresco, P. A. (2006). Fabrication and characterization of permeable degradable poly (DL-lactide-co-glycolide)(PLGA) hollow fiber phase inversion membranes for use as nerve tract guidance channels. *Biomaterials, 27*(20), 3800–3809.

Wu, C., Chen, C., Lai, J., Chen, J., Mu, X., Zheng, J., & Zhao, Y. (2008). Molecule-scale controlled-release system based on light-responsive silica nanoparticles. *Chemical Communications, 5*(23), 2662–2664.

Xhazzy, A., Naik, R. R., & Shakya, A. K. (2023). Metal–polymer nanocomposites: A promising approach to antibacterial materials. *Polymers, 15*(9), 2167.

Xie, F., Pollet, E., Halley, P. J., & Averous, L. (2014). *Advanced nano-biocomposites based on starch: Polysaccharides* (pp. 1–75). Springer International Publishing.

Xie, F., Halley, P. J., & Avérous, L. (2011). Bio-nanocomposites based on starch. In *Nanocomposites with biodegradable polymers: Synthesis, properties and future perspectives* (pp. 234–260).

Xiong, R., Grant, A. M., Ma, R., Zhang, S., & Tsukruk, V. V. (2018). Naturally-derived biopolymer nanocomposites: Interfacial design, properties and emerging applications. *Materials Science and Engineering: R: Reports, 125*, 1–41.

Xue, Y., Sant, V., Phillippi, J., & Sant, S. (2017). Biodegradable and biomimetic elastomeric scaffolds for tissue-engineered heart valves. *Acta Biomaterialia, 48*, 2–19.

Xu, P., Yang, W., Niu, D., Yu, M., Du, M., Dong, W., Chen, M., Lemstra, P. J., & Ma, P. (2020). Multifunctional and robust polyhydroxyalkanoate nanocomposites with superior gas barrier, heat resistant and inherent antibacterial performances. *Chemical Engineering Journal, 382*, 122864.

Yang, Y. Y., Shi, M., Goh, S. H., Moochhala, S. M., Ng, S., & Heller, J. (2003). POE/PLGA composite microspheres: Formation and *in vitro* behavior of double walled microspheres. *Journal of Controlled Release, 88*(2), 201–213.

Yu, X., Tang, C., Xiong, S., Yuan, Q., Gu, Z., Li, Z., & Hu, Y. (2016). Modification of collagen for biomedical applications: A review of physical and chemical methods. *Current Organic Chemistry, 20*(17), 1797–1812.

Yu, H., Guo, W., Lu, X., Xu, H., Yang, Q., Tan, J., & Zhang, W. (2021). Reduced graphene oxide nanocomposite based electrochemical biosensors for monitoring foodborne pathogenic bacteria: A review. *Food Control, 127*, 108117.

Yu, Y., Zhou, M., Zhang, W., Huang, L., Miao, D., Zhu, H., & Su, G. (2019). Rattle-type gold nanorods/porous-SiO_2 nanocomposites as near-infrared light-activated drug delivery systems for cancer combined chemo–photothermal therapy. *Molecular Pharmaceutics, 16*(5), 1929–1938.

Zare, Y., & Shabani, I. (2016). Polymer/metal nanocomposites for biomedical applications. *Materials Science and Engineering: C, 60*, 195–203.

Zhang H, Yu L, & Ding J. (2000). Roles of hydrophilic homopolymers on the hydrophobic-association-induced physical gelling of amphiphilic block copolymers in water. *Macromolecules. 41*(17), 6493–64999.

Zhang, J., Zhao, S., Zhu, Y., Huang, Y., Zhu, M., Tao, C., & Zhang, C. (2014). Three-dimensional printing of strontium-containing mesoporous bioactive glass scaffolds for bone regeneration. *Acta Biomaterialia, 10*(5), 2269–2281.

Zhang, K., & Zhao, G. (2022). An effective wound healing material based on gold incorporation into a heparin-polyvinyl alcohol nanocomposite: Enhanced *in vitro* and *in vivo* care of perioperative period. *Journal of Cluster Science, 33*(4), 1655–1665.

Zhang, R., & Ma, P. X. (1999). Poly (α-hydroxyl acids)/hydroxyapatite porous composites for bone-tissue engineering. I: Preparation and morphology. *Journal of Biomedical Materials Research: An Official Journal of the Society for Biomaterials, The Japanese Society for Biomaterials, and The Australian Society for Biomaterials, 44*(4), 446–455.

Zhang, S., Chu, Z., Yin, C., Zhang, C., Lin, G., & Li, Q. (2013). Controllable drug release and simultaneously carrier decomposition of SiO_2-drug composite nanoparticles. *Journal of the American Chemical Society, 135*(15), 5709–5716.

Zhao, H., Liu, C., Gu, Z., et al. (2020). Persistent luminescent nanoparticles containing hydrogels for targeted, sustained, and autofluorescence-free tumor metastasis imaging. *Nano Letters, 20*, 252–260.

Ziental, D., Czarczynska-Goslinska, B., Mlynarczyk, D. T., Glowacka-Sobotta, A., Stanisz, B., Goslinski, T., & Sobotta, L. (2020). Titanium dioxide nanoparticles: Prospects and applications in medicine. *Nanomaterials, 10*(2), 387.

CHAPTER 7

Nanocomposites Applications in Pharmaceutical Medicines and Biotechnological Products

LOKESH NAMA[1 #], KHUSHBOO CHOUDHARY[1 #], KRISHNA MURTI[2 #], and NITESH KUMAR[1,*]

[1]*Department of Pharmacology and Toxicology, National Institute of Pharmaceutical Education and Research, Hajipur, Vaishali, Bihar, India*

[2]*Department of Pharmacy Practice, National Institute of Pharmaceutical Education and Research, Hajipur, Vaishali, Bihar, India*

# Contributed equally
*Corresponding author.

ABSTRACT

Pharmaceutical drugs and biotechnological products both make extensive use of nanocomposites, which are materials made of a mix of nanoparticles and a matrix substance. The performance, effectiveness, and safety of numerous pharmaceutical and biotech products can be improved because of the special features offered by these nanocomposites. The wide range of applications for nanocomposites has improved the functionality, potency, and security of medicinal drugs and biotechnological products. Mechanisms for delivering drugs and numerous sophisticated medication delivery techniques are being developed using nanocomposites. These nanoparticles can be created to encapsulate medicines, proteins, or nucleic acids, protecting them from degradation and boosting their stability. In tissue engineering, where they are utilized to build scaffolds for regenerative medical applications, nanocomposites are incredibly important. Nanocomposite-based biosensors display improved sensitivity, selectivity,

Nanocomposites and Nanomaterials in Biomedical Applications. Vivek Dave, Deependra Singh, and Pallerla Naveen Reddy (Eds.)

DOI: 10.1201/9781779641236-7

and detection limits by adding nanoparticles to the sensing components. The transport of genetic material (DNA or RNA) and therapeutic proteins is enhanced by the use of nanocomposites. In conclusion, the future uses of nanocomposites in pharmaceutical drugs and biotechnological goods are set to provide revolutionary advances in combination therapy, diagnostics, gene editing, and regulatory considerations. These advancements have the potential to considerably enhance patient outcomes and transform the biotechnology and healthcare sectors.

7.1 INTRODUCTION TO NANOCOMPOSITES IN PHARMACEUTICALS AND BIOTECHNOLOGY

Numerous polymer nanocomposite materials with appropriate optical, electric, magnetic, thermal, or mechanical characteristics may be created by adding nanoparticles with length scales of 1–100 nm to polymer matrices. The term "nanocomposite" characterizes structures with nanoscale recurrence intervals between the several phases that make up the material, as well as multiphase solid materials with one, two, or three dimensions of less than 100 nm. In the future, this word may also be used to refer to porous media, colloids, gels, and copolymers, despite its widespread usage to characterize solid mixes made up of a bulk matrix and one or more nanodimensional phases with varying characteristics due to structural and chemical differences. Because of the exceptionally high surface-to-volume ratio and number density of the nanoparticles, the mass fraction of the particles injected into the polymer matrix can stay very low (on the order of 0.5–5%). In contrast to conventionally reinforced polymers, which contain macroscopic inorganic or organic fillers, polymer nanocomposites do not. Traditional polymer composites typically experience a loss of other significant qualities along with an enhancement in other properties brought on by the addition of a particular macroscopic filler. For instance, a loss in hardness, transparency, or mechanical qualities might result from a gain in stiffness, barrier properties, or fire resistance. The loss of specific qualities can be reduced or prevented with nanoparticles. In fact, spreading one kind of modified nanoparticle can concurrently improve several desired qualities. Furthermore, by embedding nanoparticles in a polymer, new physics can be created that would not be possible with macro-sized particles (Chen et al. 2011). The area at the particle surface where the bulk and particle–polymer characteristics differ is known as the particle–polymer interface, and it is critical for polymer nanocomposites because it affects both the interfacial volume and the interfacial area. Because the interface volume is so small in traditional composites, the actual properties of the polymer and fillers have a far greater influence on the composites' overall characteristics. As the particle size shrinks, the interfacial area and volume grow by orders of magnitude.

7.2 APPLICATION OF NANOCOMPOSITE

The Toyota Technology Institute created the first nanocomposite in 1987 and patented it. It was made of polyamide and phyllosilicate and had improved mechanical and thermal qualities compared with pure polymers. A new class of materials called nanocomposites has emerged to get around the restrictions of monolithic materials. Composites are made by combining two or more components. The difficulties come from controlling the elemental composition and stoichiometry in the nanocluster phase, which is one of the phases with dimensions in the nanometer range. Typically, a matrix containing a filler, such as nanoparticles, nanolayers, or nanotubes is used to create the nanocomposites. By dispersing organic nanoparticles in a polymer or lipid, an organic-organic nanocomposite is created. Mineral mixtures can be used to create inorganic-inorganic nanocomposites. A scattered nanometric inorganic filler polymer matrix is what creates organic–inorganic nanocomposites (Ray et al., 2002).

7.2.1 NANOCOMPOSITE IN CANCER THERAPY

In light of the potential to circumvent cancer cells' treatment resistance, nanocomposites have been investigated for application in drug delivery systems (DDSs). Graphene gold, chitosan palladium, gold ferrite, and graphene oxide/bismuth selenide, among others, are only a few conjugated materials that have been developed for drug administration (Heidari et al., 2016). According to a recent study by Prabhu et al., geldanamycin had unfavorable hepatotoxicity. The incorporation of geldanamycin into a polymeric nanocomposite with a superparamagnetic iron oxide base boosts anticancer breast action while lowering adverse effects. Reintroducing it into clinical use for breast cancer is promising (Prabhu et al., 2017). Dhanavel et al. created a nanocomposite utilizing chitosan and palladium in a different work on nanocomposites. For the treatment of colon cancer, capsules containing the medications curcumin and 5-fluorouracil were employed. The writers wanted to create an approach that was straightforward, adaptable, and successful. When curcumin and 5-fluorouracil were administered together, a good inhibitory impact was seen when the nanocomposite was tested *in vitro* with HT-29 cells (Dhanavel et al., 2017). A treatment technique called hyperthermia raises the body's temperature in an area afflicted by a tumor to trigger the disruption of carcinoma cells. The operating theory is based on the observation that tumor cells can be directly destroyed at temperatures between 41 and 42°C because these are narrow protective to abrupt warmth spikes than the immune cells around them (Beik et al., 2016). Additionally, the magnetic field gradient enables the tumor's nanocomposites to be vectorized for medication administration. Hybrid protein–inorganic nanoparticles, nano-graphene, and magnetic iron oxide with a silicon-coated amine-polyglycerol functional shell are all feasible. The pH of the tumor microenvironment is another issue that nanocomposite has to deal with. Cancer proliferates more quickly than healthy cells do and lacks oxygen and nutrients, which causes acidosis and a lower pH. For efficient cancer treatment, pH-sensitive nanocomposite systems, such as superparamagnetic Fe_3O_4 nanoparticles and magnetic nanoparticles based on chitosan, have been studied. Nanocomposites can penetrate biological barriers to deliver targeted intracellular drugs in an acidic tumor environment, increasing the effectiveness of therapy (Liu et al., 2014).

7.2.2 NANOCOMPOSITES IN DIAGNOSTIC OF CANCER

In the field of medical oncology, the identification of cancer is an important phenomenon, and early detection of cancer might enhance a patient's long-term prognosis. Blood tests, computed tomography, magnetic resonance imaging, and biopsies are all used in the standard method of diagnosing cancer. Recent advances in nanotechnology have improved the contrast agents for magnetic resonance imaging, enabling the quick and sensitive early identification of cancerous cells and their noninvasive management (Subramanian et al., 2015). Magnetic resonance employing iron oxides, magnetite (Fe_3O_4), maghemite (Fe_2O_3), or other ferrites is the standard method of detection by imaging. These, nevertheless, are water-insoluble. Nanocomposites have been created to enhance the image quality as a result. Commercially, nanocomposites are made from superparamagnetic iron oxides that have been coated with hydrophilic polymers made from dextran. Less toxicity and greater bioavailability are provided by the polymer coating of the formulation. Because it improves T2- and T2-weighted sequences, which measure transverse relaxation time (spin-spin), iron oxide is desirable as a contrast agent (Corot et al., 2006). Nanocomposite is made up of an antibody-coated material that can recognize particular cancer cell receptors that give out a strong contrast signal during CT scans or MRI. As soon as the nanoparticles are inside the body, the antibodies they contain attach specifically to cancer cells, essentially lighting them for scanning. Nanotechnology makes it possible to observe molecules and cells that are invisible to traditional imaging methods, making it possible to see the molecular indicators that indicate different cancer types and stages (Kanapathipillai et al., 2014). For the tumor rim, Yu et al. created a superparamagnetic iron oxide nanocomposite. Magnetic resonance imaging was used to track the treatment's development. The intravenous administration of the nanocomposite in the tumors supported the monitoring (Yu et al., 2017). A new nanocomposite gel comprised of superparamagnetic iron oxide nanoparticles that have been 2-deoxy-D-glucose functionalized has been developed for the detection of breast cancer, according to research by Shan et al. (2017).

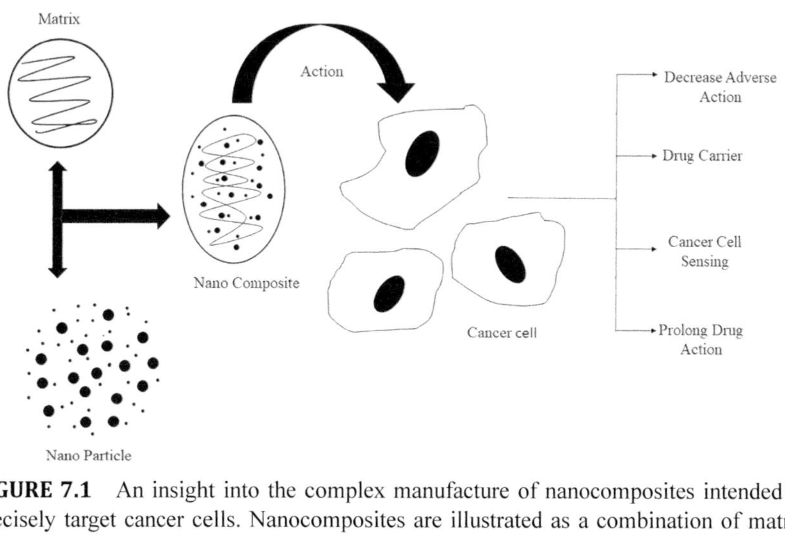

FIGURE 7.1 An insight into the complex manufacture of nanocomposites intended to precisely target cancer cells. Nanocomposites are illustrated as a combination of matrix and nanoparticle reactions, resulting in the formation of nanocomposites that interact with cancer cells and exhibit mechanistic effects. Nanocomposites reduce adverse effects, extend the duration of drug action, and enhance the capacity to detect cancer (Kanapathipillai et al., 2014).

7.2.3 NANOCOMPOSITE IN MULTIPLE SCLEROSIS THERAPY

In MS, the immune-mediated inflammatory condition that aggressively attacks the myelin and axons to significant degrees destroys myelinated axons in the central nervous system (CNS). The majority of medications that are administered systemically are unable to enter the CNS due to the physiological barriers in the CNS, notably the BBB and blood–cerebrospinal fluid barrier (BCSFB), which hinders treatment (Lu et al., 2014). Drug transport into the CNS has been widely studied using nanoparticulate carriers. Nanotechnology is often referred to as having smaller than 100 nm in size and uses nanoscale components. The CNS penetration capability of nanocomposites makes them advantageous for the first diagnosis of MS. These composites can penetrate the CNS because of their tiny size and unique surface properties. Chitosan, gelatin, and albumin are examples of natural polymers that are included in polymeric nanoparticles (PNP) composites, as are polyacrylates, polyacrylamide, and e-polycaprolactone. Many immune pathological disorders have been treated with a variety of PNP, including chitosan, polybutyl cyanoacrylate (PBCA), PCL, PLA, and chitosan. PNP, PCL, PLGA, and PLA have all received FDA approval for usage in humans (Heckman et al., 2013). By modifying the surface with antibodies, transferrin, or surfactants, these nanocomposites can be effective drug transporters into the CNS in terms of the BBB. Anti-transferrin receptor antibodies can efficiently transport drugs, such as methotrexate, nerve growth factor, and BDNF factor into the CNS when paired with PNP. For this reason, PNP composites are fascinating tools that might be applied to a range of therapeutic therapies for MS (Gao et al., 2016).

7.2.4 APPLICATION OF NANOCOMPOSITE IN GENE DELIVERY

Nanocomposites are made to preserve and enclose genetic material, such as DNA or RNA, enabling more effective and focused delivery to cells. The genetic cargo is shielded from deterioration and ensures that it reaches its destination undamaged thanks to the protective shell provided by the nanoparticles. Nanocomposites can be designed to possess certain qualities, such as surface alterations for focusing on particular cells or tissues (Keles et al., 2016). In gene therapy, where the objective is frequently to replace or fix defective genes, this level of accuracy is essential. The body's inherent defense systems may destroy genetic material before it reaches its target, which is one of several difficulties involved with gene delivery that the nanocomposite technique helps to solve (Kumar et al., 2021). The variety of nanocomposite movement that makes it beautiful. To meet particular objectives, various material combinations can be designed, whether the goal is to ensure effective delivery, reduce side effects, or improve the therapeutic efficacy of the genes. Overall, it is a fascinating field of study with positive ramifications for medical advancement. Zhao et al. looked at the possibility of using polymeric nano gene delivery systems to treat endometriosis. Their goal was to increase the effectiveness of nanoparticulate materials. The malignant cells were not affected by the placebo nanogel, but the doxorubicin (DOX)-loaded nanogels were extremely hazardous to the tumor cells (Das et al., 2016). Table 7.1 provides information on numerous formulations used and produced by multiple sources in research for targeted gene delivery to cure certain diseases, as well as information on polymeric nanocomposites in various disease states.

7.3 ENHANCED DRUG DELIVERY SYSTEMS USING NANOCOMPOSITES

A typical drug application is characterized by inadequate biodistribution, lack of selectivity, and low efficacy. Distribution of medications under rigorous control can assist in getting around these limitations and drawbacks. The medicine is delivered to the site of action using regulated DDSs, which lessens the impact of the drug on sensitive tissues and unfavorable side effects. DDSs also increase the drug's concentration in the target tissues, lowering the need for larger dosages of therapy, and shielding the medication against rapid breakdown or elimination. Nano+ composites, defined as structures with at least one dimension smaller than 100 nm, have shown great promise as drug transporters in the recent past. The nanostructures are an advantageous material for biomedical applications because of their tiny sizes and special physicochemical and biological characteristics (such as an improved reactive area and the capacity to pass cell and tissue barriers) (Nevozhay et al. 2007). To increase the absorption of poorly soluble medications, the targeting of medications to a specific place, and medication bioavailability, targeted drug administration at the site of illness can make use of nanocomposites. Successful formulations of

TABLE 7.1 Polymeric Nanocomposites in Drug and Gene Delivery.

S. No.	Type of Nanocomposites	Disease	Origin	Positive Response	References
1.	Gum ghatti bio-templated gold nanoparticles	Epilepsy, seizure, and Parkinson's	School of Pharmaceutical Education & Research (Faculty of Pharmacy), Department of Pharmaceutics, Jamia Hamdard (Hamdard University), Hamdard Nagar, New Delhi, India	Electric current-mediated neuro-signaling abnormalities were lessened by gold nanoparticles	Alam et al., (2017)
2.	(DOX)-loaded poly(lactic acid) and dextran nanogels	Tumor	Polymer Chemistry Laboratory, Department of Applied Chemistry, Indian School of Mines, Dhanbad-826004, India	Toxic impacts on tumor cells	Das et al. (2016)
3.	The PEG-SHGel NPs	Solid tumors	Department of Pharmaceutical Sciences, School of Pharmacy, Northeastern University, Boston, Massachusetts, USA	The NPs demonstrated encapsulated plasmid DNA release that was GSH concentration-dependent	Kommareddy and Amiji (2007)
4.	Poly(l-γ-glutamyl-glutamine)- paclitaxel (PGG-PTX)	Human lung cancer H460 cell line	Nitto Denko Technical Corporation, Oceanside, CA, USA	The sustained dosage of PGG-PTX is more effective than the sustained dose of PGA-PTX	Van et al. (2010)
5.	Cuboidal-shaped, poloxamer integrated, monoolein-based rosuvastatin lip cubes	Hyperlipidemia and obesity	Department of Pharmaceutics, Faculty of Pharmacy, School of Pharmaceutical Education and Research, Jamia Hamdard, New Delhi, 110062, India	Modulation of drug release improves the biopharmaceutical performance of statins	Javed et al. (2018)
6.	Polyethyleneimine-grafted chitosan oligosaccharide (CSO-PEI) with hyaluronic acid (HA) and small interfering RNA (siRNA)	Endometriosis	College of Medicine, Zhejiang University, Hangzhou, People's Republic of China	Decreased size of the endometriotic lesion. The ectopic endometrium is atrophying and deteriorating	Zhao et al. (2016)

anticancer medications utilizing nanomaterials include paclitaxel, DOX, 5-fluorouracil, and dexamethasone. Dexamethasone, a glucocorticoid with an intracellular site of action, has been encapsulated in polylactic/glycolic acid (PLGA) and polylactic acid (PLA)-based nanoparticles. A chemotherapy drug called dexamethasone has anti-proliferative and anti-inflammatory properties (Panyam and Labhasetwar, 2004). When a medication attaches to its cytoplasmic receptors, the resulting drug–receptor complex is transported to the nucleus, where it causes the production of several genes that regulate cell growth. These drug-loaded nanoparticle formulations fully stopped the growth of vascular smooth muscle cells by dispersing greater dosages of medication over a longer period (Kipp, 2004). The use of particle drug delivery technologies, including liposomes, micelles, and nanoparticles, in cancer therapy has been thoroughly investigated because of their tiny size, reduced toxicity, time-controlled release, modified pharmacokinetics, and the biological distribution of the drug. Because certain tumor cells gain resistance to several anticancer medications, chemotherapy frequently fails to completely eradicate cancer. Resistance often arises when cancer cells start to produce p-glycoprotein, a protein that can pump anticancer medications out of a cell as soon as they pass through the cell's outer membrane. According to recent studies, anticancer medications may be delivered to cells via nanoparticles without invoking the p-glycoprotein pump (Koziara et al., 2004). Human colon cancer cell line (HCT-15), paclitaxel was demonstrated to overcome drug resistance when it was encapsulated in emulsifying wax nanoparticles. Paclitaxel can be conjugated with albumin to solve the insolubility issues that it has. Abraxane, an injectable nano-suspension of paclitaxel coupled to biocompatible proteins, such as albumin, is authorized for the treatment of breast cancer. Acute hypersensitivity responses are brought on by the paclitaxel formulation solvent Cremophor-EL. Paclitaxel must be administered slowly over a few hours while patients are under pre-medication with steroids and anti-histamines to lower the chance of adverse responses (Koziara et al., 2006).

7.3.1 NANOCOMPOSITE-MEDIATED DELIVERY OF siRNA

siRNA is developing into a reliable technique for regulating gene expression with a wide range of uses. Clinical investigations utilizing nucleic acid-based treatment will necessitate major delivery system advancements.

Delivery of RNAi has been shown using quantum dots (QD). For *in vitro* RNAi delivery, PLGA and PLA-based nanocomposites have also been utilized (Shinde et al., 2007). Despite some success in the delivery of siRNA utilizing different nanomaterials, it is challenging to track their distribution and measure their transfection effectiveness in the absence of an appropriate tracking agent or marker. Developing a self-tracking, efficient transfection agent for the interference of RNA is a challenging task. Human epidermal growth factor receptor-2 (HER2/neu) siRNA was recently delivered using chitosan nanoparticles synthesized that were enclosed in QD. Because fluorescent QDs were included in the chitosan nanocomposites, this new nanocarrier enabled the tracking of the siRNA (Tan et al., 2007). Using a fluorescent microscope, it is possible to see the absorption and accumulation of nanotubes by labeling nanoparticles with a fluorescent marker, for instance, Cy-5 after nasal injection of chitosan/siRNA formulations, effective *in vivo* RNA interference was also obtained in bronchiolar epithelial cells of transgenic EGFP mice. These results demonstrate the promise of this innovative RNA-mediated treatment approach for systemic and mucosal illnesses based on chitosan (Howard et al., 2006).

7.3.2 NANOCOMPOSITE DRUG DELIVERY SYSTEM FOR TREATING CANCER

The challenge of sending diagnostic and therapeutic chemicals through the blood–brain barrier and into the brain makes brain cancer one of the most challenging tumors to diagnose and treat today. One of the most challenging diseases to treat nowadays is cancer. Numerous researchers have found that nanoparticles may deliver these medications to the brain in this way (Sumner and Kopelman, 2005). It has been suggested that apolipoprotein E aids in the transportation of drugs across the blood–brain barrier. Nanoparticles made of human serum albumin were combined with loperamide and linked to apolipoprotein E. After being injected directly into the brain, loperamide produces antinociceptive actions but does not cross the blood–brain barrier. In the tail-flick test, mice given this compound intravenously had antinociceptive effects. Naturally, the ability of this DDS to identify lipoprotein receptors is essential. Probes Encapsulated by Biologically Localized Embedding (PEBBLE), created by Kopelman and colleagues, have many uses and may contain a range of special chemicals on their surface. One target molecule that is immobilized on the surface may be the one that directs the PEBBLE toward cancer. A second agent may be used to help visualize the target with magnetic resonance imaging, and a third agent connected to the PEBBLE might deliver a deadly dose of medicine or poison to nearby cancer cells. A powerful tool against cancer can be created by combining all three functionalities into a single small polymer sphere (Steiniger et al., 2004).

7.4 NANOCOMPOSITES FOR CONTROLLED RELEASE AND TARGETED DELIVERY OF THERAPEUTIC AGENTS

The selected drug molecules should be delivered to the targeted cells or tissues via the optimal DDS, which would also release the drug under regulated circumstances. The medicine can be incorporated onto a substrate that has specific chemical, biological, and mechanical properties to control the release of the treatment. Because of their biologic compatibility, low cost of manufacture and disposal, and capacity for degradation, natural biopolymers such as chitosan provide interesting options for drug delivery applications (Kaurav et al., 2018). The drug delivery effectiveness of the nanocomposites has enhanced due to the use of such biopolymers in their synthesis, and such systems have been successfully designed to deliver pharmaceuticals in a regulated manner. A few examples are given in the following sections.

7.4.1 PROPRANOLOL CONTROL RELEASE

By using electrostatic interactions, chitosan and clays can form nanocomposites that have a promising future as cutting-edge drug delivery methods. By using spray drying, these kinds of nanocomposites can be created as microparticles. Colloidal montmorillonite and saponite clays are combined to form magnesium aluminum silicate (MAS). MAS has a wide range of uses as a suspending and stabilizing agent in pharmaceutical excipients. At dosages typically used in the pharmaceutical industry, it is regarded as nontoxic and non-irritating (Pongjanyakul et al., 2009). Numerous modifications are made to the end product as a result of the addition of MAS, including decreased microparticle swelling and a slower drug release. Even though chitosan is soluble at acidic pH levels and the microparticles had a smaller mm size range, it was demonstrated that propranolol outflow could persist for a considerable amount of time at low and neutral pH levels. It emphasizes the significance of interactions and gel formation with "drug-chitosan" and "drug-MAS" systems. The release rate also decreased as a result of the addition of trace amounts of TPP, the cross-linking agent. According to these results, chitosan: MAS microparticles have an intriguing potential for regulated drug delivery (for instance, when taken orally in capsule form) (Luo et al., 2017).

7.4.2 CONTROLLED RELEASE OF IBUPROFEN

In addition to having a significant specific surface area, graphene-based nanocarriers also stop medications from spontaneously discharging outside the target cells. To create a brand-new BTC/GPO nanocomposite drug nanocarrier. The *in situ* biosynthesis method was used to load the ibuprofen into the nanocomposites. Compared with pure BTC, the BTC/GPO nanocomposites had a greater drug storage capacity. It has been demonstrated that the pH of the release media affects drug release from IBP@BTC and IBP@ BTC/GPO. Furthermore, the produced IBP@BTC/GPO had a more sustained release behavior in comparison to IBP@BTC, indicating that GPO played a significant role in promoting the controlled release. IBP@ BTC/GPO used a non-Fickian diffusion technique to regulate drug release. The BTC/GPO nanocarrier might be an extremely effective new medicine delivery method (Manchanda and Sahoo, 2018).

7.4.3 ACECLOFENAC CONTROLLED RELEASE

Being a natural polymer, chitosan possesses a wide range of characteristics, including mucoadhesive, and it is frequently utilized in particle-based DDSs (such as nanoparticles) to release medications gradually or under control (Jana and Sen, 2017). The development of nanocomposite systems has also made use of CS's capacity to retain and release drugs over time. The size of the nanocomposites and the effectiveness of drug entrapment were both shown to be altered by variations in the polymer composition. The drug release rate from the nanocomposites may be managed by varying either the CS or the LBG. However, because the stomach's acidic pH made simply CS particles inadequate to endure the burst effect to control the drug release profiles in the gastrointestinal tract's varied pH environments and restrict the burst release of the medication in the stomach, another polymer component has to be added to the composites. By applying regulated drug delivery, this innovative nanocomposite method may be able to lessen the gastrointestinal side effects of drugs by reducing burst release (Kisfalvi et al., 2009).

7.4.4 CONTROLLED RELEASE OF METFORMIN

Metformin (MET), either by itself or along with sulfonylureas of the second generation, including gliclazide, glimperide, glipizide, and glibenclamide, is the most commonly used oral diabetic medication for type 2 diabetes mellitus (T2DM) (Li et al., 2009). The FDA-approved and well-known antidiabetic medication MET inhibits the growth of pancreatic cancer (Graham et al., 2011). Additionally, MET's effectiveness as a pancreatic cancer treatment is highlighted by the reduced growth of pancreatic cancer xenografts in nude mouse models when given to the animals. The main issues with MET are its short half-life and limited bioavailability(Li et al., 2009). Therefore, chitosan-based nanocomposites for controlled MET antidiabetic medication delivery were created to solve these issues. These substances met the criteria for a regulated release and demonstrated long-term stability in an alkaline media (Song et al., 2017).

7.5 NANOCOMPOSITES IN TARGETED DRUG DELIVERY

A technique for giving treatment to a patient that upregulates the percentage of the drug moiety in a specific body area or organ relative to other regions of the body is targeted drug administration, also known as smart drug delivery. Nanosystems are frequently used in targeted drug delivery to get around the drawbacks of traditional drug delivery. In contrast to conventional DDSs, which spread the drug throughout the body's organs, localizing, extending, targeting, and having regulated therapeutic contact with the sick tissue is the aim of a targeted DDS. Nanocomposite systems have been efficiently employed for the targeted distribution of medications; a few examples are shown here.

7.5.1 DOXORUBICIN-TARGETED DRUG DELIVERY

For the targeted administration of the anticancer medication DOX, Rasoulzadeh et al. used nanocomposites of graphene oxide and carboxymethylcellulose. The development of biodegradable carboxymethyl cellulose/

graphene oxide (CMC/GO) nanocomposite hydrogel beads for controlled release of DOX involved physical cross-linking with $FeCl_3.6H_2O$. The interaction between the stacking of GO and DOX resulted in increased capacities for loading and controlled release of the DOX from the CMC/GO nanocomposites hydrogel. The release profile of DOX from hydrogel beads showed a significant pH dependency at pH 6.8 and 7.4. Because of the strong connections between the carboxylic groups of graphene oxide and the amine groups of DOX, drug release was inhibited when the concentration of GO nanoparticles was enhanced. Conversely, drugs were released more readily from pure hydrogels than from nanocomposite hydrogels (Li et al., 2016).

7.5.2 TARGETED DRUG DELIVERY OF DOXORUBICIN IN GLIOMA

A promising technology for treating gliomas is the precise and targeted delivery of medications. To create GO@Fe_3O_4 nanocomposites, superparamagnetic Fe_3O_4 nanoparticles were chemically precipitated onto the surface of GO. The targeted delivery system Lf@GO@Fe_3O_4 was created through EDC/NHS chemistry using lactoferrin (Lf), a serum glycoprotein that transports iron. In glioma cells and vascular endothelial cells of the BBB, Lf binds to receptors that are overexpressed on the outside. A significant loading capacity and a pH-dependent release pattern were displayed by the nano delivery structure. Comparing free DOX and DOX@GO@Fe_3O_4 to Lf@GO@Fe_3O_4@DOX and C6 glioma cells, it was shown that the latter had improved intracellular transport efficiency and higher cytotoxicity. Results suggested that Lf-conjugated GO@Fe_3O_4 nanocomposites could be helpful as a therapeutic tool for the treatment of gliomas (Yoo and Park, 2004).

7.5.3 TARGET DELIVERY OF METHOTREXATE

Drug and siRNA delivery using MgAl-layered double hydroxide (MgAl-LDH) nanoparticles has significant promise. SiO_2 dot-coated layered double hydroxide (SiO_2@MgAl-LDH) nanocomposites were created using a nanodot-coating process and had good dispersibility and adjustable size for drug administration. The human osteosarcoma cell (U2OS) was treated with methotrexate (MTX), a chemotherapeutic medication, using SiO_2@MgAl-LDH nanocomposites as a nanocarrier. The positively charged LDH nanoparticles were evenly dispersed on the negatively charged SiO_2

nanodots in the ideal SiO_2@MgAl-LDH composition. The cellular uptake experiments reveal that SiO_2@MgAl-LDH nanocomposites internalized more readily and effectively inhibited the development of U2OS cancer cells. Monodispersed SiO_2@MgAl-LDH nanotechnology is a promising nonviral medication and gene delivery technique (Zhu et al., 2009).

7.6 NANOCOMPOSITES FOR TISSUE ENGINEERING AND REGENERATIVE MEDICINE

An interdisciplinary branch of study called Tissue Engineering and Regenerative Medicine (TERM) uses the idea of chemistry, biology, and engineering sciences to promote the growth, expansion, and reconstruction of impaired parts of the body. Tissue engineering is a technique that uses a combination of biomaterials, biomolecules, and cells to replace or regenerate the biological functions of tissues or organs. Scaffold fabrication techniques and scaffold biomaterials are key components in tissue engineering. As a result, new biomaterials with various compositions have been gradually investigated and developed to aid with and fulfill the requirements for tissue engineering applications.

Tissue engineering mostly focuses on bone to be repaired with the help of certain polymers. While designing these techniques, it should be considered that after repairing the tissue should not get infected, heal properly, have less pain, and should be affordable. Therefore, such biopolymers need to be selected which mimic the cellular events of the genes and ultimately repair new tissue to the injured part. The base of the nanocomposite matrix is to combine the biodegradable nanoparticles with certain fillers offering a compact property to the injured tissue and producing proliferation of tissue over there (Pina et al., 2015).

7.6.1 CHARACTERISTICS AND CRITERIA FOR OPTIMAL NANOCOMPOSITE SCAFFOLDS IN TISSUE ENGINEERING

In a comprehensive context, the perfect nanocomposite scaffold should satisfy the following requirements:

1. **Biocompatibility:** It produces a milieu that mimics the extracellular matrix (ECM), which facilitates cell proliferation, migration, and differentiation, among other physiological processes. Furthermore, its degradation should be nontoxic and capable of efficient absorption or removal from the body.

2. **Degradability:** To maintain a balanced relationship between the growth of new tissues or organs and the disintegration (resorption) of the nanocomposite framework, it should have a modifiable rate of degradation.
3. **Mechanical Properties:** It should possess ample mechanical strength to accommodate the surgical procedures and align with the specific tissue's tolerance levels. Achieving an equilibrium between porosity, stiffness, and hardness is crucial. Stiffness plays a role in guiding cell differentiation to a certain degree while preventing inadequate vascularization due to excessive porosity.
4. **Surface Characteristics:** It should offer customizable surface topography, hydrophilicity, surface energy, and related attributes. These factors enable the regulation of cell adhesion, migration, intracellular signaling, and the recruitment of cells *in vivo*.
5. **Nutrient Transport:** Scaffolds should feature a substantial network of interconnected pores to enhance the efficient conveyance of nutrients and regulatory substances.
6. **Pore Structure:** The arrangement of pores within the scaffold should align with cell morphology. Pore size should be within the crucial range, going beyond the lower bound to promote cell migration and under the upper bound to guarantee sufficient binding sites and surface area (Liu et al., 2022).

7.6.2 VARIETIES OF BIOMATERIALS EMPLOYED IN SCAFFOLD CONSTRUCTION

Over the past few decades, there has been a growing focus on the utilization of bioceramics, including hydroxyapatite (HA), zirconia, alumina, tricalcium phosphates, and bioactive glasses in the field of biomedical applications. The ability of these materials to replace and restore hard tissues, such as bone, teeth, hip joints, and bones has drawn special interest. Notably, bioceramics have demonstrated superior tissue responses compared with metals and polymers. This is primarily attributed to their excellent biocompatibility with cells, and in some cases, certain types of bioceramics such as HA and bioactive glass can directly bond with bone (Farag et al., 2023).

In general, bioceramics are categorized into two major groups based on their capacity to interact with bone tissue: bioinert ceramics (e.g., alumina and zirconia) and bioactive ceramics (e.g., HA, bioactive glasses, and bioactive glass–ceramics).

a. **Bioactive glass and bioactive glass–ceramics**

In the past few decades, bioactive glasses have become very attractive bioceramic materials for treating soft tissue injuries and bone abnormalities, indicating their place in the second and third generations of biomaterials. One of their remarkable qualities is their ability to change into HA in the body, where they make a connection with both soft tissues and bone. Hench et al. developed the first bioactive glass in the late 1960s and early 1970s. It was made by standard melting methods and was thereafter referred to as 45S5 or Bioglass.

The composition of this glass was rooted in a weight percentage (wt%) system of 45.0% SiO_2, 24.5% CaO, 24.5% Na_2O, and 6.0% P_2O_5. Remarkably, this glass demonstrated the ability to establish chemical bonds with host tissue by promoting the formation of a novel HA layer.

Since the discovery of Hench glass, numerous bioactive glasses have been developed, diverging from silicate-based compositions to include borate and phosphate formulations. Notably, borate glass has garnered significant attention as a bioactive material, owing to its comparatively reduced chemical durability and swifter transition into a HA phase in comparison to the established 45S5 glass (Kamalian et al., 2012).

Phosphate glasses, primarily founded on the Na_2O–CaO–P_2O_5 system, have found practical utility in biomedical applications. Their chemical composition closely resembles the inorganic components of bone tissue, rendering phosphate glasses effective as resorbable materials for bone treatments.

b. **Calcium phosphates**

Calcium phosphates (CaPs) represent a prominent group of bioceramic materials, finding extensive applications in the fields of dentistry, drug delivery, and bone regeneration. Their notable feature lies in their ability to seamlessly integrate with the surrounding biological tissues upon implantation, making them highly compatible with the human body. Furthermore, HA, a specific type of calcium phosphate, serves as the primary component of the inorganic matrix found in natural bone, constituting a substantial 60% of bone's composition.

Wei et al. designed scaffolds for bone tissue regeneration, incorporating a combination of HA and magnesium ammonium phosphate hexahydrate ($NH_4MgPO_4{\cdot}6H_2O$) at varying ratios, along with a hierarchical pore structure. These scaffolds exhibited porosity levels within the range of 52–78% and pore sizes ranging from 400 to 500 μm. Notably, the scaffolds demonstrated excellent cell attachment and proliferation, particularly in the case of the magnesium–calcium phosphate scaffolds, and they effectively promoted the generation of new bone tissue *in vivo* experiments (Saqaei et al., 2016).

c. **Magnesium phosphates**

In recent decades, there has been a growing fascination with magnesium phosphate ceramics (MPCs) in the realms of dentistry and bone regeneration. This heightened interest stems from the fact that MPCs offer solutions to certain constraints associated with calcium phosphate ceramics, notably in terms of biodegradability and mechanical strength. Additionally, derived MPC has shown promise for applications in orthopedics. These cements primarily rely on the formation of struvite crystals, specifically magnesium ammonium phosphate hexahydrate (MgNH4PO4·6H2O).

Considering these advantages, MPCs and their derived cements have emerged as promising candidates for facilitating the regeneration of bone tissue. Within our research group, we employed a paste extruding deposition rapid prototyping technique to create Mg-phosphate scaffolds loaded with the lysozyme drug, intended to provide both antibacterial capabilities and the promotion of hard tissue repair. The resultant scaffolds include interconnected porosity architectures, have ideal mechanical strength, are biodegradable, and have encouraged cell survival.

Furthermore, in an ongoing research effort, expanded this work by fabricating a hybrid scaffold composed of Mg-phosphate and gelatin polymer in various ratios, employing an additive manufacturing technique. Notably, the study's findings revealed that the inclusion of gelatin significantly enhanced the scaffold's mechanical strength (with a compressive strength of 16.7 ± 1.9 MPa), improved cell viability, and influenced drug release profiles.

7.6.3 CREATING BIOMIMETIC NANOCOMPOSITES UTILIZING BIOPOLYMERS AND CALCIUM PHOSPHATES

Utilizing biomimetic approaches in tissue engineering and regenerative medicine involves the design of nanocomposites with controlled geometries that mimic the natural ECM found in living organisms. These nanocomposites play a crucial role in guiding interactions between cells and the matrix, as well as facilitating cell-to-cell interactions. Nanocomposites are composed of a polymer matrix combined with nano-sized CaPs, providing the flexibility to customize the degradation and resorption rates of the matrix. Additionally, nano-sized CaPs enhance the tissue bonding properties of the polymeric materials, as well as promote cell adhesion and differentiation. Among the biomimetic nanocomposites developed for various tissue engineering applications, this discussion will focus on three-dimensional (3D) porous scaffolds, hydrogels, and nanofibrous scaffolds, which offer controlled geometries and structures and play a vital role in tissue engineering scaffolds (Pina et al., 2015).

7.6.4 SCAFFOLDS FOR CARTILAGE TISSUE ENGINEERING CRAFTED FROM 3D POROUS NANOCOMPOSITES UTILIZING CELLULOSE NANOFIBERS

Scaffolds are used in cartilage tissue engineering to replicate the characteristics of injured cartilage and generate a 3D environment that encourages the proliferation of cells and the creation of ECM. Typically, achieving a scaffold porosity with pore sizes ranging from 200 to 300 μm is considered a key criterion for effective cartilage regeneration. These porous scaffolds can be fabricated using various techniques like freeze-drying, CO_2 foaming, electrospinning, or cryogelation, employing a diverse range of polymers.

Scaffolds based on chitosan and gelatin have demonstrated advantages for supporting soft load-bearing tissues. Chitosan's positively charged surface properties have been found to promote the growth of chondrocytes, making it beneficial for cartilage repair. Gelatin, derived in part from collagen, the primary protein constituent of connective tissues, such as cartilage, skin, and bone, further enhances these properties. Additionally, a blend of these polymers can absorb water and form a hydrogel, enabling the retention of fluid within the scaffold structure, thereby mimicking the higher compression modulus observed in natural soft tissue (Naseri et al., 2016).

Nanocomposite scaffolds for cartilage regeneration were fabricated using a freeze-drying technique. This method involved the formation of porous structures from cellulose nanofibers, which were then mechanically and dimensionally stabilized using a low concentration of cross-linked chitosan/gelatin blend. It is anticipated that the pore structure developed during this process will influence both the moisture absorption and mechanical properties of the resulting scaffolds. In addition, the size and configuration of the pores allow fluid to circulate freely inside the scaffold, producing resistance forces and encouraging the growth of ECM, which are recognized as significant contributors to the load-bearing capacity under *in vivo* conditions.

The scaffold's pore structure and porosity were customized through adjustments in suspension concentrations, the ratio of fibers to the matrix, and cross-linking techniques, all to achieve the best possible mechanical properties and promote chondrocyte attachment and proliferation to produce an ECM of the highest quality. These 3D porous structures can guide cell development and act as templates for the creation of new tissue, all while facilitating the passage of nutrients and oxygen (Huang et al., 2023).

7.6.5 NANOCOMPOSITE HYDROGELS

Many strategies have been developed to integrate nanoscale techniques with traditional manufacturing processes to enhance material properties. Nanocomposite hydrogels serve as a prime illustration of the fusion between nanotechnology and biomaterial science. Hydrogels consist of interconnected polymer chains, which can be of natural or synthetic origin, linked together by cross-linkers. This results in a hydrophilic material with a gel-like macromolecular structure. These hydrogels are capable of swelling significantly, often reaching several times their initial dry weight, and can contain as much as 99% water or biological fluids. The 3D extensively hydrated porous framework can replicate the natural tissue microenvironment and is typically engineered to retain, release, or capture substances (Rafieian et al., 2019).

7.7 APPLICATIONS OF BIOCERAMICS IN SOFT TISSUE REPAIR

7.7.1 REPAIRING SOFT TISSUE INJURIES WITH BIOACTIVE GLASSES

45S5 Bioglass represents the pioneering paradigm of third-generation biomaterials, characterized by its exceptional biocompatibility, capacity to enhance tissue adhesion and growth, and the ability to activate precise cellular pathways at the genetic level. Bioglasses (BGs) are typically fabricated through processes involving either melt-quenching methods or sol–gel techniques. BGs, especially those in the phosphate system category, are designed to be resorbable. This means that their dissolution occurs upon exposure to bodily fluids within a biological environment. The release of metal ions from the structure of BGs serves specific therapeutic purposes, including the promotion of cell proliferation, the stimulation of angiogenesis, and the provision of anti-inflammatory and antibacterial effects. The structure of BGs varies, and some can exhibit mesoporous characteristics, featuring pores with diameters ranging from 2 to 50 nm. Mesoporous bioglasses (MBGs) are well-suited for effectively delivering various drugs and growth factors, which can be housed within these mesopores. Another category of BGs, known as MBG microspheres (MBGMs), possesses the unique ability to enhance the surface area, facilitating the rapid deposition of HA when exposed to bodily fluids. This distinctive feature has played a crucial role in their effectiveness within the field of tissue engineering. Various commercial formulations of BGs have been in clinical use since 1993, serving purposes such as filling periodontal bone defects, addressing orthopedic and dental needs, contributing to middle-ear surgical procedures, and even being incorporated into early prototypes of cochlear implants (Mazzoni et al., 2021).

of delicate and careful sensors for monitoring pollutants in air, water, soil, and bio-based nanocomposites, offering sustainable solutions with minimal environmental impact. This chapter emphasizes the significant contribution of nanocomposites to environmental protection by leveraging the unique properties of nanomaterials.

8.1 INTRODUCTION

The different backdrops and versatile applications of nanocomposites make them bright candidates for addressing a large range of environmental challenges, from pollution control and remediation to sustainable energy and resource management. Repeated research and innovation in nanomaterials science influence the possibility of advancing the development and deployment of nanocomposites for environmental sustainability. Nanocomposites are goods composed of a source and nanoparticles; typically, with at least one magnitude less than 100 nm (Camargo et al., 2009). In nanocomposites, the constituent materials atoms are arranged into tiny grains or clusters. Nanocomposite materials offer distinct advantages over microcomposites and monolithic materials, particularly in charge of enhanced automatic, electrical, and thermal properties. When dimensions become closer to the nanoscale, phase contact interactions greatly improve, and this is essential for improving the properties of materials (Hafeez and Hafeez, 2022). Nanoparticles exhibit great physical and chemical changes that can be tailored to fit special performance needs in various applications. Two of the primary contributing factors are an increase in the surface area-to-volume ratio and particle size. As the size of particles decreases, the surface area available for interaction with the surrounding matrix material increases exponentially. An increase in the surface area of the particle allows for the enhanced bonding between the nanoparticles and these matrixes are popular for the improvements of automatic, thermal, and electrical backdrops. Nanocomposite materials are capable of adjusting and tailoring their properties by controlling various domains, such as composition, morphology, amount, and disposal of nanoparticles within the matrix. This tunability is a significant advantage of nanocomposites, allowing for the optimization of material characteristics to meet specific application requirements. Typically, the most difficult aspects of creating nanocomposites are interfacial interactions and the uniform dispersion of particles inside the polymer matrix (Hári and Pukánszky, 2011). Nanocomposites have a large range of functions across different management like automotive corporations, the aerospace industry, electronics, packaging, construction, medical devices, energy storage and harvesting, environmental remediation, biomedical medicines and drugs, food, and agriculture. Cleanup and wastewater treatment fields have shown a great deal of interest in nanocomposites. Rapid urbanization has resulted in a significant increase in wastewater discharge, which has released a variety of contaminants into the environment and negatively impacted both the environment and human health (Hafeez and Hafeez, 2022). Exposure to indirect degradation poses a considerable risk to global health, impacting both developed and developing nations. Key areas of concern include bacterial identification and elimination, prevention of organic compound deterioration, and the detection and reduction of heavy metal ions. These efforts are crucial, as organic pollutants continue to be among the most pressing environmental threats. Tobacco smoke is a major source of air contaminants consisting of a complex mixture of harmful chemicals that pose significant health risks. Exposure to biological pollutants and unstable organic compounds leads to many health issues, such as hay fever, and respiratory conditions including asthma, as well as irritation, headaches, and neurological effects. Soil contamination from various anthropogenic sources introduces a wide range of harmful chemicals into the environment, posing significant health risks to humans and other living organisms. Water is an essential common source required for human development and applications. However, water pollution is an important global problem that poses severe hazards to human health and the environment. This pollution arises due to various artificial and organic human activities. A major cause of heavy metal pollutants in water is due to the utilization of fertilizers in agriculture. When disinfecting water, nanocomposites are utilized to either deactivate or eradicate the pathogenic germs. Nanocomposites particularly those based on materials like activated metal–organic frameworks (MOFs), carbon, or graphene have a large surface area which provides ample sites to adsorb the dye molecules from the wastewater.

Nanocomposites have the following advantages over other materials (Hafeez and Hafeez, 2022):

1. A low fill size is permitted because the ratio between surface area and volume is higher.

2. Better mechanical properties with low power loss, great ductility, and minimal initial resistance.
3. Possessing enhanced visual characteristics (light transmission is anchored by particle size).

8.1.1 CLASSIFICATION OF NANOCOMPOSITES

Nanocomposites can be divided into three classes based on the materials are as follows:

1. MMNCs or metal matrix nanocomposites
2. CMNCs or ceramic matrix nanocomposites
3. PMNCs or polymer matrix nanocomposites

8.1.1.1 METAL MATRIX NANOCOMPOSITES

Metal matrix nanocomposites (MMNCs) are one of the biggest classes of composition that are often reinforced with clay. MMNCs are advanced materials that combine a plastic or alloy source with nanosized coating particles, typically ceramics. MMNCs are engineered to synergistically combine the advantageous properties of both metals and ceramics like lower coefficients of thermal expansion and high wear resistance flexibility and fracture resolve with large specific durability and modulus. Powder metallurgy is a unique and effective method for producing metal matrix composites (MMCs), including MMNCs (Hafeez and Hafeez, 2022).

8.1.1.2 CERAMIC MATRIX NANOCOMPOSITES

Ceramic matrix nanocomposites (CMNCs) are a class of advanced goods that incorporate nanosized reinforcements into a ceramic matrix. These composites leverage the essential backdrops of ceramics, such as great hardness and thermal cohesion, and enhance them with nanoscale additives. The matrix consists of nitrides, oxides, and carbides, combined with metallic or non-metallic properties. Layers of transformation metal oxides exhibit superior backdrops compared with other metallic or organic oxides, even at nanometer scale thicknesses. This superiority arises from their exceptional phase strength and cracks resolve.

Ceramic nanocomposites have found extensive uses in different industrial sectors based on their exceptional properties, particularly their resistance to corrosion, high-temperature oxidation, and superior wear resistance compared with metals, especially in high-temperature environments (Seewagen, 2020). To improve crack resistance, heat absorption, and chemical resistance, single or multiple layers of ceramics are combined with CMNCs.

8.1.1.3 POLYMER MATRIX NANOCOMPOSITES

Polymers are widely used as matrix material in composite materials, and polymeric composite materials indeed represent a significant portion, ~ 90% of all composite materials used across various industries. Polymer matrix systems in composites can be roughly classified into three types: thermoplastics, thermosets, and elastomers. Polymers are indeed ideal materials for many applications due to their combination of advantageous properties. The particle size and distribution are of great importance in the case of nanoparticles. The inclusion of nanoparticles in polymer matrices allows for the modification and enhancement of various properties, including mechanical, electrical, optical, and thermal properties (Agboola et al., 2021).

8.2 NANOCOMPOSITES CAN ACT AS ADSORBENTS FOR CONTAMINANTS

Numerous carbonaceous substances have been employed as adsorbents in a variety of adsorption processes. Due to its high efficiency and low cost, adsorption is a useful and practical process. Due to their mechanical strength, hydraulic performance, high stability, and surface chemistry, polymeric nanocomposite adsorbents have been used as possible materials to remove various contaminants from wastewater (Pan et al., 2019). The adsorbent material's physicochemical and pore structures determine the pollutant's ability to adsorb. Nano adsorbents offer improved quick process kinetics and accessible sorption sites due to their greater surface area (Agboola et al., 2021). Adsorbents such as polymeric nanocomposites are used to separate gases, remove dyes, and for removing metal ions.

8.2.1 POLYMER NANOCOMPOSITES AS A DYE REMOVAL ADSORBENT

For 30 years, researchers have examined the mobility and distribution of dyes in water because of their possible negative effects on people, plants, animals, and aquatic life. The quality of the water is reduced when dyes leak into the water supply. According to Vijayakumar et al., certain sectors use a lot of synthetic dyes to give their products color and a lot of water, which results in a lot of colorful effluents. High oxygen demand and oxidizing agents are used to classify wastewaters based on their uneven pH hues (Vijayakumar et al., 2011). Numerous wastewater effluents present obstacles to demineralization and decolorization (Saad and Tahir, 2017). Aquatic life finds it challenging to receive adequate sunlight due to the color of the dyes (Kahlon et al., 2018). The most vital methods for removing the harmful misuse from wastewater are adsorption, precipitation, reverse osmosis, and ion exchange (Achmad et al., 2012). Because of the loss of the polymeric nanocomposites which are used as adsorbents, they are used for the elimination of dyes in the misused water. Polymeric nanocomposites have good granulometric characteristics, a large surface area, increased activated sites, and thermally and chemically they can be used as adsorbents (Chang and Juang, 2004; Tanasa et al., 2019). Nanocomposites can be classified into natural, synthetic, and organic modifications. Physical method which includes the deionized water treatment method and thermal treatment method.

8.2.2 EXAMPLES OF CONTAMINANTS THAT CAN BE REMOVED

A significant part is played by increasing the concentration of active carbon in wastewater to improve the adsorption of heavy metals. Activated carbon can be utilized as an adsorbent to absorb heavy metals found in synthetic water. The membrane with 0.5 and 0.3 wt % active carbon has a lower water flux than the membrane containing composite with 0.9 wt % active carbon, which has a higher water flux. A composite membrane containing 0.9 wt % active carbon reduced heavy metal concentrations by 2% for lead, 16% for silver, 19% for chromium, and 35% for cadmium (Khulbe and Matsuura, 2018). Polypyrrole-based adsorbents are vital as they are biocompatible, simple to synthesize, and possess redox properties that facilitate the removal of heavy metal ions from wastewater.

Heavy metals can be removed from wastewater using a variety of techniques, including exfoliated polypyrrole-organically modified clay nanocomposites, polypyrrole–graphene nanocomposites, polypyrrole–polyaniline nanofibers, ferrous ferric oxide nanocomposites, and biosorbents (Khulbe and Matsuura, 2018). Graphene and its composites are effective adsorbents when used in an aqueous solution to remove heavy metal ions and dyes. Many methods, such as adsorption, reverse osmosis, chemical precipitation, biosorption, solvent extraction, membrane filtration, and flocculation were employed to remove the lead because of the increased toxicity of lead and its importance as a heavy metal ion contaminant in geographical systems (Huang et al., 2011). Many techniques have been developed for the removal of dyes, such as degradation, filtering, adsorption, photocatalytic coagulation, biochemical degradation, and others (Pandey et al., 2017).

8.2.3 MECHANISM OF ACTION FOR WATER PURIFICATION

8.2.3.1 ADSORPTION

Adsorption is categorized as a traditional method of eliminating metal ions from contaminated water. Benefits for the adsorption process include high removal efficiency, high adsorption capacity, and producing a high-quality studied sewerage. The disadvantages include regeneration is expensive, loss of adsorbent, and limitations in using against disperse and vat dyes (Pandey et al., 2017). Adsorption is the process that occurs when molecules, ions, and atoms from a liquid, gas, or dissolved solid adhere to the surface and this process is used for the purification of the water. Activated carbon is present in the water and used for the water purification process. With its greater surface area per unit volume, this active carbon is put to use. The surface of the carbon is stuck by the water particles and those are the less soluble nonpolar particles in water, so the solute particles are not adsorbed through the flowing water.

The activity level of adsorption is based on these parameters:

1. Concentration
2. Temperature
3. Polarity of substance

Adsorption is directly proportional to concentration. When there is a high concentration of solids in water then there will be a high

in the valence band. Preventing the recombination of electrons and holes is indeed necessary to increase the efficiency of photocatalysis. To do this, scientists used a variety of techniques, including altering the surface of TiO_2 nanoparticles, doping with other elements, and creating heterojunctions with other semiconductors.

$$TiO_2 + hv \rightarrow (TiO_2) + h^+_{vb} + (TiO_2) \tag{8.1}$$

$$TiO_2 + (h^+_{vb}) + H_2O_{ads} \rightarrow (TiO_2) + HO + H^+ \tag{8.2}$$

$$TiO_2\,(hv^+_{vb}) + OH^-_{ads} \rightarrow TiO_2 + HO_{ads} \tag{8.3}$$

$$TiO_2\,(e^-_{cb}) + O_{2ads} + 2H^+ \rightarrow TiO_2 + HO_2 \rightarrow O^+_2H^+ \tag{8.4}$$

$$TiO_2\,(e^-_{cb}) + HO_2 + H^+ \rightarrow H_2O_2 \tag{8.5}$$

$$H_2O_2 + hv \rightarrow 2HO \tag{8.6}$$

$$H_2O_2 + O_2 \rightarrow HO + O_2HO^- \tag{8.7}$$

$$H_2O_2 + TiO_2(e^-_{cb}) \rightarrow HO + HO^- + TiO_2 \tag{8.8}$$

Effective confinement increases the likelihood that electrons and holes will engage in desired redox reactions as opposed to recombining and releasing energy as heat. For instance, the positive hole in the valence band of TiO_2 can react directly with adsorbed impurities to initiate oxidation reactions that break down contaminants. As shown in equations, the hole produces hydroxyl radicals ($\cdot OH$) in Eqs. 8.2 and 8.3 in Figure 8.1 by oxidizing water or hydroxide ions (OH). Oxygen is essential to the photocatalysis mechanism, especially when it comes to the photocatalytic destruction of contaminants. When oxygen (O_2) is reduced in the conduction band, superoxide ($\cdot O^{2-}$) is created (Konstantinou and Albanis, 2004). By stopping electron–hole recombination and encouraging the creation of reactive species that can react with and remove pollutants, the creation of oxygen radicals during photocatalysis is essential in accelerating the degradation of pollutants (Eq. 8.4).

The generation of superoxide radicals (O^{2-}) during photocatalysis typically occurs through the oxygen reduction process at the conduction band of the semiconductor material. The formation of hydrogen peroxide through the protonation of superoxide ions represents an additional pathway by which reactive oxygen species are generated during photocatalysis, further enhancing the efficiency of pollutant degradation (Eq. 8.5).

More hydroxyl radicals are produced during the photocatalytic process when hydrogen peroxide (H_2O_2) is utilized as an oxidizing agent. Greater pollutant degradation rates may be reached as a result of the enhanced hydroxyl radical yield, which would reduce treatment complexity and result in significant cost savings (Eq. 8.6). During photocatalysis, contaminants on the surface of TiO_2 particles can be degraded by highly reactive species such as superoxide ions (O^{2-}) and hydrogen radicals (·H). TiO_2 photocatalysis produces these two molecules (·OH and $\cdot O_2$) as its main by-products (Eqs. 8.7 and 8.8). The polluted molecules undergo continuous breakdown into smaller, less dangerous compounds during photocatalytic degradation processes. They eventually mineralize into nontoxic by-products, such as inorganic ions, water, and carbon dioxide. This thorough breakdown procedure guarantees that no sludge or leftover by-products are generated and that there is no indication of the original chemical (Al-Nuaim et al., 2022).

8.3 POLLUTANT DEGRADATION

Pollution in various forms, whether it is in the air, water, or soil, can persist for long periods, sometimes even millions of years susceptible to the type of the pollutant and the surrounding circumstances. Industry and automobile exhaust emissions are significant contributors to environmental pollution, particularly in urban and industrialized areas. Pollutants are substances that when introduced into the environment, can cause harm by disrupting the natural balance of ecosystems. There may be serious repercussions from this disturbance for both people and other living things. Numerous channels exist for pollutants to infiltrate the environment, originating from both natural and human activity (Kapil et al., 2023).

8.3.1 DIFFERENT TYPES OF POLLUTANTS

Pollutants can enter the environment in a variety of ways including human activity and natural activities as well. Pollutants come in many forms, such as pesticides, heavy metals, dyes, and polyaromatic hydrocarbons (Figure 8.2).

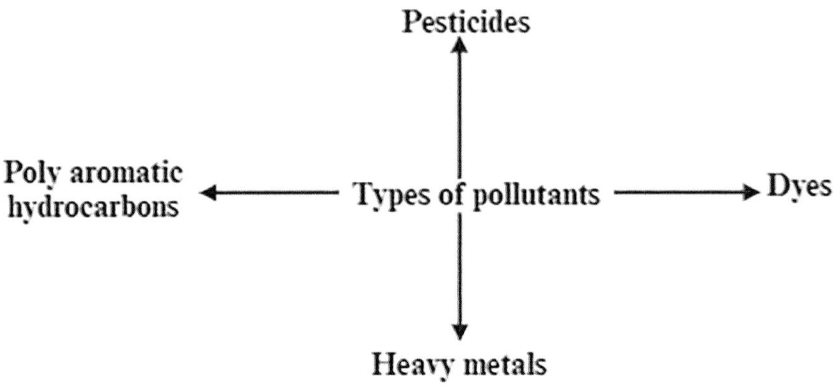

FIGURE 8.2 Different types of pollutants.

8.3.1.1 DYES

Synthetic organic chemicals called dyes are widely utilized in many industries, especially the textile industry, to give items color. On the other hand, their common use also makes them common industrial environmental pollutants. Aquatic ecosystems can be significantly impacted by synthetic dyes included in textile wastewater. One of the primary effects is the reduction of light penetration into rivers and other water bodies due to the absorption and scattering of light by the dye molecules. Industrial effluents containing dyes have the potential to dramatically raise the biochemical oxygen requirement of receiving water bodies. The amount of dissolved oxygen needed by microbes in water to decompose organic compounds is measured by the process known as biochemical oxygen demand. For instance, it has been determined that the dyes dispersed orange-1 and dispersed red-1 are mutagenic, which means they may result in genetic alterations in living things. According to classifications, dyes like Sudan I, Basic Red 9, and Crystal Violet are carcinogenic, they can cause cancer in living things. For example, Sudan I, have been outlawed in several nations because of its cancer-causing qualities (Lellis et al., 2019).

8.3.1.2 HEAVY METALS

Heavy metal pollution is a significant environmental concern stemming from various anthropogenic activities. Heavy metal pollution is a growing concern globally because of its widespread and human health, and detrimental influence on ecosystems, and economies. The increasing industrialization, urbanization, and globalization of economies have led to elevated levels of heavy metal emissions and contamination in various environmental compartments. Metallurgy and agriculture are two major industries that contribute to the discharge of harmful inorganic waste into the environment. The most common heavy metals that contaminate the environment are lead, gold, platinum, mercury, tin, cadmium, molybdenum, arsenic, cobalt, chromium, manganese, vanadium, titanium, silver, iron, copper, nickel, and zinc.

8.3.1.3 PESTICIDES

Pesticides are the chemical substances used to control, repel, or eradicate pests, which can include insects, weeds, fungi, and rodents. Pesticides serve important purposes in controlling pests and protecting crops, but can indeed pose significant risks to living organisms, food safety, and the environment (Briffa et al., 2020). Pesticides have been used in agriculture for many decades as a tool to protect livestock from pests, and crops, including weeds, insects, fungi, and pathogens. The application of pesticides helps farmers mitigate the risk of crop loss and damage caused by pests, which can significantly impact agricultural productivity and food security. Pesticide poisoning is an important social health concern globally, and the World Health Organization (WHO) has indeed highlighted its impact through various studies and reports. Pesticides contribute to a range of organic micropollutants that have negative effects on the environment. One such pesticide is dichlorodiphenyltrichloroethane (DDT).

8.3.1.4 POLYAROMATIC HYDROCARBONS

Polycyclic aromatic hydrocarbons (PAHs) are a type of persistent and extensively spread environmental contaminants that are toxic, carcinogenic, and mutagenic. As such, they pose serious health hazards. Multiple aromatic rings fused together to form polyaromatic hydrocarbons. They are indeed known for their stability and resistance to decomposition, which makes them persistent environmental pollutants. PAHs with lower molecular weights tend to be more water-soluble because they have fewer aromatic rings and thus exhibit more polar characteristics. However, as additional rings are added, their hydrophobicity increases, making them less soluble in water and more likely to adhere to organic matter or sediment in aquatic environments (Masih et al., 2010). The U.S. Environmental Protection Agency (EPA) has identified these PAHs as contaminants of concern and recommends constant monitoring to ensure their concentrations remain within acceptable limits. Naphthalene, being the simplest PAH, is often used as an indicator of the presence of other PAHs due to its early volatilization and ubiquity in the environment. Benzo[a]fluoranthene, chrysene, Phenanthrene, benzo[a]anthracene, pyrene, fluoranthene, and anthracene, are also commonly found in various environmental matrices and are associated with sources like incomplete combustion of organic materials, industrial operations, and vehicle emissions.

8.3.2 POLLUTANTS' EFFECTS ON HUMAN HEALTH

Numerous pollutants are introduced into water, air, and soil via environmental pollution, which is brought on by both human activity and natural occurrences like volcanic eruptions. These pollutants reach the human body primarily through inhalation, oral absorption, and ingestion, leading to various health issues.

8.3.2.1 RESPIRATORY HEALTH

The release of hazardous substances, such as heavy metals, nitrogen dioxide, carbon monoxide, sulfur dioxide, particulate matter, and ozone into the atmosphere is mostly caused by automobile and industry emissions. The respiratory system can be severely harmed by exposure to pollutants from car and industry emissions. This can lead to the development of upper and lower respiratory tract conditions, such as bronchitis, chronic obstructive pulmonary disease (COPD), pneumonia, and increased mucus production. By harming epithelial cells, reducing macrophage activity, and compromising mucociliary clearance, these pollutants erode the lung's natural defenses, resulting in respiratory symptoms and persistent inflammation (Shetty et al., 2023). Exposure of humans to polluted air and smoking behavior are tightly related, and both lead to respiratory problems and reduced lung capacity. Insecticide exposure has been identical to a range of respiratory symptoms, including wheezing, coughing, chest tightness, sore throat, airway inflammation, and shortness of breath. Occupational studies, such as those involving bottling plant employees, highlight the higher risk of respiratory problems in people who come in contact with pesticides at work. Cigarette smoke contains heavy elements including cadmium and lead, which seriously jeopardize respiratory health. COPD, chemical pneumonitis, and emphysema are significant respiratory disorders that are facilitated in their growth and progression by their propensity to generate inflammation, oxidative stress, and direct cytotoxicity. According to a study, people with asthma who are exposed to pesticides such as aerosols experience increased symptoms, a decline in lung function, and possibly even hyper-airway reactions. However, the exact mechanism causing these effects is yet unknown. Certain demographic groups are more susceptible to the effects of pollutants, particularly sulfates and particulates, which have been linked to increased mortality rates upon exposure. Smoking is associated with reduced lung capacity due to the harmful effects of tobacco smoke, while exposure to specific pollutants like asbestos and organic dust can lead to distinct respiratory diseases, such as asbestosis, mesothelioma, and bagassosis (Briggs, 2003).

8.3.2.2 REPRODUCTIVE HEALTH

Indirect contamination can have significant adverse effects on reproductive health. Various pollutants, including heavy metals, pesticides, endocrine-disrupting chemicals, and other toxins can interfere with reproductive processes and further cause a wide range of reproductive health problems. Exposure to environmental toxins indeed reduces the synthesis of important antioxidant enzymes like superoxide dismutase, glutathione peroxidase, and catalase, which increases oxidative stress. Male infertility is largely caused by oxidative stress, which also causes apoptosis and anomalies in germ cells. Environmental contaminants have a significant impact on reproductive function in both males and females, even at low levels of exposure (Ma et al., 2019). Acute exposure to environmental toxicants can indeed cause both apoptosis and necrosis in testicular cells, leading to significant damage to male reproductive health. Sperm exposed to environmental toxins indeed can pass genetic damage to growing embryos, resulting in early pregnancy loss and a variety of developmental issues. This transmission of genetic damage is a critical concern for reproductive health and the health of future generations. Epidemiological research supports the conclusion that exposure to particulate matter can significantly disrupt gametogenesis and reduce reproductive ability in both males and females. Exposure to particulate matter has been linked to adverse effects on gametogenesis, compromising the quality and quantity of gametes in both females and males (Carré et al., 2017). The primary mechanisms involve inflammation, oxidative stress, and endocrine disruption, leading to impaired sperm and oocyte quality, hormonal imbalances, and adverse pregnancy outcomes. Phthalate exposure is indeed associated with testosterone deficiency syndrome, which encompasses several reproductive health issues, including low sperm counts, cryptorchidism (undescended testicles), hypospadias (a disorder where the urethra opens on the underside of the penis rather than the tip), and testicular cancer (Shetty et al., 2023). The investigation suggests that toll collectors who work on highways experienced adverse effects on their reproductive health, specifically in terms of semen quality (Guven et al., 2008). Numerous consequences on reproductive and developmental health, such as aberrant sperm, infertility, spontaneous abortion, birth malformations, and developmental delay in children, have been related to specific levels of pesticide exposure (Frazier, 2007). Phthalate exposure has been implicated in various reproductive health issues, including conditions associated with testosterone deficiency syndrome (Skakkebæk et al., 2001).

8.3.2.3 CANCER

Environmental contaminants are significant contributors to cancer incidence, with the majority of cancer cases attributed to exposures in the environment. These contaminants exert their carcinogenic effects through multiple mechanisms, including deoxyribonucleic acid (DNA) damage, inflammation, oxidative stress, and hormonal disruption. Carcinogens exert their effects on multiple molecular processes within cells, disrupting normal regulatory mechanisms that control cell growth, apoptosis, DNA repair, and metabolism. These disruptions contribute to the initiation, progression, and maintenance of cancer. One major risk factor for lung cancer is occupational exposure to combustion by-products of fossil fuels (Cohen and Pope, 1995). Carcinogenesis is a multi-phase process that usually consists of three separate stages: progression, promotion, and initiation. Exposure to farming environments and occupations has been linked to a higher risk of acquiring specific cancers, such as urothelial cell carcinoma and bladder squamous cell carcinoma. Severe health hazards are associated with bisphenol A and phthalates, including the possibility of metabolic disorders, polycystic ovarian syndrome, prostate cancer, sterility, breast cancer, recurrent miscarriages, and endometrial hyperplasia. The association between bladder cancer and herbicide use among pesticide applicators, particularly with herbicides like imazethapyr and imazaquin, has been a topic of interest in epidemiological research (Koutros et al., 2016).

8.3.2.4 NEURAL HEALTH

Exposure to environmental pollutants from various sources poses significant risks to neurological health, related to the improvement and evolution of neurological disorders. Neurodevelopmental abnormalities enclose a broad spectrum of circumstances that alter the development and function of the nervous system, resulting in various cognitive, behavioral, and developmental challenges. Although neurological illnesses constitute a broad category with a variety of etiologies and clinical presentations, many of these conditions have similar pathogenic processes, including disruption of mitochondrial dysfunction, protein aggregation, blood–brain barrier, oxidative stress, and DNA damage. Pesticides, such as chlorophenoxy herbicides, ethylene bis thiocarbamates, carbamates, and organophosphates are linked to conditions, such as Alzheimer's disease, Wilson disease, amyotrophic lateral sclerosis, Parkinson's disease, and multiple sclerosis. Lead and mercury exposure can result in impaired cognitive function and memory problems (Dusek et al., 2015; Giacoppo et al., 2014). Epidemiological studies suggested a potential link between particulate matter exposure and an elevated risk of neurological conditions, including Alzheimer's disease (Haghani et al., 2020). *In vitro* research looked at the impact of biomass smoke generated from burning cow dung and has highlighted significant health concerns, particularly regarding cell inflammation (McCarthy et al., 2017). constitutes a major environmental risk factor for cardiovascular disease.

8.3.2.5 CARDIOVASCULAR HEALTH

Inhalation of particulate matter and other environmental pollutants constitutes a major environmental risk factor for cardiovascular disease. Fine and ultrafine particles, through their pro-inflammatory effects and systemic circulation, can contribute to the development of venous thromboembolism, arterial thrombosis, and plaque formation, causing acute cardiovascular issues, such as myocardial infarction, pulmonary embolism, and stroke. Indeed, there are many ways that contaminants in soil and water can harm cardiovascular health, including by causing inflammation and upsetting physiological functions. Particularly during pregnancy, pesticide exposure has been connected to many harmful cardiovascular events, such as arterial hypertension, acute myocardial infarction (heart attack), and heart failure (Shetty et al., 2023). Many diseases and deaths have been connected to pollution exposure. Cardiovascular disease is the cause of more than 60% of cases in this particular cohort (Münzel et al., 2023). These are the main reasons why people die and become disabled (Abbafati et al., 2020). Cardiovascular disorders, particularly hypertension have been related to exposure to heavy metals like lead (Rajagopalan and Landrigan, 2021).

8.4 NANOCOMPOSITES IN REMEDIATION OF WATER POLLUTION

Contaminants in water bodies give rise to a serious environmental problem called water pollution. Although the application of nanoparticles in water management shows promise in terms of eliminating impurities and enhancing water quality, it also comes with a number of real-world difficulties and hazards. Enhancing the use of nanoparticulate materials can be accomplished by developing nanocomposite materials that get benefits from the embedded nanoparticles as well as the host materials.

8.4.1 TYPES OF NANOCOMPOSITES AND THEIR USE IN WATER PURIFICATION

8.4.1.1 METAL-BASED NANOCOMPOSITE

Polymer-supported nanosilver composites are highly acclaimed for their antibacterial properties, making them valuable in various applications, including water treatment, medical devices, and textiles. More specifically, nanosilver-impregnated polyurethane and cellulose acetate have shown notable bacterial inhibition, both against Gram-negative and Gram-positive bacteria. Economically feasible microfilters for purifying drinking water have been made using nanosilver, especially in isolated or difficult-to-reach areas with limited access to clean water (Peter-Varbanets et al., 2009). Both sulfaguanidine resin (SGR) and amberlite resin (AR) beads exhibit comparable disinfection efficiency. However, there is a difference in the release of silver ions between the two types of beads, with SGR beads releasing a higher amount of silver ions compared with AR beads. The fact that the SGR beads continue to be successful in sterilizing despite this suggests that they might outlive AR beads without losing their ability to disinfect. Bimetallic nanoparticles like ruthenium—palladium, have been successfully used in wastewater treatment as reinforcements to form nanocomposites on the surface carbon matrix, particularly in cases where perchlorate is the main pollutant (Sharma et al., 2019). The synthesis of nanocomposites consisting of gold/palladium nanoparticles reinforced on TiO_2 through microemulsion methods has emerged as a promising approach for developing efficient photocatalysts.

8.4.1.2 METAL OXIDE NANOCOMPOSITE

Metal–organic nanocomposite materials are utilized with a variety of substrates, such as graphene, silica, other metal oxides, carbon nanotubes (CNTs), and polymers to eliminate various inorganic and organic impurities from both air and water. Production of alumina composites reinforced by CNTs involves incorporating CNTs into an alumina matrix, typically through a synthesis process that utilizes iron and nickel-doped energetic alumina. Additionally, Senusi et al. suggested that artificially produced titanium dioxide–zeolite nanocomposites might be highly useful for treating water, particularly for the elimination of industrial dyes (Senusi et al., 2018). The Saad et al. study concentrated on the manufacture of zinc oxide (ZnO)—chitosan nanocomposite (ZONC) to eliminate, cupric ions (Cu^{2+}), cadmium ions (Cd^{2+}), and lead ions (Pb^{2+}) from polluted water. This nanocomposite showed promising results in effectively removing these heavy metal ions (Saad et al., 2018). The study executed by Singh et al. highlights the effectiveness of boron carbide/tin oxide nanocomposites as catalysts to break down industrial dyes. The ZnO@Chitosan nanocomposite (ZONC) was created by Saad et al. using a straightforward *in situ* fabrication technique. The resulting composite showed excellent ability and quick kinetics for the cupric ions (Cu^{2+}), and cadmium ions (Cd^{2+}) to be adsorbed. According to Singh et al.'s study, boron carbide/tin oxide nanocomposites were an aid in the decomposition of commercial dyes such as Huntsman's Novacron red (NRH) (Singh et al., 2018). To remove cadmium ions (Cd^{2+}), cupric ions (Cu^{2+}), and lead ions (Pb^{2+}), from contaminated water, Saad et al. had to create a ZONC. With adsorption capacities of 135.1, 117.6, and 476.1, mg/g, respectively, the ZONC had the highest removal effectiveness at pH 6, 6.5, and 4 for Cd^{2+}, Cu^{2+}, and Pb^{2+} ions (Saad et al., 2018).

8.4.1.3 CARBON NANOCOMPOSITE

Using a magnetic multi-wall CNT nanocomposite as an adsorbent to remove cationic dyes from aqueous solutions is a creative way to remediate wastewater. An important advancement in improving the adsorption ability of organic dyes in water treatment procedures is the creation of recently generated mesopores. Mesopores with dimensions typically ranging from 2 to 50 nm, give organic compounds, such as dyes, more surface area and easily accessible locations for adsorption mesoporous nanocomposites show great promise for water purification applications because they can eliminate heavy metal species such as Cr^{6+} (chromium ions), Zn^{2+} (zinc ions), Ni^{2+} (nickel ions), Pb^{2+} (lead ions), and Cu^{2+} (copper ions) with selectivity. The ease of material preparation for these mesoporous nanocomposites allows for straightforward scale-up production, making it economically feasible while minimizing ecological impact. The incorporation of CNTs into hybrid materials, such as hybrid carbon–CNT and biochar–CNT, leads to significant improvements in their physicochemical properties. Carbon nanocomposites are composite materials made of one or more extra components mixed with carbon-based materials like CNTs. Tian et al., by their research, introduced an innovative approach to develop antibacterial agents using a synergistic nanocomposite composed of silver nanoparticles and iron oxide nanoparticles on the graphene oxide surface. This resulted in the formation of a novel nanocomposite known as graphene oxide–iron oxide nanoparticles–silver nanoparticles. The association of graphene oxide–iron oxide nanoparticles–silver nanoparticles results in an enhanced bactericidal activity of the nanocomposite compared with pure silver nanoparticles against both Gram-positive and Gram-negative bacteria alone (Tian et al., 2014).

8.4.1.4 POLYMER NANOCOMPOSITE

Polymers serve as excellent support for nanomaterials in the formation of polymer nanocomposites (PNCs) due to their unique backdrops, including adjustable porous architectures, great automatic characteristics, and chemically bound functional groups; when integrated with nanoparticles, PNCs combine the advantageous characteristics of both the nanomaterials and polymeric matrix (Zhao et al., 2012). Chitosan, being one of the most naturally abundant polysaccharides after cellulose, is particularly noteworthy for its effectiveness in pollutant removal applications (Djerahov et al., 2016). Khaydarov et al., developed and optimized nanocarbon-conjugated PNCs for the extraction of metal ions from water through a thorough investigation. This research focused on using carbon colloids as nanoparticles and polyethyleneimine as the matrix material (Khaydarov et al., 2010). Clay is an excellent matrix for the gilding of polyaniline. The announcement of a new clay nanocomposite incorporating polyaniline marks a huge step forward in the area of water treatment, especially for lead ion removal. Nithya and Sudha undertook a detailed analysis to determine if chitosan-g-poly (butyl acrylate)/bentonite nanocomposite material might be used to serve as an adsorbent to heavy metals including chromium and lead from tannery effluent. Their investigation centered on determining how well this innovative adsorbent worked to improve several important physicochemical water quality indicators, such as conductivity, total hardness, salinity, total solids, turbidity, and chemical and biological oxygen demand (BOD) (Nithya and Sudha, 2016).

8.4.1.5 MEMBRANES NANOCOMPOSITE

By capturing impurities, porous materials play a crucial function in membrane technology via filtration or adsorption. A variety of membranes, such as ultrafiltration, nanofiltration membranes, reverse osmosis, and microfiltration are commonly employed in water treatment processes. Effective water purification demands the use of highly permeable membranes, rejection, and intelligent defensive qualities, particularly when weighing cost-effectiveness and energy efficiency. Undoubtedly, nanocomposite membranes use a variety of elements to obtain improved water-filtering performance (Hernández et al., 2016; Shen et al., 2014).

8.4.2 NANOCOMPOSITES IN REMEDIATION OF AIR POLLUTION

Nanotechnology offers promising solutions for cleaning up the air through various mechanisms, including the use of nanocatalysts and filtration systems. Filters serve an important function in reducing air pollution by capturing particles and contaminants suspended in the air. The effectiveness of filters in capturing airborne particles relies on several mechanisms, including surface contact, inertial force, and electric charge. Nanofilters offer significant advantages over traditional filters due to their ability to trap particles at the nanometer scale. With pores ranging from 1 to 10 nm, nanofilters can effectively capture a large range of microorganisms, organic pollutants, and even nanoparticles that might slip through larger filters. Carbon nanotube-based membranes are known for their exceptional properties in gas separation applications, including the separation of carbon dioxide (CO_2) from other gases. Nanomembrane technology holds great promise for separating and purifying gases and pollutant vapors across various sectors, contributing to environmental protection efforts. The risk of obnoxious and hazardous spillage is a significant concern in the commercial sector, particularly in industries dealing with hazardous materials, such as chemicals, fuels, or industrial waste. The detection of small amounts of gases in commercial settings is increasingly important for various reasons, including workplace safety, environmental monitoring, and quality control. Commercial sensors play a crucial role in identifying and quantifying these gases to ensure regulatory compliance, protect human health, and optimize industrial processes. Silicon-based ultrafine particles with wireless data transmission capabilities represent a cutting-edge development in sensor technology. These sensors leverage miniaturized components and advanced microelectronics to wirelessly transmit data collected from their environment to a central server or data processing unit (Roy et al., 2021).

8.5 BIOREMEDIATION

Indeed, environmental biotechnology has a rich history, with practices like compositing and wastewater treatment being longstanding examples. Yet, recent advancements in molecular biology and ecology have ushered in new possibilities for more effective biological processes. These innovations have led to significant achievements in cleaning up polluted water and land areas. One prominent technique enabled by these advances is bioremediation. Bioremediation involves the controlled biological degradation of organic wastes to render them harmless or reduce them to levels below regulatory thresholds. Essentially, it harnesses the power of living organisms, particularly microbes, to transform indirect impurities into less harmful forms. This approach represents a sustainable and often cost-effective means of remedying environmental pollution compared with traditional methods. Bioremediation holds significant promise for cleaning up environmental contaminants, leveraging the natural abilities of microorganisms to degrade pollutants. The process of bioremediation can be enhanced through bioaugmentation, where specific microorganisms are put into the polluted site to accelerate deterioration. This technique is particularly useful in cases where indigenous microorganisms may not be sufficient or where contaminants are particularly resistant to natural degradation processes. However, bioremediation does have its limitations, especially when dealing with certain types of pollutants like chlorinated organic compounds or high aromatic hydrocarbons, which can be challenging for microorganisms to break down effectively. In such cases, the rate of cleanup can be slow or even ineffective. Notwithstanding these drawbacks, bioremediation is still a desirable choice since it may be less expensive than more conventional methods like incineration. Additionally, it provides the advantage of being less disruptive to the surrounding environment. Overall, while bioremediation holds great promise for addressing environmental contamination, it is essential to carefully assess its feasibility and effectiveness on an individual basis, taking into account the special impurities and environmental conditions involved. Additionally, Continuous research and development endeavors are advancing bioremediation methodologies and broadening their scope of use to encompass a greater variety of contaminants. The chemical structure, hydrophobicity, polarity, and other properties of pollutants, as well as the environmental factors, such as temperature, pH, and redox condition, and the characteristics of the soil, such as thickness, aggregation, and pollutant aging. All have a significant impact on the availability of contaminants and their biological degradation during environmental remediation processes. Enzymes are essential to bioremediation procedures because they catalyze the chemical changes required for the breakdown of pollutants, providing a long-lasting and ecologically benign method of remediation. Biotechnology advancements have dramatically enhanced enzyme stability, efficiency, and selectivity. These advancements heavily rely on omics technologies. Enzymes are often employed in bioremediation either as separate enzymes or as part of whole organisms as algae, fungus, or bacteria. Here is how each method works. By ensuring continuous aeration, inoculation with appropriate microbial strains, providing essential nutrition, and optimizing environmental conditions, whole-cell bioremediation can effectively harness the metabolic capabilities of microorganisms to deteriorate contaminants and promote environmental remediation. The use of individual enzymes offers advantages in terms of specificity, handling, standardization, mobility, tolerance to toxic compounds, and biodegradability. These qualities make enzyme-based bioremediation approaches highly functional and effective for the targeted elimination of impurities from various environmental matrices (Mousavi et al., 2021).

8.5.1 MICROORGANISMS USED IN BIOREMEDIATION

Microorganisms play a crucial role in maintaining biological equilibrium, particularly through their contributions to nutritional chains and ecosystem processes. Bioremediation, a marvel of environmental science, harnesses the power of living organisms, such as bacteria, algae, fungi, and yeast to restore balance to ecosystems tainted by pollutants. Microbes' remarkable adaptability enables them to grow in a wide variety of environmental circumstances, including extreme temperatures. Microorganisms exhibit remarkable adaptability, thriving in environments with extreme temperature ranges from cryogenic conditions at − 196°F (− 127°C) to near incandescent temperatures of 1200°F (649°C). Carbon is indispensable for microbial growth and metabolism, making it a fundamental nutrient in bioremediation processes. Bioremediation was carried out using microbes from diverse settings. Microbes include *Flavobacterium*, *Nitrosomonas*, *Corynebacterium*, *Bacillus*, *Arthrobacter*, *Pseudomonas*, *Xanthobacter*, *Corynebacterium*, *Mycobacterium*, *Achromobactin*, and other bacteria (Sharma et al., 2021).

8.5.1.1 AEROBIC

Degrading several kinds of environmental pollutants under aerobic circumstances is a commonly employed method. Aerobic bacteria including *Rhodococcus, Pseudomonas, Nocardia, Sphingomonas, Bacillus, Mycobacterium,* and *Flavobacterium* are highly valued for their ability to degrade a wide variety of complex organic chemicals, which makes them an invaluable resource in bioremediation initiatives (Giri et al., 2021). The aforementioned aerobic bacteria have proven capable of breaking down a variety of environmental pollutants, such as polyaromatic chemicals, alkane hydrocarbons, and pesticides. Because of their adaptable metabolism, they may use these pollutants as sources of energy and carbon, which aids in their decomposition. The growth of microorganisms in the aerobic bioremediation process is limited by the presence of oxygen.

8.5.1.2 ANAEROBIC

Anaerobic amphibious bacteria present significant prospects for the cleanup of contaminated locations because of their capacity to degrade and change pollutants into less dangerous forms, especially those affected by persistent organic pollutants and chlorinated chemicals. The anaerobic bioremediation procedure has included a variety of microorganisms, including *Pseudomonas, Aeromonas*, and sulfate-reducing bacteria. Reduction processes, in which electrons generated from the oxidation of organic substrates are used to split down the azo bonds within the dye molecules, promote the anaerobic breakdown of azo dyes. The electrochemical properties of microbes serve a significant role in the effectiveness of color removal during dye decolorization events. These properties influence the microbial interactions with dye molecules and the redox reactions involved in dye degradation. In anaerobic environments, the decolorization of dyes can lead to the formation of decolorized metabolites as intermediate or end products of microbial metabolism. External augmentation of decolorized metabolites gathered under specific conditions can indeed be a valuable approach for improving research precision and facilitating the development of more effective remediation strategies (Chen et al., 2016).

8.5.2 TYPES OF BIOREMEDIATIONS

Various methods are used based on an area's level of aeration and saturation. The term "*in situ* techniques" refers to procedures that are immediately applied to the polluted site's soil and groundwater with the least amount of environmental disruption possible. *Ex situ* procedures entail excavating or pumping polluted soil or groundwater from the location. Bioaugmentation techniques involve the deliberate addition of selected microorganisms, such as bacteria, fungi, or algae, to contaminated environments to improve the degradation of contaminants.

8.5.2.1 IN SITU BIOREMEDIATION

These strategies are frequently the most chosen ones as they minimize excavation, offer treatment on-site, and transportation of toxins (Vidali, 2001). Practices for *in situ* bioremediation fall into two categories: intrinsic and engineered. Bioventing, biosparing, and bioslurping come under engineered bioremediation.

8.5.2.1.1 Intrinsic Attenuation

Intrinsic bioremediation also known as natural attenuation, relies on the existing microbial populations and environmental conditions at contaminated sites to degrade pollutants without the need for human intervention. By creating an environment that is conducive to microbial activity, intrinsic bioremediation aims to enhance the already-occurring natural decomposition processes in the environment. Intrinsic *in situ* bioremediation includes a number of techniques aimed at enhancing the natural degradation processes occurring within the polluted environment without the need for extensive human intervention. *In situ* bioremediation was used to carry out this procedure. Over three years, *in situ* bioremediation was able to successfully reduce pollutant concentrations from 40 to 5 ppm within a legal limit. This is a noteworthy accomplishment (Bala et al., 2022). This outcome demonstrates the effectiveness of *in situ* bioremediation as an inexpensive and environmentally friendly technology for managing polluted sites.

8.5.2.1.2 Bioventing

Bioventing is an *in situ* remediation approach that enhances the natural degradation of contaminants in the soil by supplying air (primarily oxygen) and nutrients through wells. Bioventing is a remediation technique that carefully manages airflow to optimize the biodegradation of contaminants in the soil while minimizing the volatilization of pollutants into the atmosphere (Vidali, 2001). The main goal of bioventing is to aid in the biodegradation of organic pollutants by promoting the local microbiota via regulated aeration.

8.5.2.1.3 Biosparing

To do biosparing, air must be injected under pressure beneath the water table. This procedure increases the oxygen concentrations in groundwater, which speeds up the natural bacterial breakdown of pollutants. By increasing mixing in the saturated zone, biosparing improves the interaction of groundwater, soil, and oxygen. The success of biosparing relies heavily on the porosity of the soil and the biodegradability of the contaminants. Biosparing offers significant advantages because of its ease of usage and low cost of installing air injection sites having small diameters, which offers tremendous flexibility in the design and construction of the system. Biosparing is a common practice used to remove fuel and kerosene from the water supply (Bala et al., 2022).

8.5.2.1.4 Bioslurping

While soil vapor extraction, bioventing, and vacuum-assisted pumping, physically remove contaminants from the subsurface, direct oxygen delivery and biodegradation stimulation improve natural attenuation processes (Anekwe and Isa, 2021). These techniques work together to restore soil and groundwater synergistically. When it comes to cleaning up soil that has been contaminated by organic materials that are combustible or somewhat flammable, this approach works incredibly well. To repair soils polluted by combustible and highly flammable organic compounds, especially those present as free-phase liquids, this method uses a “slurp” system. Some remediation procedures adopt the strategy of employing soil moisture to limit microbial activity, especially when excessive microbial activity may impede the desired outcome. To eliminate the impurities floating on the water, bioslurping necessitates drilling 25 feet below the surface of the earth.

8.5.2.2 EX SITU BIOREMEDIATION

8.5.2.2.1 Land Farming

Land farming serves as a versatile and eco-friendly remediation method that may be effective for treating a wide variety of contaminated soils. Generally, this practice is confined to the treatment of superficial 10–35 cm of soil. Land farming has gained considerable attention as a disposal option for contaminated soil remediation due to several key advantages, including reduced experience and lasting expenses, as well as reduced cleanup obligations. Land farming, the approach to treating polluted soils often involves periodic removal and tillage of the contaminated soil. Land bioremediation using land farming technology offers a straightforward and environmentally sustainable approach to remediating polluted soil (Bala et al., 2022). Its simplicity, low capital requirements, and minimal ecological footprint make it an attractive option for addressing soil contamination while minimizing environmental impact and energy consumption.

8.5.2.2.2 Biopiles

Biopiles represent a hybrid approach that combines elements of both land farming and composting techniques. Biopiles are a more sophisticated and controlled kind of land gardening that is frequently used to alleviate surface contamination caused by petroleum hydrocarbons. Depending on the particular pollutants present and the treatment goals, biopiles are made to facilitate the growth of both native aerobic and anaerobic bacteria.

8.5.2.2.3 Bioreactor

Bioreactors are essential tools in bioremediation, providing controlled environments where microorganisms can thrive and efficiently transform contaminants into harmless by-products. The bioreactor holds the samples used for cleanup. Using bioreactors for *in situ* bioremediation offers numerous advantages over *ex situ* methods, including enhanced efficiency, increased treatment capacity, reduced site disturbance, lower costs, real-time monitoring and control, and minimized environmental impact. These benefits make bioreactor-based bioremediation an attractive option for addressing contaminated sites and restoring environmental quality.

8.6 CONCLUSION

Nanoparticles have indeed revolutionized environmental biotechnology by offering innovative solutions for pollution control, remediation, and sustainable practices. They are particularly useful in tackling environmental issues because of their unique qualities, which include a high surface area-to-volume ratio, reactivity, and adjustable surface chemistry. Examining novel and environmentally acceptable ways is necessary to meet the pressing need to remove pollutants from the environment with as little negative impact as possible, particularly in light of the shortcomings of established methods, such as coagulation, adsorption, and advanced oxidation processes. Nanocomposites which combine nanoparticles into a matrix to enhance material properties are a good way to address the new needs brought forth by scientific and technological advancements. While nanocomposites offer remarkable enhancements in material properties and applications across various fields, their processing methods pose significant challenges. These challenges present opportunities for researchers to innovate and develop solutions to overcome the issues associated with nanoscale materials. Nanocomposites offer significantly improved performance over their monolithic and micro composite counterparts, making them ideal candidates for overcoming the constraints of many conventional materials and technology. The landscape of environmental remediation and pollution control is continually evolving with the advent of innovative and eco-friendly approaches such as bioremediation, nanotechnology-based remediation, electrochemical methods, and green chemistry offer a multitude of possibilities. While some applications of these approaches are already in practice, numerous potentials remain untapped, presenting exciting new opportunities for the future. Because of its great efficiency and the nontoxic character of the synthesized materials, nano bioremediation has been a game changer in the field of environmental remediation. Diversified nanoparticles have applications in eliminating certain contaminants from their natural habitat, such as landmines and garbage spills. Natural resource depletion and industrialization are inherent to human progress. Using large- and small-scale functional active nanoparticles is a potentially comprehensive and sustainable approach to solving this grave issue. Diversified nanoparticles have demonstrated efficacy in removing certain pollutants, such as landmines and waste spills, from their native environments. Additionally, they have a negligible harmful impact on manufacturing rates, meaning they can be further studied to make them more compatible with lower production costs in the future (Roy et al., 2021). Despite their many challenges, nanocomposites have a lot of promise, particularly for specialized niche uses. Pollutants have long been a part of our environment, but their impact remains a significant and pressing concern. One of the biggest risks to human health is environmental exposure to harmful pollutants, which also contribute to significant causes of environmental diseases and deaths. The human activities that have significantly contributed to environmental degradation across the world are mining, industrialization, exploration, and urbanization. While developed countries have made significant progress in reducing pollution through knowledge, legislation, and technological innovation, both developing and developed countries must continue to share the responsibility of addressing environmental challenges.

KEYWORDS

- **bioremediation**
- **environmental remediation**
- **industrialization**
- **metal–organic frameworks**
- **nanocomposites**
- **photocatalysis**
- **pollutants**

REFERENCES

Abbafati, C., Abbas, K. M., Abbasi-Kangevari, M., Abd-Allah, F., Abdelalim, A., Abdollahi, M., Abdollahpour, I., Abegaz, K. H., Abolhassani, H., Aboyans, V., Abreu, L. G., Abrigo, M. R. M., Abualhasan, A., Abu-Raddad, L. J., Abushouk, A. I., Adabi, M., Adekanmbi, V., Adeoye, A. M., Adetokunboh, O. O. et al. (2020). Global burden of 87 risk factors in 204 countries and territories, 1990–2019: A systematic analysis for the Global Burden of Disease Study 2019. *The Lancet, 396*(10258), 1223–1249. https://doi.org/10.1016/S0140-6736(20)30752-2

Achmad, A., Kassim, J., Kim Suan, T., Che Amat, R., & Lean Seey, T. (2012). Equilibrium, kinetic and thermodynamic studies on the adsorption of direct dye onto a novel green adsorbent developed from Uncaria gambir extract. *Journal of Physical Science, 23*(1), 1–13.

Agboola, O., Fayomi, O. S. I., Ayodeji, A., Ayeni, A. O., Alagbe, E. E., Sanni, S. E., Okoro, E. E., Moropeng, L., Sadiku, R., Kupolati, K. W., & Oni, B. A. (2021). A review on polymer nanocomposites and their effective applications in membranes and adsorbents for water treatment and gas separation. *Membranes, 11*(2), 139. https://doi.org/10.3390/MEMBRANES11020139

Al-Hazmi, H. E., Luczak, J., Habibzadeh, S., Hasanin, M. S., Mohammadi, A., Esmaeili, A., Kim, S. J., Khodadadi Yazdi, M., Rabiee, N., Badawi, M., & Saeb, M. R. (2024). Polysaccharide nanocomposites in wastewater treatment: A review. *Chemosphere, 347*. https://doi.org/10.1016/J.CHEMOSPHERE.2023.140578

Al-Nuaim, M. A., Alwasiti, A. A., & Shnain, Z. Y. (2022). The photocatalytic process in the treatment of polluted water. *Chemical Papers, 77*(2), 677–701. https://doi.org/10.1007/S11696-022-02468-7

Anekwe, I. M., & Isa, Y. M. (2021). Wastewater and bioventing treatment systems for acid mine drainage–contaminated soil. *Soil and Sediment Contamination: An International Journal, 30*(5), 518–531. https://doi.org/10.1080/15320383.2020.1863909

Bala, S., Garg, D., Thirumalesh, B. V., Sharma, M., Sridhar, K., Inbaraj, B. S., & Tripathi, M. (2022). Recent strategies for bioremediation of emerging pollutants: A review for a green and sustainable environment. *Toxics, 10*(8). https://doi.org/10.3390/TOXICS10080484

Briffa, J., Sinagra, E., & Blundell, R. (2020). Heavy metal pollution in the environment and their toxicological effects on humans. *Heliyon, 6*(9). https://doi.org/10.1016/J.HELIYON.2020.E04691

Briggs, D. (2003). Environmental pollution and the global burden of disease. *British Medical Bulletin, 68*, 1–24. https://doi.org/10.1093/BMB/LDG019

Camargo, P. H. C., Satyanarayana, K. G., & Wypych, F. (2009). Nanocomposites: Synthesis, structure, properties and new application opportunities. *Materials Research, 12*(1), 1–39. https://doi.org/10.1590/S1516-14392009000100002

Carré, J., Gatimel, N., Moreau, J., Parinaud, J., & Léandri, R. (2017). Does air pollution play a role in infertility? A systematic review. *Environmental Health : A Global Access Science Source, 16*(1). https://doi.org/10.1186/S12940-017-0291-8

Chang, M. Y., & Juang, R. S. (2004). Adsorption of tannic acid, humic acid, and dyes from water using the composite of chitosan and activated clay. *Journal of Colloid and Interface Science, 278*(1), 18–25. https://doi.org/10.1016/j.jcis.2004.05.029

Chen, B. Y., Ma, C. M., Han, K., Yueh, P. L., Qin, L. J., & Hsueh, C. C. (2016). Influence of textile dye and decolorized metabolites on microbial fuel cell-assisted bioremediation. *Bioresource Technology, 200*, 1033–1038. https://doi.org/10.1016/J.BIORTECH.2015.10.011

Cohen, A. J., & Pope, C. A. (1995). Lung cancer and air pollution. *Environmental Health Perspectives, 103*(8), 219–224. https://doi.org/10.1289/EHP.95103S8219

Djerahov, L., Vasileva, P., Karadjova, I., Kurakalva, R. M., & Aradhi, K. K. (2016). Chitosan film loaded with silver nanoparticles—Sorbent for solid phase extraction of Al(III), Cd(II), Cu(II), Co(II), Fe(III), Ni(II), Pb(II) and Zn(II). *Carbohydrate Polymers, 147*, 45–52. https://doi.org/10.1016/J.CARBPOL.2016.03.080

Dusek, P., Roos, P. M., Litwin, T., Schneider, S. A., Flaten, T. P., & Aaseth, J. (2015). The neurotoxicity of iron, copper and manganese in Parkinson's and Wilson's diseases. *Journal of Trace Elements in Medicine and Biology : Organ of the Society for Minerals and Trace Elements (GMS), 31*, 193–203. https://doi.org/10.1016/J.JTEMB.2014.05.007

Frazier, L. M. (2007). Reproductive disorders associated with pesticide exposure. *Journal of Agromedicine, 12*(1), 27–37. https://doi.org/10.1300/J096V12N01_04

Giacoppo, S., Galuppo, M., Calabrò, R. S., D'Aleo, G., Marra, A., Sessa, E., Bua, D. G., Potortì, A. G., Dugo, G., Bramanti, P., & Mazzon, E. (2014). Heavy metals and neurodegenerative diseases: An observational study. *Biological Trace Element Research, 161*(2), 151–160. https://doi.org/10.1007/S12011-014-0094-5/METRICS

Giri, B. S., Geed, S., Vikrant, K., Lee, S. S., Kim, K. H., Kailasa, S. K., Vithanage, M., Chaturvedi, P., Rai, B. N., & Singh, R. S. (2021). Progress in bioremediation of pesticide residues in the environment. *Environmental Engineering Research, 26*(6). https://doi.org/10.4491/EER.2020.446

Gusain, R., Gupta, K., Joshi, P., & Khatri, O. P. (2019). Adsorptive removal and photocatalytic degradation of organic pollutants using metal oxides and their composites: A comprehensive review. *Advances in Colloid and Interface Science, 272*, 102009. https://doi.org/10.1016/J.CIS.2019.102009

Guven, A., Kayikci, A., Cam, K., Arbak, P., Balbay, O., & Cam, M. (2008). Alterations in semen parameters of toll collectors working at motorways: Does diesel exposure induce detrimental effects on semen? *Andrologia, 40*(6), 346–351. https://doi.org/10.1111/J.1439-0272.2008.00867.X

Hafeez, M., & Hafeez, M. (2022). Recent progress and overview of nanocomposites. *Nanocomposite Materials for Biomedical and Energy Storage Applications*. https://doi.org/10.5772/INTECHOPEN.102469

Haghani, A., Morgan, T. E., Forman, H. J., & Finch, C. E. (2020). Air pollution neurotoxicity in the adult brain: Emerging concepts from experimental findings. *Journal of Alzheimer's Disease, 76*(3), 773–797. https://doi.org/10.3233/JAD-200377

Hári, J., & Pukánszky, B. (2011). Nanocomposites: Preparation, structure, and properties. *Applied Plastics Engineering Handbook: Processing and Materials,* 109–142. https://doi.org/10.1016/B978-1-4377-3514-7.10008-X

Hernández, S., Saad, A., Ormsbee, L., & Bhattacharyya, D. (2016). Nanocomposite and responsive membranes for water treatment. *Emerging Membrane Technology for Sustainable Water Treatment,* 389–431. https://doi.org/10.1016/B978-0-444-63312-5.00016-4

Huang, Z. H., Zheng, X., Lv, W., Wang, M., Yang, Q. H., & Kang, F. (2011). Adsorption of lead(II) ions from aqueous solution on low-temperature exfoliated graphene nanosheets. *Langmuir*, *27*(12), 7558–7562. https://doi.org/10.1021/LA200606R/ASSET/IMAGES/MEDIUM/LA-2011-00606R_0007.GIF

Jeya Sri Lakshmi, S., Joel, C., Biju Bennie, R., Nirmal Paul Raj, A., Kumar, Y. A., & Khan, M. S. (2024). Synergistic adsorption and photocatalytic degradation of tetracycline using a Z-scheme kaolin/g-C_3N_4/MoO_3 nanocomposite: A sustainable approach for water treatment. *Journal of Environmental Management, 360*. https://doi.org/10.1016/J.JENVMAN.2024.121086

Kahlon, S. K., Sharma, G., Julka, J. M., Kumar, A., Sharma, S., & Stadler, F. J. (2018). Impact of heavy metals and nanoparticles on aquatic biota. *Environmental Chemistry Letters, 16*(3), 919–946. https://doi.org/10.1007/S10311-018-0737-4

Kapil, N., Mayani, S. V., & Gopal Bhattacharyya, K. (2023). Environmental implications of nanoceramic applications. *Results in Chemistry, 5*, 100724. https://doi.org/10.1016/J.RECHEM.2022.100724

Khaydarov, R. A., Khaydarov, R. R., & Gapurova, O. (2010). Water purification from metal ions using carbon nanoparticle-conjugated polymer nanocomposites. *Water Research, 44*(6), 1927–1933. https://doi.org/10.1016/J.WATRES.2009.11.041

Khulbe, K. C., & Matsuura, T. (2018). Removal of heavy metals and pollutants by membrane adsorption techniques. *Applied Water Science, 8*(1), 1–30. https://doi.org/10.1007/S13201-018-0661-6

Konstantinou, I. K., & Albanis, T. A. (2004). TiO_2-assisted photocatalytic degradation of azo dyes in aqueous solution: Kinetic and mechanistic investigations: A review. *Applied Catalysis B: Environmental, 49*(1), 1–14. https://doi.org/10.1016/J.APCATB.2003.11.010

Koutros, S., Silverman, D. T., Alavanja, M. C. R., Andreotti, G., Lerro, C. C., Heltshe, S., Lynch, C. F., Sandler, D. P., Blair, A., & Beane Freeman, L. E. (2016). Occupational exposure to pesticides and bladder cancer risk. *International Journal of Epidemiology, 45*(3), 792. https://doi.org/10.1093/IJE/DYV195

Lellis, B., Favaro-Polonio, C. Z., Pamphile, J. A., & Polonio, J. C. (2019). Effects of textile dyes on health and the environment and bioremediation potential of living organisms. *Biotechnology Research and Innovation, 3*(2), 275–290. https://doi.org/10.1016/J.BIORI.2019.09.001/PDF/BIORI-3-2-275.PDF

Ma, Y., He, X., Qi, K., Wang, T., Qi, Y., Cui, L., Wang, F., & Song, M. (2019). Effects of environmental contaminants on fertility and reproductive health. *Journal of Environmental Sciences (China), 77*, 210–217. https://doi.org/10.1016/J.JES.2018.07.015

Masih, J., Masih, A., Kulshrestha, A., Singhvi, R., & Taneja, A. (2010). Characteristics of polycyclic aromatic hydrocarbons in indoor and outdoor atmosphere in the North central part of India. *Journal of Hazardous Materials, 177*(1–3), 190–198. https://doi.org/10.1016/J.JHAZMAT.2009.12.017

McCarthy, C. E., Duffney, P. F., Wyatt, J. D., Thatcher, T. H., Phipps, R. P., & Sime, P. J. (2017). Comparison of *in vitro* toxicological effects of biomass smoke from different sources of animal dung. *Toxicology In Vitro : An International Journal Published in Association with BIBRA, 43*, 76–86. https://doi.org/10.1016/J.TIV.2017.05.021

Mousavi, S. M., Hashemi, S. A., Iman Moezzi, S. M., Ravan, N., Gholami, A., Lai, C. W., Chiang, W. H., Omidifar, N., Yousefi, K., & Behbudi, G. (2021). Recent advances in enzymes for the bioremediation of pollutants. *Biochemistry Research International*, *2021*(1), 5599204. https://doi.org/10.1155/2021/5599204

Münzel, T., Hahad, O., Daiber, A., & Landrigan, P. J. (2023). Soil and water pollution and human health: What should cardiologists worry about? *Cardiovascular Research*, *119*(2), 440–449. https://doi.org/10.1093/CVR/CVAC082

Nithya, R., & Sudha, P. N. (2016). Removal of heavy metals from tannery effluent using chitosan-g-poly (butyl acrylate)/bentonite nanocomposite as an adsorbent. *Textiles and Clothing Sustainability, 2*(1), 1–8. https://doi.org/10.1186/S40689-016-0018-1

Pan, B., Zhang, X., Jiang, Z., Li, Z., Zhang, Q., & Chen, J. (2019). *Polymer and polymer-based nanocomposite adsorbents for water treatment,* 93–119. https://doi.org/10.1007/978-3-030-00743-0_5

Pandey, N., Shukla, S. K., & Singh, N. B. (2017). Water purification by polymer nanocomposites: An overview. *Nanocomposites*, *3*(2), 47–66. https://doi.org/10.1080/20550324.2017.1329983

Peter-Varbanets, M., Zurbrügg, C., Swartz, C., & Pronk, W. (2009). Decentralized systems for potable water and the potential of membrane technology. *Water Research*, *43*(2), 245–265. https://doi.org/10.1016/J.WATRES.2008.10.030

Rajagopalan, S., & Landrigan, P. J. (2021). Pollution and the heart. *New England Journal of Medicine, 385*(20), 1881–1892. https://doi.org/10.1056/NEJMRA2030281/SUPPL_FILE/NEJMRA2030281_DISCLOSURES.PDF

Roy, A., Sharma, A., Yadav, S., Jule, L. T., & Krishnaraj, R. (2021). Nanomaterials for remediation of environmental pollutants. *Bioinorganic Chemistry and Applications, 2021*(1), 1764647. https://doi.org/10.1155/2021/1764647

Saad, A. H. A., Azzam, A. M., El-Wakeel, S. T., Mostafa, B. B., & Abd El-lati, M. B. (2018). Removal of toxic metal ions from wastewater using ZnO@Chitosan core-shell nanocomposite. *Environmental Nanotechnology, Monitoring & Management, 9*, 67–75. https://doi.org/10.1016/J.ENMM.2017.12.004

Saad, M., & Tahir, H. (2017). Synthesis of carbon loaded γ-Fe_2O_3 nanocomposite and their applicability for the selective removal of binary mixture of dyes by ultrasonic adsorption based on response surface methodology. *Ultrasonics Sonochemistry, 36*, 393–408. https://doi.org/10.1016/J.ULTSONCH.2016.12.010

Seewagen, C. L. (2020). The threat of global mercury pollution to bird migration: Potential mechanisms and current evidence. *Ecotoxicology, 29*(8), 1254–1267. https://doi.org/10.1007/S10646-018-1971-Z/METRICS

Senusi, F., Shahadat, M., Ismail, S., & Hamid, S. A. (2018). Recent advancement in membrane technology for water purification. *Modern Age Environmental Problems and Their Remediation*, 147–167. https://doi.org/10.1007/978-3-319-64501-8_9

Sharma, G., Kumar, A., Sharma, S., Naushad, M., Prakash Dwivedi, R., Alothman, Z. A., & Mola, G. T. (2019). Novel development of nanoparticles to bimetallic nanoparticles and their composites: A review. *Journal of King Saud University——Science*, *31*(2), 257–269. https://doi.org/10.1016/J.JKSUS.2017.06.012

Sharma, P., Pandey, A. K., Kim, S. H., Singh, S. P., Chaturvedi, P., & Varjani, S. (2021). Critical review on microbial community during *in situ* bioremediation of heavy metals from industrial wastewater. *Environmental Technology and Innovation, 24*. https://doi.org/10.1016/J.ETI.2021.101826

Shen, Y. xiao, Saboe, P. O., Sines, I. T., Erbakan, M., & Kumar, M. (2014). Biomimetic membranes: A review. *Journal of Membrane Science, 454*, 359–381. https://doi.org/10.1016/J.MEMSCI.2013.12.019

Shetty, S. S., D, D., S, H., Sonkusare, S., Naik, P. B., Kumari N, S., & Madhyastha, H. (2023). Environmental pollutants and their effects on human health. *Heliyon, 9*(9). https://doi.org/10.1016/J.HELIYON.2023.E19496

Singh, P., Kaur, G., Singh, K., Singh, B., Kaur, M., Kaur, M., Krishnan, U., Kumar, M., Bala, R., & Kumar, A. (2018). Specially designed B4C/SnO_2 nanocomposite for photocatalysis: Traditional ceramic with unique properties. *Applied Nanoscience (Switzerland), 8*(1–2), 1–9. https://doi.org/10.1007/S13204-018-0662-7/FIGURES/6

Skakkebæk, N. E., Rajpert-De Meyts, E., & Main, K. M. (2001). Testicular dysgenesis syndrome: An increasingly common developmental disorder with environmental aspects. *Human Reproduction (Oxford, England), 16*(5), 972–978. https://doi.org/10.1093/HUMREP/16.5.972

Tanasa, E., Zaharia, C., Radu, I. C., Surdu, V. A., Vasile, B. S., Damian, C. M., & Andronescu, E. (2019). Novel nanocomposites based on functionalized magnetic nanoparticles and polyacrylamide: Preparation and complex characterization. *Nanomaterials (Basel, Switzerland), 9*(10). https://doi.org/10.3390/NANO9101384

Tian, T., Shi, X., Cheng, L., Luo, Y., Dong, Z., Gong, H., Xu, L., Zhong, Z., Peng, R., & Liu, Z. (2014). Graphene-based nanocomposite as an effective, multifunctional, and recyclable antibacterial agent. *ACS Applied Materials & Interfaces, 6*(11), 8542–8548. https://doi.org/10.1021/AM5022914

Vidali, M. (2001). Bioremediation: An overview. *Pure and Applied Chemistry, 73*(7), 1163–1172. https://doi.org/10.1351/PAC200173071163/MACHINEREADABLECITATION/RIS

Vijayakumar, G., Tamilarasan, R., & Dharmendirakumar, M. (2011). Adsorption, kinetic, equilibrium and thermodynamic studies on the removal of basic dye rhodamine-B from aqueous solution by the use of natural adsorbent perlite. *Journal of Materials and Environmental Science, 3*(1), 157–170.

Zhao, S., Wang, Z., Wei, X., Zhao, B., Wang, J., Yang, S., & Wang, S. (2012). Performance improvement of polysulfone ultrafiltration membrane using well-dispersed polyaniline–poly(vinylpyrrolidone) nanocomposite as the additive. *Industrial and Engineering Chemistry Research, 51*(12), 4661–4672. https://doi.org/10.1021/IE202503P

CHAPTER 9

Application of Nanocomposites for Various Diseases

KAJAL KUMARI, TRIPTI PALIWAL, SHRUTI RICHA, KANIKA VERMA, and SWAPNIL SHARMA*

Department of Pharmacy, Banasthali Vidyapith, Banasthali, Rajasthan, India

*Corresponding author.

ABSTRACT

In recent years, the field of nanocomposites (NCs) has gained lots of attention from researchers and scientists in the context of drug delivery systems for several diseases due to their superior biomedical characteristics. NCs are multiphase materials in which one of the material dimensions should be in the range of 100 nm. The materials used for NCs are non-toxic, biocompatible, biodegradable, and environmentally benign because different polymers and nanomaterials have been integrated to fabricate NCs for drug delivery. Natural and synthetic polymers have been used for NCs. Poor drug solubility, permeability, and toxicity are now considered as a challenge, compromising medical advancements and our ability to treat diseases. To overcome these issues, one strategy could be applied, and that is, NCs. To achieve highly targeted and effective treatment, NCs were developed that only target the affected area and provide enhanced therapeutic effects. This chapter outlines the application of NC in several diseases like chronic obstructive pulmonary disease, neurological disorders, bone tissue regeneration, diabetic ulcers, and cancer.

Nanocomposites and Nanomaterials in Biomedical Applications. Vivek Dave, Deependra Singh, and Pallerla Naveen Reddy (Eds.)

DOI: 10.1201/9781779641236-9

9.1 INTRODUCTION

The term "nanocomposite" (NC) denotes a solid material comprising many phases, whereby at least one of the phases has dimensions smaller than 100 nanometers (nm). In addition, it is possible for the material to exhibit structures characterized by nano-scale periodicity between its different phases. NCs include the utilization of nanoscale constituents to create and manufacture innovative materials that demonstrate exceptional adaptability and improved physical properties. The investigation of NCs has been conducted for almost five decades, yet there is a limited number of scholarly sources that specifically discuss the significance of the processing techniques used to include organoclay into the selected plastic material. The concept of NCs was first mentioned in literature as far back as 1950, with specific mention of polyamide NCs being documented in 1976. Nevertheless, the comprehensive investigation of polymer/layered silicate clay mineral composites by Toyota researchers marked a pivotal moment in the increased scholarly and industrial exploration of NCs. Acquarulo and O'Neil (2002) did a study whereby they observed that Toyota's Central Research and Development Laboratories began research on composites consisting of polymer-layered silicate-clay minerals during the early 1980s.

Practical applications of nanoparticle-rich materials have been observed before a complete understanding of their chemical and physical characteristics. An investigation conducted by Jose-Yacaman et al. aimed to explore the underlying factors contributing to the vibrant color depth and the remarkable resilience of Maya blue paint against acids and bio-corrosion. The researchers proposed a nanoparticle mechanism as the primary explanation for these unique properties. The use of nanoscale nano-clays for the manipulation of polymer solution flow properties has been seen since the mid-1950s. They are utilized as viscosifiers in paints and as modifiers in gels, effectively functioning as thickening agents in cosmetics to ensure the uniformity of preparations. During the 1970s, polymer/clay composites gained significant attention in academic literature, as shown by their inclusion in textbooks. However, it is worth noting that the word "nanocomposites" was not widely used during this period (Alexandre and Dubois, 2000).

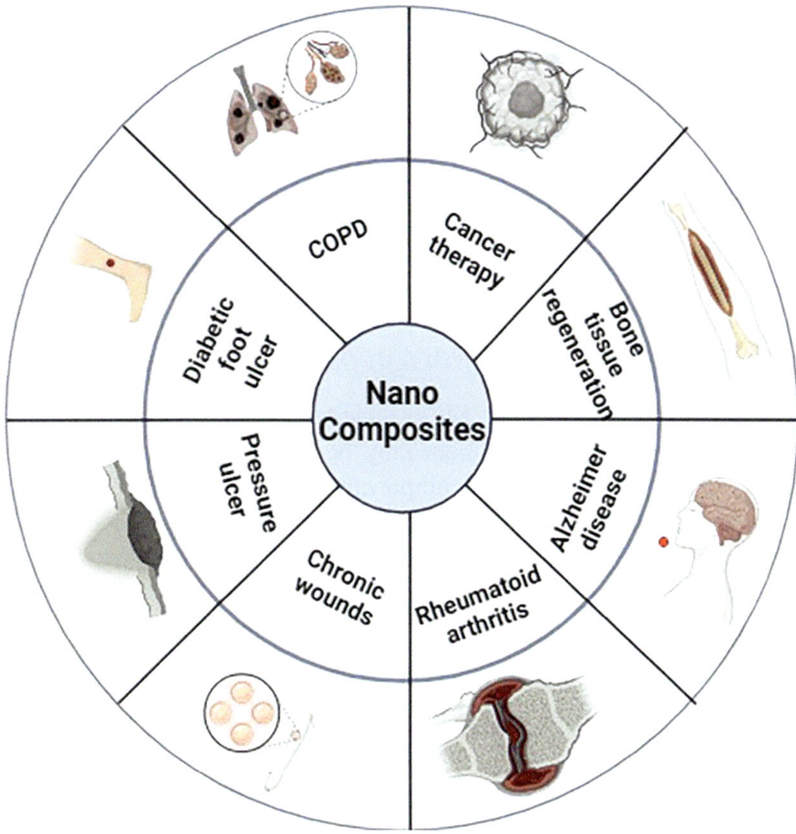

FIGURE 9.1 Application of nanocomposites for various diseases.

Because of the enormous surface area of the reinforcement, the macroscale properties of the composite material can be altered by even a small amount of nanoscale reinforcement. Carbon nanotubes (CNTs) improve electrical and thermal conductivity. Different nanoparticles may improve optical, dielectric, thermal, stiffness, strength, and wear resistance. The conventional method for integrating nano reinforcement involves dispersing it into the matrix throughout the production process. The mass fraction of inserted nanoparticles may be consistently maintained at a reasonably low level, often falling within the range of 0.5–5%. The primary cause of this phenomenon may be traced to the percolation threshold of fillers, which is notably low. This is especially true for fillers that are non-spherical in shape and have a high aspect ratio. Nanometer-thin platelets like clays and nanometer-diameter cylinders like CNTs are used in many applications. The orientation and arrangement of asymmetrical nanoparticles, the thermal property mismatch at the interface, the density of interfaces per unit volume, and the polydispersity of nanoparticles all have a significant impact on the effective thermal conductivity of NCs (Ali et al., 2019).

9.2 DIFFERENT TYPES OF NANOCOMPOSITES

9.2.1 CERAMIC MATRIX NANOCOMPOSITES

Ceramic matrix composites (CMCs) include ceramic fibers in a ceramic matrix. Carbon and carbon fibers may be ceramic components in the matrix and fibers. The primary component found in the bulk of ceramics is often generated from a group of chemicals referred to as oxides, which include nitrides, borides, and silicides. On the other hand, it is common for the secondary component to be a metallic element. For optical, electrical, magnetic, tribological, corrosion-resistant, and other protective properties, complete integration between the two components is ideal (Alimardani et al., 2019).

9.2.2 METAL MATRIX NANOCOMPOSITES

Reinforced metal matrix composites, or nanoparticle-integrated metal matrix NCs, strengthen a metal matrix. Newly popular NCs include CNT metal matrix composites (CNT-MMCs). The potential of this emerging material is very promising owing to its capacity to use the remarkable tensile strength and electrical conductivity intrinsic to CNTs (Alzahrani et al., 2018).

9.2.3 POLYMER MATRIX NANOCOMPOSITES

In a simple situation, adding nanoparticles to a polymer matrix may increase its performance by making use of the nanoscale filler's properties. The correct term is nano-filled polymer composites. The aforementioned approach has significant efficacy in producing composites with superior performance, given that there is a consistent distribution of the filler material and the nanoscale filler possesses qualities that are notably distinct or superior to those of the matrix. Thermodynamically generated phase separation counteracts the homogeneity of dispersion seen in all NCs. The aggregation of nanoscale fillers gives rise to the formation of clusters, which act as structural imperfections and ultimately lead to failure. The layer-by-layer (LbL) assembly technique entails the consecutive deposition of nanometer-scale layers composed of nanoparticles and a polymer. The selection of a suitable polyelectrolyte as the polymeric interlayer may have a substantial impact on the performance of the assembly. LBL composites have much superior performance metrics, ranging from 10 to 1000 times, compared to conventional NCs produced using extrusion or batch-mixing methods (Bahari et al., 2019; Abu Bakar et al., 2018; Balaure and Grumezescu, 2020).

9.2.4 MAGNETIC NANOCOMPOSITES

NCs exhibiting the ability to react to an external stimulus have garnered heightened attention. This phenomenon may be attributed mostly to the substantial interfacial contacts between the phases, which enhance the impact of the stimulus-response on the composite as a whole. The external stimuli might be magnetic, electrical, or mechanical. These applications benefit from magnetic NCs' electrical and magnetic response. The magnetic field penetrates deeply, affecting a larger area of the nanocomposite and improving reaction. A matrix comprising nanoparticles or nanorods may interact efficiently with a magnetic field. Materials for magnetic composite structures include Janus-type nanostructures, colloidal crystals, matrix-dispersed nanoparticles, and core–shell nanoparticles (Bengalli et al., 2019; Chaudhary et al., 2018).

9.3 ROLE OF NANOCOMPOSITE IN VARIOUS DISEASES

9.3.1 NANOCOMPOSITE IN CHRONIC OBSTRUCTIVE PULMONARY DISEASE (COPD)

Chronic obstructive pulmonary disease (COPD) is a prevalent respiratory illness that progressively diminishes lung function. The syndrome is distinguished by the occurrence of obstruction in the airways, oxidative stress, persistent inflammation, heightened production of mucus, and elevated levels of autophagy and cellular senescence. The main risk factor for COPD is smoking. Other secondary COPD risk factors include air pollution and occupational gas and fume exposure in developing countries. Antioxidative and anti-inflammatory drugs are often used to treat COPD. However, traditional antioxidant-targeted pharmaceuticals are commonly utilized to treat COPD. However, these therapies have significant limitations, including poor drug diffusion and pharmacokinetics. Recent nanotechnology and stem cell advances have changed chronic airway disease treatment. This research examines the use of antioxidants in COPD maintenance and therapy, concentrating on nanotechnology-based therapies. The above advancements include stem cell and gene therapy, which may cure chronic airway disorders, including COPD and asthma.

9.3.2 IMPORTANCE OF NANOTECHNOLOGY IN COPD

Nanoparticles are characterized as submicron-sized particles with distinctive inherent physical attributes, including diminutive dimensions, a substantial surface area-to-volume ratio, as well as qualities of biocompatibility and biodegradability. These features enable them to establish enhanced communication with adjacent tissues and cells, thereby enhancing the effectiveness of delivering drugs and genes at the cellular level. Drug delivery by nanoparticles for chronic inflammatory (CF) lung diseases offers several advantages. The methods for delivering drugs provide prolonged availability of medications and aid in circumventing airway barriers while concurrently focusing on the affected cells. An illustrative instance is the study conducted by Janib et al., whereby they showcased the efficacy of STEALTH liposome as a carrier for budesonide, a very powerful corticosteroid. The encapsulation of the medication inside the liposome facilitated regulated release, hence allowing for the sustained presence of therapeutic quantities in the lungs of rats. Moreover, this approach exhibited the added benefit of mitigating systemic toxicity. The administration of nanoparticles may also facilitate the concurrent administration of many medications for combination therapy. Zimeng et al. (2016) produced liposomes with beclomethasone and formoterol. Beclomethasone was long-lasting, whereas formoterol strengthened lung function and peripheral lung deposits without disrupting the mucociliary escalator. Thus, improved cargo-carrying nanocarriers are needed. These nanocarriers may treat COPD/emphysema.

9.3.3 MULTIFUNCTIONAL NANOCARRIERS FOR COPD: RATIONALE

9.3.3.1 SIZE

According to scientific research, the process of mucociliary clearance is strongly affected by both the charge and size of particles. Nanocarriers' ability to cross the mucus layer is a major hurdle to their application in COPD therapy. This issue arises due to the substantial thickness and high viscoelasticity of the mucus layer in individuals with COPD. Rytting et al. (2012) conducted a study whereby they observed that tiny polystyrene nanoparticles, measuring around 120 nm in diameter, exhibited significant mobility within the sputum of individuals diagnosed with cystic fibrosis. In contrast, particles of greater size, measuring around 250–270 nm in diameter, had a somewhat decreased degree of mobility. Additionally, nanoparticles may undergo surface changes like PEGylation to acquire a neutral charge.

9.3.3.2 PARTICLE SHAPE

The unique structural characteristics of particles have an impact on several biological processes, including cellular absorption, hemorheological diameter, and the ultimate destiny of the particles inside living organisms. Discoidal particles have distinct falling and migrating behaviors that often facilitate engagement with vessel walls, in contrast to spherical particles. The circulation half-life of particles is influenced by their shape. Filomicelles, micelles made of filamentous polymers, have circulation lifetimes of one week after administration, compared to two to three days for spherical particles.

Nanocarriers increase target-site drug accumulation, circulation, degradation, and customized drug delivery. It is essential for accurate and controlled pharmacological ingredient delivery. Nanocarriers and ligands that connect to specific receptors on ill cells, including breast cancer cell folate receptors, provide tailored delivery. There are several techniques to encapsulate drugs, siRNA, DNA, and peptides in nanocarriers. These approaches use molecules' physical and chemical properties. Pharmaceutical chemicals are retained better by nanocarrier surface covalent and electrostatic interactions. Variable medicine compatibility requires nanoparticle surface modifications. Production materials and surface modifications affect nanoparticle characteristics and drug release. Nanocarriers interact with intracellular and surface biomolecules. The application of biodegradable polymeric nanoparticles as an effective vehicle for sustained drug delivery has been observed due to their remarkable biocompatibility and their ability to modify their surface properties. Many studies have been conducted on their potential to provide medicinal compounds, such as pulmonary formulations for COPD, asthma, cancer, and tuberculosis (TB) medications. Numerous liposomal formulations have been investigated to deliver pharmaceutical agents to the pulmonary system. Many liposomal drug delivery techniques have FDA clearance, and many formulations are in clinical trials. The following part highlights numerous important features of nanocarrier design for pulmonary drug delivery (Din et al., 2017; Li et al., 2019; Mattu et al., 2016; Örüm and Demircioğlu, 2019; Khan et al., 2018; Nazem and Mansoori, 2011).

9.3.3.3 TARGET

Targeted *in vitro* and *in vivo* uses for nanoparticles are varied. Active and passive targeting are used to deliver nanoparticles to sick locations. Passive targeting is possible using enhanced permeability and retention (EPR). The process described entails the aggregation of nanoparticles inside tissues as a consequence of the development of disorganized and permeable blood arteries, which arise from insufficient ejection and inadequately controlled angiogenesis, particularly in the context of tumors. According to Hashizume et al., the surface charges on nanoparticles may have helped accomplish this positively charged paclitaxel-loaded liposomes function. The liposomes utilized in this work were designed to preferentially target solid tumor endothelial cells. Subsequently, these liposomes were subjected to clinical trials, including patients afflicted with advanced cancer and liver metastases.

9.3.4 NANO-THERANOSTIC FOR CHRONIC OBSTRUCTIVE PULMONARY DISEASE

Theranostics combines diagnostic and therapeutic technologies for tailored therapy. The above strategy may treat and manage CF respiratory diseases. The field of pulmonary healthcare could undergo an entirely new phase as a result of this research. For CF and COPD patients, nanotechnology in theranostics has several benefits. These benefits include the capacity to promptly detect lung inflammation and provide tailored therapeutic interventions while mitigating the occurrence of adverse medication reactions. The proposed approach involves the development of nanocarriers capable of targeting the airway and simultaneously for both therapeutic and diagnostic activities. The nanocarriers will be specifically engineered to transport therapeutic compounds while also being equipped with imaging components. This integration enables the nanocarriers to perform several roles, including cell treatment, targeted delivery, and very sensitive imaging, all inside a unified system. Prioritizing chronic obstructive lung disease management problems improves survival outcomes for CF and COPD patients (Georganopoulou et al., 2005).

Several studies have documented the use of various metal and metal oxide nanoparticles with multifunctional properties in the treatment of lung cancer. For instance, gold nanoparticles have been employed to enhance optical capabilities as well as facilitate photothermal therapy for cancer treatment. The researchers Weng et al., devised a strategy including the integration of liposomes with multifunctional CdSe/ZnS (cadmium selenide/zinc sulphide) quantum dots. This approach was used to provide targeted drug delivery for anti-cancer medications, while also serving as fluorescent imaging agents. An exemplification of this phenomena may be seen in the use of a magnetic iron oxide nanoparticle, which is amalgamated with a pharmaceutical compound and encapsulated inside a silica shell. This composite structure facilitates the transportation of a payload consisting of an MRI contrast agent and a hydrophobic drug, hence enabling the integration of diagnostic and therapeutic functionalities within a unified system. Multifunctional nanosystems for COPD require improvement despite present advances (Haes et al., 2005).

Multifunctional polymeric vesicles deliver COPD medicines through theranostic nanoparticles. These vesicles are polylactic-co-glycolic acid (PLGA). Prednisolone, corticosteroids, anti-inflammatory bronchodilators such as theophylline, and molecular probes for theranostic application in obstructive lung diseases are all transported by these vesicles. Drug- and probe-loaded nanocarriers treat COPD in a therapeutic manner.

Polyethylene glycol (PEG), polyethyleneimine (PEI), polyethylene oxide (PEO), and polyvinyl alcohol (PVA) are hydrophilic polymers with flexibility, low immunogenicity, low toxicity, and no reticuloendothelial system accumulation. The aforementioned qualities make these polymers very suitable for use as the coating agents in drug delivery applications using nanoparticles. PLGA, a copolymer, is biodegradable and biocompatible, making it appropriate for FDA-approved medicinal devices. Inhaled corticosteroids are generally used to treat asthma and other respiratory diseases. However, their pharmacological activity is short-lived. Previous studies have shown that nanoformulations of polymeric steroids with a positive charge showed increased accumulation, specifically at the site of airway inflammation. This characteristic leads to higher therapeutic benefits as compared to the administration of free steroids. One illustrative instance is the utilization of budesonide, a specific kind of polymeric micelle, which has shown efficacy in reducing the population of inflammatory cells inside the broncho-alveolar region of rat models afflicted with asthma or COPD. An alternative methodology utilizes liposomes as a nanocarrier to deliver ketofinfumarate, a mast cell stabilizer, to the specific location inside the lungs of rats by the administration of dry powder aerosols (Jahangiri et al., 2014).

9.3.5 NANOCOMPOSITE FOR CANCER THERAPY

Medical applications are being investigated for synthetic polymers, biopolymers, and NCs. Polyolefins, fluorinated polymers, polyesters, silicones, and so forth, are synthetic. In medicine, interest has been piqued by the remarkable features of NCs made of synthetic and biopolymers—hybrids of a macromolecular matrix and an organic or mineral nanofiller and other fields in recent decades. Furthermore, it explores the impact of internal and external stimuli on these materials, intending to enhance their activity and efficiency. Better outcomes are achieved by replacing traditional cancer medication carriers.

Cancer is the deadliest illness of our day, and its spread is rising. Given this dire health condition, scientists are developing innovative carriers for anticancer medicine delivery to cancer tumors (Payab et al., 2014; Wang et al., 2015; Javadzadeh et al., 2014).

Globally, a significant number of individuals get a cancer diagnosis annually, resulting in a mortality rate above 50%. The annual assessment conducted by the American Cancer Society examines the incidence and death rates of recently discovered cases (Ignjatović et al., 2013). The occurrence of cancer exhibited a stable trend among women, but it had a gradual decrease of 2% annually among males throughout the period from 2006 to 2015. In terms of cancer-related deaths, there was a fall of 1.4 and 1.8% in mortality rates for women and men, respectively, from 2007 to 2016. Expanding the application of cancer control skills to underserved populations has the potential to propel advancements in cancer cure research (Johnson and Chu, 2010).

Polymer NCs are useful in nanocarriers, tissue engineering, antimicrobials, sensors, and more. Nanofillers provide these new materials with unique features, which have sparked scientific attention. Polymer NCs are promising for disease theranostics, and nanotechnology is promising for cancer medication carriers (Johnson and Chu, 2010). Medicine seeks carriers that deliver anticancer drugs to malignant cells without damaging healthy cells (Araujo et al., 2010).

NCs are popular because of their unique design capability, eco-friendliness, ease of manufacture, and cost-effectiveness. The hybrid structural properties of composites are determined by the interplay between the nanofiller and the polymer matrix (Araújo et al., 2012). Polymers that react to stimuli may distribute drugs at a regulated pace in a stable, biologically active state (Nogueiras-Nieto et al., 2013). Novel multistimuli-responsive pharmaceuticals for theranostic application include a copolymer of N-isopropyl acrylamide-co-itaconic anhydride and a macro branched terpolymer resulting from its partial esterification with monometoxy PEG. They created NC carriers containing Fe_3O_4 nanoparticles (Xu et al., 2013).

Nanotechnology is becoming popular for diagnosing and treating various cancers. Nanocarriers have taken the place of conventional anticancer drug delivery platforms because of their nonspecificity, burst release, adverse effects, and lack of injury to normal cells. Nanocarriers enhance anticancer medication efficacy by concentrating on the target location. Few nanocarriers are clinically authorized for specific site effects (Hu et al., 2015).

9.3.6 NANOCOMPOSITE FOR BONE TISSUE REGENERATION

NC natural bone tissue is physically and biologically appropriate. Regenerating bone tissue needs biomaterials that mimic biological bone tissue. NCs are excellent for bone tissue regeneration because they offer the correct matrix environment, combine desired biological features, and provide several growth factors sequentially (Figure 9.2). No substance matches natural bone's composition, structure, and characteristics (Storm et al., 1995; Kumar et al., 2014). Nanohydroxyapeptite of graphene oxide has been proven for bone tissue regeneration. Furthermore, Purohit et al. (2020) developed a new biodegradable NC scaffold by crosslinking soy protein with hydroxyethyl cellulose, and these developed NCs improved physical and mechanical characteristics, ideal for bone tissue regeneration.

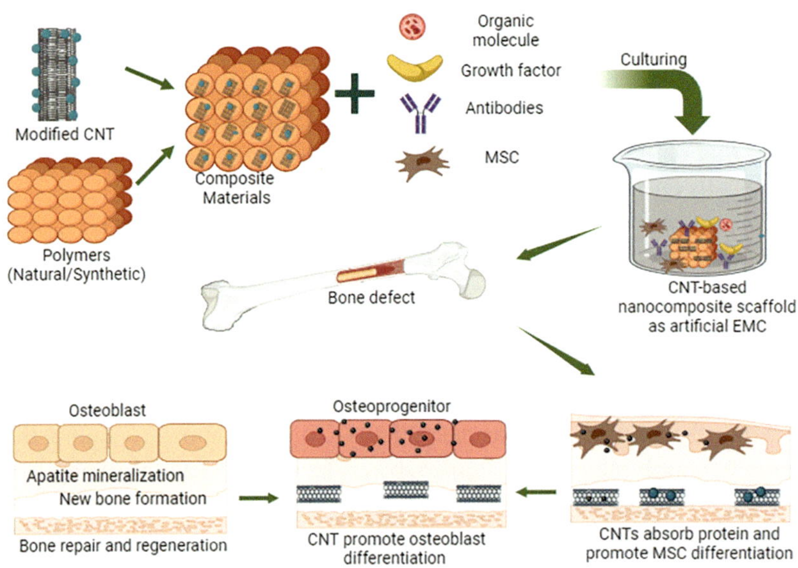

FIGURE 9.2 Role of scaffold nanocomposite in bone tissue engineering and regeneration.

9.3.7 NANOCOMPOSITE FOR COMBINATIONAL THERAPY FOR ALZHEIMER'S DISEASE (AD)

Alzheimer's is incurable and hard to diagnose. Most Alzheimer's disease (AD) medicines target amyloid-β (Aβ), although additional theranostics need consideration due to weak correlations between Aβ buildup and cognition and Phase III clinical trial failures. Recent investigations reveal that the tau pathway is intimately linked to AD symptoms, suggesting a therapeutic target. They demonstrated that these NCs decrease mitochondrial oxidative stress, tau hyperphosphorylation, and neuronal mortality *in vitro* and *in vivo* to cure AD. AD rats' memory deficiencies are greatly improved by MB-loaded CeNC/IONC/MSN-T807. According to their research, hyperphosphorylated tau-targeted multifunctional NCs could be used to treat Alzheimer's (Zhang et al., 2011; Ebadi and Rafati, 2015; Mittal et al., 2007).

We need a realistic, ultrasensitive, and specific sub-atomic detection approach to detect a mysterious illness-like promotion early. Recently discovered nanotechnology for sub-atomic biomarker detection is promising for AD diagnosis; with a realistic objective, it may identify subatomic particles *in vitro* or *in vivo* (Liu et al., 2005). Protein biomarkers are decorated to recognize DNA nanoparticles at 10–20 mol/L (10–17 mol/m^3 or attomolar) (Georganopoulou et al., 2005). The bio-barcode test matches hundreds of DNA barcodes to antibody target biomarkers using gold carrier nanoparticles (Ukmar et al., 2011). It may detect ultra-low protein biomarker levels. Decades of DNA barcodes may be replicated by polymerase chain reaction and identified by live signal transmittance to identify a biomarker unit. Researchers used a bio-standard tag test to detect oligomeric forms of amyloid β-derived diffusible ligand (ADDL) in Alzheimer's patients' cerebrospinal fluid (Caron, 2005).

A novel Alzheimer's biomarker was studied to generate ultra-sensitive, low-cost sub-nuclear biomarkers. A nano-sensor using localized surface plasmon resonance exploits trivalent silver NCs' unique optical characteristics. Each alteration around the outer nanoparticle surface alters the ambient magnetic range's refractive index. Spectroscopy can show the greatest variance in silver nanoparticle refractive index (Bhalekar et al., 2015).

9.3.8 NANOCOMPOSITES FOR THE TREATMENT OF RHEUMATOID ARTHRITIS

Recent attention has focused on nanomaterials for drug administration. These systems are more biocompatible, bioavailable, efficient, cytotoxic, controlled release, targeted drug delivery, and more than normal drug forms (Adibkia et al., 2015; Alexandre and Dubois, 2000). Recent tissue engineering and drug delivery devices utilize nanoscale ceramics-based calcium phosphates (Ali et al., 2019). Bioceramics like hydroxyapatite (HAp), $Ca_{10}(PO_4)_6(OH)_2$, are extensively used as drug carriers due to their biocompatibility, bioactivity, osteoconductivity, and osteoinductivity (Alimardani et al., 2019; Alzahrani et al., 2018). Some methods for making nanoHAp include wet chemical, hydrothermal, solid-state reaction, sol-gel, and microwave (Armstrong et al., 2017). The simplicity, low equipment cost, low processing temperature, and high product purity and consistency make wet chemical techniques better than other approaches. Tagai and Aoki introduced HAp synthesis (Hevin Rajesh, 2019). Calcium hydroxide with orthophosphoric acid makes HAp powders: $10Ca(OH)_2 + 6H_3PO_4 \rightarrow Ca_{10}(PO_4)_6(OH)_2 + 18H_2O$. TA, a synthetic corticosteroid, cures allergies, arthritis, eye problems, digestive difficulties, macular edema, and skin ailments (Abu Bakar et al., 2018). The literature says TA-loaded HA NCs have not been studied. Chemical precipitation created TA-loaded HA NCs in this investigation. Next, TA-loaded HA NCs' physicochemical and adsorption isotherms were assessed. MTT (3-[4,5-dimethylthiazol-2-yl]-2,5 diphenyl tetrazolium bromide) experiment tested TA-loaded HAp NCs' cytotoxicity. *In vivo*, the nanoformulation was tested for RA therapy in rats (Balaure and Grumezescu, 2020).

9.3.9 NANOCOMPOSITE IN CHRONIC WOUNDS AND BIOFILMS

Humans' largest organ is the integumentary system, including the skin. This organ contains bacteria, fungi, and viruses. The spatial arrangement of discrete microorganisms is contingent upon both the specific microhabitat and texture of the skin (such as burned, wet, or oily conditions), as well as the microenvironmental characteristics of the skin (including acidity levels and nutritional deficiencies) (Chaudhary et al., 2018). Biofilms are correlated with a range of human disorders. Facial, chest, and abdominal areas are prioritized. Chronic rhinosinusitis, otitis media, dental plaque, and chronic tonsillitis affect the face and cervical region. Dental plaques become apparent during the progression of the illness when a biofilm becomes sufficiently developed in the interdental and gingival regions. Dental plaque is a complex biofilm characterized by its distinct physical and functional organization (Siegel et al., 2019). Chronic wounds originate from wounds lasting beyond 45 days, making treatment difficult. The failure of the wounds to undergo healing may be ascribed to the disruption of physiological processes and the substantial influence of diverse microbiological organisms on the wound milieu. Elucidates the intricate interplay between several species, leading to the formation of bacterial biofilm-related diseases. Various forms of communication take place among the microorganisms that are linked with biofilms. Pressure, venous leg, and diabetic foot ulcers are chronic lesions. Cancer, inflammatory, and ischemic ulcers are also chronic wounds. The wound microbiology shows Serratia, Pseudomonas, Staphylococcus, and Enterobacter (Li et al., 2016).

9.3.10 NANOCOMPOSITE FOR PRESSURE ULCER

The research conducted on decubitus ulcers, also referred to as bedsores or pressure ulcers, has been limited in scope. The prevalence of this particular ulcer, typically seen in medical conditions, is higher among elderly individuals who have transitioned to long-term care facilities (LTCFs). Pressure ulcers, often referred to as bedsores, frequently develop in individuals who are subjected to prolonged immobilization. These ulcers are formed as a consequence of the pressure applied to the skin caused by friction and the continuous force imposed by the body's weight. The application of initial pressure for around 1–2 h results in cell deformation driven by pressure, ultimately leading to cell death (Thakur et al., 2017). Antibiotics have been the first-line treatment for pressure ulcers like doxycycline and amoxicillin. Since these antibiotics do not target the affected ulcer area, researchers have developed doxycycline and amoxicillin NCs loaded with magnesium aluminium double-layered hydroxide NCs, which have been found to prevent ulcer formation and improve wound healing time duration. Similarly, Chin et al. (2020) developed and identified a phomopsidione-loaded chitosan-polyethylene glycol nanocomposite hydrogel (C/PEG/Ph) as an antimicrobial dressing. Physically, the C/PEG/Ph hydrogel showed a uniform light blue color and had good attributes such as softness, flexibility, elasticity, and lack of aggregation. Fourier transform infrared (FTIR) spectroscopy analysis showed a significant shift toward lower frequencies at 1600 and 1554

cm^{-1}, which suggests successful phomopsidione incorporation into the hydrogel matrix. In the drug release assay, the hydrogel reached a plateau of release at 24 h with a cumulative release of 67.9 ± 6.4%, and no burst release effect was found throughout the experimental period. Antimicrobial testing proved that the C/PEG/Ph hydrogel provided a 99.99–100% inhibition of microbial growth against four Gram-negative bacterial species and one yeast strain. The evidence indicates that the C/PEG/Ph hydrogel has great promise as an antimicrobial dressing for pressure ulcer management.

9.3.11 NANOCOMPOSITE FOR DIABETIC ULCER

Diabetes is a long-term medical condition brought on by a problem with the metabolism of glucose. The blood viscosity of diabetic patients increases due to long-term hyperglycaemia; due to this, the elasticity of the blood vessels gets weakened, and the function of the local skin barrier also gets weakened, which causes diabetic ulcers like buttock pressure and diabetic foot ulcers. In normal individuals, the skin heals naturally after inflammation, proliferation, maturation, and hemostasis when it gets damaged. However, the diabetic patient's wound cannot possess the normal cell repair process, which results in delayed or no healing and necessitates long-term bed rest. These include bacterial infection, insufficient angiogenesis, hypoxia of the wound tissue, an imbalance in the inflammatory response surrounding the ulcer site, and other variables.

Diabetic ulcers can now be treated and managed using a variety of clinical techniques, including debridement, prompt wound dressing, blood glucose control, and infection prevention and control. Nevertheless, the efficacy of the treatment is quite low. After 120–140 days, only a small portion of diabetic ulcers heal, even with this standardized management, and the majority of patients need more advanced treatments. It has been investigated that a moist and clean environment helps in wound healing. Thus, hydrogel-based wet wound dressings have caught the interest of researchers. A number of disadvantages include poor mechanical strength, trouble regulating drug release, susceptibility to degradation, and expensive storage needs.

To overcome these limitations, researchers have found that the NC hydrogel can be proven effective for diabetic ulcers (Figure 9.3).

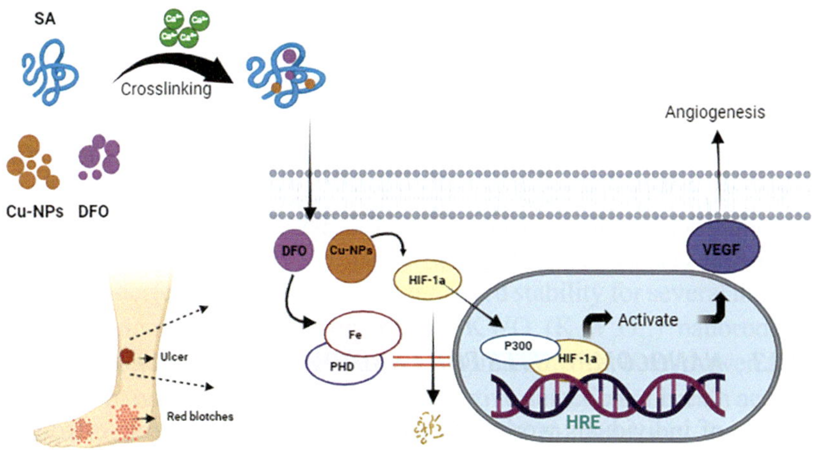

FIGURE 9.3 Nanocomposite in diabetic ulcer.

9.3.12 *NANOCOMPOSITES FOR OCULAR DISEASES*

Around the world, ocular illnesses cause vision impairment or total blindness in their victims. In 2020, research found that 43 million individuals were blind, and 596 million had distant vision impairment globally. Among the eye conditions are diabetic retinopathy, age-related macular degeneration, glaucoma, uveitis, and dry eye syndrome. All these diseases need pharmacotherapy. Conventional treatments are available for the treatment and management of ophthalmic diseases, such as eye drops. However, their uses are restricted due to insufficient bioavailability and frequent dosing. To overcome these restrictions and limitations, scientists and researchers have developed NCs for ocular diseases (Fang et al., 2021). Hyaluronic acid (HA) has been known for its unique properties like viscoelasticity, biodegradability, and biocompatibility, and it is a normal part of the tissue in the eyes as well. Because of this, HA has been proven as a carrier for ocular drug delivery systems. Latanoprost is a drug that decreases intra-ocular pressure and is used in the treatment of various ocular diseases. In 2013, Leonardus et al. developed a latanoprost HA-based nanocomposite that shows a longer sustained release of the drug when compared with liposomal formulation (Widjaja et al., 2014). The vascular endothelial growth factor (VEGF) induces angiogenetic ocular diseases. Asaf et al. (2019) developed a carbon dot NC VEGF inhibitor that has been proven to treat diabetic retinopathy and age-related macular degeneration (Shoval et al., 2019).

for the detection of dopamine in marketed preparation samples (Mathew et al., 2018). α-Syncline overexpression causes the death of dopaminergic neurons. For the inhibition of α-synuclein overexpression, gold NCs were synthesized and loaded with DNA plasmid that targets the specific cell types.

9.4 NANOCOMPOSITES FUTURE PROSPECTS

The proliferation of commercial purposes using NCs has been seeing significant growth. According to recent reports, global output is projected to surpass 600,000 tons over less than 2 years. Moreover, it is expected that this expansion will include several substantial sectors during the next 5–10 years. Mechanical advances in NC materials have enhanced their automotive and general/industrial applications. Door handles, timing belt covers, intake manifolds, engine covers, and car mirror housings are a few examples of applications. These materials are also being investigated for use in mower hoods, power tool housings, hoover cleaner impellers and blades, and cell phone and pager covers (Bengalli et al., 2019).

NCs are being studied by firms and institutes globally, according to Cutting Edge (2001). Basell USA, Bayer, Dow Chemical, Eastman Chemical, Mitsubishi Gas Chemical, Nanocor, Triton Systems, Honeywell, and RTP Co. offer NC plastics. However, the present emphasis of these endeavours mostly revolves around polyolefins or nylons. Nevertheless, it is theoretically plausible to use clay nanoparticles in any resin category.

According to Leaversuch (2001), there has been a rise in optimism regarding these new materials since they gained attention in the industry a couple of years ago. Due to growing evidence of nylon/clay NCs, polyolefin variants, and other resin matrixes and nano-fillers, exploratory activities have increased. Leaversuch stresses NCs' potential.

The appeal of NCs stems from their ability to achieve significant results in small quantities. The aforementioned product exhibits notable enhancements in its oxygen, carbon dioxide, moisture, and smell barrier characteristics. Additionally, it demonstrates heightened rigidity, strength, and heat resistance while preserving the transparency of the film and its ability to withstand impact. The manufacturing sector is now directing its attention to emerging materials and their corresponding commercial uses (Fang et al., 2021).

The field of nanotechnology is significantly transforming the realm of materials. This innovation has significant potential in the development of novel composites that exhibit improved functioning and find extensive use across several domains. The information about the manufacturing, characterization, and applications of polymer NCs aids researchers in comprehending and effectively exploiting the distinctive chemical and material principles that underlie these advanced materials.

Despite significant potential of NCs in several industrial sectors, their mainstream commercialization is hindered by both technological and economic challenges. The aforementioned factors include the influence on performance, the intricate connections between formulations, and the methods used to attain and evaluate the dispersion and exfoliation of nanofillers inside the polymer matrix. A primary hindrance to the advancement of groundbreaking NC technologies is the constrained allocation of resources toward cutting-edge equipment and the augmentation of the central research team. An area of prospective investigation for future scholarly inquiry is the broadening of nanotechnology's utilization inside diverse polymer systems. Nevertheless, it is crucial to emphasize that the effective integration of these improvements will likely need the creation of innovative compatibility protocols (Widjaja et al., 2014).

9.5 CONCLUSION

NCs are novel methods of drug administration because they are easy to manipulate non-toxic and targeting sites of action. This formulation reduces dosing frequency, enhances bioavailability, and increases patient compliance. The use of nanotechnology could completely change our way of life. This is due to the nearly limitless potential for astounding advancements in almost every industry, from diagnosis to medicine. Nano pharmaceuticals have been discussed according to the various types of NCs and their physiochemical parameters. These groups include nanocrystals, liposomes, lipid-based polymeric nanoemulsion, and protein solid-lipid nanoparticles. When it comes to treating medical issues, the use of NCs has shown to have a lot of promise for better diagnosis, treatment, and monitoring of a number of serious conditions, such as cancer, rheumatoid arthritis, COPD, diabetes, neurological disorders, ocular diseases, and diabetic ulcers.

KEYWORDS

- **biocompatible**
- **biodegradable**
- **drug delivery**
- **nanocomposites**
- **polymers**

REFERENCES

Abu Bakar, M., McKimm, J., Haque, S. Z., Majumder, M. A., & Haque, M. (2018). Chronic tonsillitis and biofilms: A brief overview of treatment modalities. *Journal of Inflammation Research, 11*, 329–337.

Acquarulo, L. A., & O Neil, C. J. (2002). Enhancing Medical Device Performance with Nanocomposite Polymers. *Medical Device and Diagnostic Industry*, *24*(5), 114–123.

Adibkia, K., Nayebi, A. M., Barzegar-Jalali, M., Hosseinzadeh, S., Ghanbarzadeh, S., & Shiva, A. (2015). Comparison of the analgesic effect of diclofenac Sodium-Eudragit® RS100 solid dispersion and nanoparticles using formalin test in the rats. *Advanced Pharmaceutical Bulletin, 5*(1), 77.

Alexandre, M., & Dubois, P. (2000). Polymer-layered silicate nanocomposites: preparation, properties and uses of a new class of materials. *Materials Science and Engineering: R,: Reports, 28*(1-2), 1–63.

Ali, K., Ahmed, B., Ansari, S. M., Saquib, Q., Al-Khedhairy, A. A., Dwivedi, S., Alshaeri, M., Khan, M. S., & Musarrat, J. (2019). Comparative *in situ* ROS mediated killing of bacteria with bulk analogue, Eucalyptus leaf extract (ELE)-capped and bare surface copper oxide nanoparticles. *Materials Science and Engineering: C, 100*, 747–758.

Alimardani, V., Abolmaali, S. S., & Borandeh, S. (2019). Antifungal and antibacterial properties of graphene-based nanomaterials: A mini-review. *Journal of Nanostructures, 9*(3), 402–413.

Alzahrani, K. E., Niazy, A. A., Alswieleh, A. M., Wahab, R., El-Toni, A. M., & Alghamdi, H. S. (2018). Antibacterial activity of trimetal (CuZnFe) oxide nanoparticles. *International Journal of Nanomedicine, 13*, 77–87.

Ama, O., Sadiq, M., Johnson, M., Zhang, Q., & Wang, D. (2020). Novel 1D/2D KWO/Ti3C2Tx nanocomposite-based acetone sensor for diabetes prevention and monitoring. *Chemosensors, 8*(4), 102.

Ananda, S. R., Kumari, L., & MV, M. (2022). Studies on room-temperature acetone sensing properties of ZnCo2O4/PPy and MnCo2O4/PPy nanocomposites for diabetes diagnosis. *Applied Physics A, 128*(8), 669.

Araújo, J., Garcia, M. L., Mallandrich, M., Souto, E. B., & Calpena, A. C. (2012). Release profile and transscleral permeation of triamcinolone acetonide loaded nanostructured lipid carriers (TA-NLC): *In vitro* and *ex vivo* studies. *Nanomedicine: Nanotechnology, Biology and Medicine, 8*(6), 1034–1041.

Araujo, J., Gonzalez-Mira, E., Egea, M. A., Garcia, M. L., & Souto, E. B. (2010). Optimization and physicochemical characterization of a triamcinolone acetonide-loaded NLC for ocular antiangiogenic applications. *International Journal of Pharmaceutics, 393*(1-2), 168–176.

Armstrong, D. G., Boulton, A. J., & Bus, S. A. (2017). Diabetic foot ulcers and their recurrence. *New England Journal of Medicine, 376*(24), 2367–2375.

Bahari, A., Roeinfard, M., Ramzannezhad, A., Khodabakhshi, M., & Mohseni, M. (2019). Nanostructured features and antimicrobial properties of Fe_3O_4/ZnO Nanocomposites. *National Academy Science Letters, 42*(1), 9–12.

Balaure, P. C., & Grumezescu, A. M. (2020). Recent advances in surface nanoengineering for biofilm prevention and control. Part I: Molecular basis of biofilm recalcitrance. Passive anti-biofoulingnanocoatings. *Nanomaterials, 10*(6), 1230.

Bengalli, R., Ortelli, S., Blosi, M., Costa, A., Mantecca, P., & Fiandra, L. (2019). *In vitro* toxicity of TiO2: SiO2 nanocomposites with different photocatalytic properties. *Nanomaterials, 9*(7), 1041.

Bhalekar, M. R., Upadhaya, P. G., Nalawade, S. D., Madgulkar, A. R., & Kshirsagar, S. J. (2015). Anti-rheumatic activity of chloroquine-SLN gel on wistar rats using complete freund's adjuvant (CFA) model. *Indian Journal of Rheumatology, 10*(2), 58–64.

Caron, J. P. (2005). Intra-articular injections for joint disease in horses. *Veterinary Clinics: Equine Practice, 21*(3), 559–573.

Chaudhary, Z., Ahmed, N., ur. Rehman, A., & Khan, G. M. (2018). Lipid polymer hybrid carrier systems for cancer targeting: A review. *International Journal of Polymeric Materials and Polymeric Biomaterials, 67*(2), 86–100.

da Silva, G. R., da Silva-Cunha, Jr. A., Behar-Cohen, F., Ayres, E., & Orefice, R. L. (2011). Biodegradable polyurethane nanocomposites containing dexamethasone for ocular route. *Materials Science and Engineering: C, 31*(2), 414–422.

Din, F. U., Aman, W., Ullah, I., Qureshi, O. S., Mustapha, O., Shafique, S., & Zeb, A. (2017). Effective use of nanocarriers as drug delivery systems for the treatment of selected tumors. *International Journal of Nanomedicine, 12*, 7291–7309.

Ebadi, A., & Rafati, A. A. (2015). Preparation of silica mesoporous nanoparticles functionalized with β-cyclodextrin and its application for methylene blue removal. *Journal of Molecular Liquids, 209*, 239–245.

Fang, G., Yang, X., Wang, Q., Zhang, A., & Tang, B. (2021). Hydrogels-based ophthalmic drug delivery systems for treatment of ocular diseases. *Materials Science and Engineering: C, 127*, 112212.

Georganopoulou, D. G., Chang, L., Nam, J. M., Thaxton, C. S., Mufson, E. J., Klein, W. L., & Mirkin, C. A. (2005). Nanoparticle-based detection in cerebral spinal fluid of a soluble pathogenic biomarker for Alzheimer’s disease. *Proceedings of the National Academy of Sciences, 102*(7), 2273–2276.

Haes, A. J., Chang, L., Klein, W. L., & Van Duyne, R. P. (2005). Detection of a biomarker for Alzheimer's disease from synthetic and clinical samples using a nanoscale optical biosensor. *Journal of the American Chemical Society, 127*(7), 2264–2271.

Hashizume, M., & Mihara, M. (2011). The roles of interleukin-6 in the pathogenesis of rheumatoid arthritis. *Arthritis, 2011*(1), 765624.

Hevin Rajesh, D. (2019). Authentication technique using ACO in WSN. *National Academy Science Letters, 42*(1), 19–23.

Hu, L., Sun, H., Zhao, Q., Han, N., Bai, L., Wang, Y., Jiang, T., & Wang, S. (2015). Multilayer encapsulated mesoporous silica nanospheres as an oral sustained drug delivery system for the poorly water-soluble drug felodipine. *Materials Science and Engineering: C, 47*, 313–324.

Ignjatović, N., Uskoković, V., Ajduković, Z., & Uskoković, D. (2013). Multifunctional hydroxyapatite and poly (D, L-lactide-co-glycolide) nanoparticles for the local delivery of cholecalciferol. *Materials Science and Engineering: C, 33*(2), 943–950.

Jahangiri, A., Davaran, S., Fayyazi, B., Tanhaei, A., Payab, S., & Adibkia, K. (2014). Application of electrospraying as a one-step method for the fabrication of triamcinolone acetonide-PLGA nanofibers and nanobeads. *Colloids and Surfaces B: Biointerfaces, 123*, 219–224.

Janib, S. M., Liu, S., Park, R., Pastuszka, M. K., Shi, P., Moses, A. S., ... & MacKay, J. A. (2013). Kinetic quantification of protein polymer nanoparticles using non-invasive imaging. *Integrative Biology, 5*(1), 183–194.

Javadzadeh, Y., Atashgah, R. B., Barzegar-Jalali, M., Soleimani, F., Mohammadi, G., Sabzevari, A., & Adibkia, K. (2014). Inclusion of piroxicam in mesoporous phosphate glass–ceramic and evaluation of the physiochemical characteristics. *Colloids and Surfaces B: Biointerfaces, 116*, 751–756.

Johnson, K. S., & Chu, D. S. (2010). Evaluation of sub-Tenon triamcinolone acetonide injections in the treatment of scleritis. *American Journal of Ophthalmology, 149*(1), 77–81.

José-Yacamán, M., Gutierrez-Wing, C., Miki, M., Yang, D. Q., Piyakis, K. N., & Sacher, E. J. T. J. (2005). Surface diffusion and coalescence of mobile metal nanoparticles. *The Journal of Physical Chemistry B, 109*(19), 9703–9711.

Kalidoss, R., Umapathy, S., & Sivalingam, Y. (2018). An investigation of GO-SnO2-TiO2 ternary nanocomposite for the detection of acetone in diabetes mellitus patient's breath. *Applied Surface Science, 449*, 677–684.

Khan, M. A., Raza, A., Ovais, M., Sohail, M. F., & Ali, S. (2018). Current state and prospects of nano-delivery systems for sorafenib. *International Journal of Polymeric Materials and Polymeric Biomaterials, 67*(18), 1105–1115.

Kumar, D., Chirravuri, S. S., & Shastri, N. R. (2014). Impact of surface area of silica particles on dissolution rate and oral bioavailability of poorly water soluble drugs: A case study with aceclofenac. *International Journal of Pharmaceutics, 461*(1-2), 459–468.

Li, L., Yang, W. W., & Xu, D. G. (2019). Stimuli-responsive nanoscale drug delivery systems for cancer therapy. *Journal of Drug Targeting, 27*(4), 423–433.

Li, N., Zhao, L., Qi, L., Li, Z., & Luan, Y. (2016). Polymer assembly: Promising carriers as co-delivery systems for cancer therapy. *Progress in Polymer Science, 58*, 1–26.

Liao, X., Yang, X., Deng, H., Hao, Y., Mao, L., Zhang, R., Liao, W., & Yuan, M. (2020). Injectable hydrogel-based nanocomposites for cardiovascular diseases. *Frontiers in Bioengineering and Biotechnology, 8*, 251.

Liu, Z., Fan, T., Zhang, W., & Zhang, D. The synthesis of hierarchical porous iron oxide with wood templates. *Microporous and Mesoporous Materials, 85*(1-2), 82–88.

Mathew, G., Dey, P., Das, R., Chowdhury, S. D., Das, M. P., Veluswamy, P., Neppolian, B., & Das, J. (2018). Direct electrochemical reduction of hematite decorated graphene oxide (α-Fe2O3@ erGO) nanocomposite for selective detection of Parkinson's disease biomarker. *Biosensors and Bioelectronics, 115*, 53–60.

Mattu, C., Silvestri, A., Wang, T. R., Boffito, M., Ranzato, E., Cassino, C., Ciofani, G., & Ciardelli, G. (2016). Surface-functionalized polyurethane nanoparticles for targeted cancer therapy. *Polymer International, 65*(7), 770–779.

Mittal, A., Kurup, L., & Mittal, J. (2007). Freundlich and Langmuir adsorption isotherms and kinetics for the removal of Tartrazine from aqueous solutions using hen feathers. *Journal of Hazardous Materials, 146*(1-2), 243–248.

Nazem, A., & Mansoori, G. A. (2011). Nanotechnology for Alzheimer's disease detection and treatment. *Insciences Journal, 1*(4), 169–193.

Nogueiras-Nieto, L., Delgado-Charro, M. B., & Otero-Espinar, F. J. (2013). Thermogelling hydrogels of cyclodextrin/poloxamerpolypseudorotaxanes as aqueous-based nail lacquers: Application to the delivery of triamcinolone acetonide and ciclopiroxolamine. *European Journal of Pharmaceutics and Biopharmaceutics, 83*(3), 370–377.

Örüm, S. M., & Demircioğlu, Y. S. (2019). One-pot synthesis and characterization of crosslinked polyphosphazene dopamine microspheres for controlled drug delivery applications. *Journal of Macromolecular Science, Part A Pure and Applied Chemistry, 56*(9), 1–6.

Obinna, O. (2020). Impact of interest rate deregulation on investment growth in Nigeria. *International Journal of Economics and Financial Issues, 10*(2), 170.

Payab, S., Jafari-Aghdam, N., Barzegar-Jalali, M., Mohammadi, G., Lotfipour, F., Gholikhani, T., & Adibkia, K. (2014). Preparation and physicochemical characterization of the azithromycin-Eudragit RS100 nanobeads and nanofibers using electrospinning method. *Journal of Drug Delivery Science and Technology, 24*(6), 585–590.

Purohit, G., Rawat, D. S., & Reiser, O. (2020). Palladium Nanocatalysts Encapsulated on Porous Silica@ Magnetic Carbon-Coated Cobalt Nanoparticles for Sustainable Hydrogenation of Nitroarenes, Alkenes and Alkynes. *ChemCatChem, 12*(2), 569–575.

Rytting, M. E. (2012). Role of L-asparaginase in acute lymphoblastic leukemia: focus on adult patients. *Blood and Lymphatic Cancer: Targets and Therapy,* 117–124.

Shoval, A., Markus, A., Zhou, Z., Liu, X., Cazelles, R., Willner, I., & Mandel, Y. Anti-VEGF-aptamer modified C-dots—a hybrid nanocomposite for topical treatment of ocular vascular disorders. *Small, 15*(40), 1902776.

Siddique, Y. H., Khan, W., Fatima, A., Jyoti, S., Khanam, S., Naz, F., Rahul, Ali, F., Singh, B. R., & Naqvi, A. H. (2016). Effect of bromocriptine alginate nanocomposite (BANC) on a transgenic Drosophila model of Parkinson's disease. *Disease Models & Mechanisms, 9*(1), 63–68.

Siddique, Y. H., Khan, W., Singh, B. R., & Naqvi, A. H. (2013). Synthesis of alginate-curcumin nanocomposite and its protective role in transgenic Drosophila model of Parkinson's disease. *International Scholarly Research Notices, 2013*, 794582.

Siegel, R. L., Miller, K. D., & Jemal, A. (2019). Cancer statistics, 2019. *CA: A Cancer Journal for Clinicians, 69*(1), 7–34.

Storm, G., Belliot, S. O., Daemen, T., & Lasic, D. D. (1995). Surface modification of nanoparticles to oppose uptake by the mononuclear phagocyte system. *Advanced Drug Delivery Reviews, 17*(1), 31–48.

Sutar, S., Patil, P., & Jadhav, J. (2022). Recent advances in biochar technology for textile dyes wastewater remediation: A review. *Environmental Research, 209,* 112841.

Thakur, S., Pramod, K. S., & Malviya, R. (2017). Utilization of polymeric nanoparticle in cancer treatment: A review. *Journal of Pharmaceutical Care Health Systems, 4*, 172.
Ukmar, T., Maver, U., Planinšek, O., Kaučič, V., Gaberšček, M., & Godec, A. (2011). Understanding controlled drug release from mesoporous silicates: Theory and experiment. *Journal of Controlled Release, 155*(3), 409–417.
Wang, L., Feng, M., Li, Y., Du, Y., Wang, H., Chen, Y., & Li, L. (2019). Fabrication of superparamagnetic nano-silica@ quercetin-encapsulated PLGA nanocomposite: Potential application for cardiovascular diseases. *Journal of Photochemistry and Photobiology B: Biology, 196*, 111508.
Wang, Y., Zhao, Q., Han, N., Bai, L., Li, J., Liu, J., Che, E., Hu, L., Zhang, Q., Jiang, T., & Wang, S. (2015). Mesoporous silica nanoparticles in drug delivery and biomedical applications. *Nanomedicine: Nanotechnology, Biology and Medicine, 11*(2), 313–327.
Widjaja, L. K., Bora, M., Chan, P. N., Lipik, V., Wong, T. T., & Venkatraman, S. S.(2014). Hyaluronic acid-based nanocomposite hydrogels for ocular drug delivery applications. *Journal of Biomedical Materials Research Part A, 102*(9), 3056–3065.
Yasir, M., & Sara, U. V. S. (2013). Preparation and optimization of haloperidol loaded solid lipid nanoparticles by Box–Behnken design. *Journal of Pharmacy Research, 7*(6), 551–558.
Xu, Q., Kambhampati, S. P., & Kannan, R. M. (2013). Nanotechnology approaches for ocular drug delivery. *Middle East African Journal of Ophthalmology, 20*(1), 26.
Zhang, Y., Zhang, J., Jiang, T., & Wang, S. (2011). Inclusion of the poorly water-soluble drug simvastatin in mesocellular foam nanoparticles: Drug loading and release properties. *International Journal of Pharmaceutics, 410*(1-2), 118–124.

CHAPTER 10

Applications of Nanocomposites for Cancer Therapy

AJAY SINGH[1,2], BIKRAM SINGH[1,2], RUHI SINGH[1], KULJIT SINGH[2,3], and BOOBALAN GOPU[1,2,*]

[1]*Pharmacology Division, CSIR – Indian Institute of Integrative Medicine, Jammu, Jammu and Kashmir, India*

[2]*Academy of Scientific and Industrial Research (AcSIR), Ghaziabad, Uttar Pradesh, India*

[3]*Infectious Diseases Division, CSIR – Indian Institute of Integrative Medicine, Jammu, Jammu and Kashmir, India*

*Corresponding author.

ABSTRACT

Cancer is a life-threatening disease, with a high incidence rate and the leading cause of mortality. Nanocomposites (NCs) have emerged as prospective therapeutic agents for cancer treatment because of their specific properties, adaptability, and cost-effectiveness. NC are multiphase nanomaterials with several benefits, such as enhanced stability, prolonged and controlled release, and biocompatibility, making them elusive targets for overcoming the limitations of traditional cancer therapies. Furthermore, NCs enable the integration of numerous capabilities within a system, enabling synergistic approaches to improve therapeutic effects. Recent advancements in NC-based theranostic platforms are under investigation for treating esophageal cancer, liver cancer, melanoma, and leukemia. This chapter explores the diverse applications of NCs in cancer therapy, imaging, photothermal therapy,

Nanocomposites and Nanomaterials in Biomedical Applications. Vivek Dave, Deependra Singh, and Pallerla Naveen Reddy (Eds.)

DOI: 10.1201/9781779641236-10

and targeted drug delivery. NC can revolutionize cancer treatment approaches and pave the way for future research and development in this exciting field.

10.1 INTRODUCTION

Cancer is a life-threatening dreadful condition that shortens life expectancy worldwide (Bray et al., 2021). Cancer remains a significant concern as the population ages due to its high incidence (19 million new cases), mortality rate (9.9 million deaths), and impetus to the quality of life and living costs (Sung et al., 2021). Despite scientific advances in oncological research, existing treatment modalities with limitations may not overcome tumor progression, multidrug resistance, selectivity, and adverse effects resulting in poor clinical outcomes. Noninvasive multimodal technologies, including photodynamic, chemodynamic, sonodynamic, photothermal, and magnetic hyperthermia therapies, have become accessible due to the growing advances in oncology research (Zhu et al., 2022). Nanoplatforms intertwined with traditional treatments like radiotherapy, surgery, and chemotherapy are synergistically applied to target cancer growth and metastasis.

Nanomaterial-based chemotherapy possesses potential prospects due to its low volume-to-surface ratio, minimal toxicity, high selectivity, better bioavailability, and efficacy (Xu et al., 2014). Nanocomposites (NCs) and their functions create a versatile framework for novel nanomaterials with diverse characteristics, notably in drug-targeting tumor cells. Custom-made NC's applicability depends on its physicochemical properties, and several metals-based NCs are applied for timely diagnosis and delivery of drugs to the tumor cells, improving the treatment efficacy (Kumar et al., 2023). Cellular uptake of NCs increased in cancer tissues, resulting in significant antitumor activity. Similarly, numerous metal-based NCs have demonstrated tremendous medical significance in cancer therapies, drug administration, and imaging. NCs enhance the versatility of the drug delivery mechanism by enabling the conjugation of pharmaceuticals to both soft materials, such as polymeric nanogel, and hard materials, such as metals (Mishra et al., 2018). A regulated or sustained therapeutic dose can be delivered to the target site using biodegradable polymer-based NCs.

In recent years, NC's applicability has drawn many scientists' attention to creating nanosized composite materials for efficient and reliable identification and therapy with better targeting efficacy and improving the quality of life. The tremendous potential of these cross-disciplinary platforms emerged as promising strategies for the formidable challenges in cancer therapeutics. Though the subject is widely researched, a few notable contributions are included in the following section for readers understanding.

10.2 CANCER THERAPY

Cancer is the leading cause of mortality rates following cardiovascular diseases, which killed 6.1 million people globally in 2019. The incidence and mortality rates differ depending on the origin of the cancer (Qamar and Communications, 2021). Nanotechnology offers potent functionality to create an advanced and new drug carrier system. Nanocarriers are expected to deliver medication molecules to specific areas where they will destroy tumor cells without harming normal cells (Huang et al., 2020). NCs have been classified into nonpolymer and polymer composites depending on the nature of the biological matrix for fillers. Polymers of natural origin contain alginate, chitosan (CS), inulin, or synthetic substances such as polylactide-co-glycolide (PLGA) and polyglycolic acid (PGA). NCs have numerous applications in cancer diagnosis and treatments, as illustrated in Figure 10.1.

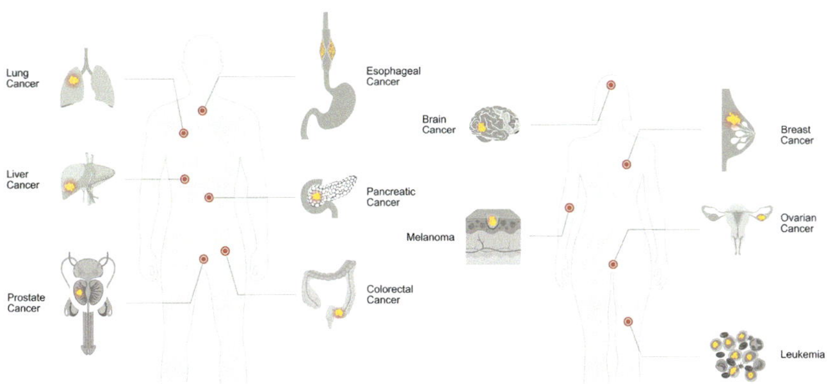

FIGURE 10.1 Nanocomposite applications in various human cancers.

Nanotechnologists manipulate ligand–receptor interactions to target cancer hallmarks. Folic acid (FA or vitamin B9) is a crucial biomolecule involved in the production of DNA, and its folic acid receptor (FAR) is found on cell surfaces. It induces cellular internalization of FA in vesicular structures *via* an endocytic mechanism. Numerous epithelial tissue malignancies, including those of the nose, brain, ovary, mammary gland, and prostate, overexpress FARs. Therefore, FA might be the targeted moiety that uses a specific interaction to deliver the anticancer medication to cell-producing FARs. Thus, folate can be used to decorate NCs. Iron oxide NC with aminated starch/ZnO coating was recently developed and designed to deliver the natural compound curcumin (CUR) using FA as a targeting molecule (Saikia et al., 2017).

10.2.1 ESOPHAGEAL CANCER

Esophageal cancer (EC) is an aggressive gastrointestinal cancer that ranks seventh most common cancer type. EC incidence rates have increased exponentially in recent years, with China accounting for more than 50% of observed cases (Rogers et al., 2022). Various approaches such as endoscopic resection, ablation, surgery, and chemo-, radio- and immunotherapy are employed to treat EC patients.

Generally, polymeric nano-formulations contain CS, a biopolymer found in fungal cell walls, employed as a drug delivery agent (Mohammed et al., 2017). The well-established polyethylene glycol (PEG)-coatings reduced opsonization and enhanced the circulating duration and effectiveness. Additionally, PEG polymer-based NCs enhanced drug delivery (Mishra et al., 2016). Results have demonstrated that the CS-sodium alginate-polyethylene glycol-crocin NCs (CSP-Cr-NCs) formulation effectively inhibited KYSE-150 cells growth in *in vitro* and exacerbated oxidative stress, reduced mitochondrial membrane potential (MMP), increased reactive oxygen species (ROS) buildup, and triggered apoptosis in these cells. The CSP-Cr-NCs additionally displayed antibacterial solid capabilities. The findings conclude that CSP-Cr-NCs might be employed as anticancer nano-formulations for EC (Hu et al., 2022).

Molybdenum-containing enzymes, such as sulfite oxidase, aldehyde oxidase, and xanthine oxidase, are essential for human metabolism. Inadequate molybdenum intake has been linked to EC (Mendel, 2009).

Numerous studies demonstrated the excellent biocompatibility of molybdenum and sulfur (MoS_2) nanoparticles (NPs). MoS_2 offers exceptional chemical and thermal resistance due to its semiconductor characteristics. MoS_2 can be loaded with diverse biomolecules, drugs, fluorescent molecules, and other nanomaterials to improve its tumor targeting, precise detection of specific biomarkers, and ability to respond precisely to the tumor milieu. This approach reduces side effects during treatment. MoS_2's outstanding photothermal conversion capabilities and near-infrared (NIR) light absorption characteristics have aided phototherapy and photoacoustic (PA) imaging techniques. MoS_2-NCs can be used for radiation treatment, immunotherapy, combination therapy, bioimaging, and microwave hyperthermia. Because of these advantages, MoS_2 NCs are usually considered extremely promising for cancer therapy and diagnosis. The study on MoS_2-NCs is still in its early phases, and many questions remain unresolved (Chen et al., 2015).

10.2.2 LIVER CANCER

NC-effective therapeutic strategies are in high demand due to increased therapeutic potential. A study was conducted to formulate and evaluate selenium/CS/PEG/allyl isothiocyanate NCs (SCPg-AI-NCs) against diethylnitrosamine (DEN)-induced hepatic carcinoma (HCC) in rat models. The results conclude that the SCPg-AI-NCs therapy reduces tumor markers with oxidative stress response. In the animal model, SCPg-AI-NCs significantly reduced pro-apoptotic proteins and increased anti-apoptotic proteins such as Bcl-2 family proteins. The therapy reduced *in vitro* cell survivability and induced cell death in $HepG_2$ cells. These findings suggested that these NCs may become a potent anticancer agent for treating liver carcinogenesis in the future (Wang et al., 2022).

A similar investigation against HCC was performed to determine the anticancer effectiveness of bio-nanomaterial CS coating iron oxide (Fe_3O_4/Cs) NC. The combination of Fe_3O_4 NPs and CS solution was subjected to gamma radiation at 20 kGy and administered to rats through intraperitoneal injections to treat DEN-induced liver cancer. In these research findings, inflammatory markers such as iNOS and TLR4 and the mTOR and MAPK signaling genes were upregulated in untreated controls. Additionally, DEN enhanced the activity of liver enzymes such as gamma-glutamyl transferase

(GGT) and alanine aminotransferase (ALT) while decreasing cleaved caspase-3 expression. The characteristics mentioned above were significantly improved following the administration of Fe_3O_4/Cs NCs (Badawy et al., 2023). Sorafenib (SOR), a tyrosine kinase inhibitor, targets the Raf family of kinases, vascular endothelial growth factor (VEGF) receptors, and platelet-derived growth factor (PDGF) receptors. Combination therapies are gaining popularity since they enhance interactions among multiple modalities to provide complementary restorative measures. A nanosystem that concurrently distributes zinc phthalocyanine and SOR to treat HCC has previously been developed (Zhang et al., 2020). This nanosystem was a cutting-edge technique to improve the therapeutic impact of SOR when HCC was treated with photodynamic therapy (PDT). For combinational photo-chemodynamic therapy in hepatic cancer, a new amphiphilic iridium complex-loading SOR was synthesized. The cleverly designed compound induced the PDT effect, encouraging cellular apoptosis, generating ROS, and inhibiting tumor proliferation and metastasis. Various cytotoxicity assays revealed that a defined ratio of iridium-based photosensitizer (IP) and SOR in iridium-based photosensitizer with Sorafineb (IPS) could improve combinational therapy efficacy while limiting the SOR adverse effects. The IPS demonstrated better biocompatibility and did not affect any organs. IPS's exceptional synergetic anticancer activity was also reconfirmed in the HCC xenograft tumor model. The combined findings offer a framework for developing an advanced medication to investigate the IPS for combinational liver cancer treatment to optimize remedial efficacy (Abuduwaili et al., 2022).

10.2.3 MELANOMA

Melanoma is a deadly skin cancer that is difficult to cure with conventional drugs and surgeries. Graphene NCs offer a larger surface area, which allows intended drugs to be distributed to target sites. Therapeutic agents are loaded using wetting behavior and ring stacking, and composite hydrogels are produced with a uniform dispersion of graphene-based NPs in media (Karimi et al., 2016). Hydrogels in a regulated and specific manner under various physiological situations aimed to treat skin cancer. Hydrogels constructed from biopolymers such as CS, bovine serum albumin, and carboxymethyl cellulose have been extensively researched as drug delivery systems (Khan et al., 2020).

Researchers developed rGO-5FU-CMARX, a novel NC hydrogel containing fluorouracil-loaded graphene oxide (GO) and carboxy methyl arabinoxylan to treat melanoma. These hydrogels inhibited the cancerous growth of U-87 cells. Their surface component delivered a substantial amount of these hydrogels under photothermal exposure. These hydrogels provided a sustained drug release when exposed to high humidity and temperature. The findings indicated that NC hydrogels could be used as biomaterials for treating and managing this lethal skin disease. These gels can be utilized in clinical settings to treat skin cancer-related infections (Nazir et al., 2021).

Also, undetectable microtumor remnants remain a worry for cancer recurrence despite notable improvements in surgical procedures after melanoma tumor removal. A research attempt was made to develop a hydrogel by mixing CS with intestinal methacrylate aided by reduced GO. The smart hydrogel can release doxorubicin (DOX) and cause localized hyperthermia in response to NIR light stimulation. These hydrogels form the polymer by adding methacryolyl that can self-reduce GO at body temperature by spreading ascorbic acid into the hydrated structure. Also, a wound-healing potential was observed when DOX was inactivated using modified GO. Additionally, an artificial *in silico* model that matched experimental findings allows the designing of therapies to increase tumor death while minimizing nonspecific injury to healthy tissues (Céspedes-Valenzuela et al., 2022). These findings might pave the way for developing multifunctional materials for cutting-edge cancer treatment applications.

A study was conducted to control tumor growth with polypyrrole and carbon dots with applications of sonosensitizer. The laboratory mice were exposed to varied carbonaceous polypyrrole (C/Ppy) concentrations and ultrasonic pulse ratios of 50 and 100% in this study. Despite being biocompatible, C/PPy inhibited cell growth by threefold when the pulse ratio of ultrasound (US) radiation was doubled. C/PPy effectively killed melanoma cancer cells as a sonosensitizer by increasing ROS generation and triggering apoptosis (Gorgizadeh et al., 2022).

10.2.4 LEUKEMIA

Many leukemic patients experience relapse despite receiving effective initial chemotherapy. As a result, alternative efforts to effectively halt and eradicate penetrative leukemic are strongly desired. Hematopoietic cells express the G-protein chemokine receptor CXCR4 (Bleul et al., 1996). Bone marrow stromal (MS-5) cells regularly release a stromal cell-derived factor ($CXCL_{12}$) for its ligand $CXCR_4$ (Tavernier-Tardy et al., 2009). Acute myeloid leukemia (AML) cell proliferation and its protection from chemotherapy-induced death may be facilitated by circulating AML cells adhering to stromal cells *via* the $CXCR_4$/$CXCL_{12}$ axis (Zeng et al., 2009). Researchers identified the $CXCR_4$-$CXCL_{12}$ axis as a prospective therapeutic target and developed a biomimetic NC utilizing a cell membrane coating approach to improve the antileukemic agents' effectiveness. An ultrathin platinum (Pt) nanozyme with the best possible catalytic characteristics was loaded into PLGA nanospheres with perfluorobromooctane (PFOB). The PFOB@PLGA@Pt core was modified using a mouse bone MS-5 cell as the shielding material and loaded with DOX to produce synergistic chemotoxicity. The results revealed that these NCs biomimic endogenous stromal cells and circulate longer to inhibit the AML cell migration through the axis mentioned and promote AML cell apoptosis (Kong et al., 2022).

A similar study confirmed that ZnO-TiO_2-CS-Amygdalin NCs induced cellular toxicity in MOLT4 (T lymphoblast leukemia) cells *via* increasing ROS and pro-apoptotic protein expression. These NC treatments drastically slowed cell growth and division in the treated groups, resulting in cell death compared to the control group (Elderdery et al., 2022b). Another study explored the biodegradability and safe elimination of lactic acid polymer and its prospective uses. Researchers investigated the potential application of novel NCs made of polylactic acid and gold as a medication delivery mechanism for cancer cells. The findings proved that the NC could effectively encapsulate daunorubicin, an anticancer drug, on their surface, and their combination showed enhanced therapeutic impacts on K562 and other leukemia cells. Notably, these novel NCs accumulate daunorubicin and may dramatically accelerate daunorubicin absorption into the leukemia cells, thus promoting the leukemic cell's cytotoxicity (Li et al., 2011).

10.2.5 BREAST CANCER

Breast cancer accounts for most diagnosed cancers and high mortality among females in varied geographical locations. Graphene-based NCs are frequently functionalized with polymers like PEG, gelatin, or CS to improve cellular absorption. This approach produces a more stable NC, which prevents the NC from releasing prematurely before reaching the target cell. Carbonaceous nanomaterials, including carbon nanotubes, GO, and their by-products, can carry anticancer agents into the nuclear region (Ayazi et al., 2020). GO was conjugated to DOX, decreased to form poly-L-lysine functionalized rGOs, and then tagged with antibodies to transport it selectively to breast cancer cells that overexpress HER2 (Zheng et al., 2016). Another approach using modified GO, FA, PEG, and adipose-derived stem cells (ADSCs) was used to promote the GO-PEG-FA NC internalization ability. After 32 days of intravenous administration in mice, this NC demonstrated minimal toxicity, and NIR radiation permitted the delivery of the NCs from ADSCs to the target site with the help of FA. Subsequent radiation is administered directly to the cancer cell to promote tumor growth (Peng et al., 2023).

Ferroptosis causes oxidative stress in cancer cells, another promising and potential cancer therapy approach. Several simple bimetallic iron-doped ZIF-8 NCs were utilized to encapsulate DOX. Due to its high pH sensitivity, the NC degrades down after internalization and liberated DOX in lysosomes promotes ROS and iron levels. GPX-4 and BCL-2 protein expression reduced with time, indicating that the NC can enhance oxidative stress in breast cancer cells by promoting ferroptosis and apoptosis (Zhong et al., 2023). Also, researchers have recently investigated the possibility of NCs doped with rare earth (RE) metals to detect metastatic breast cancer remains in distant organs. It was determined that customized RE-doped NCs might be utilized to perform optical imaging studies in pre-clinical breast cancer models (Pierce et al., 2017).

10.2.6 LUNG CANCER

CS-based nanomedicines usually exhibit high levels of bioactivity, biocompatibility, and polycationicity and excellent biodegradability (Cheung et al., 2015). Because of the flexibility of their structures, multifunctional NCs with excellent solubility and steadiness for administering drugs to cancer cells are possible (Naikoo et al., 2021).

CS oligosaccharide (CSO)-modified paclitaxel (PTX)-loaded liposomes was developed using synthetic P123 polymers (CP20 and CP50). CSO-modified liposomes enhanced PTX release while reducing leakage in the endosome/lysosome environment. Better cellular absorption and widespread distribution of CP50-LSs significantly inhibited A549 cell growth. Liposomes are more reactive toward the tumor's collagen protein accumulated on the CP50, enabling A549 cells to internalize them. These NCs showed higher therapeutic efficacy, with a 90% tumor inhibition rate in the mouse model. These results demonstrated that these NCs had the best therapeutic efficacy and less toxicity among the liposomes examined (Miao et al., 2021).

Distinct preparatory procedures have been utilized to generate alginate-based nanocarrier systems. Huang's group effectively combined sodium with alginate-coated magnetite and silver nanoparticles (Ag-NPs) painted on it to treat human lung carcinoma. The synthesized NC demonstrated extremely low cell viability, high cellular death, and low survivability in a dose-dependent manner while being reversed in the normal endothelial cell line. The significant anticancer ability against prevalent cell lines may be attributed to its antioxidant potential since it blocked half of the DPPH molecules (Huang et al., 2022).

Since they are in the micro-range, drug-loaded NPs and excipients, primarily sugars, are combined to form NCs suitable for deep lung absorption. The sugar dissolves in the alveoli fluid, releasing the primary NPs. Drying processes like spray-drying transform PLGA NCs into inhalable NCs utilizing sugar carriers. Quick drying with higher thermal applications results in smaller particle sizes. The unusually low melting point (45°C) of PLGAs causes particle aggregation and poor water dispersibility. Trehalose and lactose have higher melting points than PLGA, making it possible to fabricate NCs with the required conditions when the inlet temperature is accurately controlled (Tomoda et al., 2009). Polymeric NCs made of amphiphilic folate-PEG-hydrophobically modified dextran (F-PEG-HMD) micelles were created in the presence of specific carriers. Temozolomide combined with F-PEG-HMD dextran was dried and sprayed with carriers such as leucine and mannitol to form an inhalable dry powder. These NCs demonstrated profound lung accumulation due to fine particle size (RoSièRe et al., 2015).

10.2.7 PROSTATE CANCER

Prostate cancer (PC) is the most devastating cancer that endangers men's lives globally. It is challenging to make an early diagnosis of PC due to the limitations of the available diagnostic techniques and infrastructure (Appleton et al., 2015).

Aptamers are short, single-stranded artificial DNA that can be synthesized using the SELEX technique. Aptamers are being researched as cutting-edge targeting ligands for early diagnosis and effective PC therapy because they selectively bind to the molecule by establishing complementary spatial configurations. Prostate-specific membrane antigen (PSMA) protein, an integral glycoprotein, has been suggested as a promising target for detecting and controlling PC (Ghosh and Heston, 2004). In 2010, Savory et al. identified a DNA aptamer against PSMA using genetic algorithms. Various strategies employ PSMA biosensing based on aptamer. A new generation of PSMA detection sensors that utilize the EATR technique and aptamer identification has been developed. Combining two functional DNA probes (PSMA-aptamer and a Taqman probe coupled with quencher and fluorophore), this extremely sensitive and economical aptasensor provided an outstanding signal, revealing its significant promise for practical assay (Dai et al., 2020). Furthermore, aptamer-NCs-based diagnosis of PC can be performed using ultrasound imaging, MRI, and by using nuclear and multimodal imaging (Feng et al., 2022).

Vanillin was utilized as a reducer and stabilizer to develop a simple and ecologically friendly process for synthesizing green platinum nanoparticles (PtNPs) using GO. The findings showed that GO-PtNPs, as compared to GO or PtNPs, generated a more potent dose-dependent impact and triggered cellular apoptosis with disrupted membrane integrity and lactate dehydrogenase leakage. GO-PtNPs increased oxidative stress with nitric oxide, malondialdehyde, and carbonylated protein levels. GO-PtNPs triggered cell death through the regulation of apoptosis genes. These findings bolstered the hypothesis that these GO-PtNPs may impair cell survival by causing DNA damage and genomic toxicity, altering the gene expression involved in DNA damage and repair, and elevating modified nucleotide levels (Gurunathan et al., 2019). This research could advance the development of biologically synthesized biocompatible and non-toxic agents, one of the promising methods for treating cancer.

Developing a highly effective MRI agent aids in rapid and precise cancer diagnosis. A research team developed a novel gadolinium oxide (GdO) and iron oxide (IO) NC as an MRI contrast agent using the hydrothermal synthesis approach. Its surface was then coupled with a tumor-targeting ligand (PDGFB-PEG), resulting in a tumor-aiming probe for the precise diagnosis of prostatic cancer. Furthermore, with high biocompatibility, targeted IO/GdO accumulated in tumor cells more than pIO/GdO without

TABLE 10.1 List of Nanocomposite-Based Therapeutic Agents and Their Mechanisms on Cancer Types.

Type of Cancer	Type of Nanocomposites (NCs)	Element Used	Mechanism of Action	Reference
Esophageal cancer	CSP-Cr-NCs,	Sodium	↓MMP, ↑ROS, apoptosis	Hu et al. (2022)
	MoS_2	molybdenum, sulphur		Mendel (2009)
	Cu_2O/PEDOT:PSS/ZnO	Copper, zinc, oxygen		Tseng et al. (2020)
	SnO_2-SAPG- Carvacrol	Tin, oxygen		Jiang et al. (2022)
	DOX-Au	Gold		Moawad et al. (2023)
Liver cancer	SCPg-AI-NCs	Selenium	Apoptosis, ↑ROS,apoptosis, oxidative stress	Wang et al. (2022)
	Fe_3O_4/Cs	Iron, oxygen, cesium		Badawy et al. (2023)
	Zinc phthalocyanine-Sorafenib	Zinc		Zhang et al. (2020)
	DBT–CSNPs	Titanium		Shaban et al. (2021)
Melanoma	ZnO/CeO_2	Zinc, cerium, oxygen	↑ROS, apoptosis	Alabyadh et al. (2022)
	Graphene nanocomposites	Graphene oxide		Karimi et al. (2016)
	rGO-5FU-CMARX			Nazir et al. (2021)
	SISMA-ChiMA-rGO-DOX NCs			Céspedes-Valenzuela et al. (2022)
	C/PPy	Carbon		Gorgizadeh et al. (2022)
	Zn-Cur NC	Zinc	↑ROS, low pH	Deshpande et al. (2022)
	Mn-Fe_2O_4/C	Manganese, iron, oxygen		Gorgizadeh et al. (2020)
Leukemia	PFOB@PLGA@Pt	Platinum	↑Caspase-3, 8, 9↓MMP, ↑ROS,	Kong et al. (2022)
	ZnO-TiO_2-chitosan-amygdalin	Zinc, oxygen, titanium		Elderdery et al. (2022b)
	ZnO-TiO_2-chitosan-farnesol			Elderdery et al. (2022c)
	CuO-TiO_2-chitosan-escin			Elderdery et al. (2022a)

TABLE 10.1 *(Continued)*

Type of Cancer	Type of Nanocomposites (NCs)	Element Used	Mechanism of Action	Reference
Breast cancer	Poly-L-lysine-GO NC	Graphene oxide	↑ROS, ↑Bcl-2, ↓MMP, ↑ROS, apoptosis	Zheng et al. (2016)
	GO-PEG-FA			Peng et al. (2023)
	Fe-doped ZIF-8 - rare earth (RE) doped NCs	Iron		Zhong et al. (2023)
	AuPtPd/rGO	Gold, platinum, palladium		Dong et al. (2018)
Lung cancer	CSO-PTX loaded liposomes	–	↑Cellular absorption, ↑Cytotoxicity,	Miao et al. (2021)
	Fe_3O_4/Alg-Ag	Iron, oxygen, silver		Huang et al. (2022)
	PLGA	NIL		Tomoda et al. (2009)
	F-PEG-HMD	Folate		RoSièRe et al. (2015)
Prostate cancer	PSA - EATR amplification	–	↑ROS, ↑NO, ↑MDA, Apoptosis, ↑8-oxo-(d)G	Feng et al. (2022)
	GO-Pt	Platinum, oxygen	Cell death	Gurunathan et al. (2019)
	IO/GdO-PDGFB	Gadolinium, iron, oxygen		Li et al. (2022)
Brain cancer	Dox/alg-Fe_3O_4	Iron, oxygen	BBB penetrating and homotypic targeting effect	Su and Cheng (2015)
	wFIONs			Kang et al. (2023)
	CS/CMC/GQDs/ZnO@QUR	Zinc, oxygen		Ostovar et al. (2023)
Colorectal cancer	Au/Ag-CQD-NC,	Silver, gold	↓IL-6, ↓IL-1β, ↓TNF-α, apoptosis	Liu et al. (2020)
	Rv *in situ* gelling films	–	Cytotoxicity	Md et al. (2021)
	Cu_2O_2@$CaCO_3$@HA	Copper, calcium, oxygen		Chang et al. (2020)

TABLE 10.1 *(Continued)*

Type of Cancer	Type of Nanocomposites (NCs)	Element Used	Mechanism of Action	Reference
Pancreatic cancer	Fe_3O_4@DOP	Iron, oxygen	↑ ROS, ↑ photothermal responses	Dong et al. (2022)
	TiO_2/C	Titanium, oxygen, carbon	Cellular death	Cao et al. (2021)
	GEM-POSS-SQ	–		Abudukelimu et al. (2022)
	IONPs-PEG-MCC triple scabs	Iron oxide		Zou et al. (2020)
Ovarian cancer	L-arginine/ Fe_3O_4	Iron, oxygen	↓MMP, ↑ROS, apoptosis	Xu et al. (2021)
	rGO-Ag	Graphene oxide, silver		Choi et al. (2018)
	Fe_3O_4@CS-Starch/Cu	Iron, oxygen, copper		Ji et al. (2022)
	AZG-BSA	Gold, zinc		Naskar et al. (2018)

PDGFB-PEG, demonstrating its remarkable tumor-targeting capacity in *in vivo*. Finally, systemic administration of PDGFB-pIO/GdO improved the diagnostic signal in animals, resulting in better quality images and effective orthotopic prostatic cancer diagnosis. This work suggested PDGF-B tagged IO/GdO NCs as a rationally built nanoplatform, suggesting promise for PC diagnostic detection by MRI contrast enhancement (Li et al., 2022).

10.2.8 BRAIN CANCER (GLIOBLASTOMA)

Gliomas, the most aggressive and infiltrative brain tumor, are untreatable with existing therapies. A patient with a glioma has a less than 1.5 years average survival period after diagnosis. Although several anticancer drugs kill glioma cells *in vitro*, their antitumor effects in *in vivo* studies are primarily constrained by their poor solubility and short elimination half-lives. US-FDA-approved Fe_3O_4 (iron oxide) NPs and alginate were combined to develop a novel NC with encapsulated DOX to address the above issues. The Dox/alg-Fe_3O_4 NPs destroyed tumor cells without harming healthy nontumor cells, according to both *in vitro* and *in vivo* testing data (Su and Cheng, 2015). Similarly, Li et al. (2010) developed polyoxalate nanocarriers made with CS for sustained release of cisplatin.

Radiotherapy and systemic chemotherapy are frequently ineffective postoperative therapies for glioblastoma multiforme (GBM). Tumor cells that have deeply absorbed into the brain parenchyma cannot be surgically eliminated, posing a substantial challenge to treating GBM. Scientists developed an injectable hydrogel NC and an *in vivo* technique to treat deeply infiltrative GBM cells after surgery. An injectable hydrogel NC solution containing drug-loaded micelles with water-dispersible ferrimagnetic iron oxide nanocubes (wFIONs) was injected into multiple sites in the brain parenchyma to create soft intracortical drug reservoirs. wFIONs generated heat in alternating magnetic fields (AMFs) to accelerate micelle formation, drug release, and diffusion, allowing deep drug penetration. The micelles are designed to release the drug in an acidic tumor microenvironment. This technique may be useful for individuals with untreatable GBM and may improve the effectiveness of multimodal therapies or traditional therapeutic modalities (Kang et al., 2023).

Brain tumor-focused delivery of diagnostic contrast agents is still a challenging task. Nanoscale contrast agents are employed for targeted diagnosis of glioblastoma using available imaging modalities. Using bio-orthogonal click reactions, the endothelial integrin receptor-targeting peptide cyclic RGD decorated onto the NC surface provides a BBB-penetrating and homotypic targeting effect in the brain tumor microenvironment. Cellular and animal experiments confirmed the better-targeting results of cRGD-labeled brain tumor cell membrane coating outcomes (Duan et al., 2020). This study illustrates how a surface-modified cell membrane could be used as a theranostic strategy to circumvent delivery barriers in brain tumors.

10.2.9 COLORECTAL CANCER

Solid tumors are still treated with surgery followed by adjuvant chemotherapy. Targeted therapies involving NPs, immunotherapy, hormone therapy, and stem cell transplantation met adverse side effects, such as the possibility of metastatic tumor recurrence, intolerable cytotoxicity, unrestricted invasions into tissues, and exceptionally low bioavailability (AbuMousa et al., 2018).

Some recent developments in nano-bioengineering include liposomes, cubosomes, targeted NPs, -rods, -sheets, and quantum dots, which have significantly improved theranostics while lowering the adverse effects of cytotoxic drugs. A gold/silver coupled carbon quantum dot NCs (Au/Ag-CQD-NC) utilizing an effortless mechanism for photothermal cancer treatment (León Félix et al., 2019). The effectiveness of the hybrid NC was confirmed in *in vivo* studies, along with the role of epigenetic regulation, which could facilitate the translation of laboratory findings to clinical settings (Liu et al., 2020).

Resveratrol (Rv) *in situ* gelling films for long-acting sustained-release (SR) oral NCs were proposed for treating colorectal cancer (CRC). Rv-Soy protein (Rv-Sp) wet granules enclosed in sodium alginate (NA) dry film were assessed for drug encapsulation effectiveness, swelling studies, pH-drug release profile, and storage duration impacts. The optimized formulation demonstrated its anticancer activity against CRC cells. Rv NC formulation significantly enhanced Bcl-2-associated protein X (BAX) and decreased the expression of B-cell lymphoma 2, interleukins (IL-1 beta and IL-6), and tumor necrosis factor-alpha (TNF-alpha) in HCT-116 cells. These findings imply that long-acting Rv NC gels could be an effective therapy option for CRC (Md et al., 2021). Native CS and its derivatives (trimethyl CS, carboxymethyl CS, N-succinyl CS, PEGylated CS, hyaluronic acid (HA)-coupled CS, thiolated CS, and FA conjugated CS) have been converted into NC/NPs as drug carriers for colon cancer therapeutics (Bose and Wong, 2018).

Re-sponsive oncotherapy is the rational design of NCs triggered by the tumor-associated microenvironment (TAMs). Overexpressed endogenous H_2S levels and a mildly acidic microenvironment are common physiological characteristics of CRC (Chang et al., 2020). Researchers prepared TAMs-activated Cu_2O@$CaCO_3$@HA (CCH) NCs to accumulate in tumor cells more efficiently than normal tissue. In the acidic TAMs, the outer $CaCO_3$ mineralized layer disintegrates and releases Ca^{2+}, which causes the calcification of cancer cells. The mesoporous Cu_2O quickly disintegrated into ultrasmall $Cu_{13}S_{16}$ NPs without protecting the $CaCO_3$ shell when in contact with endogenous H_2S. $Cu_{13}S_{16}$ NPs expressed potent therapeutic behavior on laser radiation and were eliminated *via* urine, thereby deterring long-term toxicity. CCH-based synergistic therapy can transform "cold" tumors into "hot" tumors by reprogramming TAMs from pro-tumoral M2 to tumoricidal M1 phenotype. The CCH-based synergistic therapy may elicit potent immunogenicity responses by combining with CD47 blockage, leading to primary tumor ablation and suppression of cancer recurrence and metastasis (Chang et al., 2020).

10.2.10 PANCREATIC CANCER

Diagnosing pancreatic cancer using imaging and biopsy is challenging as they need expensive equipment and better interpretation. The detection sensitivity of pancreatic cancer cells both in biofluids and on tissues is increased by combining hybridization chain reaction (HCR) amplification and dopamine-coated Fe_3O_4 nanoparticles (Fe_3O_4@DOP NPs) to form free 6-carboxyfluorescein (FAM) labeled DNA (H1-FAM/H2-FAM). Furthermore, the pancreatic cancer cells fluoresced much more intensely than the control cell lines HPDE-C7 and HepG-2, which express less mucin1 protein. The versatility of the HCR amplification system was employed to detect the presence of cancer cells in pancreatic tissue in clinical settings (Dong et al., 2022).

A novel polyhedral oligomeric silsesquioxane (POSS)-based nanoplatform (GEM-POSS-SQ) facilitated the efficient grafting of the chemotherapeutic and photothermal agent, gemcitabine and squaraine, respectively, into POSS for combination therapy. The resulting NPs produced a substantial synergistic chemo-photothermal effect, particularly in chemoresistant PANC-1 pancreatic cancer cells, and inhibited SQ aggregation. Confocal laser scanning microscopy revealed nuclear damage and NIR-triggered rupture of cell membranes. Intra-tumoral administration of GEM-POSS-SQ NPs under NIR irradiation significantly slowed the PANC-1 tumor growth in mice with no observable adverse effects with a significant change in TUNEL- and Ki-67-positive cells in isolated tumor tissue (Abudukelimu et al., 2022).

The prognosis of pancreatic ductal adenocarcinoma (PDAC) depends on an accurate diagnosis and prompt therapy. Antibodies are critical therapeutic agents, whereas magnetic iron oxide nanoparticles (IONPs) are excellent MRI contrast agents. In a study, a novel NC was developed to diagnose and treat PDAC by fusing triple single-chain antibodies (scAbs) and IONPs by coupling the surface of polyethylene glycol-modified IONPs (IONPs-PEG) with triple scAbs, creating the IONPs-PEG-MCC triple scAbs NC. After a detailed investigation, it was found that the NCs were nontoxic to normal PDAC and could specifically bind to MUC4/CEACAM6/CD44v6-expressing PDAC cells. The triple scAbs IONPs-PEG-MCC NC also displayed an effective anti-pancreatic cancer result by reducing the MRI T2-weighted signal intensity in biological studies (Zou et al., 2020).

10.2.11 OVARIAN CANCER

Ovarian cancer (OC) is a common invasive malignant tumor that affects the female reproductive system. OC is diagnosed at an advanced stage and is prone to spreading to the pelvic and abdominal cavities (Kuroki and Guntupalli, 2020). Magnetic nanocomposite (MNC) materials are being explored for comprehensive cancer therapy due to their distinctive properties absent in other materials. A study demonstrated a low-cost and eco-friendly L-arginine-modified MNC (L-arginine/Fe_3O_4) fabrication method. The proposed NC was incorporated with gold nanoparticles (Au NPs), using sodium borohydride as a reductant agent. A detailed characterization and assessment showed that NC decreased the malignant ovarian cell line viability (Xu et al., 2021).

R-phycoerythrin (RPE) was used in the biomolecule-mediated synthesis of reduced GO-silver NCs (rGO-Ag). The resulting NC was well characterized and analyzed in OC stem cells (OvCSCs) and found that rGO-Ag was toxic to OvCSC's cell viability, as evidenced by increased ROS generation, MMP reduction, and apoptotic expression. In addition, salinomycin plus rGO-Ag at extremely low doses increased apoptosis five times more than each therapy and selectively destroyed OvCSCs and sensitized tumor cells (Choi et al., 2018). Conjugated bio-nanomaterials with magnetic cores have recently piqued researchers' interest as a viable substrate for next-generation biological composites. Recent research reported a CS-starch-based NC supported with iron oxide and copper (Fe_3O_4@CS-Starch/Cu) as an anticancer agent in ovarian cell lines (Ji et al., 2022).

We illustrated that the underlying mechanism by which various NCs kill various cancer cells is identical (Figure 10.2). Better drug delivery and simple intracellular absorption of NC-coated drugs result in efficient cell killing. The disruption of the mitochondrial membrane integrity induced by the NC-coated drugs leads to a considerable loss of MMP, indicating the release of SMAC and cytochrome c (Cyt-C). ROS generation is

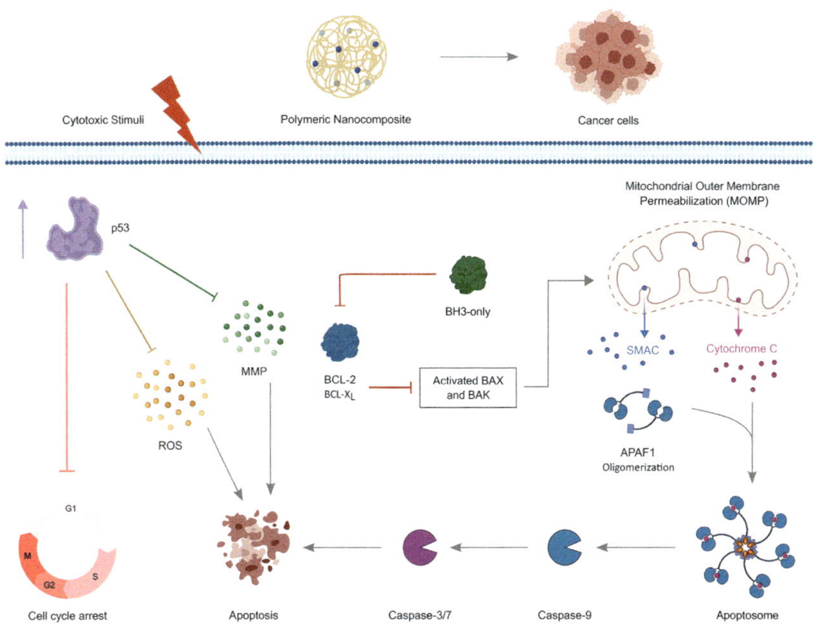

FIGURE 10.2 Illustration of the fundamental process by which a nanocomposite results in cell death.

primarily produced in the mitochondria of human cells and is associated with MMP breakdown. Numerous studies have shown decreased anti-apoptotic proteins, including p53, p21, and p27, and increased levels of pro-apoptotic proteins like BAX, BAK, and caspase 3. All these mechanisms result in cell cycle arrest and cell death (Hu et al., 2022; Mendel, 2009; Wang et al., 2022; Karimi et al., 2016; Céspedes-Valenzuela et al., 2022; Elderdery et al., 2022b; Feng et al., 2022; Xu et al., 2021; Tseng et al., 2020; Shaban et al., 2021).

10.3 METHOD OF APPLICATION IN CANCER TREATMENT

Cancer is linked with several pathogenic processes. Its limitations are drug resistance, stem-like cell growth, and poor selectivity/specificity. Several NC-based approaches have been developed to improve cytotoxicity and restricted selectivity while increasing drug capacity and bioavailability. NCs incorporate NPs into a matrix material, typically on a nanoscale. These NPs can be programmed to carry drugs, target specific cancer cells, enhance imaging, or have other therapeutic functions.

10.3.1 NANOCOMPOSITE-BASED CHEMOTHERAPY

Chemotherapy's limited solubility in water, short biological half-lives, inability to target specific organs, and the emergence of multidrug resistance deter its beneficial effects. These agents can be delivered *via* stimuli-responsive systems instead of existing methods to address the limitations. Because of a precise stimulus response impact, these delivery systems can control drug release and reduce injury to healthy tissue (Seifu and Nath, 2019).

Polymer NCs are biocompatible and biodegradable. Insoluble biopolymers such as collagen, elastin, CS, keratin, and silk can be transformed into soluble derivatives through chemical and enzymatic hydrolysis. CS is preferred due to its nontoxicity, high charge density, biocompatibility, and mucoadhesion characteristics (Sionkowska et al., 2016). Also, CS NCs can improve retention duration, allowing them to release drug molecules for a more extended period, making them the ideal drug delivery system. CUR is the principal curcuminoid found in turmeric and administered as a chemotherapeutic agent using CS as a drug carrier. Its low solubility makes it challenging to use as a medication. Poly (alkyl-2-oxazoline) (PAOx) amphiphilic block copolymers were developed for effective distribution at the cancer tumor site. The CUR-loaded PAOx (100 nm-sized) NPs showed excellent stability and enhanced CUR solubility in water. In vitro release experiments revealed their possible significance as a suitable nanocarrier to overcome CUR's constraints (Raveendran et al., 2017).

10.3.2 NANOCOMPOSITE-BASED PHOTODYNAMIC THERAPY

PDT, a phototherapic technique that combines light with a photon sensitizer, is touted as a promising anticancer strategy because of its restricted invasion, lower systemic toxicity, reduced drug aversion, and more precise tumor targeting capabilities (Li et al., 2020). PDT comprises a step-by-step treatment procedure that kills the cancerous and precancerous cells after light activation. Specific photosensitizers (PS) in an oxygen-rich tumor microenvironment are activated by light radiation. Singlet oxygen ($^1O_2$) and ROSes are released upon activation and kill cancer cells. The wavelength's value of light radiation determines the depth of that light's penetration. Scintillating nanoparticles (ScNP) and light wavelengths with PS allow greater penetration into tumor tissues (Wilson and Hepatology, 2002).

PS such as copper–cysteamine (Cu–Cy), cerium-titanium oxide (TiO_2:Ce), octahedral molybdenum cluster compound $(n\text{-}Bu_4N)_2(Mo_6I_8(OCOCF_3)_6$ and various types of scintillating NPs such as Y_2O_3:Eu, $Y_3Al_5O_{12}$:Eu, $NaGdF_4$:Gd^{3+}, and $LiGa_5O_8$:Cr were commonly used. Sun et al. (2020) investigated photosensitizer conjugates called gold clustoluminogens (AIE-Au) with x-ray radiation (1 Gy dose) to develop effective low-dose XPDT. AIE-Au significantly absorbed X-rays and produced hydroxyl radicals that damaged DNA, enhancing the radiotherapy's effects. Glutathione-protected gold clusters (GCs) aggregation also boosted the X-ray excited luminescence. AIE-Au clustoluminogens activated the rose bengal (RB) PS, converted X-rays into optical luminescence oxidized lipid membranes, and stimulated ROS generation.

Guo et al. (2018) used alcohol dehydrogenase-1 (ADH-1), HA, and mesoporous titanium dioxide (MTN) NP conjugates to deliver the drug molecules into tumor tissues. These conjugates had an average size of roughly 110 nm. HA and ADH-1 are linked to the surface of the MTN. N-cadherin's activity was blocked by ADH-1 acting as an antagonist, potentially disrupting the tumor vasculature. Additionally, MTN produced more ROS generation when exposed to X-ray radiation at 0.15–0.35 mA and 60–75 kV.

10.3.3 NANOCOMPOSITE-BASED SONODYNAMIC THERAPY

Sonodynamic therapy (SDT) is an emerging approach to the site-specific and non-invasive removal of solid tumors. SDT combines sonosensitizers with low-intensity ultrasound. Target tissues are sensitized using a non-toxic chemical agent and exposed to a relatively low-intensity ultrasound. SDT uses ultrasonic stimulation to activate sonosensitizers, which release energy and produce ROS that kill tumor cells.

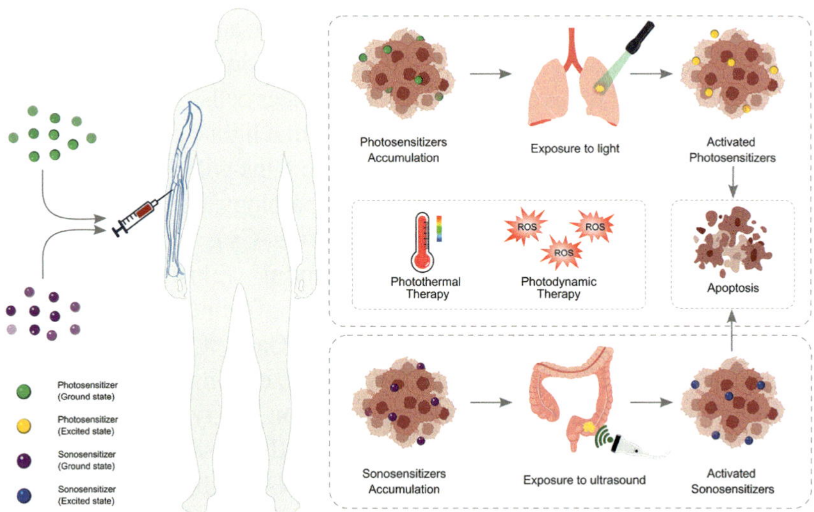

FIGURE 10.3 Mechanism of photodynamic and sonodynamic therapy of nanocomposites.

The presence of a hypoxic environment in pancreatic tumors reduces the effectiveness of oxygen-dependent type I SDT. Cao et al. (2021) developed a TiO_2/C NC, a stable and hypoxia-tolerant sonosensitizer with a carbon structure produced from a metal-organic framework (MOF).TiO_2/C showed excellent stability under repeated ultrasonic irradiation, hypoxia tolerance, strong biocompatibility, and produced superoxide radicals, leading to effective type I SDT. TiO_2/C significantly induced tumor cell apoptosis through SDT-induced DNA damage (Cao et al., 2021). The general mechanism by which PDT and SDT cause cell death is illustrated in Figure 10.3.

10.4 FUTURE PERSPECTIVES

The complex and diverse kinds of cancer continue to provide one of the most critical challenges in modern medicine, necessitating novel and focused therapeutic strategies. NCs have recently emerged as a potential new field in the search for more potent cancer therapies. These nanoscale materials, which combine NPs or nanomaterials with a matrix material, have the potential to fundamentally alter how cancer is identified, diagnosed, and treated.

NCs have unique characteristics that may revolutionize drug delivery and be used to advance cancer therapy. They provide better imaging, precise drug delivery, and the possibility for a theranostic approach. Additionally, NCs may solve some enduring issues in cancer treatment, such as drug resistance and collateral damage to healthy tissue. This investigation into the potential uses of NCs in cancer treatment focuses on cutting-edge breakthroughs and intriguing new research that could completely alter the field of oncology. This discussion offers insights into the exciting prospects and challenges of employing NCs as a powerful weapon against cancer while acknowledging the dynamic nature of scientific research.

Moreover, NC applications as contrast agents in MRI and CT scans are promising. This technique increases the likelihood of a successful course of treatment by enabling early detection and precise diagnosis of cancer. NPs with radiation-absorbing molecules increase the susceptibility of cancer cells to radiation treatment while minimizing damage to neighboring healthy tissues. NPs could transport drug molecules directly to the tumor location or locally heat the tumor *via* photothermal therapy. Drug resistance in the treatment of cancer may be addressed *via* NCs. They can be designed to deliver a combination of drugs to target specific drug resistance mechanisms.

Soon, we can expect more NC-based cancer therapies to enter clinical trials and be approved for clinical use. Regulatory approvals and commercialization will depend on successful testing and safety profiles. It is important to remember that the development and applications of NCs for cancer treatment are complex processes that may encounter several challenges, including issues with safety, production scale, and regulatory constraints. NCs have great promise in cancer treatment as research advances; nevertheless, it may take some time before they are widely used in clinical settings.

KEYWORDS

- **cancer therapy**
- **cancer types**
- **chemotherapy**
- **esophageal cancer**
- **nanocomposites**
- **nanoparticles**

REFERENCES

Abudukelimu, S., Wei, G., Huang, J., Zhao, G., Wei, L., Cui, W., Lu, M., & Yao, W. (2022). Polyhedral oligomeric silsesquioxane (POSS)-based hybrid nanocomposite for synergistic chemo-photothermal therapy against pancreatic cancer. *Chemical Engineering Journal, 442*, 136124.

Abuduwaili, W., Wang, X., Huang, A.-T., Sun, J.-L., Xu, R.-C., Zhang, G.-C., Liu, Z.-Y., Wang, F., Zhu, C.-F., Liu, T.-T., et al. (2022). Iridium complex-loaded sorafenib nanocomposites for synergistic chemo-photodynamic therapy of hepatocellular carcinoma. *ACS Applied Materials and Interfaces, 14*(33), 37356–37368.

AbuMousa, R. A., Baig, U., Gondal, M. A., AlSalhi, M. S., Alqahtani, F. Y., Akhtar, S., Aleanizy, F. S., & Dastageer, M. A. (2018). Photo-catalytic killing of HeLa cancer cells using facile synthesized pure and Ag loaded WO3 nanoparticles. *Scientific Reports, 8*(1), 15224.

Alabyadh, T., Albadri, R., Es-Haghi, A., Yazdi, M. E. T., Ajalli, N., Rahdar, A., & Thakur, V. K. (2022). ZnO/CeO2 nanocomposites: Metal-organic framework-mediated synthesis, characterization, and estimation of cellular toxicity toward liver cancer cells. *Journal of Functional Biomaterials, 13*(3), 139.

Appleton, L., Wyatt, D., Perkins, E., Parker, C., Crane, J., Jones, A., Moorhead, L., Brown, V., Wall, C., & Pagett, M. (2015). The impact of prostate cancer on men's everyday life. *European Journal of Cancer Care, 24*(1), 71–84.

Ayazi, H., Akhavan, O., Raoufi, M., Varshochian, R., Motlagh, N. S. H., & Atyabi, F. (2020). Graphene aerogel nanoparticles for *in situ* loading/pH sensitive releasing anticancer drugs. *Colloids and Surfaces B: Biointerfaces, 186*, 110712.

Badawy, M. M., Abdel-Hamid, G. R., & Mohamed, H. E. (2023). Antitumor activity of chitosan-coated iron oxide nanocomposite against hepatocellular carcinoma in animal models. *Biological Trace Element Research, 201*(3), 1274–1285.

Bleul, C. C., Fuhlbrigge, R. C., Casasnovas, J. M., Aiuti, A., & Springer, T. A. (1996). A highly efficacious lymphocyte chemoattractant, stromal cell-derived factor 1 (SDF-1). *Journal of Experimental Medicine,184*(3), 1101–1109.

Bose, A., & Wong, T. W. (2018). Oral colon cancer targeting by chitosan nanocomposites. In *Applications of nanocomposite materials in drug delivery* (pp. 409–429). Elsevier.

Bray, F., Laversanne, M., Weiderpass, E., & Soerjomataram, I. (2021). The ever-increasing importance of cancer as a leading cause of premature death worldwide. *Cancer, 127*(16), 3029–3030.

Cao, J., Sun, Y., Zhang, C., Wang, X., Zeng, Y., Zhang, T., & Huang, P. J. (2021). Tablet-like TiO2/C nanocomposites for repeated type I sonodynamic therapy of pancreatic cancer. *Acta Biomaterialia, 129*, 269–279.

Céspedes-Valenzuela, D. N., Sánchez-Rentería, S., Cifuentes, J., Gómez, S. C., Serna, J. A., Rueda-Gensini, L., Ostos, C., Muñoz-Camargo, C., & Cruz, J. C. (2022). Novel photo-and thermo-responsive nanocomposite hydrogels based on functionalized rGO and modified SIS/Chitosan polymers for localized treatment of malignant cutaneous melanoma. *Frontiers in Bioengineering and Biotechnology, 10*, 947616.

Chang, M., Hou, Z., Jin, D., Zhou, J., Wang, M., Wang, M., Shu, M., Ding, B., Li, C., & Lin, J. (2020). Colorectal tumor microenvironment-activated bio-decomposable and metabolizable Cu2O@ CaCO3 nanocomposites for synergistic oncotherapy. *Advanced Materials, 32*(43), 2004647.

Chen, Y., Tan, C., Zhang, H., & Wang, L. (2015). Two-dimensional graphene analogues for biomedical applications. *Chemical Society Reviews, 44*(9), 2681–2701.

Cheung, R. C. F., Ng, T. B., Wong, J. H., & Chan, W.Y. (2015). Chitosan: An update on potential biomedical and pharmaceutical applications. *Marine Drugs, 13*(8), 5156–5186.

Choi, Y.-J., Gurunathan, S., & Kim, J.-H. (2018). Graphene oxide–silver nanocomposite enhances cytotoxic and apoptotic potential of salinomycin in human ovarian cancer stem cells (OvCSCs): A novel approach for cancer therapy. *International Journal of Molecular Sciences, 19*(3), 710.

Dai, J., Su, Y., Zhong, S., Cong, L., Liu, B., Yang, J., Tao, Y., He, Z., Chen, C., & Jiang, Y. (2020). Exosomes: Key players in cancer and potential therapeutic strategy. *Signal Transduction and Targeted Therapy, 5*(1), 145.

Deshpande, S. S., Veeragoni, D., Rachamalla, H. K., & Misra, S. (2022). Anticancer properties of ZnO-Curcumin nanocomposite against melanoma cancer and its genotoxicity profiling. *Journal of Drug Delivery Science and Technology*, 75, 103703.

Dong, Q., Jia, X., Wang, Y., Wang, H., Liu, Q., Li, D., Wang, J., & Wang, E. (2022). Sensitive and selective detection of Mucin1 in pancreatic cancer using hybridization chain reaction with the assistance of Fe3O4@ polydopamine nanocomposites. *Journal of Nanobiotechnology, 20*(1), 94.

Dong, W., Ren, Y., Bai, Z., Yang, Y., Wang, Z., Zhang, C., & Chen, Q. (2018). Trimetallic AuPtPd nanocomposites platform on graphene: Applied to electrochemical detection and breast cancer diagnosis. *Talanta, 189*, 79–85.

Duan, Y., Wu, M., Hu, D., Pan, Y., Hu, F., Liu, X., Thakor, N., Ng, W. H., Liu, X., Sheng, Z. Zheng, H., & Liu, B. (2020). Biomimetic nanocomposites cloaked with bioorthogonally labeled glioblastoma cell membrane for targeted multimodal imaging of brain tumors. *Advanced Functional Materials, 30*(38), 2004346.

Elderdery, A. Y., Alhamidi, A. H., Elkhalifa, A. M., Althobiti, M. M., Omer, N. E., Alsugoor, M. H., Alsuhaymi, N., Atebien, E. M., Hamza, S. M., Alzahrani, B., et al. (2022a). Synthesis, characterization, and antimicrobial and antiproliferative effects of CuO-TiO2-chitosan-escin nanocomposites on human leukemic MOLT4 cells. *Nanomaterials, 12*(21), 3753.

Elderdery, A. Y., Alzahrani, B., Alanazi, F., Hamza, S. M., Elkhalifa, A. M., Alhamidi, A. H., Alabdulsalam, A. A., Mohamedain, A., Kumar, S. S., & Mok, P.L. (2022b). Amelioration of human acute lymphoblastic leukemia (ALL) cells by ZnO-TiO2-Chitosan-Amygdalin nanocomposites. *Arabian Journal of Chemistry, 15*(8), 103999.

Elderdery, A. Y., Alzahrani, B., Hamza, S., Mostafa-Hedeab, G., Mok, P. L., & Subbiah, S. K. (2022c). Synthesis of zinc oxide (ZnO)-titanium dioxide (TiO 2)-chitosan-farnesol nanocomposites and assessment of their anticancer potential in human leukemic MOLT-4 cell line. *Bioinorganic Chemistry and Applications, 2022*, 5949086.

Feng, X., Zhu, Y., Wang, F., Guo, T., Dou, X., Lin, M., & Tian, W. (2022). The aptamer functionalized nanocomposite used for prostate cancer diagnosis and therapy. *Journal of Nanomaterials, 2022*, 1–17.

Ghosh, A., & Heston, W. D. W. (2004). Tumor target prostate specific membrane antigen (PSMA) and its regulation in prostate cancer. *Journal of Cellular Biochemistry, 91*(3), 528–539.

Gorgizadeh, M., Behzadpour, N., Salehi, F., Daneshvar, F., Vais, R. D., Nazari-Vanani, R., Azarpira, N., Lotfi, M., & Sattarahmady, N. (2020). A MnFe2O4/C nanocomposite as a novel theranostic agent in MRI, sonodynamic therapy and photothermal therapy of a melanoma cancer model. *Journal of Alloys and Compounds, 816*, 152597.

Gorgizadeh, M., Zahraie, N., Behzadpour, N., Vais, R. D., & Sattarahmady, N. (2022). Carbon dots/polypyrrole nanocomposite as a sonosensitizer for treatment of melanoma tumor. *Applied Physics A, 128*(8), 738.

Guo, Z., Zheng, K., Tan, Z., Liu, Y., Zhao, Z., Zhu, G., Ma, K., Cui, C., Wang, L., & Kang, T. (2018). Overcoming drug resistance with functional mesoporous titanium dioxide nanoparticles combining targeting, drug delivery and photodynamic therapy. *Journal of Materials Chemistry B, 6*(46), 7750–7759.

Gurunathan, S., Jeyaraj, M., Kang, M.-H., & Kim, J.-H. (2019). Graphene oxide–platinum nanoparticle nanocomposites: A suitable biocompatible therapeutic agent for prostate cancer. *Polymers (Basel), 11*(4), 733.

Hu, A., Alarfaj, A. A., Hirad, A. H., Veeraraghavan, V. P., Surapaneni, K. M., Hussein-Al-Ali, S. H., Natarajan, N., & Elayappan, P. K. (2022). Chitosan-sodium alginate-polyethylene glycol-crocin nanocomposite treatment inhibits esophageal cancer KYSE-150 cell growth via inducing apoptotic cell death. *Arabian Journal of Chemistry, 15*(20), 103844.

Huang, J., Guo, J., Zhu, J., & Zou, X. (2022). Supported silver nanoparticles over alginate-modified magnetic nanoparticles: Synthesis, characterization and treat the human lung carcinoma. *Journal of Saudi Chemical Society, 26*(1), 101393.

Huang, L., Huang, J., Huang, J., Xue, H., Liang, Z., Wu, J., & Chen, C. (2020). Nanomedicine–a promising therapy for hematological malignancies. *Biomaterials Science, 8*(9), 2376–2393.

Ji, N., Dong, C., & Jiang, J. (2022). Evaluation of antioxidant, cytotoxicity, and anti-ovarian cancer properties of the Fe3O4@ CS-Starch/Cu bio-nanocomposite. *Inorganic Chemistry Communications, 140*, 109452.

Jiang, G., Mohideen, A. P., Seshadri, V. D., & Rengarajan, T. (2022). Biosynthesized tin oxide-sodium alginate-polyethylene glycol-carvacrol nanocomposite shows anticancer activity on esophagus squamous carcinoma cells. *Process Biochemistry, 121*, 403–412.

Kang, T., Cha, G. D., Park, O. K., Cho, H. R., Kim, M., Lee, J., Kim, D., Lee, B., Chu, J., Koo, S., et al. (2023). Penetrative and sustained drug delivery using injectable hydrogel nanocomposites for postsurgical brain tumor treatment. *ACS Nano, 17*(6), 5435–5447.

Karimi, M., Eslami, M., Sahandi-Zangabad, P., Mirab, F., Farajisafiloo, N., Shafaei, Z., Ghosh, D., Bozorgomid, M., Dashkhaneh, F., & Hamblin, M. R. (2016). pH-Sensitive stimulus-responsive nanocarriers for targeted delivery of therapeutic agents. *Wiley Interdisciplinary Reviews: Nanomedicine and Nanobiotechnology, 8*(5), 696–716.

Khan, M. U. A., Raza, M. A., Razak, S. I. A., Kadir, M. R. A., Haider, A., Shah, S. A., Mohd Yusof, A. H., Haider, S., Shakir, I., & Aftab, S. (2020). Novel functional antimicrobial and biocompatible arabinoxylan/guar gum hydrogel for skin wound dressing applications. *Journal of Tissue Engineering and Regenerative Medicine, 14*(10), 1488–1501.

Kong, F., He, H., Bai, H., Yang, F., Ma, M., Ning, G., & Zhang, Y. (2022). A biomimetic nanocomposite with enzyme-like activities and CXCR4 antagonism efficiently enhances the therapeutic efficacy of acute myeloid leukemia. *Bioactive Materials, 18*, 526–538.

Kumar, S., Shukla, M. K., Sharma, A. K., Jayaprakash, G. K., Tonk, R. K., Chellappan, D. K., Singh, S. K., Dua, K., Ahmed, F., Bhattacharyya, S., et al. (2023). Metal-based nanomaterials and nanocomposites as promising frontier in cancer chemotherapy. *MedComm (2020), 4*(2), e253.

Kuroki, L., & Guntupalli, S. R. (2020). Treatment of epithelial ovarian cancer. *BMJ, 371*, m3773.

León Félix, L., Sanz, B., Sebastián, V., Torres, T., Sousa, M. H., Coaquira, J., Ibarra, M., & Goya, G. F. (2019). Gold-decorated magnetic nanoparticles design for hyperthermia applications and as a potential platform for their surface-functionalization. *Scientific Reports, 9*(1), 4185.

Li, J., Chen, C., Wang, X., Gu, Z., & Chen, B. (2011). Novel strategy to fabricate PLA/Au nanocomposites as an efficient drug carrier for human leukemia cells in vitro. *Nanoscale Research Letters, 6*, 1–8.

Li, Z., Huang, P., He, R., Lin, J., Yang, S., Zhang, X., Ren, Q., & Cui, D. (2010). Aptamer-conjugated dendrimer-modified quantum dots for cancer cell targeting and imaging. *Materials Letters, 64*(3), 375–378.

Li, X., Lovell, J. F., Yoon, J., & Chen, X. (2020). Clinical development and potential of photothermal and photodynamic therapies for cancer. *Nature Reviews Clinical Oncology, 17*(11), 657–674.

Li, Q., Xiao, J., Zhang, C., Wang, P., Li, W., Wang, Y., Ran, L., Qin, X., Yan, M., Qian, J., et al. (2022). PDGF-B conjugating mesoporous IO/GdO nanocomposites for accurate diagnosis of orthotopic prostatic cancer through T1-T2 dual-modal MRI contrast enhancement. *Materials Today Advances, 16*, 100278.

Liu, F, Wang, X.-D., & Du, S.-Y. (2020). Production of gold/silver doped carbon nanocomposites for effective photothermal therapy of colon cancer. *Scientific Reports, 10*(1), 7618.

Md, S., Abdullah, S., Alhakamy, N. A., Alharbi, W. S., Ahmad, J., Shaik, R. A., Ansari, M. J., Ibrahim, I. M., & Ali, J. (2021). Development, optimization, and in vitro evaluation of novel oral long-acting resveratrol nanocomposite *in situ* gelling film in the treatment of colorectal cancer. *Gels, 7*(4), 276.

Mendel, R. R. (2009). Cell biology of molybdenum. *Biofactors, 35*(5), 429–434.

Miao, Y.-Q., Chen, M.-S., Zhou, X., Guo, L.-M., Zhu, J.-J., Wang, R., Zhang, X.-X., & Gan, Y. (2021). Chitosan oligosaccharide modified liposomes enhance lung cancer delivery of paclitaxel. *Acta Pharmacologica Sinica, 42*(10), 1714–1722.

Mishra, P., Nayak, B., & Dey, R. K. (2016). PEGylation in anti-cancer therapy: An overview. *Asian Journal of Pharmaceutical Sciences, 11*(3), 337–348.

Mishra, D. K., Yadav, K. S., Prabhakar, B., & Gaud, R. (2018). Nanocomposite for cancer targeted drug delivery. In *Applications of nanocomposite materials in drug delivery* (pp. 323–337). Elsevier.

Moawad, M., Fahmy, A. M., Elsherbeni, E.-S. A., Youssef, A. M., & El-Ghannam, G. (2023). Enhancing the efficiency of doxorubicin-conjugated with gold nanocomposites for active targeting and laser treatment for human esophagus cancer cells. *Egyptian Journal of Chemistry, 66*(10), 17–22.

Mohammed, M. A., Syeda, J. T., Wasan, K. M., & Wasan, E. K. (2017). An overview of chitosan nanoparticles and its application in non-parenteral drug delivery. *Pharmaceutics, 9*(4), 53.

Naikoo, G. A., Awan, T., Hassan, I. U., Salim, H., Arshad, F., Ahmed, W., Asiri, A. M., & Qurashi, A. (2021). Nanomaterials-based sensors for respiratory viral detection: A review. *IEEE Sensors Journal, 21*(16), 17643–17656.

Naskar, A., Bera, S., Bhattacharya, R., Roy, S. S., & Jana, S. (2018). Effect of bovine serum albumin immobilized Au–ZnO–graphene nanocomposite on human ovarian cancer cell. *Journal of Alloys and Compounds,* 734, 66–74.

Nazir, S., Khan, M. U. A., Al-Arjan, W. S., Abd Razak, S. I., Javed, A., & Kadir, M. R. A. (2021). Nanocomposite hydrogels for melanoma skin cancer care and treatment: In-vitro drug delivery, drug release kinetics and anti-cancer activities. *Arabian Journal of Chemistry, 14*(5), 103120.

Ostovar, S., Pourmadadi, M., & Zaker, M. A. (2023). Co-biopolymer of chitosan/ carboxymethyl cellulose hydrogel improved by zinc oxide and graphene quantum dots nanoparticles as pH-sensitive nanocomposite for quercetin delivery to brain cancer treatment. *International Journal of Biological Macromolecules, 253*(Pt 4), 127091.

Peng, Z., Chang, Q., Xing. M., & Lu, F. (2023). Active hydrophilic graphene oxide nanocomposites delivery mediated by adipose-derived stem cell for elevated photothermal therapy of breast cancer. *International Journal of Nanomedicine, 18,* 971–986.

Pierce, M. C., Higgins, L. M., Ganapathy, V., Kantamneni, H., Riman, R. E., Roth, C. M., & Moghe, P. V. (2017). Rare-earth doped nanocomposites enable multiscale targeted short-wave infrared imaging of metastatic breast cancer. In *Proceedings of SPIE - The International Society for Optical Engineering* (pp. 75–81).

Qamar, S. U. R. (2021). Nanocomposites: Potential therapeutic agents for the diagnosis and treatment of infectious diseases and cancer. *Colloid and Interface Science Communications, 43*, 100463.

Raveendran, R., Mullen, K. M., Wellard, R. M., Sharma, C. P., Hoogenboom, R., & Dargaville, T. R. (2017). Poly (2-oxazoline) block copolymer nanoparticles for curcumin loading and delivery to cancer cells. *European Polymer Journal, 93*, 682–694.

Rogers, J. E., Sewastjanow-Silva, M., Waters, R. E., & Ajani, J. A. (2022). Esophageal cancer: Emerging therapeutics. *Expert Opinion on Therapeutic Targets, 26*(2), 107–117.

RoSièRe, R., Gelbcke, M., Mathieu, V., Van Antwerpen, P., Amighi, K., & Wauthoz N. (2015). New dry powders for inhalation containing temozolomide-based nanomicelles for improved lung cancer therapy. *International Journal of Oncology, 47*(3), 1131–1142.

Saikia, C., Das, M. K., Ramteke, A., & Maji, T. K. (2017). Evaluation of folic acid tagged aminated starch/ZnO coated iron oxide nanoparticles as targeted curcumin delivery system. *Carbohydrate Polymers,157*, 391–399.

Savory, N., Abe, K., Sode, K., & Ikebukuro, K. (2010). Selection of DNA aptamer against prostate specific antigen using a genetic algorithm and application to sensing. *Biosensors and Bioelectronics, 26*(4), 1386–1391.

Seifu, M. F., & Nath, L. K. (2019). Polymer-drug conjugates: Novel carriers for cancer chemotherapy. *Polymer-Plastics Technology and Materials, 58*(2), 158–171.

Shaban, N. Z., Yehia, S. A., Awad, D., Shaban, S. Y., & Saleh, S. R. (2021). A titanium (IV)–dithiophenolate complex and its chitosan nanocomposite: Their roles towards rat liver injuries *in vivo* and against human liver cancer cell lines. *International Journal of Molecular Science, 22*(20), 11219.

Sionkowska, A., Kaczmarek, B., Lewandowska, K., Grabska, S., Pokrywczyńska, M., Kloskowski, T., & Drewa, T. (2016). 3D composites based on the blends of chitosan and collagen with the addition of hyaluronic acid. *International Journal of Biological Macromolecules, 89*, 442–448.

Su, C.-H., & Cheng, F.-Y. (2015). *In vitro* and *in vivo* applications of alginate/iron oxide nanocomposites for theranostic molecular imaging in a brain tumor model. *RSC Advances, 5*(109), 90061–90064.

Sun, W., Luo, L., Feng, Y., Cai, Y., Zhuang, Y., Xie, R. J., Chen, X., & Chen, H. (2020). Aggregation-induced emission gold clustoluminogens for enhanced low-dose X-ray-induced photodynamic therapy. *Angewandte Chemie, 132*(25), 10000–10007.

Sung, H., Ferlay, J., Siegel, R. L., Laversanne, M., Soerjomataram, I., Jemal, A., & Bray, F. (2021). Global cancer statistics 2020: GLOBOCAN estimates of incidence and mortality worldwide for 36 cancers in 185 countries. *CA: A Cancer Journal for Clinicians, 71*(3), 209–249.

Tavernier-Tardy, E., Cornillon, J., Campos, L., Flandrin, P., Duval, A., Nadal, N., & Guyotat, D. (2009). Prognostic value of CXCR4 and FAK expression in acute myelogenous leukemia. *Leukemia Research, 33*(6), 764–768.

Tomoda, K., Ohkoshi, T., Hirota, K., Sonavane, G. S., Nakajima, T., Terada, H., Komuro, M., Kitazato, K., & Makino, K. (2009). Preparation and properties of inhalable nanocomposite particles for treatment of lung cancer. *Colloids and Surfaces B: Biointerfaces, 71*(2), 177–182.

Tseng, K.-W., Hsiao, Y.-P., Jen, C.-P., Chang, T.-S., & Wang, H.-C. (2020). Cu2O/PEDOT: PSS/ZnO nanocomposite material biosensor for esophageal cancer detection. *Sensors (Basel), 20*(9), 2455.

Wang, Y., Li, L., Ma, T., Cheng, X., & Liu, D. (2022). Design, synthesis, and apoptosis-promoting effect evaluation of chalcone derivatives containing aminoguanidine units. *Anticancer Agents in Medicinal Chemistry, 22*(11), 2116–2124.

Wilson, B. C. (2002). Photodynamic therapy for cancer: Principles. *Canadian Journal of Gastroenterology and Hepatology*, *16*, 393–396.

Xu, J., Li, L., Zhang, J., & Min, Y. (2021). Decorated of Au NPs over L-arginine-modified Fe3O4 nanoparticles as a novel nanomagnetic composite for the treatment of human ovarian cancer. *Arabian Journal of Chemistry, 14*(8), 103283.

Xu, J.-J., Zhao, W.-W., Song, S., Fan, C., & Chen. H.-Y. (2014). Functional nanoprobes for ultrasensitive detection of biomolecules: An update. *Chemical Society Reviews, 43*(5), 1601–1611.

Zeng, Z, Shi, Y. X., Samudio, I. J., Wang, R.-Y., Ling, X., Frolova, O., Levis, M., Rubin, J. B., Negrin, R. R., Estey, E. H., et al. (2009). Targeting the leukemia microenvironment by CXCR4 inhibition overcomes resistance to kinase inhibitors and chemotherapy in AML. *Blood, 113*(24), 6215–6224.

Zhang, GC, Liu J, Yu XN, Deng Y, Sun Y, Liu TT, Dong L, Zhu CF, Shen XZ, Zhu, J. M., Weng, S. Q., & Li, Y. (2020). Bismuth-based mesoporous nanoball carrying sorafenib for computed tomography imaging and synergetic chemoradiotherapy of hepatocellular carcinoma. *Advanced Healthcare Materials, 9*(21), 2000650.

Zheng, X. T., Ma, X. Q., & Li, C. M. (2016). Highly efficient nuclear delivery of anti-cancer drugs using a bio-functionalized reduced graphene oxide. *Journal of Colloid and Interface Science, 467*, 35–42.

Zhong, Y., Peng, Z., Peng, Y., Li, B., Pan, Y., Ouyang, Q., Sakiyama, H., Muddassir, M., & Liu, J. (2023). Construction of Fe-doped ZIF-8/DOX nanocomposites for ferroptosis strategy in the treatment of breast cancer. *Journal of Materials Chemistry B, 27*.

Zhu, R., Zhang, F., Peng, Y., Xie, T., Wang, Y., & Lan, Y. (2022). Current progress in cancer treatment using nanomaterials. *Frontiers in Oncology, 12*, 930125.

Zou, J., Chen, S., Li, Y., Zeng, L., Lian, G., Li, J., Chen, S., Huang, K., & Chen, Y. (2020). Nanoparticles modified by triple single chain antibodies for MRI examination and targeted therapy in pancreatic cancer. *Nanoscale, 12*(7), 4473–4490.

CHAPTER 11

Application of Nanocomposites in 3D and 4D Printing of Nanomaterials in Biomedical Research

PALLERLA NAVEEN REDDY[1,2], DEEPENDRA SINGH[3], VIVEK DAVE[1], and RUCHI TANDON[2,*]

[1]Department of Pharmacy, School of Health Sciences, Central University of South Bihar, Gaya, India

[2]Translational Health Science and Technology Institute, Faridabad, India

[3]University Institute of Pharmacy, Pt. Ravishankar Shukla University, Raipur, India

[*]Corresponding author.

ABSTRACT

Nanocomposites are a new class of promising materials having a broad range of uses in drug delivery systems, biomedical research, and other areas that are made up of nanomaterials combined with a matrix material. In the context of 3D and 4D printing technologies, this chapter offers an in-depth review of the uses of nanocomposites for developing intricate biomedical devices, implantable gadgets, drug delivery systems, dental prostheses, and biological sensors. The combination of printing techniques and nanocomposites has opened up new opportunities for precise and tailored drug delivery systems that have higher therapeutic efficacy. The properties of nanocomposites emphasize their special qualities, such as improved drug-loading capacity and mechanical strength. This chapter

Nanocomposites and Nanomaterials in Biomedical Applications. Vivek Dave, Deependra Singh, and Pallerla Naveen Reddy (Eds.)

DOI: 10.1201/9781779641236-11

explores how nanocomposites, in the context of the versatility of 3D and 4D printing, have made it possible to fabricate scaffolds for tissue engineering and patient-specific drug delivery devices, as well as to create complex structures that can encapsulate drugs, growth factors, or therapeutic nanoparticles, ensuring precise spatial control over drug release. Additionally, nanocomposites that have stimuli-responsive components within them have the ability to change form or release drugs in response to external triggers like pH, temperature, or light. On-demand medicines and personalized medicine have enormous potential because of this innovative approach to drug administration. This chapter concludes by showcasing the incredible ability of nanocomposites to transform medication delivery systems via 3D and 4D printing technology.

11.1 INTRODUCTION

Nanocomposites are evolving into materials that hold great promise for drug delivery systems and biomedical research. These materials represent a significant advancement in material science, particularly in biomedical research for designing targeted and sustained release drug delivery systems reduce adverse effects while increasing therapeutic efficacy (Caballero-Aguilar et al., 2020). These materials incorporate the characteristics of nanoparticles incorporated in a polymer matrix which led to unique functionalities that are better than those of individual components (Rahmani et al., 2022). These nanoengineered materials have several beneficial qualities, such as increased mechanical strength, better biocompatibility, and customized functions, which make them vital for a wide range of biomedical applications.

Simultaneously, advances in 3D and 4D printing technology have completely changed the healthcare industry by making it possible to fabricate intricate structures with unprecedented levels of precision and personalization (Deshmukh et al., 2020; Wang et al., 2017). 3D printing is also called as additive manufacturing. Using digital designs, 3D printers build objects with three dimensions layer by layer. It provides previously unattainable control over internal architecture, design, and material properties, and enables the creation of personalized implants, tissue models, and drug delivery systems (Richards et al., 2013). Dynamic methods for administering medication can be made even more flexible with 4D printing, which gradually adds a new level of functionality and changes its shape in response to external stimuli (Zeng et al., 2023). The combination of 3D and 4D printing technologies with nanocomposites has great potential for biomedical research. Through the utilization of nanomaterial diversity and printing technique's accuracy, scientists may create complex structures with customized properties for a range of uses, including tissue engineering and medication delivery.

For biological applications, incorporating nanocomposites into 3D and 4D printing has many benefits (Figure 11.1) (Archer et al., 2018; Rahman et al., 2020). It includes:

1. Personalized drug delivery systems with adjustable release kinetics that precisely target particular tissues or cells.
2. Enhanced biocompatibility and bioactivity, which promote cellular adhesion, proliferation, and tissue regeneration.
3. Precise control over material qualities such as porosity, mechanical strength, and degradation rate to replicate real tissues and organs.

By providing novel approaches for controlled drug release, tailored medicine, and on-demand therapeutic interventions, this type of technology not only enriches the field of tissue engineering but also improves research on drug delivery systems (Goyanes et al., 2014). To better understand how nanocomposites can change healthcare paradigms, this chapter examines their different applications in both 4D and 3D printing of drug delivery systems and in biomedical research.

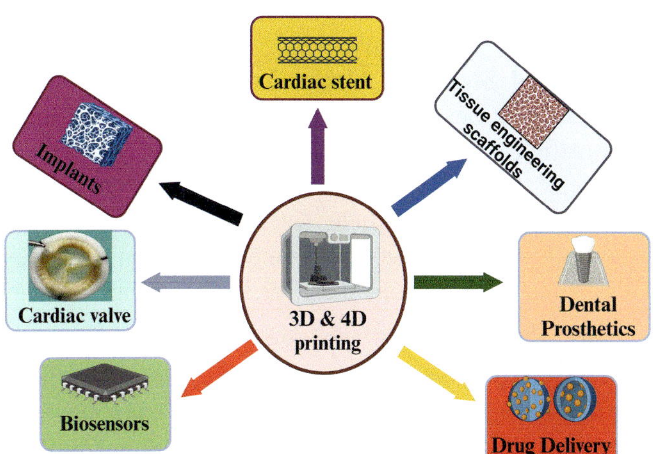

FIGURE 11.1 Applications of nanocomposites in 3D and 4D printing.

11.2 FUNDAMENTALS OF NANOCOMPOSITES

Composite matrixes containing nanomaterials known as nanocomposites are made of a host material or matrix that has been reinforced with nanoparticles or nanofillers. These nanoparticles, which normally have size between 1 and 100 nm, are dispersed throughout the matrix to give the composite material required special characteristics.

Depending on the intended use, nanocomposites might have different compositions. Through the manipulation of the nanophase's composition, distribution, shape, and size, researchers can customize the overall properties of the nanocomposite for certain applications in biomedicine. Polymers, metals, ceramics, and their alloys are often utilized as matrix materials. The required processing conditions, desired mechanical characteristics, and biocompatibility all plays an integral part in the matrix material selection and characteristics of nanocomposites can be precisely engineered to comply with particular requirements in biomedical research, such as:

1. **Biocompatibility:** It is possible to develop nanocomposites that are specifically designed to resemble the biological environment, which will foster proliferation, cell adhesion, and tissue integration. Techniques for altering nanocomposite surfaces, such as functionalization with bioactive compounds or peptides, improve biocompatibility, interact better with biological systems, reduce the body's immune response, and facilitate interaction with live tissues.
2. **Mechanical Strength:** Mechanical behavior of the composite material can be significantly enhanced by the inclusion of nanoparticles. By strengthening the matrix and enhancing its resistance to deformation and fracture, nanoparticles serve as reinforcing agents. Particle size, distribution, and orientation control permit the exact adjustment of mechanical features, such as by using carbon nanotubes, for load-bearing applications like scaffolds or bone implants.
3. **Porosity:** To promote tissue growth, waste elimination, and nutrient diffusion, nanocomposites can be engineered with regulated porosity. Porous features promote cellular infiltration and vascularization by replicating the internal structure of real tissues

and organs. For tissue engineering applications, porous features in the matrix may enhance cell adherence and development while also increasing surface area. Precise control over distribution, pore size, and shape is made possible by methods including additive manufacturing, phase separation, and sacrificial templating.

4. **Thermal and Electrical Conductivity:** Improved thermal and electrical conductivity in nanocomposites can lead to new applications in biosensing, thermal control, and electrostimulation to promote muscle or nerve growth. Incorporating conductive nanoparticles, such as graphene or carbon nanotubes, enhances the composite material's electrical signaling and heat dissipation properties.
5. **Biological Properties:** Bioactive compounds, genes for targeted drug delivery, biosensing, growth factors, or medications to impart particular biological features can all be used to functionalize nanocomposites (Allo et al., 2012). Surface alterations facilitate localized medication delivery, wound healing, and tissue regeneration by allowing for the regulated release of therapeutic substances.

11.2.1 MATERIALS FOR 4D AND 3D PRINTING APPLICATIONS THAT CAN BE UTILIZED FOR THE DEVELOPMENT OF NANOCOMPOSITES

Complex biomedical constructions can be created with intriguing new possibilities through 4D/3D printing. Nanocomposites are especially well suited for these methods because of their printability and adaptability. Nanocomposites provide flexible platforms for creating innovative materials with specialized features for 3D/4D printing and biomedical research because they allow for precise control over composition, structure, and processing parameters.

1. **Polymer-Based Nanocomposites:** For 3D printing, adding nanoparticles to polymer matrices improves their thermal stability, mechanical strength, and processability. Polycaprolactone, polylactic acid (PLA), polymer-clay, polymer-graphene, and polymer-metal oxide nanocomposites are examples of common

polymer-nanoparticle combinations. Hydrogels can also be reinforced with different types of nanoparticles for improved characteristics (Bose and Saha, 2003).

2. **Ceramic-Based Nanocomposites:** Ceramic matrices reinforced with nanoparticles exhibit high strength, heat resistance, wear resistance, and biocompatibility for biomedical applications. Ceramic-hydroxyapatite nanocomposites, ceramic-metal nanocomposites, and ceramic-carbon nanotube composites are some examples. However, printability is challenging, and further research is ongoing (Bose and Saha, 2003).

11.3 NANOCOMPOSITES IN 3D PRINTING

Nanocomposite's improved qualities and adaptability have made them stand out among potential materials for 3D printing applications. When working with nanocomposite inks or filaments, several 3D printing technologies can be employed, each has its own benefits and is also utilized in drug delivery systems and biomedical research. The best 3D printing method for nanocomposites is determined by some parameters, such as:

1. **Characteristics of the Nanocomposite:** The concentration, shape, and size of nanoparticles within composite material can influence printability.
2. **Desired Resolution and Feature Size:** Techniques like stereolithography offer a higher resolution for intricate structures, while fused deposition modeling might be sufficient for less detailed applications.
3. **Biocompatibility Requirements:** Direct ink writing is specifically designed for biocompatible materials, making it ideal for tissue engineering applications.

11.3.1 OVERVIEW OF 3D PRINTING TECHNOLOGIES SUITABLE FOR NANOCOMPOSITES

11.3.1.1 STEREOLITHOGRAPHY

Stereolithography (SLA) utilizes photopolymerization to create objects layer by layer from a liquid resin containing nanoparticles into desired

3D structures that harden when exposed to UV light (Kim et al., 2010). The process of stereolithography involves spatially regulating a liquid photopolymerizable resin's solidification. On a computer-driven building platform, a three-dimensional object is built layer by layer using a computer-controlled laser beam or digital light projection (Lu et al., 2006). In order to achieve this, the many layers that need to be made are defined by numerical analysis of the object's design (Provin and Monneret, 2002). According to Griffith and Halloran, complex three-dimensional objects are created through the progressive polymerization of layers with various lighting patterns (Griffith and Halloran, 1996). Due to its high resolution and capacity for producing detailed structures, this technology is well-suited for applications that call for smooth surface finish and fine features, such as complex scaffolds and microfluidic devices (Li et al., 2021; Melchels et al., 2010). Nanocomposite resins can be formulated to include nanoparticles dispersed within the liquid resin, allowing for the fabrication of nanocomposite-based objects with precise control over material properties (Bleach et al., 2002). However, a limitation of stereolithography is the limited selection of commercially available nanocomposite resins compatible with the specific laser wavelength used (Dhariwala et al., 2004), and additionally, these nanocomposite networks become brittle at very high crosslink densities, which limits their utility (Gurr et al., 2008).

11.3.1.2 FUSED DEPOSITION MODELING

Thermoplastic filaments are extruded through a nozzle that is heated to fabricate the desired shape, which subsequently hardens layer-by-layer by using fused deposition modeling printer (Ahn et al., 2002). Because of its price, ease of usage, and adaptability, fused deposition modeling is often used. By distributing nanoparticles throughout thermoplastic polymers, nanocomposite filaments can be developed for the printing of shapes based on nanocomposite with better functional and mechanical characteristics (Seng et al., 2020). Fused deposition modeling is compatible with a range of nanocomposite filaments and provides material diversity. However, compared to stereolithography, fused deposition modeling often has a poorer resolution, which restricts the ability to create extremely intricate structures (Iqbal Hussain, 2021).

11.3.1.3 DIRECT INK WRITING

Direct ink writing also sometimes referred to as robocasting or extrusion-based 3D printing, is a technique that uses a syringe or nozzle to apply viscous inks or pastes onto a substrate. This technique makes tissue engineering applications possible by precisely depositing materials to create complex structures (Zhang et al., 2021). By dispersing nanoparticles in a polymer matrix or carrier fluid, nanocomposite inks can be created, allowing for the creation of structures based on nanocomposite technology that have specific qualities like high strength, conductivity, or biocompatibility. Carefully adjusting the bio ink's viscosity and rheological characteristics is necessary for achieving smooth extrusion from the printing nozzle and retention of the intended shape after deposition. The bio ink's printability is still challenging.

These 3D printing instruments are utilized for a wide range of biological applications because they each have special advantages and capabilities for working with nanocomposite materials. Scientists are consistently designing and advancing novel 3D printing methodologies exclusively tailored for bio-fabrication employing nanocomposites. With these developments, the limits of current techniques will be addressed and the full potential of nanocomposites to create intricate and useful biological structures will be realized.

11.4 APPLICATION OF NANOCOMPOSITES IN IMPLANTABLE DEVICES BY 3D PRINTING

The introduction of 3D-printed nanocomposites has brought up a dramatic change in the field of implanted devices. Personalized implants and complex drug delivery systems are only two examples of implantable devices that could be developed with the help of nanocomposite-based 3D printing. The creation of drug delivery devices with exact control over drug release kinetics and spatial distribution is made possible by nanocomposite-based 3D printing. These gadgets offer personalized medicine strategies by being able to be customized to each patient's unique requirements. Nanocomposite-based 3D printing techniques can be used to construct implantable drug-delivery implants, such as customized drug-eluting stents and implants for targeted drug delivery to certain tissues or organs.

11.4.1 IMPROVED BIOCOMPATIBILITY AND OSSEOINTEGRATION IN PATIENT-SPECIFIC IMPLANTS

Conventional implants frequently encounter difficulties in orthopedic applications, such as inadequate osseointegration (bone bonding) and rejection. Orthopedic applications can benefit from patient-specific implants with improved osseointegration and biocompatibility because of 3D-printed nanocomposites. These implants minimized the possibility of implant rejection or failure by exactly matching the anatomy of the patient. They also encourage greater tissue integration. Implants made of nanocomposite materials can be designed to resemble the mechanical characteristics of native bone tissue, which will encourage bone fusion and stability over the span of time. Implants tailored to each patient's anatomy can be designed and 3D printed using medical imaging data, such as CT scans. As a result, implant fit is improved and surgical complexity is decreased (Rengier et al., 2010). Moreover, by adding bioactive nanoparticles such as hydroxyapatite (HA), a significant component of bone to the implant material can strengthen and speed up osseointegration by enhancing bone cell proliferation and adhesion to the implant surface (Turnbull et al., 2018). Because of its chemical resemblance to biological forms, hydroxyapatite is recognized for its established and expected characteristics, including osteoconductive characteristics, bioactivity, absence of immunological response, and biocompatibility. Considering its basic properties, low cost, and ease of manufacturing, hydroxyapatite is a suitable material for the development of implants and scaffolds. As a targeted medication carrier, it can also be used to treat numerous bone disorders (Ghiasi et al., 2019). One study from 2022 showed how well hydroxyapatite microspheres reinforced polylactic-co-glycolic acid (PLGA) could be utilized to create 3D-printed scaffolds for bone tissue engineering. The study also showed how these scaffolds could enhance bone regeneration (Wei et al., 2022). The use of hydroxyapatite coatings in bone implants is accepted as a means of strengthening the implant material. Hydroxyapatite is utilized in the field of implants that serves as a composite material for strength in conjunction with other materials (Fendi et al., 2024; Huang et al., 2021; Khalili and Naji, 2020).

For example, a study carried out in 2022 developed poly (organophosphazene)-based thermo-responsive nanocomposites (TNC) carrying growth factors (bone morphogenetic protein-2 and transforming growth factor beta-1). The TNC bioink formulation produced the smoothest filaments during extrusion at 25°C, which led to the best 3D printing results. At 37°C body temperature, the printed bioink exhibited enough rigidity, and the 3D structure was preserved without the need for extra chemical crosslinking procedures after printing. The implant was effectively placed in a model with a calvarial bone deficiency. Thick, newly produced bone successfully replaced the defect area, and the implanted TNC bioink scaffold deteriorated over six weeks (Kim et al., 2023). In another study, poly(vinyl alcohol) was innovatively blended into PLA to fabricate composite filaments, and the fabricated macroporous scaffolds were subjected to a gas foaming process. This led to the creation of microporosities scaffolds which help in bone tissue repair (Song et al., 2018).

In a study by W. Wang et al., composite scaffolds made of nano-hydroxyapatite (n-HA) and PLA were 3D printed using fused deposit modeling. The study showed that the PLA/n-HA composites were excellent candidates for 3D printing, yielding scaffolds with mechanical strength that could be adjusted based on the ratio of n-HA components. The composite's osteogenic induction capabilities and biocompatibility outperformed those of pure PLA scaffolds. A viable method for fixing large bone lesions is provided by this combination of PLA and n-HA (Wang et al., 2021). Senatov et al., utilized fused filament fabrication to develop composite filaments consisting of polylactide/15 wt% hydroxyapatite for 3D printed scaffolds. It has been shown that 3D-printed PLA/15%HA porous scaffolds with a 98% shape recovery rate can be used as self-fitting implants to replace small bone defects because of the shape memory effect (Senatov et al., 2016).

In an experiment conducted by Y. J. Wang et al., water-based biodegradable shape memory polyurethane with polyethylene oxide/gelatin and Superparamagnetic iron oxide nanoparticles showed improved printability, shape memory characteristics, and cell viability. The polyurethane was printed using a low-temperature fused deposition manufacturing 3D printer. Furthermore, the osteogenesis of the human mesenchymal stem cells in the scaffolds, as well as the secretion and deposition of collagen and calcium, were facilitated by the release of superparamagnetic iron oxide nanoparticles from the polyurethane-based 3D printed scaffolds. This novel scaffold has the potential to be beneficial in minimally invasive surgery as a customized bone substitute for regeneration of bone tissue (Wang et al., 2018).

A PLA scaffold reinforced with graphene at different weight percentages (0.1, 0.2, and 0.3) was developed by Belaid et al. by 3D printing. Fused deposit modeling was employed to produce the scaffolds. To aid with bone regeneration, these scientists designed scaffolds with a 70% infill and a porous interconnected network with pore diameters of roughly 300 μm. Compared to the control PLA scaffolds, the graphene-incorporated scaffolds had a more hydrophilic and rougher surface. Tensile tests showed that the mechanical properties enhanced with the application of graphene. Composite scaffolds were superior than pure PLA scaffolds in terms of stimulating cell proliferation and mineralization, which may have resulted in the development of bone, in addition to being biocompatible (Belaid et al., 2020).

11.4.2 3D-PRINTED CARDIAC VALVES AND STENTS USING NANOCOMPOSITES

Heart valves and stents are life-saving implants, but they sometimes have durability issues and need to be replaced in order to treat complications like blood clot development that arise after surgery. For the development of stents and heart valves, nanocomposites provide improved durability and drug delivery capabilities (Cherian et al., 2021). Incorporating nanoparticles into implant material increased its mechanical strength and corrosion resistance, prolonging the device's lifespan and lowering the need for repeat operations. Therapeutic agents can be delivered to surrounding tissue using drug-eluting stents and heart valves made of nanocomposites. This can reduce inflammation, encourage tissue regeneration, and prevent restenosis or thrombosis (Kim et al., 2019; Ooi et al., 2007).

A 2017 study investigated the use of graphene oxide and poly(ε-caprolactone) (PCL) in the 3D printing of biocompatible nanocomposite stents. The findings pointed to enhanced mechanical characteristics and the possibility of regulated drug delivery, paving the way for stents of the future generation. For example, graphene nanoplatelets are added to Poly(ε-caprolactone) to form nanocomposites, which are then combined with Niclosamide for restenosis and coagulation and phytic acid (IP6, inositol phosphate). Utilizing data from volume-rendered CT images of the targeted coronary heart clot, computer-aided design, fused deposit modeling, and stents printed from poly(ε-caprolactone) are utilized to produce nanocomposites that exhibited superior mechanical properties and also no toxicity when tested on the HUVEC cell line (Misra et al., 2017).

11.4.3 PROSPECTS OF 3D PRINTED NANOCOMPOSITE SCAFFOLDS IN TISSUE ENGINEERING AND NEURAL IMPLANTS

A patient's quality of life can be greatly impacted by neurological conditions. TAMAY et al., suggest that 3D printed nanocomposite scaffolds have potential uses in tissue engineering and brain implants (Tamay et al., 2019). By imitating the composition and characteristics of neural tissue, these 3D-printed nanocomposite scaffolds offer an environment that is favorable for the formation of brain cells and the outgrowth of axons (Mobaraki et al., 2020). According to Koppler et al., this may lead to better treatment outcomes for neurodegenerative illnesses and spinal cord injuries (Koffler et al., 2019). The mechanical and biochemical cues of the native tissue can be mimicked by nanocomposite-based scaffolds, facilitating integration with the host tissue and functional recovery. This has potential applications in the treatment of epilepsy and Parkinson's disease (Saylam et al., 2021). A 2022 study, showed the possibility for future neural implants by effectively demonstrating the viability of 3D printed scaffolds employing graphene-poly-L-lactic acid (PLLA) scaffolds to promote cell alignment and differentiation, but research in this field is still ongoing (Gasparotto et al., 2022).

11.5 USE OF NANOCOMPOSITES IN 3D PRINTED DENTAL PROSTHETICS

To create stronger, lighter, and more aesthetically pleasing dental crowns, bridges, implants, and orthodontic aligners, 3D printing with nanocomposites has intriguing prospects for the future development of dental prosthetics.

11.5.1 CREATING STRONG AND LIGHTWEIGHT DENTAL CROWNS AND BRIDGES

Conventional bridges and crowns are bulky and prone to chipping. 3D printing can be used to develop strong and lightweight dental crowns and bridges using nanocomposite materials (Cristache and Totu, 2021). The mechanical properties of the finished product—crowns and bridges that can withstand biting forces and resist wear and tear much better than traditional materials—are improved by the incorporation of nanoparticles into the resin or filament. Additionally, durability is ensured while reducing overall weight, improving patient comfort by reducing pressure on the jaw (Lauvahutanon et al., 2015). Better fit and comfort are achieved by precisely fabricating dental prosthesis using 3D printing, which is customized to each patient's unique anatomy. Furthermore, 3D printing's digital designs allow for the production of complex patterns and textures for enhanced aesthetics. The mechanical characteristics of conventional crowns and 3D-printed nanocomposite crowns were examined in a 2023 study. According to the findings, the 3D-printed nanocomposite crowns demonstrated strength that was on par with or even better than the original, paving up the possibilities for patients to have lighter but more robust crowns (Rizzante et al., 2023).

11.5.2 NANOCOMPOSITE-BASED DENTAL IMPLANTS WITH ENHANCED FUNCTIONALITY

Long-term dental implant success depends on robust osseointegration, or the connection of the implant with the bone. Dental implants made of nanocomposite materials have better functionality, especially in terms of osseointegration and appearance. In order to increase stability for extended periods and lower the chance of implant failure, bioactive nanoparticles such as hydroxyapatite can be added to the implant material to encourage bone cell proliferation and integration with the surrounding tissue (Turnbull et al., 2018). Additionally, prosthesis with better aesthetics can be developed by engineering nanocomposites to resemble the color and translucency of real teeth. This is especially crucial for dental implants in the anterior area, as patient satisfaction is greatly influenced by aesthetics. A 2021 study explored the potential of 3D printed nanocomposite materials with hydroxyapatite for dental implant applications. The study highlighted the possibility of achieving improved mechanical properties and biocompatibility, paving the way for more natural-looking and functional implants (Han et al., 2021).

11.5.3 CUSTOMIZED ORTHODONTIC ALIGNERS USING NANOCOMPOSITES

Traditional orthodontic treatments often involve uncomfortable metal braces. 3D printing with nanocomposites also presents the possibility of creating customized orthodontic aligners. By utilizing digital scans of the patient's teeth, personalized aligners can be designed and 3D printed using nanocomposite materials. These aligners offer a more comfortable and discreet alternative to traditional orthodontic treatments, such as braces. Nanocomposite materials can provide the necessary strength and flexibility required for orthodontic aligners, ensuring optimal performance and durability throughout the treatment process. Nanocomposites can be formulated to be transparent, similar to existing clear aligners, while potentially offering improved strength and clarity. This could enhance patient comfort and potentially improve treatment efficiency (Mallik et al., 2020). While this application is still under development, the ability to 3D print customized aligners with potentially superior properties holds promise for the future of orthodontics.

11.6 USE OF NANOCOMPOSITES IN DRUG DELIVERY SYSTEMS

The emergence of 3D-printed nanocomposite structures is bringing about a major change in the medication delivery industry. Targeted and customized treatment plans are now possible because of the novel drug delivery systems which are nanocomposite-based 3D. According to Rahman et al., 3D printing using nanocomposite technology presents novel approaches to drug delivery systems. This technology allows for exact control over drug release kinetics and spatial distribution, hence opening up new avenues for customized and targeted treatment plans (Rahman et al., 2024). This is an example of the best way to use nanocomposite structures for long-term, targeted medication administration.

11.6.1 CONTROLLED POROSITY FOR SUSTAINED AND TARGETED DRUG RELEASE

Traditional drug administration methods often face challenges like unintended side effects and difficulty in maintaining therapeutic drug levels. Nanocomposite structures fabricated using 3D printing can be designed with controlled porosity, allowing for the incorporation of drugs within the porous network. This controlled porosity enables the targeted and sustained release of drugs, where the porosity and structure can be changed to modify the rate and duration of drug release, reducing the requirement for repeated dosing. By adjusting the nanocomposite scaffold's surface characteristics and pore size and distribution, specific drug release profiles suitable for various therapeutic applications can be achieved, such as tissue regeneration, cancer treatment, and antimicrobial therapy (Adepu and Khandelwal, 2018). Localized drug delivery can minimize systemic side effects often associated with traditional oral medications. In 2018, a research study revealed that controlled porosity scaffolds made of 3D-printed nanocomposite materials might be used to administer Cefazolin locally to surgery site infections. The study highlighted the potential for this approach to reduce systemic antibiotic use and improve treatment efficacy (Visscher et al., 2018).

11.6.2 DESIGN OF NANOCOMPOSITE FILAMENTS FOR FUSED DEPOSITION MODELING PRINTING

The most often utilized 3D printing method is fused deposition modeling (FDM) (Goyanes et al., 2014). For FDM printing, nanocomposite filaments can be designed to encapsulate and release drugs in reaction to particular stimuli like light, pH, or temperature. Under the right circumstances, drug release can be triggered by functional nanoparticles added to the filament matrix, such as pH-sensitive or thermoresponsive polymers. For example, thermo-responsive polymers releases drugs upon reaching body temperature, targeting the desired site of action (Jahanbakhshi and Shahrousvand, 2021), whereas pH-sensitive polymers release their payload only in specific acidic or basic environments, mimicking the pH variations within the body (Reyes-Ortega, 2014), and Light-sensitive nanoparticles can be incorporated into the filament, allowing for controlled drug release using external light sources, this offers the potential for precise spatial and temporal control over medicine delivery (Tao et al., 2020). Nanocomposite filaments that include stimuli-responsive nanoparticles can react to variations in the surrounding environment, enabling medication release at the target spot on-demand. This controlled drug release mechanism enhances therapeutic efficacy while minimizing systemic side effects. Cardoso et al., produced graphene-PLA filament biosensing systems using FDM. The PLA-graphene surface was utilized to create the enzymatic biosensor, which measures glucose in blood plasma. For the analysis of plasma, the biosensor showed appropriate recovery values (90–105%) and a limit of detection of 15 μM L−1, with intra and inter day precision < 5%. Furthermore, after surface treatment, the 3D-printed sensor showed improved electrochemical properties for the immediate identification of uric acid and nitrite (Cardoso et al., 2020).

11.6.3 POTENTIAL FOR 4D PRINTED NANOCOMPOSITE DRUG DELIVERY SYSTEMS

Drug delivery devices made of 4D-printed nanocomposite materials offer an additional level of functionality by releasing or altering their structure in response to external stimuli. When these dynamic systems are exposed to external stimuli, like changes in pH or temperature or biological signals, they can go through predefined transformations. 4D printed drug delivery systems can display programmable behaviors, such as shape deformation, self-folding, or controlled release, by incorporating shape-memory polymers or stimuli-responsive nanoparticles into the nanocomposite matrix, this allows for precise spatiotemporal control over drug delivery (Sajjad et al., 2024). Furthermore, by utilizing stimuli-responsive materials, 4D printed systems can deliver drugs when exposed to certain environmental changes like a rise in temperature or a shift in pH, further enhancing targeted therapy. For instance, a scaffold implanted for bone regeneration could gradually degrade and release growth factors as new bone tissue forms (Tajvar et al., 2023). These developments in 3D printing using nanocomposite materials have the potential to completely transform drug delivery systems by providing tailored and adaptable treatments that increase patient outcomes.

11.7 APPLICATIONS OF 3D PRINTED NANOCOMPOSITES IN BIOLOGICAL SENSORS

The development of biological sensors for the detection of disease, real time biomarker monitoring, and customized medicine techniques has advanced significantly due to the integration of nanocomposites with 3D printing (Xu et al., 2017).

11.7.1 3D PRINTED NANOCOMPOSITE BIOSENSORS

Traditional biosensors often face limitations like rigidity and difficulty in customization, whereas 3D printed nanocomposite biosensors offer the capability of flexibility, customization, and real-time monitoring of biomarkers and physiological parameters. These biosensors are typically composed of nanocomposite materials that exhibit high sensitivity and selectivity towards target biomarkers (Su et al., 2023), this allows for precise detection of disease markers or physiological changes. Nanoparticles embedded within the sensor matrix facilitate enhanced detection through surface plasmon resonance, electrochemical properties, or fluorescence. The utilization of 3D printing technology facilitates the precise manufacturing of sensor structures with adaptable shapes, thus augmenting their sensitivity and accuracy (Palmara et al., 2021). A 2018 study showed how to develop glucose biosensors by employing specially made enzyme ink for 3D printing on tattoo paper using the direct ink writing process, and modifying regular carbon ink with Prussian blue as an electron transfer mediator. Compared to traditional screen-printing techniques, the printed sensors showed improved sensitivity, specificity, and lower material consumption (Nesaei et al., 2018).

11.7.2 CUSTOMIZED BIOSENSORS FOR DISEASE DETECTION

3D printed nanocomposite scaffolds can be utilized to create customized biosensors for disease detection. These scaffolds offer a three-dimensional environment in which biomolecules can bind to target analytes selectively like DNA probes or antibodies can be immobilized in those spaces. Biosensors can be designed with high specificity and sensitivity for detecting a variety of diseases, including metabolic disorders, cancer, and infectious diseases (Lin et al., 2021; Morales and Halpern, 2018). Customized biosensors have the potential to facilitate prompt disease detection, better patient outcomes, and thereby enabling timely intervention. Damiati et al., reported on the development of an efficient acoustic and hybrid electrochemical biosensor printed using a 3D printer for continuous identification of HepG2 liver cancer cells (Damiati et al., 2017). The biosensors look into the relationship between recombinant proteins and CD133, a tumor biomarker that is highly expressed in cells that cause liver cancer. According to Chinen et al., tumor markers are often proteins that cells produce in larger amounts in response to cancer (Chinen et al., 2015). These markers can help determine prognosis, direct treatment, and assist in detecting cancer in its early stages (Vargas and Harris, 2016).

11.7.3 NANOCOMPOSITE-BASED SENSORS FOR PERSONALIZED MEDICINE AND DETECTION OF DISEASE AT THEIR EARLY STAGES

Nanocomposite-based sensors offer the potential for personalized medicine and detection of disease at their early-stage approaches. By detecting biomarkers at the early stages of disease progression, these sensors enable timely interventions and personalized treatment strategies (Krause et al., 2013). The versatility of nanocomposite materials allows for the development of sensors with tunable properties, such as sensitivity, selectivity, and response time, catering to individual patient needs. In addition to providing real-time data on an individual's health status, 3D printed biosensors can be designed for constant tracking of physiological indicators or biomarkers, which enables early identification of potential health issues. This allows for the development of individualized treatment plans that are specific to each individual's particular requirements. This individualized approach to managing illness can potentially lower healthcare expenses while enhancing the outcomes for patients.

11.7.4 POTENTIAL FOR IMPLANTABLE BIOSENSORS USING 3D PRINTED NANOCOMPOSITES

Implantable biosensors utilizing 3D printed nanocomposites offer a minimally invasive solution for continuous monitoring of biomarkers and physiological parameters. These biosensors can be integrated into medical devices or implanted directly inside the body to provide continuous data on health status. Nanocomposite materials with biocompatible and bioresorbable properties enable the fabrication of implantable biosensors that minimize immune response and tissue rejection. By leveraging 3D printing technology, these biosensors can be customized to fit specific anatomical sites and patient requirements (Khosravani and Reinicke, 2020).

11.8 NANOCOMPOSITES IN 4D PRINTING

4D printing is by adding the element of time into the fabrication process to 3D printing. 4D printing utilizes materials that can alter shape or function in response to external variables like humidity, light, temperature, or pH, whereas 3D printing uses layers of materials to create static objects. The transition from 3D to 4D printing enables the production of dynamic frameworks with the ability to self-repair, self-assemble, or adjust to their surroundings over some time.

Nanocomposites plays a crucial part in enabling the 4D printing of dynamic structures by imparting stimuli-responsive properties to the printed materials. Nanoparticles embedded within the composite matrix can respond to external stimuli, triggering changes in the material's shape, mechanical properties, or functionality. For instance, incorporating shape-memory alloy nanoparticles can enable printed structures to change shape upon exposure to heat (Khalid et al., 2022). By incorporating different types of nanoparticles into the composite material, structures can be designed that undergo pre-programmed transformations in response to specific stimuli. This responsiveness is essential for achieving programmable and controllable transformations in 4D printed objects. According to studies by Choong et al., on the correlation between curing depth and energy density in SLA, the material's shape memory capabilities deteriorate as cross-linker content but curing depth gets better (Choong et al., 2017). Using SLA, Zarek et al. were able to successfully create a range of intricate 4D structures, and they were able to confirm that the printed object's shape memory characteristics held up to thermal stimulation (Zarek et al., 2016). Lu et al. developed a magnetic force-assisted SLA through which magnetic fillers can be disseminated in a particular pattern and are controlled in resinous liquid with the aid of a magnetic field from the outside (Lu et al., 2017). This opens the door for the use of magnetically driven smart polymer composites in the biomedical sector through 4D printing enabling the creation of a wide range of innovative sophisticated structures. A study carried out by Chowdhury et al. utilized both tert-Butyl Acrylate, diethylene glycol diacrylate, and phenylbis (2,4,6-trimethylbenzoyl) phosphine oxide, and nanoparticles made up of graphene were dispersed at a slow pace into the resin and printed by micro stereolithography into U shape and after printing, these shape memory polymers demonstrated a considerable increase in recovery time, a decrease in surface roughness, and greater tensile stress, without observable changes to strain fixity. Therefore, it is discovered that the addition of graphene particles of 0.3% yields the most advantageous set of features and that all other qualities—aside from tensile stress—begin to decrease with additional graphene addition (Chowdhury et al., 2021).

11.8.1 USES OF 4D PRINTING WITH NANOCOMPOSITE MATERIALS IN BIOMEDICAL RESEARCH

4D printing with nanocomposite materials enables the creation of implants with morphing shapes that gradually adjust to the anatomy of the patient. By undergoing controlled alterations in response to physiological circumstances, these implants can improve patient comfort and biocompatibility. Nanocomposites-based 4D printed medication delivery devices allow for the regulated release of therapeutic drugs in response to particular stimuli (Malachowski et al., 2014). Nanoparticles embedded within the drug delivery device can respond to physiological changes, like temperature variations or pH, triggers to release of the drug at the target site. For instance, a hydrogel implant could release antibiotics only when exposed to bacterial infection markers (Devnarain et al., 2021; Trachsel et al., 2020; Wang et al., 2014). Hydrogels based on nanocomposite technology developed through 4D printing offer customizable characteristics in the field of regenerative medicine and tissue engineering. Similar to the dynamic nature of biological tissues, these hydrogels can alter their stiffness or form in response to external stimuli. Nanocomposite-based implants and microneedles fabricated using 4D printing technology incorporate drugs within the composite matrix for controlled drug delivery. These devices release drugs in a programmable manner at specific rates upon encountering a change in surrounding conditions like pH or temperature, ensuring precise dosing and localized treatment (Che et al., 2024). Nanocomposite-based 4D printing holds immense potential for advancing biomedical research and clinical applications, offering innovative solutions for drug delivery systems, personalized medicine, and tissue engineering.

For example, Wan et al. incorporated carbon nanotubes (CNTs) to poly (D,L-lactide-co-trimethylene carbonate) (PLMC) in an attempt to enable PLMC-based 4D-printed structures and their capacity to transform shape when exposed to an electrical charge. The 4D-printed structure showed excellent shape changing behavior under 25 V in less than 16 s. After that, PLMC was converted into crosslinking networks (c-PLMC) via a reaction between UV light and hydrogen abstraction. After UV crosslinking, the c-PLMC/CNT's solvent resistance and mechanical strength were significantly enhanced, but its volume electrical conductivity and actuation temperature remained mostly unchanged (Wan et al., 2019).

According to Zhang et al., 4D printed porous stent for customized bone repair, PLA/Fe_3O_4 composite filaments are fabricated by an FDM printer. The stent could revert to its original form in 100 s after being exposed to a magnetic field, and it could be customized for various defect sizes. This implies that since varied bone sizes may be produced to meet the particular deficiency, the 4D structure may be utilized for tailored bone healing (Zhang et al., 2019). At human body temperature, the 4D printed cartilage scaffold by Deng et al., showed good mechanical properties, shape memory efficiency, and biocompatibility. Nanohydroxyapatite was added to a shape memory polyurethane matrix to accomplish this. According to Deng et al., the 4D-printed cartilage scaffold presents a novel method for the personalized and less invasive treatment of cartilage defects. A scaffold can be developed from a readily insertional form to a deployed form that matches the defect (Deng et al., 2023).

Gold nanoparticles and nano-hydroxyapatite were added to a shape memory polyurethane composite that was 4D printed to create a computed tomography-developable orbital stent for ocular therapy. To support the ocular tissue more precisely and steadily, an orbital stent was printed into a 4D bionic honeycomb pore structure, and an outer shape that matched the ocular coloboma was developed. The bionic idea and computed tomography reconstruction method served as the foundation for this design. Research by Deng et al. demonstrated better efficacy of the composite in the accurate less invasive surgical treatment of enophthalmos as compared to traditional orbital implants (Deng et al., 2022). Lin et al. constructed a polyethylene glycol/shape memory PLA-based 4D-printed shape-retaining biocomposite via melt extrusion and fabricated it using fused deposit modeling for constructing an intestinal stent with a wave-like architecture. The biocomposite effectively reduced the possibility of tissue damage by exhibiting stress strain characteristics that were exactly like those of living tissues. Among the stent's advantages were its biocompatibility, biodegradability, customized design, high flexibility, and ability to complete the shape memory operation at 42 to 43°C. Furthermore, scientists used the

been processed using fused deposit modeling, but SLA appears to be a more promising technology than any other 3D printing process because of its high resolution and the manner in which the laser interacts with graphene. Graphene powder has been widely employed in SLS and FDM processes for printed items and nanocomposite prototypes. The improved mechanical, thermal, and other physical properties of these nanoparticles were demonstrated. Similar techniques have been used to create intricate, multipurpose, high-resolution 3D shapes using 4D printing. A fundamental link between material properties and processing parameters has been discovered to be significant in all of these 3D and 4D techniques and research is needed soon on novel 3D and 4D fabrication techniques, adaptable printing materials, processing techniques, and a wide range of applications. Numerous 3D printing techniques, including direct ink writing, stereolithography, and fused deposition modeling, are been effectively modified to handle nanocomposite materials. With these approaches, one can fabricate scaffolds for the application of tissue engineering, sophisticated drug delivery systems, and implants tailored to the needs of individual patients. Even with all the tremendous potential, further research and development is needed to address issues with printability, dispersion of nanoparticles, and other health hazards. However, these obstacles are being actively addressed by current research. Nanocomposites have enormous potential for personalized treatment, on-demand drug delivery, and the development of cutting-edge biomedical devices that can greatly enhance patient outcomes and care in the future of 3D and 4D printing. Nanocomposites have the potential to completely change the fields of biomedical engineering and drug delivery by improving printing methods, maximizing material qualities, and resolving safety issues.

KEYWORDS

- **additive manufacturing**
- **fused deposit modeling**
- **implants**
- **personalized medicine**
- **3D printing**

REFERENCES

Adepu, S., & Khandelwal, M. (2018). Broad-spectrum antimicrobial activity of bacterial cellulose silver nanocomposites with sustained release. *Journal of Materials Science*, *53*(3), 1596–1609. https://doi.org/10.1007/S10853-017-1638-9/METRICS.

Ahn, S. H., Montero, M., Odell, D., Roundy, S., & Wright, P. K. (2002). Anisotropic material properties of fused deposition modeling ABS. *Rapid Prototyping Journal*, *8*(4), 248–257. https://doi.org/10.1108/13552540210441166/FULL/XML.

Allo, B. A., Costa, D. O., Dixon, S. J., Mequanint, K., & Rizkalla, A. S. (2012). Bioactive and biodegradable nanocomposites and hybrid biomaterials for bone regeneration. *Journal of Functional Biomaterials*, *3*(2), 432. https://doi.org/10.3390/JFB3020432.

Archer, C. R., Sargeant, R., Basak, J., Pilling, J., Barnes, J. R., & Pointon, A. (2018). Characterization and validation of a human 3D cardiac microtissue for the assessment of changes in cardiac pathology. *Scientific Reports*, *8*(1), 1–15. https://doi.org/10.1038/s41598-018-28393-y.

Belaid, H., Nagarajan, S., Teyssier, C., Barou, C., Barés, J., Balme, S., Garay, H., Huon, V., Cornu, D., Cavaillès, V., & Bechelany, M. (2020). Development of new biocompatible 3D printed graphene oxide-based scaffolds. *Materials Science and Engineering: C*, *110*, 110595. https://doi.org/10.1016/J.MSEC.2019.110595.

Bleach, N. C., Nazhat, S. N., Tanner, K. E., Kellomäki, M., & Törmälä, P. (2002). Effect of filler content on mechanical and dynamic mechanical properties of particulate biphasic calcium phosphate—polylactide composites. *Biomaterials*, *23*(7), 1579–1585. https://doi.org/10.1016/S0142-9612(01)00283-6.

Bose, S., & Saha, S. K. (2003). Synthesis and characterization of hydroxyapatite nanopowders by emulsion technique. *Chemistry of Materials*, *15*(23), 4464–4469. https://doi.org/10.1021/CM0303437/ASSET/IMAGES/MEDIUM/CM0303437N00001.GIF.

Caballero-Aguilar, L. M., Silva, S. M., & Moulton, S. E. (2020). Three-dimensional printed drug delivery systems. *Engineering Drug Delivery Systems*, 147–162. https://doi.org/10.1016/B978-0-08-102548-2.00006-8.

Cardoso, R. M., Silva, P. R. L., Lima, A. P., Rocha, D. P., Oliveira, T. C., do Prado, T. M., Fava, E. L., Fatibello-Filho, O., Richter, E. M., & Muñoz, R. A. A. (2020). 3D-Printed graphene/polylactic acid electrode for bioanalysis: Biosensing of glucose and simultaneous determination of uric acid and nitrite in biological fluids. *Sensors and Actuators B: Chemical*, *307*, 127621. https://doi.org/10.1016/J.SNB.2019.127621.

Che, Q. T., Seo, J. W., Charoensri, K., Nguyen, M. H., Park, H. J., & Bae, H. (2024). 4D-printed microneedles from dual-sensitive chitosan for non-transdermal drug delivery. *International Journal of Biological Macromolecules*, *261*, 129638. https://doi.org/10.1016/J.IJBIOMAC.2024.129638.

Cherian, A. M., Nair, S. V., Maniyal, V., & Menon, D. (2021). Surface engineering at the nanoscale: A way forward to improve coronary stent efficacy. *APL Bioengineering*, *5*(2). https://doi.org/10.1063/5.0037298.

Chinen, A. B., Guan, C. M., Ferrer, J. R., Barnaby, S. N., Merkel, T. J., & Mirkin, C. A. (2015). Nanoparticle probes for the detection of cancer biomarkers, cells, and tissues by fluorescence. *Chemical Reviews*, *115*(19), 10530–10574. https://doi.org/10.1021/ACS.CHEMREV.5B00321/ASSET/ACS.CHEMREV.5B00321.FP.PNG_V03.

Choong, Y. Y. C., Maleksaeedi, S., Eng, H., Wei, J., & Su, P. C. (2017). 4D printing of high performance shape memory polymer using stereolithography. *Materials and Design, 126*, 219–225. https://doi.org/10.1016/J.MATDES.2017.04.049.

Chowdhury, J., Anirudh, P. V., Karunakaran, C., Rajmohan, V., Mathew, A. T., Koziol, K., Alsanie, W. F., Kannan, C., Balan, A. S. S., & Thakur, V. K. (2021). 4D printing of smart polymer nanocomposites: Integrating graphene and acrylate based shape memory polymers. *Polymers, 13*(21), 3660. https://doi.org/10.3390/POLYM13213660.

Cristache, C. M., & Totu, E. E. (2021). 3D printing-processed polymers for dental applications. In T. J. Gutiérrez (Eds.), *Reactive and functional polymers volume three: Advanced materials* (pp.141–164). Cham: Springer. https://doi.org/10.1007/978-3-030-50457-1_7.

Damiati, S., Küpcü, S., Peacock, M., Eilenberger, C., Zamzami, M., Qadri, I., Choudhry, H., Sleytr, U. B., & Schuster, B. (2017). Acoustic and hybrid 3D-printed electrochemical biosensors for the real-time immunodetection of liver cancer cells (HepG2). *Biosensors and Bioelectronics, 94*, 500–506. https://doi.org/10.1016/J.BIOS.2017.03.045.

Deng, Y., Yang, B., Zhang, F., Liu, Y., Sun, J., Zhang, S., Zhao, Y., Yuan, H., & Leng, J. (2022). 4D printed orbital stent for the treatment of enophthalmic invagination. *Biomaterials, 291*, 121886. https://doi.org/10.1016/J.BIOMATERIALS.2022.121886.

Deng, Y., Zhang, F., Liu, Y., Zhang, S., Yuan, H., & Leng, J. (2023). 4D printed shape memory polyurethane-based composite for bionic cartilage scaffolds. *ACS Applied Polymer Materials, 5*(2), 1283–1292. https://doi.org/10.1021/ACSAPM.2C01833/SUPPL_FILE/AP2C01833_SI_001.PDF.

Deshmukh, K., Muzaffar, A., Kovářík, T., Křenek, T., Ahamed, M. B., & Pasha, S. K. K. (2020). Fundamentals and applications of 3D and 4D printing of polymers: Challenges in polymer processing and prospects of future research. *3D and 4D Printing of Polymer Nanocomposite Materials: Processes, Applications, and Challenges*, 527–560. https://doi.org/10.1016/B978-0-12-816805-9.00017-X.

Devnarain, N., Osman, N., Fasiku, V. O., Makhathini, S., Salih, M., Ibrahim, U. H., & Govender, T. (2021). Intrinsic stimuli-responsive nanocarriers for smart drug delivery of antibacterial agents-An in-depth review of the last two decades. *Wiley Interdisciplinary Reviews. Nanomedicine and Nanobiotechnology, 13*(1), e1664. https://doi.org/10.1002/WNAN.1664.

Dhariwala, B., Hunt, E., & Boland, T. (2004). Rapid prototyping of tissue-engineering constructs, using photopolymerizable hydrogels and stereolithography. *Tissue Engineering, 10*(9–10), 1316–1322. https://doi.org/10.1089/TEN.2004.10.1316.

Fendi, F., Abdullah, B., Suryani, S., Raya, I., Tahir, D., & Iswahyudi, I. (2024). Hydroxyapatite based for bone tissue engineering: innovation and new insights in 3D printing technology. *Polymer Bulletin, 81*(2), 1097–1116. https://doi.org/10.1007/S00289-023-04794-6/METRICS.

Gasparotto, M., Bellet, P., Scapin, G., Busetto, R., Rampazzo, C., Vitiello, L., Shah, D. I., & Filippini, F. (2022). 3D printed graphene-PLA scaffolds promote cell alignment and differentiation. *International Journal of Molecular Sciences, 23*(3), 1736. https://doi.org/10.3390/IJMS23031736.

Ghiasi, B., Sefidbakht, Y., & Rezaei, M. (2019). Hydroxyapatite for biomedicine and drug delivery. *Advanced Structured Materials, 104*, 85–120. https://doi.org/10.1007/978-3-030-10834-2_4.

Goyanes, A., Buanz, A. B. M., Basit, A. W., & Gaisford, S. (2014). Fused-filament 3D printing (3DP) for fabrication of tablets. *International Journal of Pharmaceutics*, *476*(1–2), 88–92. https://doi.org/10.1016/J.IJPHARM.2014.09.044.

Griffith, M. L., & Halloran, J. W. (1996). Freeform fabrication of ceramics via stereolithography. *Journal of the American Ceramic Society*, *79*(10), 2601–2608. https://doi.org/10.1111/J.1151-2916.1996.TB09022.X.

Gurr, M., Hofmann, D., Ehm, M., Thomann, Y., Kubier, R., & Mülhaupt, R. (2008). Acrylic nanocomposite resins for use in stereolithography and structural light modulation based rapid prototyping and rapid manufacturing technologies. *Advanced Functional Materials*, *18*(16), 2390–2397. https://doi.org/10.1002/ADFM.200800344.

Han, Y., Wei, Q., Chang, P., Hu, K., Okoro, O. V., Shavandi, A., & Nie, L. (2021). Three-dimensional printing of hydroxyapatite composites for biomedical application. *Crystals*, *11*(4), 353. https://doi.org/10.3390/CRYST11040353.

Huang, Z., Wan, Y., Zhu, X., Zhang, P., Yang, Z., Yao, F., & Luo, H. (2021). Simultaneous engineering of nanofillers and patterned surface macropores of graphene/hydroxyapatite/polyetheretherketone ternary composites for potential bone implants. *Materials Science and Engineering. C, Materials for Biological Applications*, *123*, 111967. https://doi.org/10.1016/J.MSEC.2021.111967.

Iqbal Hussain. (2021). *A comparative analysis of fused deposition modeling and stereolithography 3D printers*.

Jahanbakhshi, M., & Shahrousvand, M. (2021). Stimuli-responsive polymers as smart drug delivery systems. *Modeling and Control of Drug Delivery Systems*, *2021*, 67–77. https://doi.org/10.1016/B978-0-12-821185-4.00015-4.

Khalid, M. Y., Arif, Z. U., Noroozi, R., Zolfagharian, A., & Bodaghi, M. (2022). 4D printing of shape memory polymer composites: A review on fabrication techniques, applications, and future perspectives. *Journal of Manufacturing Processes*, *81*, 759–797. https://doi.org/10.1016/J.JMAPRO.2022.07.035.

Khalili, V., & Naji, H. (2020). Developing a mechanochemical surface pretreatment to increase the adhesion strength of hydroxyapatite electrophoretic coating on the NiTi alloy as a bone implant. *Surface and Coatings Technology*, *397*, 125985. https://doi.org/10.1016/J.SURFCOAT.2020.125985.

Khosravani, M. R., & Reinicke, T. (2020). 3D-printed sensors: Current progress and future challenges. *Sensors and Actuators A: Physical*, *305*, 111916. https://doi.org/10.1016/J.SNA.2020.111916.

Kim, J., Choi, H. S., Kim, Y. M., & Song, S. C. (2023). Thermo-responsive nanocomposite bioink with growth-factor holding and its application to bone regeneration. *Small*, *19*(9), 2203464. https://doi.org/10.1002/SMLL.202203464.

Kim, K., Yeatts, A., Dean, D., & Fisher, J. P. (2010). Stereolithographic bone scaffold design parameters: Osteogenic differentiation and signal expression. *Tissue Engineering Part B: Reviews*, *16*(5), 523–539. https://doi.org/10.1089/TEN.TEB.2010.0171.

Kim, T. H., Lee, J. H., Ahn, C. B., Hong, J. H., Son, K. H., & Lee, J. W. (2019). Development of a 3D-printed drug-eluting stent for treating obstructive salivary gland disease. *ACS Biomaterials Science and Engineering*, *5*(7), 3572–3581. https://doi.org/10.1021/ACSBIOMATERIALS.9B00636/SUPPL_FILE/AB9B00636_SI_001.PDF.

Koffler, J., Zhu, W., Qu, X., Platoshyn, O., Dulin, J. N., Brock, J., Graham, L., Lu, P., Sakamoto, J., Marsala, M., Chen, S., & Tuszynski, M. H. (2019). Biomimetic 3D-printed scaffolds for spinal cord injury repair. *Nature Medicine*, *25*(2), 263–269. https://doi.org/10.1038/S41591-018-0296-Z.

Krause, C. E., Otieno, B. A., Latus, A., Faria, R. C., Patel, V., Gutkind, J. S., & Rusling, J. F. (2013). Rapid microfluidic immunoassays of cancer biomarker proteins using disposable inkjet-printed gold nanoparticle arrays. *ChemistryOpen*, *2*(4), 141–145. https://doi.org/10.1002/OPEN.201300018.

Lauvahutanon, S., Takahashi, H., Oki, M., Arksornnukit, M., Kanehira, M., & Finger, W. J. (2015). In vitro evaluation of the wear resistance of composite resin blocks for CAD/CAM. *Dental Materials Journal*, *34*(4), 495–502. https://doi.org/10.4012/DMJ.2014-293.

Li, J., Zhang, Y., Wang, P., Wang, G., Liu, Y., Liu, Y., & Li, Q. (2021). Selectively metalizable stereolithography resin for three-dimensional DC and high-frequency electronics via hybrid additive manufacturing. *ACS Applied Materials and Interfaces*, *13*(19), 22891–22901. https://doi.org/10.1021/ACSAMI.1C01199/SUPPL_FILE/AM1C01199_SI_003.PDF.

Lin, C., Huang, Z., Wang, Q., Zou, Z., Wang, W., Liu, L., Liu, Y., & Leng, J. (2023). Mass-producible near-body temperature-triggered 4D printed shape memory biocomposites and their application in biomimetic intestinal stents. *Composites Part B: Engineering*, *256*, 110623. https://doi.org/10.1016/J.COMPOSITESB.2023.110623.

Lin, Y., Bariya, M., & Javey, A. (2021). Wearable biosensors for body computing. *Advanced Functional Materials*, *31*(39), 2008087. https://doi.org/10.1002/ADFM.202008087.

Lu, Guo, P., & Pan, Y. (2017). Magnetic-field-assisted projection stereolithography for three-dimensional printing of smart structures. *Journal of Manufacturing Science and Engineering, Transactions of the ASME*, *139*(7), 071008.

Lu, Y., Mapili, G., Suhali, G., Chen, S., & Roy, K. (2006). A digital micro-mirror device-based system for the microfabrication of complex, spatially patterned tissue engineering scaffolds. *Journal of Biomedical Materials Research Part A*, *77A*(2), 396–405. https://doi.org/10.1002/JBM.A.30601.

Malachowski, K., Breger, J., Kwag, H. R., Wang, M. O., Fisher, J. P., Selaru, F. M., & Gracias, D. H. (2014). Stimuli-responsive theragrippers for chemomechanical controlled release. *Angewandte Chemie International Edition*, *53*(31), 8045–8049. https://doi.org/10.1002/ANIE.201311047.

Mallik, A., Malhotra, Y., Badhan, A., Malhotra, P. U., & Rasila, D. (2020). Current trends in orthodontics. *International Healthcare Research Journal*, *4*(8), RV1–RV4. https://doi.org/10.26440/IHRJ/0408.11285.

Melchels, F. P. W., Feijen, J., & Grijpma, D. W. (2010). A review on stereolithography and its applications in biomedical engineering. *Biomaterials*, *31*(24), 6121–6130. https://doi.org/10.1016/J.BIOMATERIALS.2010.04.050.

Misra, S. K., Ostadhossein, F., Babu, R., Kus, J., Tankasala, D., Sutrisno, A., Walsh, K. A., Bromfield, C. R., & Pan, D. (2017). 3D-printed multidrug-eluting stent from graphene-nanoplatelet-doped biodegradable polymer composite. *Advanced Healthcare Materials*, *6*(11), 1700008. https://doi.org/10.1002/ADHM.201700008.

Mobaraki, M., Ghaffari, M., Yazdanpanah, A., Luo, Y., & Mills, D. K. (2020). Bioinks and bioprinting: A focused review. *Bioprinting*, *18*, e00080. https://doi.org/10.1016/J.BPRINT.2020.E00080.

Morales, M. A., & Halpern, J. M. (2018). Guide to selecting a biorecognition element for biosensors. *Bioconjugate Chemistry*, *29*(10), 3231–3239. https://doi.org/10.1021/ACS.BIOCONJCHEM.8B00592/ASSET/IMAGES/MEDIUM/BC-2018-005923_0007.GIF.

Nesaei, S., Song, Y., Wang, Y., Ruan, X., Du, D., Gozen, A., & Lin, Y. (2018). Micro additive manufacturing of glucose biosensors: A feasibility study. *Analytica Chimica Acta*, *1043*, 142–149. https://doi.org/10.1016/J.ACA.2018.09.012.

Ooi, O. C., Mullany, C. J., & Rihal, C. S. (2007). Revascularization options for ischemic heart disease: Coronary artery bypass grafting and percutaneous coronary intervention. In E. M. Antman, M. M. Givertz, M. E. Josephson, J. de Lemos, S. Oparil, & F. M. Sacks (Eds.), *Cardiovascular therapeutics: A companion to braunwald's heart disease* (pp. 157–177). Philadelphia: W.B. Saunders. https://doi.org/10.1016/B978-1-4160-3358-5.50014-0.

Palmara, G., Frascella, F., Roppolo, I., Chiappone, A., & Chiadò, A. (2021). Functional 3D printing: Approaches and bioapplications. *Biosensors and Bioelectronics*, *175*, 112849. https://doi.org/10.1016/J.BIOS.2020.112849.

Provin, C., & Monneret, S. (2002). Complex ceramic-polymer composite microparts made by microstereolithography. *IEEE Transactions on Electronics Packaging Manufacturing*, *25*(1), 59–63. https://doi.org/10.1109/TEPM.2002.1000484.

Rahman, Md. H., Liza, N. Y., Hossain, K. R., Kalambhe, D. R., Shyeed, Md. A., & Noor, D. H. (2024). Additive manufacturing in nano drug delivery systems. *Pharmaceutical Science Advances*, *2*, 100036. https://doi.org/10.1016/J.PSCIA.2024.100036.

Rahman, Z., Charoo, N. A., Kuttolamadom, M., Asadi, A., & Khan, M. A. (2020). Printing of personalized medication using binder jetting 3D printer. In *Precision medicine for investigators, practitioners and providers* (pp. 473–481). Elsevier. https://doi.org/10.1016/B978-0-12-819178-1.00046-0.

Rahmani, S., Maroufkhani, M., Mohammadzadeh-Komuleh, S., & Khoubi-Arani, Z. (2022). Polymer nanocomposites for biomedical applications. In *Fundamentals of bionanomaterials* (pp. 175–215). Netherlands: Elsevier. https://doi.org/10.1016/B978-0-12-824147-9.00007-8.

Rengier, F., Mehndiratta, A., Von Tengg-Kobligk, H., Zechmann, C. M., Unterhinninghofen, R., Kauczor, H. U., & Giesel, F. L. (2010). 3D printing based on imaging data: Review of medical applications. *International Journal of Computer Assisted Radiology and Surgery*, *5*(4), 335–341. https://doi.org/10.1007/S11548-010-0476-X.

Reyes-Ortega, F. (2014). pH-responsive polymers: Properties, synthesis and applications. In M. R. Aguilar & J. S. Román (Eds.), *Smart polymers and their applications* (pp. 45–92). Woodhead Publishing. https://doi.org/10.1533/9780857097026.1.45.

Richards, D. J., Tan, Y., Jia, J., Yao, H., & Mei, Y. (2013). 3D printing for tissue engineering. *Israel Journal of Chemistry*, *53*(9–10), 805. https://doi.org/10.1002/ijch.201300086.

Rizzante, F., Bueno, T., Guimarães, G., Moura, G., Teich, S., Furuse, A., & Mendonça, G. (2023). Comparative physical and mechanical properties of a 3D printed temporary crown and bridge restorative material. *Journal of Clinical and Experimental Dentistry*, *15*(6), e464. https://doi.org/10.4317/JCED.60507.

Sajjad, R., Chauhdary, S. T., Anwar, M. T., Zahid, A., Khosa, A. A., Imran, M., & Sajjad, M. H. (2024). A review of 4D printing – Technologies, shape shifting, smart polymer based materials, and biomedical applications. *Advanced Industrial and Engineering Polymer Research*, *7*(1), 20–36. https://doi.org/10.1016/J.AIEPR.2023.08.002.

Saylam, E., Akkaya, Y., Ilhan, E., Cesur, S., Guler, E., Sahin, A., Cam, M. E., Ekren, N., Oktar, F. N., Gunduz, O., Ficai, D., & Ficai, A. (2021). Levodopa-loaded 3D-printed poly (lactic) acid/chitosan neural tissue scaffold as a promising drug delivery system for the treatment of Parkinson's disease. *Applied Sciences, 11*(22), 10727. https://doi.org/10.3390/APP112210727.

Senatov, F. S., Niaza, K. V., Zadorozhnyy, M. Y., Maksimkin, A. V., Kaloshkin, S. D., & Estrin, Y. Z. (2016). Mechanical properties and shape memory effect of 3D-printed PLA-based porous scaffolds. *Journal of the Mechanical Behavior of Biomedical Materials*, *57*, 139–148. https://doi.org/10.1016/J.JMBBM.2015.11.036.

Seng, C. T., A/l Eh Noum, S. Y., A/l Sivanesan, S. K., & Yu, L. J. (2020). Reduction of hygroscopicity of PLA filament for 3D printing by introducing nano silica as filler. *AIP Conference Proceedings*, *2233*(1), 020024. https://doi.org/10.1063/5.0001927/611159.

Song, P., Zhou, C., Fan, H., Zhang, B., Pei, X., Fan, Y., Jiang, Q., Bao, R., Yang, Q., Dong, Z., & Zhang, X. (2018). Novel 3D porous biocomposite scaffolds fabricated by fused deposition modeling and gas foaming combined technology. *Composites Part B: Engineering*, *152*, 151–159. https://doi.org/10.1016/J.COMPOSITESB.2018.06.029.

Su, S., Wang, J., Parupelli, S. K., & Desai, S. (2023). The 3D printing of nanocomposites for wearable biosensors: Recent advances, challenges, and prospects. *Bioengineering, 11*(1), 32. https://doi.org/10.3390/BIOENGINEERING11010032.

Tajvar, S., Hadjizadeh, A., & Samandari, S. S. (2023). Scaffold degradation in bone tissue engineering: An overview. *International Biodeterioration and Biodegradation*, *180*, 105599. https://doi.org/10.1016/J.IBIOD.2023.105599.

Tamay, D. G., Usal, T. D., Alagoz, A. S., Yucel, D., Hasirci, N., & Hasirci, V. (2019). 3D and 4D printing of polymers for tissue engineering applications. *Frontiers in Bioengineering and Biotechnology*, *7*, 453880. https://doi.org/10.3389/FBIOE.2019.00164/BIBTEX.

Tao, Y., Chan, H. F., Shi, B., Li, M., & Leong, K. W. (2020). Light: A magical tool for controlled drug delivery. *Advanced Functional Materials*, *30*(49), 2005029. https://doi.org/10.1002/ADFM.202005029.

Trachsel, L., Romio, M., Grob, B., Zenobi-Wong, M., Spencer, N. D., Ramakrishna, S. N., & Benetti, E. M. (2020). Functional nanoassemblies of cyclic polymers show amplified responsiveness and enhanced protein-binding ability. *ACS Nano*, *14*(8), 10054–10067. https://doi.org/10.1021/ACSNANO.0C03239.

Turnbull, G., Clarke, J., Picard, F., Riches, P., Jia, L., Han, F., Li, B., & Shu, W. (2018). 3D bioactive composite scaffolds for bone tissue engineering. *Bioactive Materials*, *3*(3), 278–314. https://doi.org/10.1016/J.BIOACTMAT.2017.10.001.

Vargas, A. J., & Harris, C. C. (2016). Biomarker development in the precision medicine era: Lung cancer as a case study. *Nature Reviews Cancer, 16*(8), 525–537. https://doi.org/10.1038/nrc.2016.56.

Visscher, L. E., Dang, H. P., Knackstedt, M. A., Hutmacher, D. W., & Tran, P. A. (2018). 3D printed Polycaprolactone scaffolds with dual macro-microporosity for applications in local delivery of antibiotics. *Materials Science and Engineering: C*, *87*, 78–89. https://doi.org/10.1016/J.MSEC.2018.02.008.

Wan, X., Zhang, F., Liu, Y., & Leng, J. (2019). CNT-based electro-responsive shape memory functionalized 3D printed nanocomposites for liquid sensors. *Carbon*, *155*, 77–87. https://doi.org/10.1016/J.CARBON.2019.08.047.

Wang, W., Zhang, B., Li, M., Li, J., Zhang, C., Han, Y., Wang, L., Wang, K., Zhou, C., Liu, L., Fan, Y., & Zhang, X. (2021). 3D printing of PLA/n-HA composite scaffolds with customized mechanical properties and biological functions for bone tissue engineering. *Composites Part B: Engineering*, *224*, 109192. https://doi.org/10.1016/J.COMPOSITESB.2021.109192.

Wang, X., Jiang, M., Zhou, Z., Gou, J., & Hui, D. (2017). 3D printing of polymer matrix composites: A review and prospective. *Composites Part B: Engineering*, *110*, 442–458. https://doi.org/10.1016/J.COMPOSITESB.2016.11.034.

Wang, Y. J., Jeng, U. S., & Hsu, S. H. (2018). Biodegradable water-based polyurethane shape memory elastomers for bone tissue engineering. *ACS Biomaterials Science and Engineering*, *4*(4), 1397–1406. https://doi.org/10.1021/ACSBIOMATERIALS.8B00091/SUPPL_FILE/AB8B00091_SI_001.PDF.

Wang, Y., Shim, M. S., Levinson, N. S., Sung, H. W., & Xia, Y. (2014). Stimuli-responsive materials for controlled release of theranostic agents. *Advanced Functional Materials*, *24*(27), 4206–4220. https://doi.org/10.1002/ADFM.201400279.

Wei, J., Yan, Y., Gao, J., Li, Y., Wang, R., Wang, J., Zou, Q., Zuo, Y., Zhu, M., & Li, J. (2022). 3D-printed hydroxyapatite microspheres reinforced PLGA scaffolds for bone regeneration. *Biomaterials Advances*, *133*, 112618. https://doi.org/10.1016/J.MSEC.2021.112618.

Xu, Y., Wu, X., Guo, X., Kong, B., Zhang, M., Qian, X., Mi, S., & Sun, W. (2017). The boom in 3D-printed sensor technology. *Sensors*, *17*(5), 1166. https://doi.org/10.3390/S17051166.

Yan, S., Zhang, F., Luo, L., Wang, L., Liu, Y., & Leng, J. (2023). Shape memory polymer composites: 4D printing, smart structures, and applications. *Research*, *6*, 0234. https://doi.org/10.34133/RESEARCH.0234/ASSET/9CD33FCE-BD4E-4924-841E-6FCB3564943B/ASSETS/GRAPHIC/RESEARCH.0234.FIG.015.JPG.

Zarek, M., Layani, M., Cooperstein, I., Sachyani, E., Cohn, D., & Magdassi, S. (2016). 3D printing of shape memory polymers for flexible electronic devices. *Advanced Materials*, *28*(22), 4449–4454. https://doi.org/10.1002/ADMA.201503132.

Zeng, C. J., Liu, L. W., Zhao, W., Liu, Z. X., Xin, X. Z., Liu, Y. J., & Leng, J. S. (2023). Mechanical properties and shape memory behavior of 4D printed functionally graded cellular structures. *Science China Technological Sciences*, *66*(12), 3522–3533. https://doi.org/10.1007/S11431-023-2475-3/METRICS.

Zhang, B., Chung, S. H., Barker, S., Craig, D., Narayan, R. J., & Huang, J. (2021). Direct ink writing of polycaprolactone/polyethylene oxide based 3D constructs. *Progress in Natural Science: Materials International*, *31*(2), 180–191. https://doi.org/10.1016/J.PNSC.2020.10.001.

Zhang, F., Wang, L., Zheng, Z., Liu, Y., & Leng, J. (2019). Magnetic programming of 4D printed shape memory composite structures. *Composites Part A: Applied Science and Manufacturing*, *125*, 105571. https://doi.org/10.1016/J.COMPOSITESA.2019.105571.

CHAPTER 12

Biomedical Applications of Various Herbal Nanocomposites

SUMAIYA QASIM and SURAJPAL VERMA*

School of Pharmaceutical Sciences, Delhi Pharmaceutical Sciences and Research University (DPSRU), Puspvihar, New Delhi, India

*Corresponding author.

ABSTRACT

Nanotechnology, a rapidly developing field, has the potential to revolutionize many industries by improving material properties through the use of nanoscale physicochemical properties. Materials with at least one component at nanoscale dimensions, or nanocomposites, provide a framework for developing novel materials with enhanced characteristics. These composites integrate organic, inorganic, or polymeric materials into metal or nonmetal matrices, yielding materials tailored for specific applications. This chapter explores the biosynthesis of various nanocomposites using biopolymers, biological agents such as bacteria and fungi, and plant extracts. Herbal nanocomposites are used in fields like diagnostics, testing, pharmaceuticals, and textiles. Nanotechnology offers promise to improve absorption and decrease toxicity in herbal medicine, hence, improving therapeutic efficacy. Despite its traditional safety and efficacy, challenges persist in selecting suitable nanoformulations, polymers, and extraction methods, alongside concerns regarding safety data and stability. The integration of nanotechnology into herbal medicine holds promise for addressing these challenges and unlocking new therapeutic avenues resulting in improved bioavailability, specificity, and reduced toxicity, which expands the scope of herbal medications. By utilizing suitable carriers, plant extracts can be

Nanocomposites and Nanomaterials in Biomedical Applications. Vivek Dave, Deependra Singh, and Pallerla Naveen Reddy (Eds.)

DOI: 10.1201/9781779641236-12

administered with enhanced physicochemical properties via nanocomposites. Although the area of herbal nanocomposites has a lot of promise, more study and advancement are required before clinical application and the benefits in biomedical disciplines could be achieved.

12.1 INTRODUCTION

The rapidly growing field of nanotechnology promises significant innovation and progress in different sectors. The slight changes in materials at the nanoscale using physicochemical properties help in the improved properties of novel material. The improved properties are the major reason behind the growth of this field, which is helping in fields of communication, biotechnology, agriculture, environmental remedies, material science, chemistry, pharmaceuticals and phytopharmaceuticals, and so forth (Piyal Mondal et al., 2020). Materials that have at least one component with nanoscale dimensions are called nanocomposites. A nanocomposite material is composed of many phases, each of which has at least one, two, or three nanometer-sized dimensions. Phase interfaces are produced when material dimensions are reduced to the nanoscale (Omanović-Mikličanin, et al., 2020). These composites are designed to yield cutting-edge materials with improved characteristics. Different organic, inorganic, or polymeric materials can be included into metal or nonmetal matrices that create materials having specific characteristics for particular nanocomposites. Biosynthesis of zinc oxide nanoparticles (NPs) can be achieved by a range of biological agents, comprising bacteria, fungus, and plant extracts (Dar et al., 2022).

Nowadays, more people are using herbal medications since they may treat a wide range of illnesses with less side effects and more effective therapeutic results (Bhokare et al., 2016). Herbal medications are often sold in conventional dose forms, but in recent years, numerous effective methods—particularly those involving nanotechnology—have been developed to improve the therapeutic effects and delivery effectiveness of extracts (Shah et al., 2021). Herbal medications are often sold in conventional dose forms, but in recent years, numerous effective methods—particularly those involving nanotechnology—have been developed to improve the therapeutic effects and delivery effectiveness of extracts (Dar et al., 2021). As the research advances, we anticipate witnessing further transformative breakthroughs in the years ahead.

12.2 NANOCOMPOSITE

When two or more distinct constituent materials each with a significant characteristic (physical or chemical) combined naturally or by engineering, to create a new substance with superior properties than original materials in a specific finished structure, known as a composite (Sen, 2020). Nanocomposite is made up of many phases, at least one of which has a nanometer-scale size in the range of 1–100 nm, and they can be NPs, nanoclays, nanofibers, nanotubes, and so forth. The synergy between various constituents is the aim that is accomplished by the nanoscale-phase process (Omanović-Mikličanin, et al., 2020). A small quantity of nanofillers is used in the technology of nanocomposites. Nanofillers can be sheet-like, spherical, or tube-like, their macroscopic features can be greatly altered while maintaining the essential features by their inclusion as nanoscale organic and inorganic components are combined resulting in better characteristics and overcome defects like greater mechanical attributes modulus and strength, permeability to gases, water, and hydrocarbons, thermal and dimensional stability, flame retardancy, optical qualities, chemical resistance, magnetic, dielectric, magnetic, and electrical properties (Kosmidou et al., 2008).

12.2.1 CLASSIFICATION OF NANOCOMPOSITES

Based on the matrix nanocomposites can be classified into three groups (Figure 12.1):

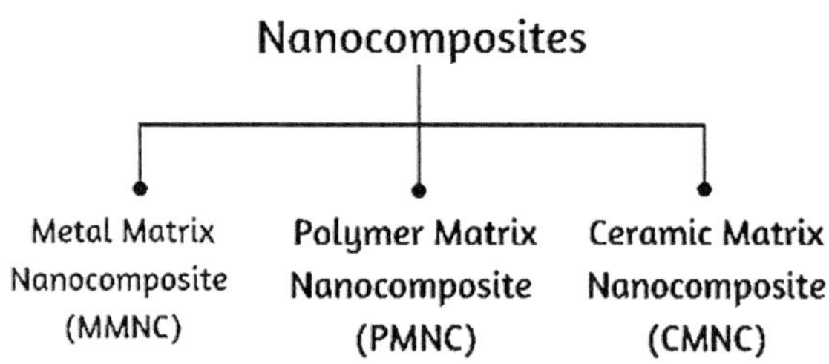

FIGURE 12.1 Classification of nanocomposites based on matrix.

12.2.1.1 METAL MATRIX NANOCOMPOSITE (MMNC)

Metal matrix nanocomposites (MMNCs) are multiphase materials made of an alloy or ductile metal matrix that has been seeded with nanoscale reinforcing particles (Khandoker et al., 2011). They can be characterized by high ductility, plasticity, strength, and so forth. Al, Fe, Mg, W, Pb, and Sn metals can be used as metal matrix for MMNCs (Omanović-Mikličanin, et al., 2020). Preparation of MMNCs can be done by various methods, including spray pyrolysis (Dermenci et al., 2014), chemical methods (e.g., sol–gel process) (Roy et al., 1993), liquid infiltration (Contreras et al., 2004), and so forth. MMNC have wide industrial applications as catalysts, electronic packaging, industrial high speed machinery, magnetic devices, and much more (Malaki et al., 2019).

12.2.1.2 POLYMER MATRIX NANOCOMPOSITE (PMNC)

In polymer nanocomposites, the NPs or nanofillers are scattered throughout the polymer matrix. They can be further classified as 1-D (Smart et al., 2006), 2-D, and 3-D polymers. High damage tolerance, flame retardancy, barrier properties, and high-specific modulus make them the most widely used nanocomposite. *In situ* intercalative polymerization (Anandhan and Bandyopadhyay, 2011), sol–gel method (Hussain et al., 2006), melt intercalation (Vaia et al., 1995), and so forth can be used for PMNC's synthesis. PMNCs are most commonly used in the preparation of nanocomposites for the pharmaceutical and mechanical industry.

12.2.1.3 CERAMIC MATRIX NANOCOMPOSITE (CMNC)

In ceramic matrix, nanocomposites consist of more than one phase in which at least a phase has nanoscale dimensions. In order to improve chemical stability, wear resistance and thermal stability ceramic phases are intentionally added resulting in combined mechanical, optical, chemical, and magnetic capabilities. Spray pyrolysis, chemical methods, and vapor techniques can be used in the processing of CMNCs. Ceramic-ceramic-based nanocomposites proved to be extremely beneficial in the area of artificial joint implants and in dentistry (Vaia and Wagner, 2004).

12.2.2 HERBAL NANOCOMPOSITE

In recent years, nanocomposite technology has been widely used in the development of herbal nanocomposites in cosmetics, the textile industry, pharmaceuticals, food, and chemical industry. Composites are created through the integration of two or more distinct materials with distinct characteristics. Together, these many materials provide the distinctive characteristics of the composite, which are the individual qualities of the constituents. Studies related to herbal compounds like aloe vera, resveratrol, neem, and various plant extract nanocomposites prepared with suitable biopolymers like chitosan (CS), alginate, and metal oxide showed promising results (Rajendran et al., 2012). By combining these materials, nanocomposites exhibit a blend of properties derived from their individual constituents. In the pharmaceutical industry, herbal nanocomposites offer opportunities for the development of advanced drug delivery systems. By encapsulating herbal extracts within nanomaterial matrices, these formulations facilitate targeted delivery of bioactive compounds, leading to enhanced therapeutic outcomes and reduced side effects. Moreover, the nanoscale dimensions of these carriers allow for precise control.

12.3 BIOPOLYMERS USED IN NANOCOMPOSITE PREPARATION

12.3.1 CHITOSAN

Natural polysaccharide CS is derived from chitin, the second-most abundant biopolymer with distinct structural and physiological properties that is frequently found in the shells and exoskeletons of various crustaceans. CS is a potent chelating agent that readily forms complexes with heavy and transition metals. Their special blend of qualities, such as their biodegradability, biocompatibility, and antibacterial activity, makes them the perfect polymer for use in a variety of industrial settings, such as textiles, agriculture, food science, pharmaceuticals, and biomedical science (Wang et al., 2005).

12.3.2 ALGINATE

Alginate, which is mostly made of calcium and sodium alginate, is a significant algal biopolymer that is mostly formed from bacteria and brown seaweed. It has gained a lot of popularity since it is inexpensive, readily available, biocompatible (much like glycosaminoglycan in the extracellular matrix; ECM), and biodegradable. Due to the abundance of carboxyl and hydroxyl functional groups on its structural chain, alginate is primarily an unbranched polyanionic polymer that can be modified to form gels, films, NPs/spheres, fibers, and micro particles. Sodium (Na) and calcium (Ca) alginate have a wide range of physicochemical features and uses due to their chemical diversity, mucoadhesion, sol–gel transition, hemostaticity, good bio- and cytocompatible behavior, and biodegradable qualities (Zia et al., 2020).

12.3.3 GELATINE

Gelatine is a basic protein and polypeptide mixture that is derived from a variety of animals. It has no discernible color or flavor. It is appealing for many applications due to its many functional qualities, which include excellent film-forming ability, gelling properties, emulsifying and foaming capabilities, and biocompatible and biodegradable qualities. Furthermore, the phase behavior of diluted and semidiluted solutions of gelatine can be readily modified by external conditions due to the denatured and distinct amino acid sequence. Mechanical characteristics of gelatine can be improved with the incorporation of nanofillers such as clay, carbon nanotubes, graphene oxide, hydroxyapatite, calcium carbonate, NPs, and so forth (Yang et al., 2021).

12.3.4 ALBUMIN

The primary protein in plasma, albumin, is primarily responsible for binding water, cations (such as Ca^{2+}, Na^{+}, and K^{+}), fatty acids, hormones, and medication molecules within the body. It has a significant impact on how plasma's osmotic pressure is regulated and how well plasma proteins buffer acids from bases (Engelhart and Moughamian, 1968). Nano drug delivery system included with albumin polymer helps in the accumulation of drugs at tumor or inflammatory tissue sites. Albumin stabilizes metal ions efficiently which helps in the development of novel drug delivery systems; external factors can be used for the same like magnetic and electric fields, temperature, mechanical forces, and so forth (Xu et al., 2020).

12.4 APPLICATIONS OF HERBAL NANOCOMPOSITE (Table 12.1 and Figure 12.2)

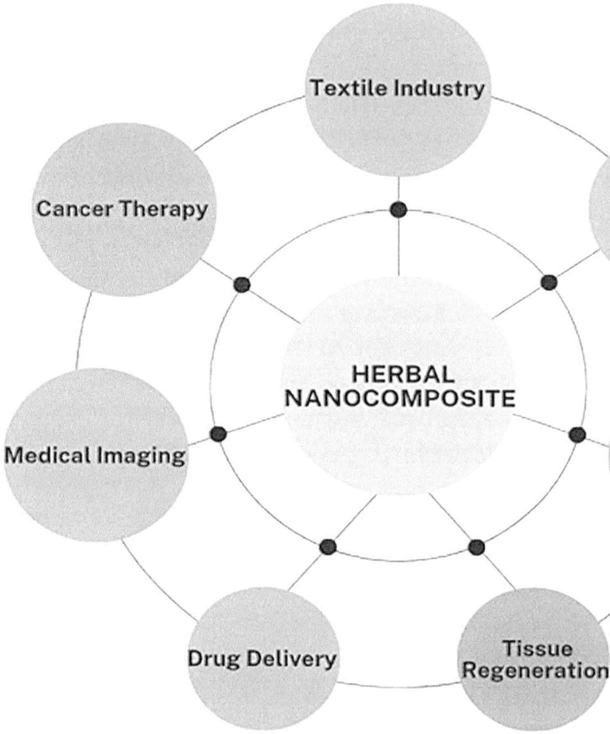

FIGURE 12.2 Applications of herbal nanocomposites.

12.4.1 OVARIAN CANCER THERAPY

In recent years, many drug delivery approaches have been tested for cancer treatment. The primary goal of these approaches is to increase therapeutic effectiveness and reduce toxicity through selective tumor targeting, particularly through the use of nano-carriers to precisely deliver drug payloads to malignant tumor regions via the intravenous route. Because of the specific anatomical structures of tumor tissue (e.g., leaky vesicle, immature lymphatic, etc.), the distribution of these nano-carriers with specific properties (e.g. surface charge, particle size, etc.) in tumor tissue can be ameliorated through an enhanced permeability and retention effect.

However, this passive approach to tumor targeting with an EPR effect has inherent limitations. To improve tumor-targeting effectiveness, active tumor-targeting strategies (i.e., attachment of target ligands onto nano-carriers) are developed. Alongside single ingredients, small synthetic chemicals or whole extracts derived from bio products can be utilized for the tumor treatment. When compared to the use of tiny synthetic compounds or biologics, the delivery of these active components obtained from natural products reduces adverse effects and demonstrates equal therapeutic effectiveness.

Using solvent evaporation and emulsification techniques (Figure 12.3), the ethanol extract of *Angelica gigas Nakai* (AGN) and resveratrol (RSV)-based NPs were formed. The emulsification and solvent evaporation operations altered the crystalline properties of AGN Ex and RSV, resulting in amorphous states for AGN/RSV NPs and AGN NPs. AGN extract served as both an anticancer agent and a medicinal matrix of NPs. At normal physiological pH, sustained release profiles (spanning 5 days) of RSV and decursin (D) and decursinol angelate (DA), the typical markers of AGN extract, were identified. AGN/RSV NPs shown greater antiproliferation potentials in ovarian cancer (SKOV-3) cells compared to AGN NPs, AGN extract, and AGN NPs + RSV (Nam et al., 2018).

12.4.2 FOR MEDICAL TEXTILE

12.4.2.1 ANTIBACTERIAL ACTIVITY

Fabric sets were treated with nanocomposites comprising herbal extracts from *Achyranthes aspera*, *Senna auriculata,* and CS. The study assessed the antibacterial effectiveness by using the EN ISO 20645 standard, whereas fabric durability was evaluated by the AATCC–124 test method. Physical characteristics such as air permeability, abrasion resistance, and tensile strength were scrutinized according to ASTM D 737–1996, AATCC 119–2004, and ASTM D 5035–2006, respectively. By using HET-CAM, no skin irritation was observed by checking the biocompatibility of the treated fabric. Compared to traditionally treated fabrics, the nanocomposite-treated fabrics exhibited larger inhibitory zones against *Escherichia coli* (31 mm) and *Staphylococcus aureus* (29 mm). Moreover, they demonstrated favorable durability, physical properties, and biocompatibility. The incorporation of CS and herbal extracts enhanced

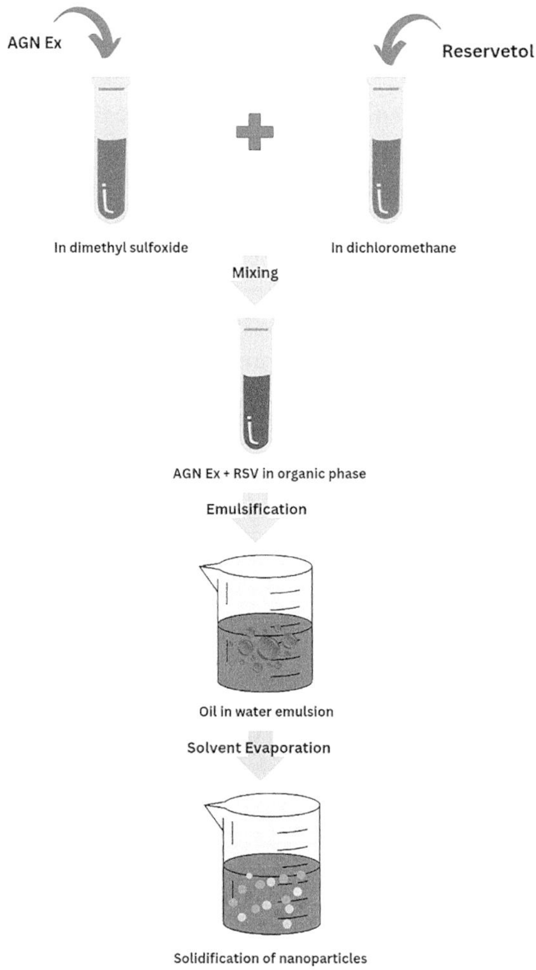

FIGURE 12.3 Preparation of AGN/RSV nanoparticles.

the antibacterial activity and overall quality of the finished fabrics, making them suitable for various applications requiring both functionality and comfort (Chandrasekar et al., 2014).

12.4.2.2 ANTI-OXIDANT, ANTI-UV, AND ANTI-BACTERIAL

Using the wet processing approach, or solution-based synthesis, various herbal NPs were created from the leaves of the *Azadirachta indica,*

Acalypha indica, Tridax procumbens, Piper betle, and *Aloe vera* plants. The goal was to create biocompatible nanomaterials with exceptional therapeutic qualities. The effectiveness of herbal NPs was ascertained by comparing and contrasting the different herbal NPs based on their functional and physico-chemical characteristics when applied to cotton textiles. NPs of *T. procumbens* have greater antioxidant activity. The CS polymer-prepared herbal nanocomposites made from *Aloe vera* demonstrated protection (UPF = 62.3). The *A. indica* nanocomposite-coated cotton fabrics demonstrated effective results against *S. aureus* (35.62 mm) and *E. coli* (33.13 mm) due to its superhydrophobic qualities (154.5°) and higher antibacterial properties. Research helps identify the best ways to process herbal NPs and improve their UV protection, antioxidant, antibacterial, and self-cleaning properties for use in textile and biomedical applications (Subramani et al., 2020).

12.4.2.3 CARRIER FOR PLANT EXTRACT

Plant extracts with biopolymers like sodium alginate as an alternative to clinical drugs and the phytochemicals found in natural plant extracts that possess properties such as antifungal, antibacterial, anticancer, antimicrobial, and other diseases found to be useful in various biomedical applications. More investigation and testing are required to determine the effectiveness and dosage of the extract; however, they may be used as a stand-in for clinical drugs to help treat or suppress specific diseases.

Alginates are often made of various bacterial strains and kelp or brown algae. Alginate is a linear polysaccharide having an anionic character. Its average molecular weight is 216.121 g/mol and its chemical formula is ($C_6H_7NaO_6$). An alginate's capacity to form an ionic gel is its most crucial property. Ionized alginates are created mainly by electrostatic interactions between cations and anions, negatively charged carboxyl groups of the alginate moiety, resulting in polyelectrolyte complexes. In the egg-box model, the gelation mechanism consists of coordination and chelation. During this process, the alginate units attach to the polyvalent metal ion in conjunction with these cross-linked agents' hydrogen bonding interactions with the atoms of oxygen on the G blocks of the two nearby polymer chains. Biopolymers such as alginate cross-linked with Ca^{2+} also have a vast application in nerve and tissue regeneration, bone, 3D printing, drug delivery systems, and 3D printing. Particularly because of their magnetic properties, iron oxide NPs have been used to create alginate-based nanocomposite materials that have the potential to improve drug delivery systems. A highly effective and nontoxic medication known as a nano-hybrid was created by combining a biopolymer nanocomposite with plant extracts. Therefore, it can be said that plant extract can be carried by an alginate bases nanocomposite to help boost its biological effectiveness (Figure 12.4).

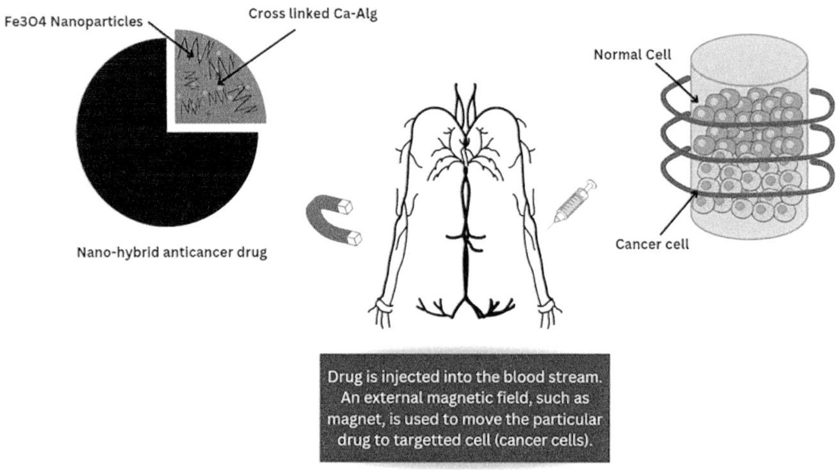

FIGURE 12.4 Working principle of plant-based nanocomposite drugs in drug delivery for cancer treatment.

12.4.3 *ANTI-MICROBIAL ACTIVITY*

Applications for metal oxide NPs are numerous, from soil stabilization to solar cells and biomedicine. Various metals have been used to synthesize nanocomposites, and their uses have been studied in agriculture and other industries. Silver is commonly used in skin ointments because of its innate propensity to serve as an antibacterial agent and NPs of zinc oxide are found to be the most effective antibacterial and antimicrobial agent. Recently, plant materials have been used in NP synthesis, and this approach has proven to be a simple and cost-effective way to prepare nanomaterials. Because of the surface morphology, small particle size, and surface morphology, the synthesized material has a large surface-to-volume ratio, as well as its size distribution. For antibacterial and antimicrobial activity, the best option is considered as NPs of zinc oxide.

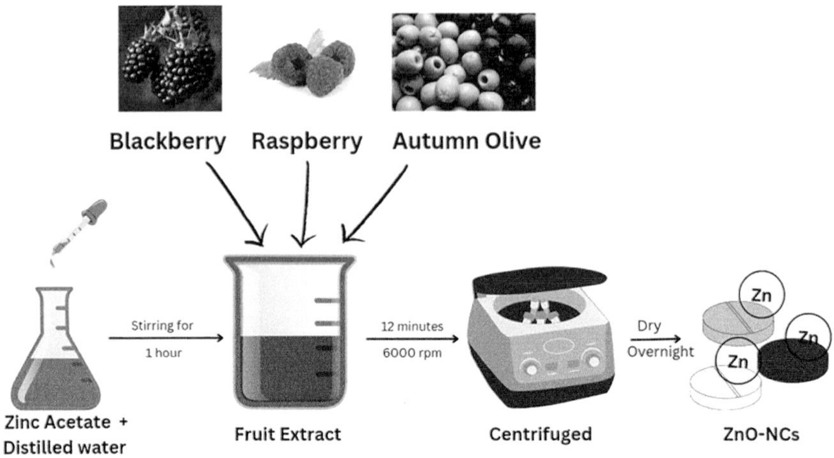

FIGURE 12.5 Synthesis of zinc oxide nanocomposite using fruit extracts.

Zinc oxide nanocomposites were made via an environmentally friendly method. By using a simple technique for creating a ZnO nanocomposite from the extract of *Rubus idaeus, Rubus fruticosus,* and *Elaeagnus umbellata*, hazardous chemicals were not used in the production of these nanocomposites. Instead of using hazardous reduction and stabilization chemicals, fruit extract was utilized. Using Debye-Scherrer's equation, the average particle sizes of the synthesized ZnO nanocomposites were determined to be 102.47, 102.97, and 151.99 nm. The synthesized nanocomposites showed significant antibacterial properties against the bacterial strains of *E. coli* (341), *S. aureus* (345B), and *Pseudomonas aeruginosa* (5994 NLF) (Dar et al., 2022) (Figure 12.5).

12.4.4 ANTI-BACTERIAL ACTIVITY

Instead of employing plant extract, microorganisms like fungi can be used to synthesize NPs, which effectively reduce the metals. Plant-mediated synthesis of metal nanocomposites has become notably more popular in recent years. The fact that plant-mediated NP synthesis proceeds more quickly than microorganism-based NP synthesis is one important factor. ZnO nanopowders can be generated via a variety of techniques, such as spray pyrolysis, hydrothermal techniques, precipitation techniques, sol–gel techniques, microwave-assisted techniques, thermal breakdown of organic precursors, electrodeposition, and the utilization of micro emulsions, as previously reported.

Plant extracts from *Terminalia* and *Aloe vera* were used to create zinc oxide nanocomposites. The synthesis of NPs was carried out with less harmful chemical production and in an environmentally sustainable manner. Several analytical methods, including FTIR, UV-VIS, and SEM, were used to characterize the nanocomposites. The generated NPs demonstrated stretching and bending at predicted specific wave numbers in FTIR experiments, indicating the production of nanocomposites. This was caused by a variety of functional groups. SEM pictures demonstrated the morphology of the NP and validated the structure of the nanocomposite. When tested against two bacterial strains, *E. coli,* and *Burkholderia stabilis*, the nanocomposites demonstrated effective antibacterial activity (Dar et al., 2021).

12.4.5 HEPATIC CARCINOMA

Green synthesis was used to create the *Nigella sativa* seed extract silver (Ag) nanocomposites, which were then tested against HepG2 cells, a kind of hepatocellular cancer. In addition to being an environmentally friendly and economical process with low production costs, the green chemistry for Ag nanocomposites (AgNCs) synthesis has many benefits for the impending industrial scale-up of production. It also does not require any kind of polymer for stabilization or reduction, surfactant, or co-surfactant, nor does it require sophisticated equipment for synthesis. Evaluation of synthesized AgNCs completed based on zeta potential, UV spectroscopy, DLS, TEM, FTIR, and *in vitro* release studies. According to the HepG2 cell viability assay and morphological changes, *N. sativa* AgNCs in a dose-dependently manner efficiently limits cell proliferation. Based on the investigation of *in vitro* cell survival and morphological changes in HepG2 cells, it was suggested that *N. sativa* AgNCs potentially suppress the proliferation of HepG2 cells in a concentration-dependent manner. These exhibit strong anticancer efficacy against human hepatocellular carcinoma, as demonstrated by the fragmentation of apoptotic nuclei and their build-up in ROS generation, all while causing no harm to normal cells. Thus, these have the potential to serve as an effective anticancer medicine for hepatocellular carcinoma (Vijian et al., 2022).

12.4.6 ORTHOPEDIC IMPLANTS

Implant-associated infection is one of the most serious side effects of orthopedic surgery. A suitable coating on the implant can successfully prevent biofilm formation, even though the implant interface may be a surface that is susceptible to bacterial colonization. In order to stop this, henna and thyme extract were added to a nano-fibrous hydrogel based on CS and polyethylene oxide (PEO). By using two-nozzle electrospinning, CS/PEO/thymus and CS/PEO/henna extract nanofibers were created. The ideal proportion of CS to PEO to herbal extract was evaluated, the fibers demonstrated a fine inhibitory zone against both gram-positive bacteria and gram-negative bacteria when employed as a coating on orthopedic screws or by themselves. There is no evidence of any negative impact of the coatings on red blood cells that could cause hemolysis. The results of the test—cytocompatibility exhibited that the produced nano-fibrous in tests using fibroblast cells, the covering proved to be completely biocompatible. According to the study, thyme extract-loaded CS/PEO and henna coatings show promise for usage in various medical settings, including implantation coating and wound dressing, where they can inhibit bacterial adherence, growth, and infection in general (Usmani et al., 2019).

12.4.7 LUNG CANCER THERAPY

Due to its systemic side effects and poor solubility, etoposide (ETP) has limited use as a prospective lung cancer treatment. The study suggested using inhaled boronate-targeted human serum albumin (HAS) nanocomposites to deliver ETP and berberine (BER), a herbal remedy, together for localized lung cancer treatment. In order to improve drug solubility and make it easier for the drug to be encapsulated within the hydrophilic albumin NPs, ETP was first pre-formulated as phospholipid complex (EPC). Second, for synergistic effect in lung cancer therapy BER and EPC co-loaded with great efficiency in HAS NPs. The NPs displayed a sequential drug release pattern and suitable size around 200 nm. Moreover, conjugation of HSA NPs and aminophenylboronic acid (APBA) resulted in enhanced internalization into A549 lung cancer cells and cytotoxicity, when compared to sialic acid residues binding to nontargeted NPs or free drugs expressed by cancer cells. Better antitumor efficacy of inhalable nanocomposites is demonstrated by *in vivo* investigations in lung cancer animal models. The alternative strategy for systemic delivery of ETP and

BER in lung cancer therapy is offered by inhalable APBA-HSA nanocomposites (Kharat et al., 2021).

12.4.8 CATALYTIC EFFECT

Fe_3O_4/ZrO_2 and nano ZrO_2 nanocomposite were prepared by co-precipitation and sol–gel methods, respectively. Then, by using *Centaurea cyanus* flower extract, AgNPs were mixed with them in eco-friendly and mild conditions with stabilizing and reducing agents. The synthesized nanocomposites were characterized by using FT-IR, VSM, XRD, EDS, and FE-SEM techniques. The findings of the experiment verified that, AgNPs with a diameter of 30–90 nm formed on the surface of supports at room temperature. After that, the synthesized nanocomposite's catalytic activity was studied in the reduction of methyl orange (MO) and 4-nitrophenol (4-NP). $Ag/Fe_3O_4/ZrO_2$ nanocomposite, which is magnetically recoverable, can be recycled multiple times without experiencing a significant reduction in catalytic activity (Elgohary et al., 2018).

12.4.9 TISSUE ENGINEERING

Cartilage tissue damage is accompanied by problems with mobility. Tissue engineering is a young field that has the potential to repair or regenerate damaged tissues. It is essential to have a suitable scaffold that promotes cell adhesion, recruitment, proliferation, and differentiation. In order to enhance the hydrogel's qualities, NPs are included in its matrix. To strengthen the hydrogel matrix, fillers such as organic/polymeric and inorganic NPs (clay, hydroxyapatite, graphene, and metallic NPs) can be employed to form a nanocomposite. Because of its sol-gel-phase transition at body temperature, chitosan/glycerophosphate (CS/GP), hydrogel is a thermosensitive hydrogel that is a biological material with significant applications in tissue engineering (Rostami-Vartooni et al., 2019).

12.4.10 PACKAGING TO INCREASE SHELF LIFE

Biopolymers—plant-derived cellulose, starch, agar, and carnauba; animal-derived whey protein, gelatin, casein, and beeswax; and microorganism-derived dextrans, xanthan, pullulan, bacterial cellulose, and polylactic acids—have all been thoroughly researched and used as biodegradable

TABLE 12.1 Nanocomposites and Their Applications.

S. No.	Nanocomposite and Extract	Method Used for Preparation	Applications	References
1	Ethanolic extract of *Angelica gigas Nakai* and resveratrol	Nanocrystal concept (emulsification and solvent evaporation method)	Ovarian cancer therapy	Nam et al., 2018
2	*Senna auriculata* and *Achyranthes aspera* extract with the solution of chitosan	Coacervation method	Antibacterial activity for textile fabric	Chandrasekar et al., 2014
3	Chitosan with *P. betle, A. indica, Az. indica,* and *T. procumbens*	Wet chemical preparation	UV blocking activity – textile fabrics	Subramani et al., 2020
4	Calcium ions crosslinking with biopolymers like alginate and plant extracts with biopolymer nanocomposite.	Polymer crosslinking with divalent cation	3D printing, nerve and tissue regeneration of bone, drug delivery system, Anti-cancer activity	Dewi et al., 2022
5	ZnO nanocomposites synthesized by *Rubus fruticosus* (blackberry, *Rubus idaeus* (raspberry), and *Elaeagnus umbellata* (autumn olive)	Green synthesis	Antimicrobial activity against *Staphylococcus aureus* (*345B*) bacteria, *Escheracia coli* (*341*) (Gram positive bacteria) and gram-negative *Pseudomonas aeruginosa* (*5994NLF*,	Dar et al., 2022
6	*Aloe vera* and *Terminalia* plants extract to synthesis ZnO nanocomposite	Green Synthesis	Antibacterial activity (against *Burkholderia stabilis* and *Escherichia coli*)	Dar et al., 2021
7	*Nigella sativa* extract with silver nanocomposite	Green synthesis	Anticancer effect against hepatocellular carcinoma	Vijian et al., 2022
8	Henna and thyme extract filled in Chitosan/polyethylene oxide nanocomposite	Electrospinning method	Antibacterial coatings of orthopaedic implants	Usmani et al., 2019
9	Boronate-targeted inhalable nanocomposite	Desolvation method	Etoposide delivery and berberine (herbal drug), for lung cancer treatment	Kharat et al., 2021
10	Aqueous *Centaurea cyanus* flower extract with $Ag/Fe_3O_4/ZeO_2$ nanocomposite	Sol-gel and co-precipitation methods	Reduction of organic pollutants from wastewater	Elgohary et al., 2018

substitutes for synthetic plastic packaging. Reinforcement of these biopolymers with nanomaterials such as zinc oxide NPs, cellulose nanofiber, zinc oxide, alginate NPs improved barrier properties, physico-chemcal, mechanical, antimicrobial, and antioxidant activities. The most popular technique for creating these films, including blends and hybrids, is solution casting. Studies have reported a shelf life of 14–40 days, by the applications of nanocomposite coatings on cut fruits depending on the type of fruit (Asadi et al., 2018).

12.5 ADVANTAGES AND DISADVANTAGES

The pursuit of enhancing targeted delivery and efficacy while minimizing adverse effects has led to the exploration of novel materials and formulations. Herbal nanocomposites represent a cutting-edge intersection of traditional herbal medicine and nanotechnology, holding immense promise in revolutionizing drug delivery and therapeutic efficacy. By leveraging the nanoscale dimensions of these carriers, herbal nanocomposites achieve precise control over drug loading and release kinetics, enhancing bioavailability and ensuring targeted delivery to specific tissues or cells, resulting in composites with unprecedented properties including, improved stability, solubility, and therapeutic potency. This synergy amplifies the therapeutic benefits of herbal remedies, unlocking their full potential in treating various ailments. Moreover, the nanocomposite platform enables sustained release of bioactive compounds, prolonging therapeutic effects and reducing the frequency of administration, thereby enhancing patient compliance. Beyond enhancing drug delivery, herbal nanocomposites also offer advantages in terms of safety and biocompatibility. The use of natural herbal extracts reduces the risk of adverse effects associated with synthetic drugs, while the biodegradable nature of nanomaterials ensures compatibility with biological systems (Khan et al., 2018; Puglia et al., 2019; Kaplan, 2022; Okamoto, 2013).

Despite their potential benefits, herbal nanocomposites present several challenges and disadvantages that must be addressed for their widespread implementation in healthcare. Firstly, standardizing herbal products for use in nanocomposites proves challenging due to the variability in natural sources and extraction methods, leading to inconsistencies in composition and efficacy.

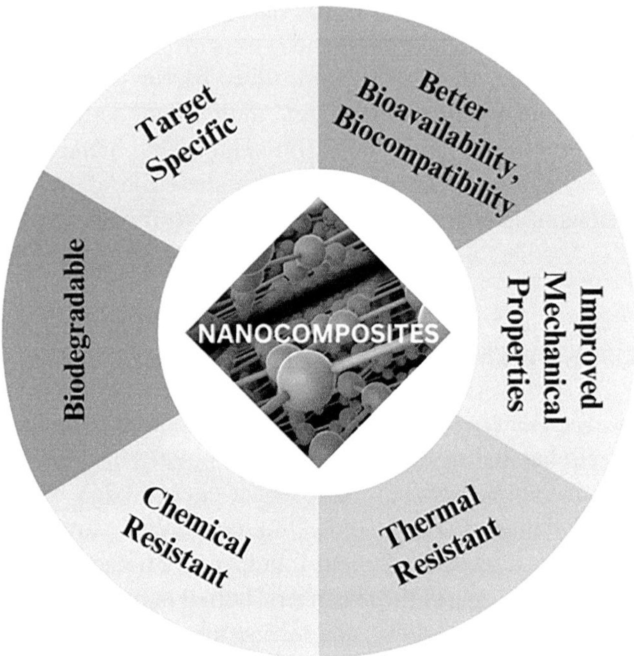

FIGURE 12.6 Advantages of herbal nanocomposites.

Moreover, the use of raw herbal materials can introduce microbial and heavy metal contamination, posing significant health risks to consumers and complicating quality control measures. The lack of available stability data for herbal nanocomposites hinders their regulatory approval and clinical translation. Without comprehensive stability profiles, it is difficult to assess the shelf-life and long-term viability of these formulations, raising concerns about their safety and effectiveness over time. Furthermore, the preparation of herbal nanocomposites is often an expensive and complex process, involving intricate extraction techniques and specialized equipment. These factors contribute to the overall cost of production, limiting accessibility to patients, particularly in resource-constrained settings. Another challenge is the propensity of herbal nanocomposites to aggregate, leading to reduced bioavailability and compromised therapeutic efficacy. Aggregation can occur during formulation or storage, resulting in uneven distribution of active ingredients and diminished performance of the composite material. Moreover, scaling up the production of herbal nanocomposites from laboratory to industrial scale poses significant technical hurdles. The intricate nature of the extraction process and the need for precise control over particle size and distribution make large-scale manufacturing both costly and logistically challenging (Maghsoudi et al., 2020; Bhardwaj and Arora, 2018).

In essence, herbal nanocomposites represent a ground-breaking approach to healthcare, blending centuries-old herbal wisdom with modern nanotechnology to create advanced therapeutic solutions hold promise as innovative therapeutic agents, they are not without limitations.

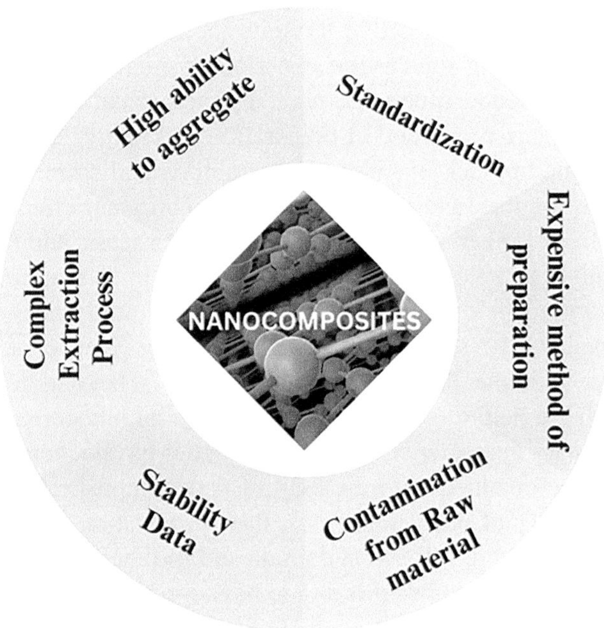

FIGURE 12.7 Disadvantages of herbal nanocomposites.

Collaborative efforts between researchers, regulators, and industry stakeholders are needed to overcome these challenges and pave the way for the development of safe, effective, and accessible herbal nanocomposite-based therapies. As research in this field progresses, these innovative systems hold the potential to address unmet medical needs, offering safer, more effective treatments with enhanced patient outcomes. The journey from traditional herbal medicine to state-of-the-art nanocomposites exemplifies the evolution of medicine, marking a new chapter in the quest for improved health and well-being. As research continues to evolve, we anticipate witnessing the development of novel formulations, improved manufacturing techniques, and expanded applications for herbal nanocomposites, paving the way for a brighter, healthier, and more sustainable future.

12.6 CONCLUSION

Research on herbal medicine with improved bioavailability and less toxicity may benefit greatly from the integration of nanotechnology and herbal medicine. In comparison to the traditional uses of medicine, it is relatively safe and effective. Despite the advances, there are a few challenges in this field such as the selection of suitable nanoformulation, polymer for phytoconstituent or plant extract, isolation of the herbal constituent, expensive method of preparation and optimization, difficulty in process scale up, lack of safety data, stability, and so forth. Inspite of these challenges, the field provided us with significant effective results, with the help of further studies, modification in process, and formulation selection, might be able to overcome the situation in coming future.

The emerging field of nanotechnology has its advantages, including better bioavailability, specificity, and less toxicity. The scope of herbal medications with the help of nanotechnology is beyond reach. Plant extracts with the help of suitable carriers can be administered in order to have the primary characteristics as well as better physiochemical properties that can be obtained with the help of nanocomposites. This chapter covered the basics of nanocomposites, their preparation, and application in the biomedical field. It is a vast domain and further research and development is required to get benefits and to have a clinical application.

KEYWORDS

- **herbal**
- **nanocomposite**
- **nanoparticles**
- **phytopharmaceuticals**
- **polymer**
- **shelf-life**

REFERENCES

Anandhan, S., & Bandyopadhyay, S. (2011). Polymer nanocomposites: From synthesis to applications. In J. Cuppoletti (Ed.), *Nanocomposites and polymers with analytical methods*. http://dx.doi.org/10.5772/17039.

Asadi, N., Alizadeh, E., Salehi, R., Khalandi, B., Davaran, S., & Akbarzadeh, A. (2018). Nanocomposite hydrogels for cartilage tissue engineering: A review. *Artificial Cells, Nanomedicine, and Biotechnology*, *46*(3), 465–471. https://doi.org/10.1080/21691401.2017.1345924.

Bhardwaj, A., & Arora, R. (2018). *Nanoformulations: A novel approach against hypoxia*. Elsevier Inc. https://doi.org/10.1016/B978-0-12-813999-8.00012-4.

Bhokare, S., Dongaonkar, C., Lahane, S. V. B. P., Sawale, V. S., & Thombare, M. S. (2016). Herbal novel drug delivery: A review. *World Journal of Pharmaceutical Sciences*, *5*(8), 593–611. https://doi.org/10.20959/wjpps20168-74.

Chandrasekar, S., Vijayakumar, S., & Rajendran, R. (2014). Application of chitosan and herbal nanocomposites to develop antibacterial medical textile. *Biomedicine and Aging Pathology*, *4*(1), 59–64. https://doi.org/10.1016/j.biomag.2013.10.007.

Contreras, A., López, V. H., & Bedolla, E. (2004). Mg/TiC composites manufactured by pressureless melt infiltration. *Scripta Materialia*, *51*(3), 249–253. https://doi.org/10.1016/j.scriptamat.2004.04.007.

Dar, A., Rehman, R., & Mohyuddin, A. (2022). Efficacy of various types of berries extract for the synthesis of ZnO nanocomposites and exploring their antimicrobial potential for use in herbal medicines. *BioMed Research International, 2023*. https://doi.org/10.1155/2023/9815767.

Dar, A., Rehman, R., Zaheer, W., Shafique, U., & Anwar, J. (2021). Synthesis and characterization of ZnO-nanocomposites by utilizing aloe vera leaf gel and extract of terminalia arjuna nuts and exploring their antibacterial potency. *Journal of Chemistry*, *2021*. https://doi.org/10.1155/2021/9448894.

Dermenci, K. B., Genc, B., Ebin, B., Olmez-Hanci, T., & Gürmen, S. (2014). Photocatalytic studies of Ag/ZnO nanocomposite particles produced via ultrasonic spray pyrolysis method. *Journal of Alloys and Compounds*, *586*, 267–273. https://doi.org/10.1016%2Fj.jallcom.2013.10.004.

Dewi, M. K., Chaerunisaa, A. Y., Muhaimin, M., & Joni, I. M. (2022). Improved activity of herbal medicines through nanotechnology. *Nanomaterials*, *12*(22), 4073. https://doi.org/10.3390/nano12224073.

Elgohary, M. M., Helmy, M. W., Abdelfattah, E. Z. A., Ragab, D. M., Mortada, S. M., Fang, J. Y., & Elzoghby, A. O. (2018). Targeting sialic acid residues on lung cancer cells by inhalable boronic acid-decorated albumin nanocomposites for combined chemo/herbal therapy. *Journal of Controlled Release*, *285*, 230–243. https://doi.org/10.1016/j.jconrel.2018.07.014.

Engelhart, M. D., & Moughamian, H. (1968). Book reviews: Book reviews. *Educational and Psychological Measurement*, *28*(2), 619–620. https://doi.org/10.1177/001316446802800256. (Original work published 1968).

Hussain, F., Hojjati, M., Okamoto, M., & Gorga, R. E. (2006). Review article: Polymer-matrix nanocomposites, processing, manufacturing, and application: An overview. *Journal of Composite Materials*, *40*(17), 1511–1575. https://doi.org/10.1177/0021998306067321.

Kaplan, A. (2022). The nanocomposites designs of phytomolecules from medicinal and aromatic plants: Promising anticancer-antiviral applications. *Beni-Suef University Journal of Basic and Applied Sciences*, *11*, 17. https://doi.org/10.1186/s43088-022-00198-z.

Khan, I., Bahuguna, A., Kumar, P., Bajpai, V. K., & Kang, S. C. (2018). In vitro and in vivo antitumor potential of carvacrol nanoemulsion against human lung adenocarcinoma A549 cells via mitochondrial mediated apoptosis. *Scientific Reports*, *8*(1), 712–714. https://doi.org/10.1038/s41598-017-18644-9.

Khandoker, N., Hawkins, S. C., Ibrahim, R., Huynh, C. P., & Deng. F. (2011). Tensile strength of spinnable multiwall carbon nanotubes. *Procedia Engineering*, *10*, 2572–2578. https://doi.org/10.1016/j.proeng.2011.04.424.

Kharat, Z., Sadri, M., & Kabiri, M. (2021). Herbal extract loaded chitosan/PEO nanocomposites as antibacterial coatings of orthopaedic implants. *Fibers and Polymers*, *22*(4), 989–999.

Kosmidou, T. V., Vatalis, A. S., Delides, C. G., Logakis, E., Pissis, P., & Papanicolaou, G. C. (2008). Structural, mechanical and electrical characterization of epoxy-amine/carbon black nanonocomposites. *Express Polymer Letters*, *2*(5), 364–372. https://doi.org/10.3144/expresspolymlett.2008.43.

Maghsoudi, S., Shahraki, B. T., Rabiee, N., & et al. (2020). Burgeoning polymer nano blends for improved controlled drug release: A review. *International Journal of Nanomedicine*, *15*, 4363–4392. https://doi.org/10.2147/IJN.S252237.

Malaki, M., Xu, W., Kasar, A. K., & et al. (2019). Advanced metal matrix nanocomposites. *Metals*, *9*(3), 330. https://doi.org/10.3390/met9030330.

Nam, S., Lee, S. Y., Kang, W. S., & Cho, H. J. (2018). Development of resveratrol-loaded herbal extract-based nanocomposites and their application to the therapy of ovarian cancer. *Nanomaterials*, *8*(6), 384–396. https://doi.org/10.3390/nano8060384.

Okamoto, M. (2013). Polymer Nanocomposites. Eng, *4*(1), 257–279. https://doi.org/10.3390/eng4010028.

Omanović-Mikličanin, E., Badnjević, A., Kazlagić, A., & Hajlovac, M. (2020). Nanocomposites: A brief review. *Health Technology*, *10*, 51–59. https://doi.org/10.1007/s12553-019-00380-x.

Piyal Mondal, A., Anweshan, A., & Purkait, M. K. (2020). Green synthesis and environmental application of iron-based nanomaterials and nanocomposite: A review. *Chemosphere*, *259*, 127509. https://doi.org/10.1016/j.chemosphere.2020.127509.

Puglia, C., Santonocito, D., Musumeci, T., & et al. (2019). Nanotechnological approach to increase the antioxidant and cytotoxic efficacy of crocin and crocetin. *Planta Medica*, *85*(3), 258–265. https://doi.org/10.1055/a-0732-5757.

Rajendran, R., Radhai, R., Balakumar, C., Ahamed, H. A. M., Vigneswaran, C., & Vaideki, K. (2012). Synthesis and characterization of neem chitosan nanocomposites for development of antimicrobial cotton textiles. *Journal of Engineered Fibers and Fabrics*, *7*(1), 136–141. https://doi.org/10.1177/155892501200700116.

Rostami-Vartooni, A., Moradi-Saadatmand, A., Bagherzadeh, M., & Mahdavi, M. (2019). Green synthesis of $Ag/Fe_3O_4/ZrO_2$ nanocomposite using aqueous *Centaurea cyanus* flower extract and its catalytic application for reduction of organic pollutants. *Iranian Journal of Catalysis*, *9*(1), 27–35.

Roy, S., Das, D., Chakravorty, D., & Agrawal, D. C. (1993). Magnetic properties of glass-metal nanocomposites prepared by the sol-gel route and hot pressing. *Journal of Applied Physics*, *74*, 4746–4749. https://doi.org/10.1063/1.354344.

Sen, M. (2020). Nanocomposite materials. In *Nanotechnology and the environment* (pp. 1–12). https://doi.org/10.5772/intechopen.93047.

Shah, S. M. A., Nisar, Z., Nisar, J., Akram, M., Ghotekar, S. K., & Oza, R. (2021). Nanobiomedicine: A new approach of medicinal plants and their therapeutic modalities. *Journal of Materials and Environmental Science*, *12*(1), 1–14.

Smart, S. K., Cassady, A. I., Lu, G. Q., & Martin, D. J. (2006). The biocompatibility of carbon nanotubes. *Carbone New York*, *44*, 1034–1047. https://doi.org/10.1016/j.carbon.2005.10.011.

Subramani, K., Saha, R., Palanisamy, S., & et al. (2020). Wet chemical preparation of herbal nanocomposites from medicinal plant leaves for enhanced coating on textile fabrics with multifunctional properties. *SN Applied Sciences*, *2*(4), 1–14. https://doi.org/10.1007/s42452-020-2459-z.

Usmani, A., Mishra, A., Jafri, A., Arshad, M., & Siddiqui, M. A. (2019). Green synthesis of silver nanocomposites of nigella sativa seeds extract for hepatocellular carcinoma. *Current Nanomaterials*, *4*(3), 191–200. https://doi.org/10.2174/2468187309666190906130115.

Vaia, R. A., Jandt, K. D., Kramer, E. J., & Giannelis, E. P. (1995). Kinetics of polymer melt intercalation. *Macromolecules*, *28*(24), 8080–8085. https://doi.org/10.1021/ma00128a016.

Vaia, R. A., & Wagner, H. D. (2004). Framework for nanocomposites. *Materials Today*, *7*(11), 32–37. https://doi.org/10.1016/S1369-7021(04)00506-1.

Vijian, R. S., Yusefi, M., & Shameli, K. (2022). Plant extract loaded sodium alginate nanocomposites for biomedical applications: A review. *Journal of Nanoscience and Nanotechnology*, *6*(1), 14–30. https://doi.org/10.37934/jrnn.6.1.1430.

Wang, X., Du, Y., Fan, L., Liu, H., & Hu, Y. (2005). Chitosan-metal complexes as antimicrobial agent: Synthesis, characterization and structure-activity study. *Polymer Bulletin*, *55*(1–2), 105–113. https://doi.org/10.1007/s00289-005-0414-1.

Xu, K., Zhang, J., Xue, W., & et al. (2020). Albumin-stabilized manganese-based nanocomposites with sensitive tumor microenvironment responsivity and their application for efficient SiRNA delivery in brain tumors. *Journal of Materials Chemistry B*, *8*(7), 1507–1515. https://doi.org/10.1039/c9tb02341k.

Yang, Z., Chaieb, S., & Hemar, Y. (2021). Gelatin-based nanocomposites: A review. *Polymer Reviews*, *61*(4), 765–813. https://doi.org/10.1080/15583724.2021.1897995.

Zia, F., Salman, M., Ali, M., & et al. (2020). Alginate-based bionanocomposites. *Bionanocomposites Green Synthesis and Applications Micro and Nano Technologies*, *2020*, 173–205. https://doi.org/10.1016/B978-0-12-816751-9.00008-8.

CHAPTER 13

Applications of Various Nanocomposites Used for Antimicrobial Activity, Tissue Engineering, Biosensors, and Dental

DIKSHA KUMARI[1,2#], PARAMPREET KOUR[1#], VISHWANI JAMWAL[1,2#], BOOBALAN GOPU[2,3], and KULJIT SINGH[1,2,*]

[1]*Infectious Diseases Division, CSIR - Indian Institute of Integrative Medicine, Jammu, Jammu and Kashmir, India*

[2]*Academy of Scientific and Innovative Research (AcSIR), Ghaziabad, Uttar Pradesh, India*

[3]*Pharmacology Division, CSIR - Indian Institute of Integrative Medicine, Jammu, Jammu and Kashmir, India*

#Equal contribution
*Corresponding author.

ABSTRACT

Nanocomposites (NCs) are one of the most important materials of modern technology which comprises at least two components, one with the continuous matrix phase and another uncontinuous reinforcement material to attain the best features from each component. As state of the art, these materials have been explored for their wide spectrum of applicability in the field of medicine, the pharmaceutical industry, drug discovery, and the energy industry. NCs have proved to be a breakthrough for the design of bio-sensing devices to improve the performance of detection platforms by increasing the electrochemical reactivity of biomolecules and can endorse the electron-transfer reactions of proteins. Moreover, in dentistry,

Nanocomposites and Nanomaterials in Biomedical Applications. Vivek Dave, Deependra Singh, and Pallerla Naveen Reddy (Eds.)

DOI: 10.1201/9781779641236-13

NCs have been in use for their extravagant properties such as improved mechanical stability, bone growth, and bacterial control for better oral hygiene. Thus, in the present chapter, we have highlighted the applications of these NCs as antimicrobial agents where we have discussed the role of different NCs-based materials in curbing different bacterial and parasitic infections. The readers will also get acquired the role of these miraculous particles as biosensors, in dentistry and the field of tissue engineering. Altogether, the use of NCs has emerged as a way forward for their effective role in pharmacological research.

13.1 INTRODUCTION

Nanotechnology is a rapidly growing field of research consisting of nano-dimension particles with multidimensional research applications, innovation, and industrial usage (Hulla et al., 2015; Malik et al., 2023). Nanoparticles (NPs) are known to be the building blocks of nanotechnology. These are the solid particulates with size ranging from 10 to 1000 nm (Okpala, 2013; Yetisgin et al., 2020). In the recent era of scientific discoveries, the utilization of NP-based technology has paced with leaps and bounds (Singh et al., 2023). In particular, nanocomposites (NCs) are natural or engineered solid nanomaterials that are formed when two or more different constituent materials, with their significant features, are integrated to incarnate a new substance with superior properties to original materials (Dutt et al., 2023). The idea behind designing these unique materials is to improve their quality such as stiffness, strength, low coefficient of expansion, resistance against fatigue, ease in manufacturing complex shapes, simple repair of damaged structures, corrosion resistance, and so forth, without causing substantial impartment or the loss of originality. However, the properties of NCs are greatly influenced by the size scale of their component phases and the degree of mixing between the two phases.

Broadly, NC materials are classified into three categories based on their matrix materials: ceramic matrix (CMNCs), polymer matrix (PMNCs), and metal matrix (MMNCs) nanocomposites (Omanović-Mikličanin et al., 2020). CMNCs contain matrix and fibers embedded in a ceramic matrix that leads to increased durability and reduced brittleness. These types of NCs are widely used in the electronic industry, ferroelectric properties, high-performance catalysts, data storage technology, solid-state laser media, optical amplifiers, and so forth (Li et al., 2020). Another type, PMNCs, is formed by NPs rooted in polymer matrix. These NCs resulted in a synergistic effect of both components and thus stand out from the other two. They come up with enhanced electronic, mechanical, ionic, chemical properties, and so forth, that open up the doors for a wide spectrum of applications in various scientific arenas such as power tool housing, fuel cells, solar cells, health care, and so forth (Hussain et al., 2006). The third type is MMNCs in which nanomaterials are coated in metal or alloy matrix. These types of NCs are primarily of great value in the display of screens, electronics, equipment, LED, storage devices, and biomedicines (Gupta et al., 2016). An overview of various applications of NCs is shown in Figure 13.1.

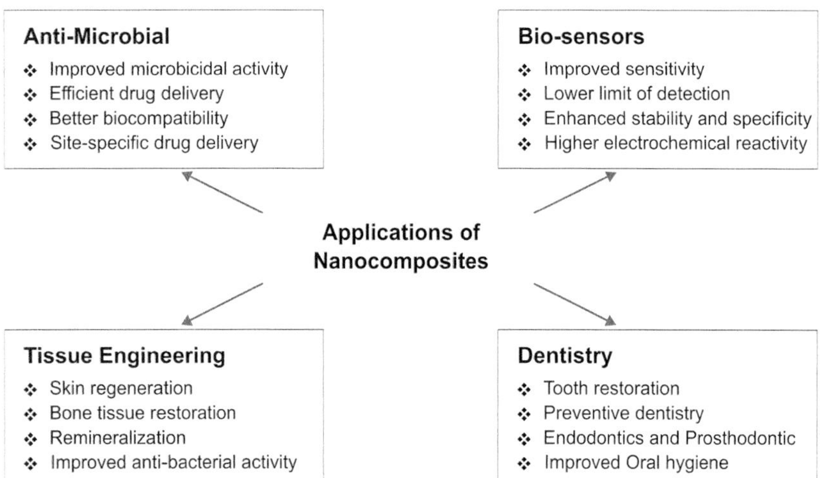

FIGURE 13.1 Various applications of nanocomposites.

NCs have shown tremendous importance in the biomedical, and food industries, biosensors, tissue engineering, textile applications, drug improvement, and development. Several NCs such as silver chitosan NCs, carbon nanotubes NCs, metal oxide NCs, and nanofibers are known to combat antimicrobial drug resistance, overcoming toxicity and targeted drug delivery. Matharu et al. (2018) in their comprehensive review demonstrated various possible antimicrobial activities of NCs along with discussing their composition, and physical and chemical properties. These miraculous materials are also known for their effectiveness as biosensors that augment their detection performance by increasing the electrochemical reactivity of biomolecules. By incorporating the nanostructures in the biosensor, the advantages of improved sensitivity and specificity can be further achieved. Some prominent examples of nano-biosensors include PEDOT/RGO, NiNPs, Au-PANI, and so forth (Baiju, 2020). Besides these, studies have reported the vitality of NCs in the field of tissue engineering and dentistry. The NCs are molded in such a way that it mimics the natural biological milieu of the body making them desirable for tissue engineering. In this context, Bramhill et al. (2017) elucidated the usage of NCs in tissue repair, adjusting degradation as well as in regeneration processes. With respect to dental science, NCs are used for fighting antimicrobial oral hygiene, tooth restoration, dental endodontics, and much more (Sreenivasalu et al., 2022). The present chapter acquainted the readers with the plethora of applicability of NCs-based technology, namely, antimicrobial agents, antiparasitic, role in tissue engineering, dentistry, and as potential biosensors. A detailed study of all the aforementioned parts is elaborated in further sections of this chapter.

13.2 NANOCOMPOSITES AS ANTIMICROBIAL AGENTS

13.2.1 TUBERCULOSIS

Tuberculosis (TB) is a long-lasting clinical infection caused by the most devastating pathogen *Mycobacterium tuberculosis* challenging millions of lives every year. This disease remains a daunting global health concern and requires urgent attention and effective therapeutic drugs to combat its dreadful smash on public health (Hatae et al., 2023). There are various anti-mycobacterial drugs available now, but the duration of first-line drug treatment for TB patients lasts from 3 to 24 months depending upon the type of TB (Perveen et al., 2022; Saifullah et al., 2013). These currently available treatment options against TB surely help in controlling the promulgation of active TB, but the multidrug-resistant and extensive drug-resistant *M. tuberculosis* strains are still a global health concern demanding immediate and effective drugs with less hepatotoxicity and short-term drug regimens (Jafari et al., 2020).

Nanotechnology has acquired special attention in the field of infectious diseases by contributing its nano-therapeutics application for disease treatment, especially in the case of TB. The unique advantage provided by lipid-based NPs such as high biocompatibility, immense stability, and efficient drug delivery system has compelled modern medicine to shift toward nanotechnology-based treatment options. These NPs encapsulate various anti-TB drugs that have the advantage of sustained drug release capacity at the target site (Hatae et al., 2023). Nowadays, metallic nanoparticles (MeNPs) are credited as antitubercular agents that disturb the integrity of bacterial cell membranes, employing these NPs both as a diagnostic and as drug delivery tool. Recent studies have revealed that various metal-based NPs such as mixed nanocrystals of silver and zinc oxide (ZnO) contribute to the effective clearance of *M. tuberculosis H37Rv* and MDR-TB in THP-1 cell lines (Jafari et al., 2020). An interesting study conducted by Punjabi et al. (2018) synthesized the silver nanoparticles (AgNPs) and zinc nanoparticles (ZnNPs) having dimensions of 40 and 60 nm. It was observed that AgNPs were the most potent antimicrobial agents against Methicillin-resistant *Staphylococcus aureus* (MRSA), Vancomycin-resistant *Enterococcus* (VRE), and *Pseudomonas aeruginosa* with minimum inhibitory concentration (MIC) ranging from 1.25 to 5 mg/mL. Also, it was noticed that both AgNPs and ZnNPs were active against *M. tuberculosis* and MDR strains having MIC of 1.25 mg/mL. Another exciting study conducted by Saifullah et al. (2013) prepared NCs by adopting a coprecipitation and ion exchange strategy. They characterized and intercalated para-aminosalicylic acid (PASA), an effective anti-TB agent, into aluminum and zinc-layered double hydroxides (ZLDH). PASA drugs were incorporated as alternate monolayers between ZLDH layers with an estimated drug loading of 22.8% for PASA-D prepared by the direct method and 16.6% for PASA-I manufactured by the indirect method. Cell viability studies were performed for these PASA-D and PASA-I-based NCs against human lung fibroblast MRC-5 and 3T3 mouse fibroblast cells for 72 h and were observed that these NC formulations were completely safe for various cell lines at least for 72 h. Thus, these NCs can be used as efficient drug delivery carriers for antitubercular drugs with controlled drug release properties without any cytotoxicity. Besides lipid-based nanocarriers, polysaccharides are also greatly employed for target-based drug delivery approaches. Another research conducted by Mehnath et al. (2019) synthesized biodegradable three-dimensional drug delivery NCs that are sericin-chitosan-doped maleate gellan gum (MA-GG) formed by the esterification process. Natural silk sericin was grafted over this 3D MA-GG structure. These polysaccharide-based Maleate Gellan Gum-Silk-Sericin-Chitosan

(MA-GG-SS-CS) NCs were encapsulated with antitubercular drugs such as rifampicin and pyrazinamide for efficient and sustained drug release of >80% for a long duration of 120 h. This acidic environment-based drug release at pH 4.0 showed high antimycobacterial activity by disrupting bacterial cell membranes. These findings prove that NCs are promising antimicrobial agents to combat drug resistance trends and are exciting options for advanced treatment strategies against TB.

13.2.2 ESKAPE PATHOGENS

The term ESKAPE comprises six highly virulent and opportunistic nosocomial pathogens exhibiting multidrug resistance that is a looming threat to public health. These pathogens include *Enterococcus faecium, Staphylococcus aureus, Klebsiella pneumonia, Acinetobacter baumannii, Pseudomonas aeruginosa*, and *Enterobacter species* (Mulani et al., 2019). *Enterococci* represent a group of gram-positive bacteria producing a variety of serious infections such as urinary tract infections (UTIs), soft tissue infections, intra-abdominal infections, and endocarditis. *Staphylococci* is a group of gram-positive bacteria producing a broad range of clinical infections that are capable of causing acute as well as chronic infections due to its inherent biofilm-producing ability (Shinu et al., 2022). *Klebsiella pneumonia* is a gram-negative bacterium that is responsible for various UTIs and catheter-associated UTIs. Wound or surgical site infections are another major problem caused by these bacteria which accounts for about 13% of all infections associated with *K. pneumonia* (Martin and Bachman, 2018). *A. baumannii* is an opportunistic pathogen and nonfermentative. It mainly colonizes the respiratory tract, pleural cavity, urinary tract, surgical sites, and blood which gives rise to various life-threatening infections, particularly in the case of immuno-compromised patients. *P. aeruginosa* is a gram-negative, facultative anaerobe that causes serious respiratory infections accountable for about 10% of nosocomial infections (Ramos-Martín and D'Amelio, 2023). *Enterobacter* species represents gram-negative, nonfastidious bacteria that cause a broad range of serious problems such as urinary and respiratory infections, septicaemia, meningitis, and so forth (Mancuso et al., 2021; Murugaiyan et al., 2022). Reckless use of unprescribed antibiotics has provoked the emergence of antimicrobial resistance that places a heavy burden on hospital sectors demanding efficient therapeutic drug.

Nanotechnology has been used widely in the medicinal field for diagnostic as well as therapeutic purposes. Another major role of this technology has gained the attention of many researchers for effectively targeting and destructing dreadful pathogens belonging to the ESKAPE family. A recent study conducted by Maruthapandi et al. (2020) evaluated the antimicrobial properties of polyaniline (PANI) composite against two gram-negative pathogens *K. pneumonia* and *P. aureginosa* (Figure 13.2). In this study, CuO, TiO_2, and SiO_2 were incorporated in PANI using sonochemical method. The antibacterial effect of these PANI composites was tested against two ESKAPE pathogens along with CuO, SiO_2, and TiO_2 taken as a control to determine the effect of PANI on bacterial growth. It was observed that both PANI-CuO and PANI-TiO_2 inhibited *P. aureginosa* growth after 6 h of treatment with these composites, whereas PANI- SiO_2 needed 12 h to completely eradicate bacterial growth. On the other hand, pure oxide particles (control) exhibited no bactericidal effect on *P. aureginosa* growth. In the case of *K. pneumonia*, it was observed that PANI-TiO_2 suppressed growth after exposure of 6 h, and the PANI-SiO_2 effect was highlighted after 12 h of composite exposure. Various experimental studies have revealed that acidic pH-induced PANI binds to the lipopolysaccharide layer of the bacterial membrane. This binding event compromised bacterial activities by changing membrane potential and eventually causing cell damage. This study suggested the essential therapeutic role of polymer-based composite against various nosocomial pathogens that are majorly responsible for antimicrobial resistance against the various available treatment options. These findings have gained special attention to utilize various polymers such as polypyrrolepropylic acid and poly(N-acetyl tyramine) for synthesizing polymer-based composite grafted with MeNPs toward the development of a new class of therapeutic drugs against various gram-negative pathogens.

Another interesting study conducted by Shehabeldine et al. (2023) synthesized copper oxide nanoparticles (CuONPs) from the cell filtrate of *Penicillium chrysogenum*. The agar diffusion test-based antibacterial activity of CuONPs was tested against *S. aureus* and *Escherichia coli* and reported the inhibition zone of 11 ± 0.57 and 16 ± 0.53 mm at 100 μg/mL of CuONPs, respectively. Also, resazurin-based studies revealed that both gram-positive and gram-negative bacteria were susceptible to CuONPs with reported MIC of 3.12 and 12.5 μg/mL for *E. coli* and *S. aureus*. Furthermore, it was observed that the minimum bactericidal concentration (MBC) value of CuONPs was 25 μg/mL for killing more than 99% of *E. coli*, whereas this MBC raised to 50 μg/mL for a similar antimicrobial effect on *S. aureus*. These results reflect that these CuONPs were more susceptible to *E. coli* as compared to *S. aureus*. This study concludes the cruciality of fungal-based CuONPs for having promising therapeutic potential against gram-negative bacteria that is providing a ray of hope for drug development against dreadful bacterial infections in the future. A similar approach was carried out by Stojkovska et al. (2019) who synthesized alginate hydrogels having AgNPs and honey. Antibacterial evaluation of alginate-based AgNPs was performed against a panel of broad-spectrum gram-negative and gram-positive bacteria along with 13 multiresistant clinical isolates of *A. baumannii, P. aeruginosa*, and *S. aureus*. It was observed that there was a significant decrease in bacterial colonies after 1 h for all the tested strains of *P. aeruginosa* and *S. aureus* whereas strains of MRSA remained unaffected. Surprisingly, at the released silver concentration of about 9 μg/mL, these hydrogel-based NCs exhibited extraordinary bactericidal potency against the standard as well as various clinical strains, except only three hospital strains in which the bactericidal effect was not detected. These results satisfied the antibacterial potency of hydrogel-based NCs in combating serious hospital-associated infections caused by these multi-drug-resistant bacterial strains.

Silva-Santos et al. (2016) explored the potency of silver NCs as a therapeutic agent against various nosocomial pathogens causing serious health issues. They adopted a green chemistry process to prepare silver-NCs using a fermentative medium of *Xanthomonas* spp. to synthesize xanthan gum polymer. The antimicrobial potency of these synthesized NCs was tested against a group of gram-positive and gram-negative bacteria, namely, *E. coli, E. faecalis, P. aeruginosa*, and *S. aureus* along with multi-drug-resistant clinical strains of *P. aeruginosa* and *A. baumannii.* It was found that these nanocomposites were highly susceptible to all the bacterial strains with the reported zone of inhibition ranging from 10.0 ± 1.0 mm to 12.9 ± 0.8 mm having 500 mg/mL of silver-NCs. This finding concludes that silver-NCs synthesized by one-step methodology can be a promising strategy to avoid the growth of multi-drug-resistant strains of *P. aeruginosa* and *A. baumannii*. Similarly, Li et al. (2021) designed multifunctional silver-NCs that have excellent antibacterial properties against carbapenem-resistant *A. baumannii* (CRAB). They constructed IPM@AgNPs-PEG-NOTA-NCs acting as a pH-sensitive nanodrug delivery system. Antibacterial assessment of imipenem and silver-NCs against CRAB revealed that the

MICs of imipenem and silver-NCs were 256 and 150 μg/mL, respectively. Further, it was observed that SH-PEG-NOTA alone was unable to decrease bacterial growth but when a combination of silver-NCs at 50 μg/mL and SH-PEG-NOTA with 60 μg/mL loaded with imipenem about 148 μg/mL was tested against CRAB, it exhibited outstanding antibacterial effect as compared to imipenem alone. Moreover, it was observed that IPM@AgNPs-PEG-NOTA-NCs were acting as a resistance-reversal agent by chelating the zinc of metallo-β-lactamases forming stable metal complexes. As a result, the nonhydrolyzed β-ring of imipenem is devoted to contributing as an antibacterial agent. Also, antibiofilm studies revealed that the IPM@AgNPs-PEG-NOTA-NCs were confirmed to be 2.24 times more suppressive than imipenem alone. Real-time quantitative polymerase chain reaction studies confirmed the extraordinary effect of NCs in the downregulation of the biofilm-related gene, ompA, involved in bacterial attachment to host cells promoting biofilm establishment. Furthermore, *in vivo* studies revealed that there was a significant decrease in neutrophils, lymphocytes, and white blood cells after treating CRAB-infected mice as compared to imipenem. Also, decreased level of C-reactive protein, IL-6, and TNF-α was observed. These extraordinary results support the therapeutic efficiency of IPM@AgNPs-PEG-NOTA-NCs as an alternative pH-sensitive nano-drug in combating CRAB-associated infections.

Another interesting approach was carried out by Luthfiah et al. (2021) to synthesize TiO_2-SiO_2-NCs using a photocatalytic technique against *S. aureus* and *P. aeruginosa*. The antibacterial activity of these NCs was tested by both optical densities and dilution test methods for qualitative and quantitative analysis, respectively. The outstanding antibacterial activity was evaluated by 1:0.5 TiO_2-SiO_2-NCs having inhibition percentages of 98.69 and 97.44% for *S. aureus* and *P. aeruginosa*, respectively. It was found that the incorporation of SiO_2 into TiO_2 enhanced the photocatalytic activity of TiO_2 as an antibacterial particle. These findings provide a therapeutic agent that has excellent antibacterial properties synthesized from eco-friendly natural material. A similar approach was performed by Natarajan (2015) who designed epoxy/Ag-TiO_2 polymer NCs as an antibiofilm agent against the pre-established biofilm of *S. aureus* and *E. coli*. It was found that many functional inorganic nano-filters, especially TiO_2 embedded in polymer matrix, were known to have an antibacterial nature having Ag^+ ions that provided combinational biocidal properties of both silver and photocatalytic properties of TiO_2. In this study, the antibiofilm property of epoxy/Ag-TiO_2 polymer NCs was studied against *S. aureus* and *E. coli* under UV irradiation. It was observed that OD_{600} of crystal violet elated from the biofilm of both bacterial pathogens ranged from 0.121 to 2.8 which justified that biofilm inhibition was more prominent in the case of *E. coli* as compared to *S. aureus*. Furthermore, to validate the ability of TiO_2/Ag-TiO_2 on NCs surface for photo-killing properties, the experiment was carried out under both dark and irradiated conditions. It was found that the biofilm inhibition was more prominent under UV-irradiated conditions with 56% and 77% inhibition for epoxy/TiO_2 and epoxy/Ag-TiO_2 in the case of *E. coli*, whereas for *S. aureus* the biofilm inhibition percentage was 43% for epoxy/TiO_2 and 67% inhibition was found for epoxy/Ag-TiO_2 NCs, respectively. These results concluded that these epoxy/TiO_2 and epoxy/Ag-TiO_2-based NCs under UV-irradiated conditions exhibited powerful antibiofilm activity against both gram-positive and gram-negative bacteria. All these findings conclude that NCs-based therapeutic technology plays a significant role in combating the dreadful smash of ESKAPE pathogen-associated infections on public health.

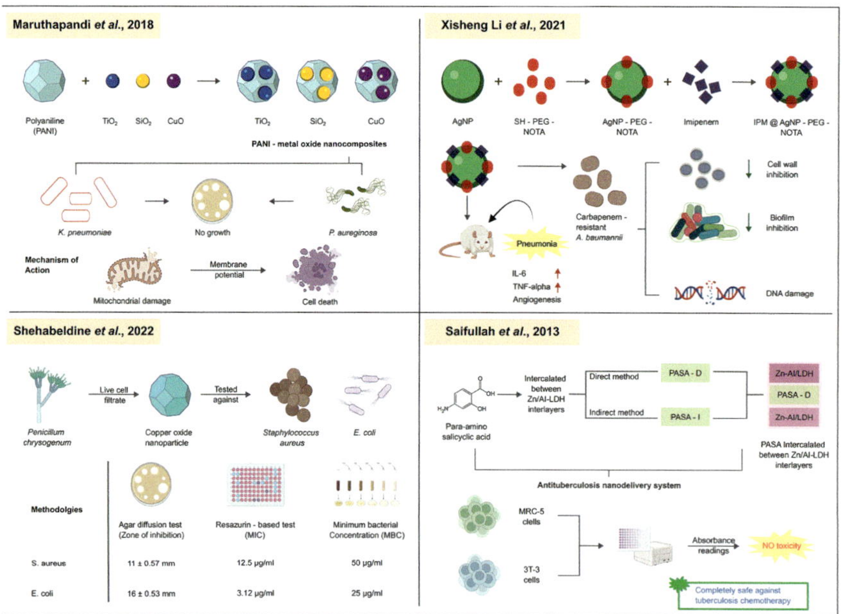

FIGURE 13.2 Schematic representations of experimental investigations demonstrating the importance of nanocomposites in increasing antibacterial activity.

Source: Prepared using BioRender.com.

13.2.3 LEISHMANIASIS

Leishmaniasis, a tropically neglected disease, ranks second to malaria in the cause of parasitic mortality and morbidity (Kumari et al., 2022; Singh et al., 2016). This zoonotic disease is prevalent in underdeveloped and developing countries, endemic to geographically distinct areas of Asia, Europe, the Middle East, Northern Africa, South-eastern Mexico, and other parts of South America (Kumari et al., 2021; Kumari and Singh, 2022). Unicellular protists belonging to the genus *Leishmania,* grouped in the family Trypanasomatidae and order Kinetoplastida, are the causative agents of this debilitating disease. First-line drugs for the treatment of leishmaniasis include pentavalent antimonials, followed by amphotericin B, miltefosine, and paromomycin as second-line drugs of choice (Gopu et al., 2023).

To overcome the existing therapeutic barriers against leishmaniasis, Bahraminejad et al. (2023) prepared layered double hydroxides (LDHs), so-called anionic clay nanocarriers using the coprecipitation method. They intercalated amphotericin B into the interlayer space of LDHs containing divalent cation Zn and trivalent Al to prepare Amp-Zn/Al-LDH NCs. Finally, the prepared Zn-Al LDHs were characterized, and their antileishmanial effects were evaluated against *Leishmania major* using *in vitro* and *in silico* models. Results of *in vitro* and *in silico* assays revealed that Amp-Zn/Al-LDH NCs had promising inhibitory effects on the growth of *L. major* parasites. A comparison of IC_{50} values concluded that clinical amastigotes were more sensitive to NCs than promastigotes. Intercalation of amphotericin B with NCs provided a higher potent effect against *Leishmania* parasites compared to the conventional form of the drug as it was found that positively charged Amp-Zn/Al-LDH NCs were taken up more quickly by macrophages *via* endocytosis compared to amphotericin B. Metallic and metallic oxide nanoparticles including ZnO, silver (Ag^{2+}), silver oxide (AgO), and Ag-doped ZnO (Ag/ZnO) are associated with a wide range of biological properties such as antimicrobial, antitumor, anti-inflammatory, antidiabetic, antioxidant, antiviral, and immuno-modulatory properties. Due to their immense biological properties, Barbosa et al. (2022) developed new NCs with ZnO and different percentages of AgO. Their findings suggested that AgO nanocrystals at 65 and 68% inhibited extra and intracellular replication of *L. braziliensis*. Ag-ZnO/65%AgO NCs presented a high selectivity index by modulating the activation and expression of surface receptors. They also led to the production of cytokine by human peripheral blood mononuclear cells toward a pro-inflammatory phenotype, pointing them as a promising option for *L. braziliensis* treatment. Encouraged by the diversified applications of carbon materials, Kumar et al. (2018) unraveled the potential of carbon dots (CDs) and gallium-doped carbon dots (Ga@CDs) against leishmaniasis. By ultrasonication, they formulated an ointment with CDs and Ga@CDs against *L. donovani.* Fluorescing dots made the ointment detectable in cells and tissues during cell imaging, drug targeting, and delivery applications *in vivo*. The prepared ointment was found to possess higher leishmanicidal activity with a minimal concentration of 30 µg/mL for CDs/Ga@CDs, compared with its commercial counterpart. *In vitro* and *in vivo* experiments performed with mice and healthy human blood cells confirmed their biocompatibility and nontoxic nature.

Alishahi et al. (2020) prepared a nanofibrous mat of polyethylene oxide, gelatin, poly (vinyl alcohol), and chitosan *via* an electrospinning technique. The main objective behind the synthesis was to prepare a topical drug delivery system that can efficiently deliver glucantime to the site of cutaneous *Leishmania* infection. Prepared drug-loaded nanofibers were screened for antileishmanial activity against *Leishmania* promastigotes and for cytotoxicity against fibroblasts by flow cytometry and indirect MTT methods. Experimental results indicated that fabricated nanofibers significantly killed the promastigotes up to 78% and also did not possess any cytotoxicity toward fibroblast cells. Drug-loaded nanofibers exhibited antileishmanial activity and killed parasites in a dose-dependent manner. Apart from this, Kumar et al. (2016) developed miltefosine-encapsulated PLGA–PEG nanoparticles to deliver the drug into the infected macrophage cells for the treatment of visceral leishmaniasis. Nano-encapsulated miltefosine was found to significantly inhibit amastigotes in the splenic tissue as compared to conventional miltefosine and amphotericin B. Improved efficiency of nano-composed drugs both *in vitro* and *in vivo* study was due to their increased contact surface area and a significant reduction in size as compared to conventional ones. Gelvez et al. (2021) conducted a specific analysis to show that AgNP-PVP-meglumine antimoniate NCs reduce the *L. amazonensis* infection in macrophages *in vitro.*

13.2.4 MALARIA

Malaria is a severe disease caused by a protozoan parasite belonging to the genus *Plasmodium*. The disease is transmitted by the bite of the female

Anopheles mosquito and remains the leading cause of mortality around the world. More than 120 subspecies of *Plasmodium* exist, of which only five cause malarial infections in humans (Talapko et al., 2019). It infects around half a billion people worldwide and is endemic in 86 tropical and subtropical countries, encompassing all of Africa as well as large areas of Southeast Asia, the Eastern Mediterranean, the Western Pacific, and the Americas. The first- and second-line treatments recommended by the World Health Organization (WHO) include single drugs such as chloroquine, lumefantrine, primaquine, mefloquine, or in combination with artemisinin or its derivatives (Chaves et al., 2022).

Abinaya et al. (2018) synthesized novel zinc oxide NCs (ZnO-NCs) using bacterial exopolysaccharides (EPS) obtained from *Bacillus licheniformis*. This nanoparticle synthesized *via* the coprecipitation method was then tested for its antibacterial and larvicidal toxicity against malaria and Zika virus vectors. A 100 μg/mL dosage of NPs considerably obstructed the growth of gram-negative as well as gram-positive bacteria, whereas a concentration of 75 μg/mL was found to inhibit the biofilm growth of *Candida albicans*. Moreover, at very low doses, these EPS-ZnO NPs led to complete killing of *Anopheles stephensi* and *Aedes aegypti* mosquito's larvae. In addition, the histological studies displayed the significant disruption of cells and tissues in the mid-gut of treated larvae. Impressed by the immense biological properties of collagen/polymer-based biomaterials, Murugan et al. (2017) synthesized chitosan-silver nanocomplex (Ch-AgNPs) along with its modified approach (Figure 13.3). They combined chitosan with a noble metal Ag to form the bio-encapsulated chitosan-Ag nanocomplex (BNC). These (BNC) along with classic (Ch-AgNPs) were then evaluated for their potential against parasites and vectors. Along with this the toxicity of larvae and pupae of *A. stephensi*, and cytotoxicity against breast cancer cells (MCF-7) were also tested. From the results obtained, BNC was found to be effective against *A. stephensi* both in the laboratory and in the field as compared to chitosan alone. BNC also reduced the longevity and fecundity of *A. Stephensi*, making it a promising candidate to initiate the Bax/Bcl-2/cytochrome c/caspase-3 signaling pathway for enhancing the apoptosis of MCF-7 cells.

Solairaj and Rameshthangam (2017) synthesized chitin nanoparticles (CNP), AgNP, and a-chitin/silver NCs (CNP/AgNCs) and evaluated their antibacterial, antifungal, and mosquito larvicidal potential. Biological assays revealed that the CNP/AgNCs have an improved antimicrobial effect on the bacterial growth (*K. Pneumonia, P. vulgaris*, and *S. aureus*) as well as fungi (*T. viridae, C. albicans, A. alternate*, and *A. niger*). Larvicidal assays were carried out with all three nanoparticles (CNP, AgNP, and CNP/AgNCs) against the first, second, third, and fourth instar larvae of *A. Aegypti*. Results suggested that CNP/AgNCs showed the lowest LC_{50} and LC_{90} values and 100% mortality than CNP and AgNP. CNP exhibited weak larvicidal activity against all the instars, while AgNP showed LC_{50} at 18.23 ppm for the first instar. Another study performed by Tian et al. (2022) made use of the nanodelivery system to effectively target the brain to cure cerebral malaria, a life-threatening neurological complication caused by *Plasmodium falciparum*. In the present study, glucose transporter 1 (GLUT1) at the blood–brain barrier was used as a desirable target for the design of targeted materials and the preparation of brain-targeted liposomes. Through molecular dynamics simulation, it was found that the cholesterol-undecanoate-glucose was a preferred targeting conjugate beneficial to the deciphering of brain-targeted liposomes by GLUT1. They previously reported that artesunate (ATS) in combination with ligustrazine

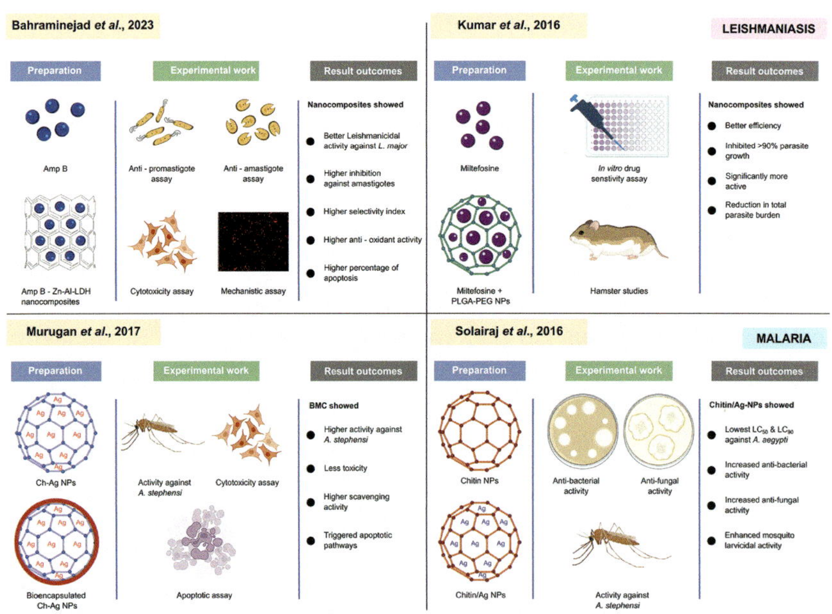

FIGURE 13.3 Schematic representations of experimental studies showing the significance of the nanocomposites in improving antileishmanial and antimalarial activities.

Source: Prepared using BioRender.com.

hydrochloride (TMP) improved the efficacy of anticerebral malaria in a mouse model. In this study, ATS and TMP were coencapsulated into liposomes to evaluate their anticerebral malaria efficacy and brain target ability. Brain-targeted liposomes (na-ATS/TMP@lipoBX) were found to significantly enhance the brain fluorescence intensity in mice as compared to nontargeted liposomes without glucose modification. *In vivo* pharmacokinetic analysis revealed that brain-targeted liposomes (na-ATS/TMP@ lipoBX) could pave their way to brain in two ways *viz* systemic circulation and nasal-brain pathways producing significant antimalarial effects with enhanced biosafety.

13.3 NANOCOMPOSITES FOR TISSUE ENGINEERING

In this ever-growing period of science and technology, extensive research has been conducted in the field of NCs by closely impersonating the biological milieu as well as a natural matrix. The characteristics of NCs are manipulated such that the functional necessities of every cell and tissue can be met. Thus, NCs-based tissue engineering has proven to be a breakthrough therapeutic technique for repairing damaged tissue or organs. Tissue engineering is the branch of science dealing with cure of people who lost their organs or restoring damaged tissues. In this regard, recent studies regarding the exploration of NCs in tissue engineering have been emphasized. Khan et al. (2022) developed the photo-assisted Arabinoxylan, the nanosystem (Zn@rGO), and nanohydroxyapatite polymeric NCs ARX-g-(Zn@rGO)/HAp using free radical polymerization and by the fabrication of porous bioactive scaffolds using the freeze–drying technique. It was observed that the higher the Zn content, the larger the pore size. Further, the hydrophobic nature changes to hydrophilic with the upsurge in contact time. Maximum swelling in deionized water and biodegradation in PBS were seen in AZG-4. Furthermore**,** the collaborative effect of (Zn@rGO)-HAp and polymeric contents displayed antibacterial against both gram-negative and gram-positive bacteria strains, whereas anticancer activity (U87 cell lines) was seen majorly in AZG-4. Thus, the study revealed that the bioactive scaffold AZG-4 would be a potential biomaterial to repair and regenerate defective bone tissue in medical applications.

In the era of medication, soft tissue engineering has been a potential wound-healing strategy to treat chronic wounds. In this regard, Nandhakumar et al. (2022) studied the *in vitro* effectivity of bioactive glass

(BG) Na-free, GO and BG/GO, and BG (Na-free)/GO NCs due to excellent biocompatibility and granulation capability. An *in vitro* experiment on wound-healing mechanism revealed that the BG (Na-free) and BG displayed improved wound healing in comparison to BG (Na-free)/GO, BG/GO, and GO which may be due to the nanosheet structure of GO leading to inhibition of cellular growth. The ROS measurement using DPPH further elucidated that 45S5 BG and GO materials significantly scavenge the free radical by 80% in BG (Na-free), 60% in BG, 65% in BG (Na-free)/GO, 80% in BG/GO, and 70% for GO at a dosage of 70 mg/mL. The results also showed that BG and BG (Na-free) composites are better for early stages of wound repairment; however, GO composites have considerable stability along with anti-inflammatory actions. Altogether, findings unveil the promising role of 45S5 BG and GO-NCs for curing chronic or diabetic wound healing (Figure 13.4). Similarly, Alam et al. (2023) aimed to produce hybrid bio-NCs by combining collagen (Col), *Nigella sativa* (Ns) oil, and chitosan (Cs) with varied ratios by a bilayer green electrospinning on polyvinyl chloride (PVA) layer. It was observed that water-soluble Col (above 60% solution) failed to produce fibers due to the surface tension of the solution. However, Col (50% solution) with equal amounts of Ns and Cs formed continuous fibers of a mean diameter of 384 ± 27 nm and 79% porosity. Additionally, fabricated bio-NCs $Col_{20}Ns_{10}Cs_{10}$ have demonstrated excellent absorbing capabilities along with slow-drying properties, imitating the dermal conditions and thus, allowing proper gaseous exchange. Furthermore, the synergistic effects of Ns and Cs have shown enhanced antibacterial activity, while native Col recovers tissue damage. In the *in vivo* rabbit model, approximately 84% wound closure was achieved in only 10 days, which is faster (1.5 times) than the control model indicating its tissue regeneration efficacy. So, as a whole, established hybrid bio-NCs displayed potential as a material for skin tissue engineering.

Tissue engineering is the multidisciplinary field of designing a beneficial product that utilizes the combination of matrix scaffolds with viable human cell systems, for the repair or regeneration of damaged tissues. One study by Bagheri and group (Valizadeh et al., 2022) aimed to develop a formulation of polyhedral oligomeric silsesquioxane-polycaprolactone (POSS-PCL) using the electrospinning method. The cytotoxicity analysis on mesenchymal stem cells (MSCs) in the presence of POSS-PCL nanofiber displayed a safe profile along with reducing cellular stress, better cell growth, angiogenesis, and bone differentiation concerning control (PCL alone). Further, the upregulation of RUNX2, Osteocalcin, and VEGF factor indicated bone development and angiogenesis, whereas no expression of BAX and Nrf2 gene implies that the POSS-PCL scaffold is nontoxic and does not cause any oxidative stress. Thus, the study concluded that POSS-PCL nanofiber scaffold can be potentially utilized in applications of tissue engineering.

In another interesting, by utilizing a computational approach, Almalki et al. (2023) evaluated multicomponent biomaterial Poly lactic-co-glycolic acid (PLGA) and Chitosan (CH) with Zirconium dioxide (ZrO_2) nanotube using density functional theory (DFT) approach. First, the interaction between the PLGA and CH monomers and ZrO_2 demonstrated that the binding affinity for PLGA and CH monomers on the surface of ZrO_2 is higher due to its shorter equilibrium interval. The molecular dynamics simulation studies discovered that on interaction with PLGA and CH on the ZrO_2 surface, there was 47 folds increase of their Young's modulus of PLGA-ZrO_2 and 27 folds in case CH-ZrO_2 which further determined that the mechanical properties of PLGA and CH are upgraded by the addition of ZrO_2 to the polymer matrix. Therefore, the study suggested that PLGA-ZrO_2 NCs could be a promising candidate in bone regeneration and other tissue engineering fields. A recent study by Canales et al. (2023) prepared the poly (lactic acid)-based calcium oxide nanoparticles PLA/n-CaO fibers (10 and 20 wt% incorporation) using an electrospinning technique. The result showed that the formed fibers have significant homogeneity as well as porosity along with increased diameter at 20 weight percentage. The fibers further presented a hydrophobic nature, decreased tensile strength, and improved rigidity. The PLA/n-CaO fibers have also displayed the capability of making a hydroxyapatite layer after 1-week post immersion in SBF solution. Furthermore, bactericidal activity PLA/n-CaO with 20 wt% demonstrated a 30 and 11% reduction in the viability of *S. aureus* and *E. coli*, respectively. Additionally, there was no cellular toxicity when tested against the bone marrow ST-2 cell line. However, the induced cellular adhesion, proliferation, differentiation, and upregulation of alkaline phosphatase level after a week of exposure to PLA/n-CaO in comparison with PLA was observed. Moreover, *in vivo* assessment affirms the biocompatibility of formed formulations. The significant promotions of connective tissue formation along with the reduced formation of the material's fibrous encapsulation are suggestive of the healing process. Altogether, the study highlighted that PLA/n-CaO NCs with their antimicrobial, bioactive, osteoblastic, and biocompatible features could be ideal therapeutics for bone regeneration in the future.

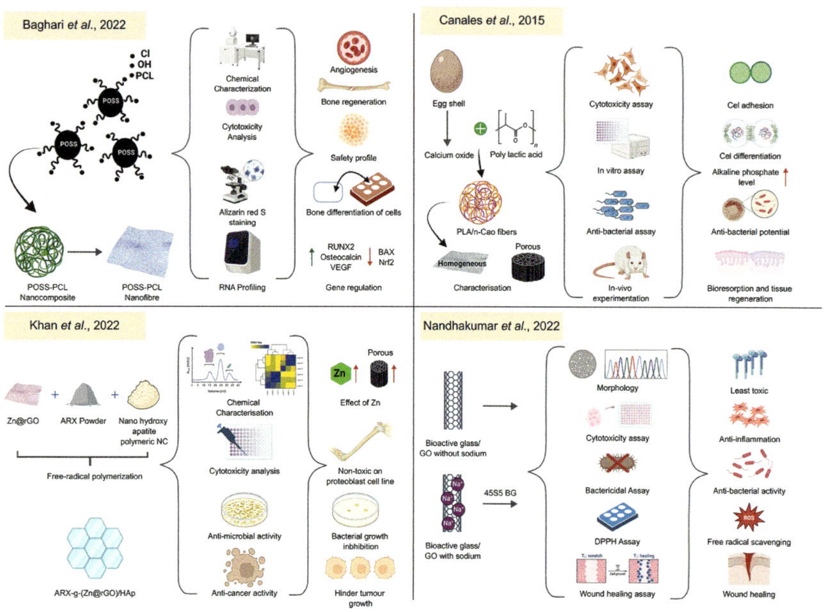

FIGURE 13.4 Pictorial representations of experimental results demonstrating the importance of nanocomposites in tissue engineering applications.

Source: Prepared using BioRender.com.

13.4 NANOCOMPOSITES FOR BIOSENSORS

Biosensor, a subset of chemical sensors, is a device-based technology that makes use of biomolecules of interest to measure molecules with biological significance. This advanced technology detects analytes with complex chemical structures, by binding soluble analyte targets with a complementary ligand (biological recognition element) immobilized on the sensor surface (Dakshayini et al., 2019). Conventionally, these highly sensitive, selective, and cost-effective analytical devices incorporate biological components with a physicochemical detector to quantitatively and qualitatively detect a wide range of analytes such as drugs, proteins, antibodies, enzymes, and nucleic acids. They have a very low limit of detection and can detect the concentration of biomolecules in nano and picomolar, making it an approach of paramount importance in the medicines and healthcare industry for estimating physiological as well as metabolic parameters (Khan et al., 2020).

Recent years have witnessed the exploitation of different nanomaterials for the design of highly efficient biosensors. Biosensors are doped with nanomaterials due to their multifunctional and novel properties and explored to reduce limits of detection along with high selectivity (Jiang and Wu, 2019). This synergistic combination of biosensors with nanotechnology has made sensing more effective and has produced sensors that are more stable and biocompatible. A wide range of nanomaterials have been used as an intermediate substrate material between ligand and electrode surfaces. These hybrid materials obtained on a special combination of two or more nano-dimensional entities include carbon nanotubes, graphene, metal oxides, chitosan conducting polymers, and so forth. Because of its outstanding properties, chitosan has been used as an excellent immobilization matrix for biological recognition elements. This natural polysaccharide is known to have a wide spectrum of features such as good mechanical strength, biocompatibility, availability, high permeability, cost effectiveness, and toxicity. Chitosan NCs-based biosensors have been used for various bio-detection applications. These chitosan biosensors have illustrated improved sensitivity, better selectivity, and stability for the detection of various targets such as glucose, proteins, DNAs, and small biomolecules to bacteria (Jiang and Wu, 2019).

Due to their unique conducting nature, conducting polymers have emerged as fascinating NCs materials. Polypyrrole, polythiophene, polyacetylene, PANI, polycarbazole, poly(*o*-methoxy aniline), and polyanisidine are different conducting polymers having applications in biosensors (Kaur et al., 2021). For better detection capability of biosensors, these conducting polymers are copolymerized with different nanomaterials (graphene, carbon, metals) and biopolymers (chitosan, chitin, gelatin, cellulose, and proteins), each component adding its property for combined enhancement of biosensing applications and medical devices (Naseri et al., 2018). Due to their ability to rationalize the size and shape on nanoscale at biosensing platforms, these conducting polymers are used by various research groups in tissue engineering, bio-electrodes, drug delivery, and biosensors to detect biological and synthetic moieties. Apart from this, polypyrrole is highly compatible with cell and biological tissues, whereas PANI has exhibited good biocompatibility with selective cell lines (Shoaie et al., 2019).

Unique physical properties including high specific surface area, carrier mobility, good electrical conductivity, flexibility, and optical transparency have spectacularly accelerated research on fabricated low-cost electrode material known as Graphene. Graphene along with its oxygenated derivatives (graphene oxide and reduced graphene oxide) have significantly contributed as nanomaterials in the field of biosensors. The presence of oxygenated functional groups has made these 2D nanosheets strongly hydrophilic, facilitating chemical functionalization (Ye and Lin, 2023). Graphene in combination with other inorganic nanoparticles has been greatly used to detect a wide variety of biological analytes such as glucose, H_2O_2, cholesterol, cancer biomarkers, food pathogens, heavy metal ions, and pesticides, with notable sensitivity, low detection limits, and enhanced stability (Krishnan et al., 2019). Apart from these, gold nanoparticles have gained interest as NCs material due to several intriguing properties like high surface-to-volume ratio, high surface energy, good electrical conductivity, high stability, and so forth. Their compact size makes them excellent platforms for the detection and identification of pathogens in native biological samples and for retaining the bioactivity of immobilized biomolecules (Lednický and Bonyár, 2020; Mobed et al., 2020). Metal oxides also play an enhancing role in immobilization and stability of enzymes. Sensing electrodes are coated with advanced bimetallic nanomaterials to greatly enhance the specificity, sensitivity, and biosensing capability of biosensors (Stephanie et al., 2021).

Making use of this NCs-based biosensor technology, Phukon et al. (2014) fabricated a biosensor for the analysis of artemisinin in bulk and spiked human serum. The biosensor was designed with natural polymer polyhydroxyalkanoate-gold nanoparticle mounted on an indium-tin oxide glass plate. Under diffused reaction conditions, the electro-catalytic reduction of artemisinin was monitored by using horse-radish peroxidase enzyme adsorbed on the electrode surface. The current study proposed a good analytical alternative for the detection of antimalarial drug artemisinin in human body fluids as compared to conventional chromatographic methods and further revealed an increase in differential pulse voltammetric peak current for artemisinin along with greater sensitivity of fabricated biosensor to artemisinin. Mobed et al. (2020) developed a NC-based biosensor for fast, rapid, and accurate detection of *Leishmania* in patients. They designed a novel ultra-sensitive DNA-based biosensor based on surface plasmon resonance immunosensor to detect C1 protein in *L. infantum*. They combined cysteamine with a poly (amidoamine) dendrimer (PAMAM(G4)) and made a layer-by-layer assembly on a gold surface to immobilize the protein. With this thiolated probe immobilized on the surface of AuNPs, the group was able to detect C1 protein in *L. infantum* with higher selectivity, sensitivity, and suitable linear range.

Khoder and Korri-Youssoufi (2020) highlighted the ability of nanowire polypyrrole (new-PPy), formed by electro polymerization with template-free, as a platform for amperometric detection of genomic DNA of *M. tuberculosis*. Properties of nw-PPy such as hydrophilic character, good conductivity, and large surface area improved the sensitivity of DNA sensing with a detection limit of 0.36 atomolar. This nanostructured polypyrrole served as a transducer and helped in the improvement of DNA immobilization and detection by providing chemical and electronic properties. Their study added to the fact that these new-PPy-sensitive E-DNA biosensors could be used as a platform for the detection of various biomolecules in complex samples involving specific bioreceptors. Singh et al. (2013) employed Graphene oxide (GO)-Chitosan (CHI) NCs for the fabrication of DNA-based electrochemical biosensors for rapid and sensitive diagnosis of typhoid. The biosensor was prepared by covalent immobilization of a *Salmonella typhi*-specific DNA probe to distinguish between complementary, noncomplementary, and one-base mismatch sequences. Prepared NCs biosensor showed detection limit of 10 fM in phosphate buffer and 100 fM in serum samples. It is also attributed to excellent electrochemical activity due to enhanced electron transfer by graphene oxide and good biocompatibility for DNA immobilization by chitosan.

13.5 NANOCOMPOSITES FOR DENTAL

In recent times, nanotechnology has momentously evolved from Richard Feynman's theory to practical aspects of science including dentistry. The nanomaterials, in combination with nanotechnology, led to exploration of NCs as nano-dental adducts in prosthodontics, orthodontics, regeneration, and so forth. Also, the drug-release capabilities of the nano-dental adduct offer an additional edge to nano-dentistry over their traditional counterparts. The dental practitioners aim for the prophylactic therapeutics of dental caries and periodontal disease along with developing exceedingly efficient bio-mimicking supportive NCs for dental restorative materials.

Thus, in the following section, we discuss the recent research publications retrieved from PubMed, Web of Science, and Google Scholar which elucidate the potential applications of NCs in the field of dentistry.

Hamouda et al. (2023) demonstrated the effectiveness of Ulva/Ag/cellulose NCs isolated from *Ulva lactuca,* a green alga against *Streptococcus mutans* ATCC 25175 and *Lactobacillus acidophilus* CH-2. Various characterization parameters were utilized to elucidate the physical and chemical features of Ulva/nano-cellulose and Ulva/Ag/cellulose NCs. Scanning electron microscopy (SEM) analysis shows that nano-cellulose contains filaments, whereas Ulva/Ag cellulose NCs are fibrous in nature. On the other side, transmission electron microscopy (TEM) results determined the diameter of 17–28 nm and 18.59–38.11 nm for Ulva/nano-cellulose and Ulva/Ag cellulose NCs, respectively. Further, the crystalline structure of both was validated by X-ray powder diffraction (XRD). The evaluation of the zeta potential of Ulva/Ag cellulose NCs and Ulva/nano-cellulose concluded that the former was more negative in charge than the latter. Also, the antibacterial effect of both against *S. mutans* ATCC 25175 and *L. acidophilus* CH-2 discovered the increased affectivity of Ulva/Ag/cellulose NCs in comparison to Ulva/nano-cellulose. Moreover, Ulva/Ag/cellulose NCs incorporated with fluoride displayed improved antibacterial activity than without fluoride. Thus, the study concludes the potential of Ulva/Ag/cellulose NCs as an economic and safe antibacterial agent to cure dental decay by blending in toothpaste and adding to tooth filler. Similarly, Hu et al. (2022) in their analysis showed that 2D MXene Ti3C2Tx nanosheets incorporated in epoxy resin displayed extravagant mechanical strength and roughness along with improved antibacterial potential. The study noted that the bactericidal activity was enhanced when illuminated in natural light in comparison to the dark, which may be due to the high photo-thermal efficiency possessed by MXene. Moreover, the cytotoxicity against normal oral cell lines indicated that MXene Ti3C2Tx nanosheets are a promising candidate to cure a wide spectrum of dental problems without causing significant damage to host cells. As a whole, MXene/resin nanosheets could be a novel state of the art with long-lasting antibacterial and high-performance capabilities to improve dental health and oral hygiene.

In another relevant study by Kazemi et al. (2023), the protein-repellent and antibacterial features of PEG-modified 3-methacryloxypropyltrimethoxy-silane silica nanoparticles (PMS) as potential dental resin were assessed. The experimental investigation showed that there was a 28%

reduction in the protein absorption capacity of PMS containing 20 wt% PMS nanoparticles when compared with the MPS-modified silica at the same weight percentage. Further, the increase in flexure strength and hardness was observed while analyzing the mechanical properties of PEG-modified silica NCs relative to normal PEG-modified silica. The study also reported that PMS NCs displayed 13-folds higher bactericidal potential against *Streptococcus mutans* ATCC 35668 than that of MPS alone along with significantly inhibiting the biofilm development. Thus, to sum up, the results clearly stated that PEG-modulated silica NCs have improved flexural strength, protein-repellent, and antibacterial properties that have potential applications in dental restoration. In one of the interesting studies, Kanak et al. (2023) evaluated poly lactic acid (PLA) nano-hydroxyapatite (n-HAp), that is, PLA-nHAP NCs as a potential antibacterial drug carrier in dentistry. The degradation experiment showed that PLA-nHAP degrades at a slower rate in deionized water as compared to phosphate buffer saline. Further, safety profile assessment on Vero cells and BHK-21 cell lines affirms that it was nontoxic as well as biocompatible. Furthermore, the examination of gentamicin-encapsulated NCs during *in vitro* studies claimed that the sustained drug release pattern was increased with the rise in pH such that at pH of 5.5, 80% of the drug was released, whereas at pH of 7.4, the sustained drug release reduced to 50%. The study altogether suggests that fabricated drug-loaded electrospun PLA-nHAp NCs with their ability of continuous drug release have greater applicability in the treatment of antibacterial infections in dental care and orthopedics.

Similarly, Taymour et al. (2022) evaluated polyether-ether-ketone/bioglass (PEEK/BG) and polyether-ether-ketone/forsterite (PEEK/FT) as nano-fillers for dental implants. An inclusive characterization was performed to assess the morphology, mechanical properties crystalline behavior, and bioactive features of these biomaterials. The SEM analysis of PEEK revealed the homologous surface having uniform distribution within the matrix without aggregation, whereas there was increased roughness in the case of NCs. Further, it was observed that 20 wt% forsterite was the most suitable NCs formulation compared with bioglass-based NCs that provide improved elasticity, strength, and micro-hardness. More importantly, PEEK NCs also exhibited the ability to precipitate calcium and phosphate bone minerals on their surface overall indicating their utilization as a potential alternative for metal implant material in dental and orthopedic applications. More recently, Maleki Dizaj et al. (2023) aimed to examine the role of antimicrobial gelatin-curcumin-coated NCs (GC-NCs) in thwarting peri-implantitis in dental implants and its stability in curing abutment. The experimentation reported that GC-NCs are nontoxic when assessed on dental pulp stem cells. Further, the drug dissolution study of curcumin release was conducted at intervals of 1, 30, and 60 days after coating inside the simulated body fluid solution. The results presented that there was substantial release of the drug till the 30th day and the healing persisted a month. However, more investigation is mandatory to decipher its physicochemical, mechanical, and antimicrobial facets. It is thought-provoking to note that NCs-based materials have also contributed to bone reconstruction and dental rehabilitation after trauma or tumor resection which have been a critical challenge for maxillofacial surgeons. In this context, an *in vivo* examination in a pig model was performed by Unnikrishnan et al. (2022) to demonstrate the bone formation potential of NCs fibrous scaffold (silica-nanohydroxyapatite-gelatin strengthened with poly L-lactic acid yarns-CSF) for delayed titanium implantation. The autograft extracted from the iliac crest was taken as a control for comparison. The study revealed that there was complete attachment and vascularized bone formation within half a year of the experiment. Also, the biological features, namely, bone maturation and remodeling as well as angiogenesis are similarly conserved as in that of autograft control. Moreover, CSF also displayed improved osseointegration properties in titanium dental implants in the later stages of development. Thus, it can be deduced that CSF-based NCs can be a way forward approach for dental rehabilitation and bone regeneration processes.

13.6 CONCLUSION

In today's modern era, nanotechnology has utilized the mechanics to control and manipulate the particle size as well as structure in the desired nano range for accomplishing the envisioned purposes. In previous times, the utilization of nanoscience was restricted to physico-chemical engineering, but with the evolution of time, various biological scientists recognized its endless benefits and thus explored it in their respective fields. NCs possess an excellent antimicrobial potency and therapeutic efficacy that greatly contributed to combating serious hospital-associated infections caused by multidrug-resistant bacterial strains. These miraculous materials have also been reported to enhance the leishmanicidal and larvicidal potential of conventional therapy along with improving their drug delivery system. Interestingly, NCs' ability to mimic the characteristics of natural tissues to a certain extent has also made them desirable for tissue engineering and a wide spectrum of dental applications. In tissue engineering applications, nanoscale topological characteristics influence cell adhesion, survival, proliferation, and differentiation. Various nanomaterials such as silver, gold, chitosan, and so forth are amalgamated to fabricate a scaffold that can retain the biological activities of nanomaterials and prevent particle aggregation. In dentistry, nanomaterials favored the usage of nano dental adducts in prosthodontics, regeneration, orthodontics, and so forth. NCs-based bio-detectors have completely revolutionized the field of biosensing. The conducting polymers are copolymerized with different nanomaterials to develop highly specific, selective, sensitive, and biocompatible biosensing devices. These exclusive features of NCs have made it stand out from other conventional materials. However, NCs come up with major drawbacks such as complex synthesis procedures, least cost affectivity, and scarce knowledge of toxicity caused in humans. Despite the above-mentioned hurdles, scientific communities are now focusing on creating less expensive methods to synthesize nanoparticle scaffolds that fit in the regulatory framework along with significantly reducing the associated toxicity concerns. Advances in manufacturing capabilities such as electrospinning, sheet deposition methods, and so forth have allowed the development of more novel NCs with greater capabilities to regulate both physio-chemical properties. All of these concerns need to be addressed before we see NCs as desirable contributors to the future of science and technology.

KEYWORDS

- **antimicrobial agents**
- **drug resistance**
- **ESKAPE pathogens**
- **leishmaniasis**
- **nanocomposites**
- **tuberculosis**

REFERENCES

Abinaya, M., Vaseeharan, B., Divya, M., Sharmili, A., Govindarajan, M., Alharbi, N. S., Kadaikunnan, S., Khaled, J. M., & Benelli, G. (2018). Bacterial exopolysaccharide (EPS)-coated ZnO nanoparticles showed high antibiofilm activity and larvicidal toxicity against malaria and Zika virus vectors. *Journal of Trace Elements in Medicine, 45*, 93–103.

Alam, M. R., Alimuzzaman, S., Shahid, M. A., Fahmida-E-Karim, & Hoque, M. E. (2023). Collagen/Nigella sativa/chitosan inscribed electrospun hybrid bio-nanocomposites for skin tissue engineering. *Journal of Biomaterials Science, Polymer Edition, 34*, 1517–1538.

Alishahi, M., Khorram, M., Asgari, Q., Davani, F., Goudarzi, F., Emami, A., Arastehfar, A., & Zomorodian, K. (2020). Glucantime-loaded electrospun core-shell nanofibers composed of poly (ethylene oxide)/gelatin-poly (vinyl alcohol)/chitosan as dressing for cutaneous leishmaniasis. *International Journal of Biological Macromolecules, 163*, 288-297.

Almalki, A. H., Belal, A., Farghali, A. A., Mahmoud, R., Mustafa, F., & Abd El-Mageed, H. (2023). Electronic, mechanical, and thermal properties of zirconium dioxide nanotube interacting with poly lactic-co-glycolic acid and chitosan as potential agents in bone tissue engineering: insights from computational approaches. *Journal of Biomolecular Structure and Dynamics, 41*, 1–13.

Bahraminejad, S., Pardakhty, A., Sharifi, I., Keyhani, A., Salarkia, E., & Ranjbar, M. (2023). Synthesis and physicochemical characterization of Zn–Al layered double hydroxides (LDHs) as a delivery system for amphotericin B: *In vitro* and *in silico* antileishmanial study. *Heliyon, 9*(4), e15308.

Baiju, J. (2020). Polymer nanocomposite-based electrochemical sensors and biosensors. In G. Morteza Sasani & D. Soumen (Eds.), *Nanorods and nanocomposites* (Ch. 9). IntechOpen.

Barbosa, R. M., Obata, M. M. S., Neto, J. R. dC., Guerra, R. O., Borges, A. V.B. E., Trevisan, R. O., Ruiz, L. C., Bernardi, Jd. M., Oliveira-Scussel, A. C. dM., & Tanaka, S. C. S. V. (2022). Development of Ag-ZnO/AgO Nanocomposites Effectives for Leishmania braziliensis Treatment. *Pharmaceutics, 14*(12), 2642.

Bramhill, J., Ross, S., & Ross, G. (2017). Bioactive nanocomposites for tissue repair and regeneration: A review. *International Journal of Environmental Research and Public Health, 14*(1), 66.

Canales, D., Moyano, D., Alvarez, F., Grande-Tovar, C. D., Valencia-Llano, C. H., Peponi, L., Boccaccini, A. R., & Zapata, P. A. (2023). Preparation and characterization of novel poly (lactic acid)/calcium oxide nanocomposites by electrospinning as a potential bone tissue scaffold. *Biomaterials Advances, 153*, 213578.

Chaves, J. B., de Moraes, B. P. T., Ferrarini, S. R., da Fonseca, F. N., Silva, A. R., & Gonçalves-de-Albuquerque, C. F. (2022). Potential of nanoformulations in malaria treatment. *Frontiers in Pharmacology, 13*, 999300.

Dakshayini, B., Reddy, K. R., Mishra, A., Shetti, N. P., Malode, S. J., Basu, S., Naveen, S., & Raghu, A. V. (2019). Role of conducting polymer and metal oxide-based hybrids for applications in ampereometric sensors and biosensors. *Microchemical Journal, 147*, 7–24.

Dutt, Y., Pandey, R. P., Dutt, M., Gupta, A., Vibhuti, A., Vidic, J., Raj, V. S., Chang, C.-M., & Priyadarshini, A. (2023). Therapeutic applications of nanobiotechnology. *Journal of Nanobiotechnology, 21*(1), 1–32.

Gélvez, A. P. C., Diniz Junior, J. A. P., Brígida, R. T. S. S., & Rodrigues, A. P. D. (2021). AgNP-PVP-meglumine antimoniate nanocomposite reduces Leishmania amazonensis infection in macrophages. *BMC Microbiology, 21*(1), 1–14.

Gopu, B., Kour, P., Pandian, R., & Singh, K. (2023). Insights into the drug screening approaches in leishmaniasis. *International Immunopharmacology, 114*, 109591.

Gupta, P., Kumar, D., Quraishi, M., & Parkash, O. (2016). Metal matrix nanocomposites and their application in corrosion control. In *Advances in nanomaterials* (pp. 231–246).

Hamouda, R. A., Qarabai, F. A., Shahabuddin, F. S., Al-Shaikh, T. M., & Makharita, R. R. (2023). Antibacterial activity of ulva/nanocellulose and Ulva/Ag/cellulose nanocomposites and both blended with fluoride against bacteria causing dental decay. *Polymers, 15*(4), 1047.

Hatae, A. C., Roque-Borda, C. A., & Pavan, F. R. (2023). Strategies for lipid-based nanocomposites with potential activity against Mycobacterium tuberculosis: microbial resistance challenge and drug delivery trends. *OpenNano, 13,* 100171.

Hu, Y., Xu, Z., Pu, J., Hu, L., Zi, Y., Wang, M., Feng, X., & Huang, W. (2022). 2D MXene Ti3C2Tx nanosheets in the development of a mechanically enhanced and efficient antibacterial dental resin composite. *Frontiers in Chemistry, 10*, 1090905.

Hulla, J., Sahu, S., & Hayes, A. (2015). Nanotechnology: History and future. *Human Experimental Toxicology, 34*(12), 1318–1321.

Hussain, F., Hojjati, M., Okamoto, M., & Gorga, R. E. (2006). Polymer-matrix nanocomposites, processing, manufacturing, and application: An overview. *Journal of Composite Materials, 40*(17), 1511–1575.

Jafari, A., Nagheli, A., Foumani, A. A., Soltani, B., & Goswami, R. (2020). The role of metallic nanoparticles in inhibition of Mycobacterium tuberculosis and enhances phagosome maturation into the infected macrophage. *Oman Medical Journal, 35*(6), e194.

Jiang, Y., & Wu, J. (2019). Recent development in chitosan nanocomposites for surface-based biosensor applications. *Electrophoresis, 40*(16-17), 2084–2097.

Kanak, N. A., Shahruzzaman, M., Islam, M. S., Takafuji, M., Rahman, M. M., & Kabir, S. F. (2023). Fabrication of electrospun PLA-nHAp nanocomposite for sustained drug release in dental and orthopedic applications. *Materials, 16*(10), 3691.

Kaur, G., Kaur, A., & Kaur, H. (2021). Review on nanomaterials/conducting polymer based nanocomposites for the development of biosensors and electrochemical sensors. *Polymer-Plastics Technology Materials Science Engineering C, 60*(5), 504–521.

Kazemi, M., Navarchian, A. H., & Ahangaran, F. (2023). Effects of silica surface modification with silane and poly (ethylene glycol) on flexural strength, protein-repellent, and antibacterial properties of acrylic dental nanocomposites. *Dental Materials, 39*(10), 863–871.

Khan, M. U. A., Al-Arjan, W. S., Ashammakhi, N., Haider, S., Amin, R., & Hasan, A. (2022). Multifunctional bioactive scaffolds from ARX-g-(Zn@ rGO)-HAp for bone tissue engineering: *In vitro* antibacterial, antitumor, and biocompatibility evaluations. *ACS Applied Bio Materials, 5*(11), 5445–5456.

Khan, M., Hasan, M., Hossain, S., Ahommed, M., & Daizy, M. (2020). Ultrasensitive detection of pathogenic viruses with electrochemical biosensor: State of the art. *Biosensors and Bioelectronics 166*, 112431.

Khoder, R., & Korri-Youssoufi, H. (2020). E-DNA biosensors of M. tuberculosis based on nanostructured polypyrrole. *Materials Science Engineering C, 108*, 110371.

Krishnan, S. K., Singh, E., Singh, P., Meyyappan, M., & Nalwa, H. S. (2019). A review on graphene-based nanocomposites for electrochemical and fluorescent biosensors. *RSC Advances, 9*(16), 8778–8881.

Kumar, V. B., Dolitzky, A., Michaeli, S., & Gedanken, A. (2018). Antiparasitic ointment based on a biocompatibile carbon dot nanocomposite. *Applied Nano Materials, 1*(4), 1784–1791.

Kumar, R., Sahoo, G. C., Pandey, K., Das, V., Topno, R. K., Ansari, M. Y., Rana, S., & Das, P. (2016). Development of PLGA–PEG encapsulated miltefosine based drug delivery system against visceral leishmaniasis. *Materials Science Engineering C, 59*, 748–753.

Kumari, D., Mahajan, S., Kour, P., & Singh, K. (2022). Virulence factors of Leishmania parasite: Their paramount importance in unraveling novel vaccine candidates and therapeutic targets. *Life Sciences, 306*, 120829.

Kumari, D., Perveen, S., Sharma, R., & Singh, K. (2021). Advancement in leishmaniasis diagnosis and therapeutics: An update. *European Journal of Pharmacology, 910*, 174436.

Kumari, D., & Singh, K. (2022). Exploring the paradox of defense between host and Leishmania parasite. *International Immunopharmacology, 102*, 108400.

Lednický, T., & Bonyár, A. (2020). Large scale fabrication of ordered gold nanoparticle–epoxy surface nanocomposites and their application as label-free plasmonic DNA biosensors. *ACS Applied Materials & Interfaces, 12*(4), 4804–4814.

Li, X., Gui, R., Li, J., Huang, R., Shang, Y., Zhao, Q., Liu, H., Jiang, H., Shang, X., & Wu, X. (2021). Novel multifunctional silver nanocomposite serves as a resistance-reversal agent to synergistically combat carbapenem-resistant Acinetobacter baumannii. *ACS Applied Materials Interfaces, 13*(26), 30434–30457.

Li, C., Sun, W., Lu, Z., Ao, X., & Li, S. (2020). Ceramic nanocomposite membranes and membrane fouling: A review. *Water Research, 175*, 115674.

Luthfiah, A., Permana, M. D., Deawati, Y., Firdaus, M. L., Rahayu, I., & Eddy, D. R. (2021). Photocatalysis of nanocomposite titania–natural silica as antibacterial against Staphylococcus aureus and Pseudomonas aeruginosa. *RSC Advances, 11*(61), 38528–38536.

Maleki Dizaj, S., Torab, A., Kouhkani, S., Sharifi, S., Negahdari, R., Bohlouli, S., Fattahi, S., & Salatin, S. (2023). Gelatin–Curcumin nanocomposites as a coating for implant healing abutment: *In vitro* stability investigation. *Clinics and Practice, 13*(1), 88–101.

Malik, S., Muhammad, K., & Waheed, Y. (2023). Nanotechnology: A revolution in modern industry. *Molecules, 28*(2), 661.

Mancuso, G., Midiri, A., Gerace, E., & Biondo, C. (2021). Bacterial antibiotic resistance: The most critical pathogens. *Pathogens, 10*(10), 1310.

Martin, R. M., & Bachman, M. A. (2018). Colonization, infection, and the accessory genome of *Klebsiella pneumoniae*. *Frontiers in Cellular and Infection Microbiology, 8*, 4.

Maruthapandi, M., Saravanan, A., Luong, J. H., & Gedanken, A. (2020). Antimicrobial properties of the polyaniline composites against Pseudomonas aeruginosa and Klebsiella pneumoniae. *Journal of Functional Biomaterials, 11*(3), 59.

Matharu, R. K., Ciric, L., & Edirisinghe, M. (2018). Nanocomposites: Suitable alternatives as antimicrobial agents. *Nanotechnology, 29*(28), 282001.

Mehnath, S., Sithika, M. A. A., Arjama, M., Rajan, M., Praphakar, R. A., & Jeyaraj, M. (2019). Sericin-chitosan doped maleate gellan gum nanocomposites for effective cell damage in Mycobacterium tuberculosis. *International Journal of Biological Macromolecules, 122*, 174–184.

Mobed, A., Mehri, P., Hasanzadeh, M., & Mokhtarzadeh, A. (2020). Binding of Leishmania spp with gold nanoparticles supported polyethylene glycol and its application for the sensitive detection of infectious photogenes in human plasma samples: A novel biosensor. *Journal of Molecular Recognition, 33*(7), e2839.

Mulani, M. S., Kamble, E. E., Kumkar, S. N., Tawre, M. S., & Pardesi, K. R. (2019). Emerging strategies to combat ESKAPE pathogens in the era of antimicrobial resistance: a review. *Frontiers in Microbiology, 10*, 539.

Murugaiyan, J., Kumar, P. A., Rao, G. S., Iskandar, K., Hawser, S., Hays, J. P., Mohsen, Y., Adukkadukkam, S., Awuah, W. A., & Jose, R. A. M. (2022). Progress in alternative strategies to combat antimicrobial resistance: Focus on antibiotics. *Antibiotics, 11*(2), 200.

Murugan, K., Jaganathan, A., Suresh, U., Rajaganesh, R., Jayasanthini, S., Higuchi, A., Kumar, S., & Benelli, G. (2017). Towards bio-encapsulation of chitosan-silver nanocomplex? Impact on malaria mosquito vectors, human breast adenocarcinoma cells (MCF-7) and behavioral traits of non-target fishes. *Journal of Cluster Science, 28*, 529–550.

Nandhakumar, M., Thangaian, D. T., Sundaram, S., Roy, A., & Subramanian, B. (2022). An enduring *in vitro* wound healing phase recipient by bioactive glass-graphene oxide nanocomposites. *Scientific Reports, 12*(1), 16162.

Naseri, M., Fotouhi, L., & Ehsani, A. (2018). Recent progress in the development of conducting polymer-based nanocomposites for electrochemical biosensors applications: A mini-review. *The Chemical Record, 18*(6), 599–618.

Natarajan, K. (2015). Antibiofilm activity of epoxy/Ag-TiO2 polymer nanocomposite coatings against Staphylococcus aureus and Escherichia coli. *Coatings, 5*(2), 95–114.

Okpala, C. C. (2013). Nanocomposites–An overview. *International Journal of Engineering Research and Development, 8*(11), 17–23.

Omanović-Mikličanin, E., Badnjević, A., Kazlagić, A., & Hajlovac, M. (2020). Nanocomposites: A brief review. *Health and Technology, 10*, 51–59.

Perveen, S., Kumari, D., Singh, K., & Sharma R. (2022). Tuberculosis drug discovery: Progression and future interventions in the wake of emerging resistance. *European Journal of Medicinal Chemistry, 229*, 114066.

Phukon, P., Radhapyari, K., Konwar, B. K., & Khan, R. (2014). Natural polyhydroxyalkanoate–gold nanocomposite based biosensor for detection of antimalarial drug artemisinin. *Materials Science Engineering C, 37*, 314–320.

Punjabi, K., Mehta, S., Chavan, R., Chitalia, V., Deogharkar, D., & Deshpande, S. (2018). Efficiency of biosynthesized silver and zinc nanoparticles against multi-drug resistant pathogens. *Frontiers in Microbiology, 9*, 2207.

Ramos-Martín, F., & D'Amelio, N. (2023). Drug resistance: An incessant fight against evolutionary strategies of survival. *Microbiology Research, 14*(2), 507–542.

Saifullah, B., Hussein, M. Z., Hussein-Al-Ali, S. H., Arulselvan, P., & Fakurazi, S. (2013). Antituberculosis nanodelivery system with controlled-release properties based on para-amino salicylate–zinc aluminum-layered double-hydroxide nanocomposites. *Drug Design, Development and Therapy, 7*, 1365–1375.

Shehabeldine, A. M., Amin, B. H., Hagras, F. A., Ramadan, A. A., Kamel, M. R., Ahmed, M. A., Atia, K. H., & Salem, S. S. (2023). Potential antimicrobial and antibiofilm properties of copper oxide nanoparticles: Time-kill kinetic essay and ultrastructure of pathogenic bacterial cells. *Applied Biochemistry & Biotechnology, 195*(1), 467–485.

Shinu, P., Mouslem, A. K. A., Nair, A. B., Venugopala, K. N., Attimarad, M., Singh, V. A., Nagaraja, S., Alotaibi, G., & Deb, P. K. (2022). Progress report: Antimicrobial drug discovery in the resistance era. *Pharmaceuticals, 15*(4), 413.

Shoaie, N., Daneshpour, M., Azimzadeh, M., Mahshid, S., Khoshfetrat, S. M., Jahanpeyma, F., Gholaminejad, A., Omidfar, K., & Foruzandeh, M. (2019). Electrochemical sensors and biosensors based on the use of polyaniline and its nanocomposites: A review on recent advances. *Microchimica Acta, 186*, 1–29.

Silva Santos, K., Barbosa, A. M., da Costa, L. P., Pinheiro, M. S., Oliveira, M. B. P. P., & Padilha, F. F. (2016). Silver nanocomposite biosynthesis: Antibacterial activity against multidrug-resistant strains of Pseudomonas aeruginosa and Acinetobacter baumannii. *Molecules, 21*(9), 1255.

Singh, K., Ahlawat, S., Kumari, D., Matlani, U., Singh, M., Kaur, T., & Rao, A. (2023). Safety of Nanoparticles: Emphasis on Antimicrobial Properties. In *Biomedical Applications Toxicity of Nanomaterials* (Vol. 1, pp. 425–458).

Singh, K., Garg, G., & Ali, V. (2016). Current therapeutics, their problems and thiol metabolism as potential drug targets in leishmaniasis. *Current Drug Metabolism, 17*(9), 897–919.

Singh, A., Sinsinbar, G., Choudhary, M., Kumar, V., Pasricha, R., Verma, H., Singh, S. P., & Arora, K. (2013). Graphene oxide-chitosan nanocomposite based electrochemical DNA biosensor for detection of typhoid. *Sensors and Actuators B: Chemical, 185*, 675–684.

Solairaj, D., & Rameshthangam, P. (2017). Silver nanoparticle embedded α-chitin nanocomposite for enhanced antimicrobial and mosquito larvicidal activity. *Journal of Polymers the Environment, 25*, 435–452.

Sreenivasalu, P. K. P., Dora, C. P., Swami, R., Jasthi, V. C., Shiroorkar, P. N., Nagaraja, S., Asdaq, S. M. B., & Anwer, M. K. (2022). Nanomaterials in dentistry: Current applications and future scope. *Nanomaterials, 12*(10), 1676.

Stephanie, R., Kim, M. W., Kim, S. H., Kim, J.-K., Park, C. Y., & Park, T. J. (2021). Recent advances of bimetallic nanomaterials and its nanocomposites for biosensing applications. *Trends in Analytical Chemistry, 135*, 116159.

Stojkovska, J., Petrovic, P., Jancic, I., Milenkovic, M. T., & Obradovic, B. (2019). Novel nano-composite hydrogels with honey effective against multi-resistant clinical strains of Acinetobacter baumannii and Pseudomonas aeruginosa. *Applied Microbiology & Biotechnology, 103*, 8529–8543.

Talapko, J., Škrlec, I., Alebić, T., Jukić, M., & Včev, A. (2019). Malaria: The past and the present. *Microorganisms, 7*(6), 179.

Taymour, N., Fahmy, A. E., Gepreel, M. A. H., Kandil, S., & El-Fattah, A. A. (2022). Improved mechanical properties and bioactivity of silicate based bioceramics reinforced poly (ether-ether-ketone) nanocomposites for prosthetic dental implantology. *Polymers, 14*(8), 1632.

Tian, Y., Zheng, Z., Wang, X., Liu, S., Gu, L., Mu, J., Zheng, X., Li, Y., & Shen, S. (2022). Establishment and evaluation of glucose-modified nanocomposite liposomes for the treatment of cerebral malaria. *Journal of Nanobiotechnology, 20*(1), 1–22.

Unnikrishnan, P., Iyer, S., Manju, V., Reshmi, C., Menon, D., Nair, S. V., & Nair, M. (2022). Nanocomposite fibrous scaffold mediated mandible reconstruction and dental rehabilitation: An experimental study in pig model. *Biomaterials Advances, 133*, 112631.

Valizadeh, H., Dindar-Safa, K., & Zarghami, N. (2022). Fabricating a robust POSS-PCL nanofiber scaffold for nesting of mesenchymal stem cells: Potential application in bone tissue regeneration. *Journal of Biological Engineering, 16*(1), 1–14.

Ye, C., & Lin, C.-T. (2023). Graphene-based nanocomposites for biosensors. *Microchimica Acta, 190*(5), 183.

Yetisgin, A. A., Cetinel, S., Zuvin, M., Kosar, A., & Kutlu, O. (2020). Therapeutic nanoparticles and their targeted delivery applications. *Molecules, 25*(9), 2193.

CHAPTER 14

Future Aspects and Trends of Nanocomposites in Various Biomedical Applications

PRIYA JHA[1], VIVEK DAVE[1,*], VAIBHAV VERMA[1], and DEEPENDRA SINGH[2]

[1] *Department of Pharmacy, School of Health Sciences, Central University of South Bihar, Gaya, Bihar, India*

[2] *University Institute of Pharmacy, Pt. Ravishankar Shukla University, Raipur, Chhattisgarh, India*

[*] Corresponding author.

ABSTRACT

This chapter explores the growing field of nanocomposites in biomedical applications, combining nanotechnology and materials science to create new possibilities. It underscores their pivotal role in transforming the delivery of medication, enabling precise and controlled drug release, fostering the adoption of combination therapies, and surmounting biological barriers. Furthermore, nanocomposites significantly alter the administration of vaccines. In addition to medication delivery, they hold great significance in tissue engineering, providing customized solutions for the repair and regeneration of various tissues, such as cardiac, bone, skin, and neurological tissues. Wound healing is a significant area of focus, and various nanocomposites show promise in improving wound treatment effectiveness and speeding up the healing process. The incorporation of nanocomposites into biosensing technologies marks the beginning of a new era characterized by extremely

Nanocomposites and Nanomaterials in Biomedical Applications. Vivek Dave, Deependra Singh, and Pallerla Naveen Reddy (Eds.)

DOI: 10.1201/9781779641236-14

sensitive and specific detection systems, which have great potential for advancements in healthcare diagnostics and monitoring. Nevertheless, it is crucial to successfully navigate through regulatory obstacles in order to guarantee the ethical and secure incorporation of nanocomposites in biomedical applications. This involves taking into account many issues, such as product compatibility, environmental effects, and long-term consequences. Nanocomposites in biomedicine have a promising future, driven by continuous research, interdisciplinary collaborations, and breakthroughs in manufacturing techniques. We expect these developments to revolutionize healthcare and enhance patient outcomes.

14.1 INTRODUCTIONS TO NANOCOMPOSITE

A composite is a substance composed of many components. Nanocomposite materials, abbreviated as NCs, are composed of a matrix that provides support for one or more fillers. These filler materials can take the form of particles, sheets, fibers, and so forth, as shown in Figure 14.1 and have dimensions less than 100 nm. Additionally, they have an elevated surface-to-volume ratio. Nanocomposites exhibit enhanced characteristics, including decreased mass, heightened stiffness, more significant longevity, and increased resistance to abrasion. Composite materials with these properties exhibit enhanced recyclability, superior flame retardancy, reduced porosity, and improved electrical conductivity compared to conventional composite materials. The use of nanofillers, such as carbon nanotubes, in composites enhances their durability and resistance to damage and fracture while preserving their surface quality and transparency. Enhanced characteristics, such as reduced ability to allow gasses, water, and hydrocarbons to pass through; improved thermal endurance and resistance to fire, as well as decreased smoke production; and enhanced chemical resistance, surface aesthetics, electrical insulation, and optical properties, are important factors to consider in nanocomposites that contain additives in addition to filler components. Nonporous nanocomposites are highly durable, strong, and recyclable, which is why they are employed in vacuum packs, food and drink packaging, and medical equipment protection (Singh et al., 2021).

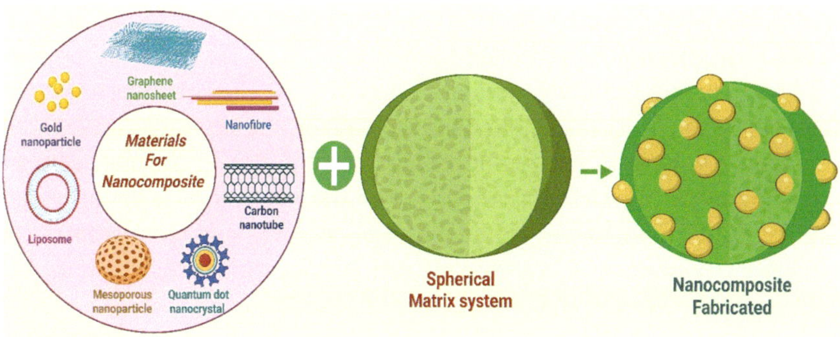

FIGURE 14.1 Fabrication of nanocomposite through embedding nanomaterials into a matrix system.

The emergence of nanoparticles (NPs) in society was driven by increasing socioeconomic demand, innovations in science and technology, and the implementation of sophisticated characterization methods. Nanotechnology has piqued the attention of all scientific sectors due to the multifunctionality, greater performance, and extraordinary sustainability qualities of NPs. The advancement of technology in characterizing and conducting scientific research has resulted in the creation of easily manipulated, versatile engineered materials referred to as nanocomposites (Gupta et al., 2023). Nanocomposites' special qualities and adaptability make them essential for use in biological applications. With its ability to precisely manage medication administration, medicinal substances may be released in a focused and sustained manner, reducing adverse effects and enhancing treatment effectiveness. In addition, nanocomposites can function as bioactive scaffolds for tissue technology to promote adhesion of cells, cell proliferation, and differentiation, consequently encouraging regenerative medicine procedures. They also make sensitive and targeted bioimaging modalities possible, which helps with early illness identification and monitoring. Additionally, nanocomposites are promising for biosensing applications because they provide quick and precise diagnostic tests. All things considered, nanocomposites customized qualities make them indispensable for developing individualized medication, boosting patient care, and achieving better overall healthcare results (Alghamdi, 2024).

Due to their unique qualities and versatility, nanocomposites find extensive and ongoing use in biomedical applications. Nanocomposites are playing a critical role in various important areas of biomedicine, particularly in systems to deliver medications. By offering precise control over the rate at which medications are released, nanocomposites facilitate the precise delivery of medications to particular tissues or cells. They have the ability to encapsulate medications, shield them from deterioration, and release them gradually, increasing therapeutic efficacy and lowering adverse effects. Nanocomposites function as scaffolds for tissue regeneration in tissue design and the field of regenerative medicine by providing a biocompatible and bioactive environment that promotes cell development, proliferation, and differentiation. We can create them to stimulate tissue regeneration within various organs and tissues, mimicking the extracellular matrix. Diagnostic imaging nanocomposites serve as contrast elements in several types of diagnostic imaging modalities. They improve imaging resolution, making it easier to accurately localize and identify diseases like cancer in a timely manner. These biosensors and bioimaging nanocomposites are important for making pathogen and biomolecule sensors that are very sensitive and selective. Advanced bioimaging approaches allow molecular analysis of biological processes. Implantable Tech Stents, orthopedic implants, and prostheses utilize nanocomposites. They are more biocompatible and operate better over time because they have mechanical properties like actual tissues, reduce rejection, and promote tissue integration (Zhang et al., 2015). Theranostics: Nanocomposites combine medicinal and diagnostic qualities, enabling personalized therapy. They simplify real-time treatment response assessment and patient-specific therapy modification. Nanocomposite dressings and scaffolds promote cell migration, proliferation, and tissue regeneration to treat wounds. Additional antimicrobials help prevent infections and speed healing. Nanocomposites minimize systemic toxicity and enable tumor-targeted delivery of anticancer drugs. Functionalizing them with targeted ligands improves tumor medication accumulation and cancer cell attachment (Bardania et al., 2023).

14.2 APPLICATIONS OF NANOCOMPOSITE IN DRUG DELIVERY SYSTEMS

14.2.1 CONTROLLED AND TARGETED DRUG RELEASE

We can design nanocomposites to systematically release pharmaceuticals through controlled methods like diffusion, degradation, or stimuli-responsive mechanisms. The implementation of this regulated release guarantees the most favorable medication concentrations at the intended location, minimizing adverse effects and enhancing the treatments effectiveness. By functionalizing nanocomposites with targeting ligands, it becomes possible to deliver medications specifically to desired tissues or cells (Hassan et al., 2021); some examples of trending nanocomposites are shown in Table 14.1. This targeted delivery approach reduces the amount of medicine that enters the bloodstream and enhances the absorption of the drug at the desired site of effect. As a result, this leads to better therapeutic outcomes and fewer adverse effects in nontargeted areas of the body.

A smart MXene@Hydrogel nanocomposite enables controlled drug release *via* NIR responsiveness. Agarose and MXene concentrations, irradiation intensity, and exposure time modulate drug release kinetics. The system achieves reversible softening and gelling, functioning as an integer control mechanism for accurate medication delivery and guaranteeing optimal blood levels for clinical therapy by near-infrared (NIR) induced cancer cell apoptosis (Dong et al., 2021). According to Gao et al. (2016), nanocomposites made of graphene oxide (GO) are very good at mixing with water and can hold a lot of drugs, which makes them perfect for drug delivery. GO was replaced with carboxymethyl chitosan, fluoroscein isothiocyanate, and lactobionic acid to make a system for delivering drugs that targets specific areas. The mixture took in the cancer-fighting drug doxorubicin, worked well with liver cancer cell lines, and killed the cells within 24 h of being mixed (Jayalekshmi et al., 2018). Under ultrasound and pH conditions, Cur-loaded ZnFe2O4@mZnO and ZnFe2O4@mZnO-Ox-p-g-C3N4 nanocarriers achieve cumulative drug release rates of 70.28 and 56.05% at pH 5.5 and 66.24 and 45.40% at pH 7.4, respectively, indicating their efficacy in dual-targeted delivery to cancerous tissues (Rostami et al., 2021). The GO-Ca-Alg2/CPT nanocomposite effectively delivers 52.39% of the drug camptothecin (CPT) through controlled drug release. Moreover, it demonstrated potent anticancer activity against HeLa cancer cells, indicating its targeted efficacy. Concentrations exceeding 62.5 μg/mL exhibited significant cytotoxicity against cancer cells, highlighting its potential as an oncological therapy that utilizes a precise drug delivery mechanism. One of the research objectives was to explore the use of polyamide-disulfide NPs as carriers for pharmaceutical administration methods. Amlodipine was selected as a representative drug, leading to the development of amlodipine-polymer nanocomposites. The findings indicated that these nanocomposites exhibited a particle size of around

130 nm on average. Furthermore, the amlodipine-polymer composite demonstrated a regulated release of amlodipine, releasing approximately 85% of the medication within a 24-h timeframe (Ahmed et al., 2022). It was discovered that the iron glycerolates layer of MNP@Fe(III)Glyc remains stable, while the silica glycerolates layer of MNP@Fe(III)/SiGlyc undergoes substantial hydrolysis, reaching 76.4% within 24 h at 25°C. Importantly, these artificially created materials exhibited no harmful effects on living cells *in vitro*, according to MTT-assay experiments. This research suggests potential applications in designing magnetic nanocomposites for controlled-release medicine delivery systems (Khonina et al., 2023).

14.2.2 COMBINATION THERAPY

Nanocomposites enable the simultaneous delivery of multiple medicines or therapeutic agents. This allows combination therapy to have synergistic effects, overcome drug resistance, or target different pathways involved in disease development. A nano nanocomposite made up of selenium, ampicillin, and the β-lactamase inhibitor 3-carboxyphenylboronic acid (Pba) lets the drugs be released in a controlled and targAt a concentration of 30 μg/mL, this nanocomposite effectively blocks the activity of β-lactamase, which makes ampicillin more effective against *E. coli* (R) bacteria that are resistant to ampicil. Wang et al. (2021) achieved this by suppressing the AcrAB-TolC efflux pump system and removing reactive oxygen species (ROS), both in laboratory conditions (*in vitro*) and in living organisms (*in vivo*).

14.2.3 CROSSING BIOLOGICAL BARRIERS

Nanocarriers have the capability to traverse biological barriers such as the blood–brain barrier (BBB) or the mucosal barrier, enabling the transport of drugs to areas of the body that were previously difficult to reach, such as the central nervous system. When given to healthy mice or a model of a mouse with a brain tumor, 1PA (pazopanib) loaded polymer covalent organic framework (COF) nanocomposites containing polyethylene glycol (PEG) with an average molecular weight of 350, carboxymethyl cellulose (CCM), and 3-aminopropyl)triethoxysilane (APTES) can cross the BBB. They reduce the number of renal cells in the tumor lesions and enhance the effects of antiangiogenesis and apoptosis (Zhang et al., 2020). The drug delivery method uses bubbles to temporarily break down the BBB. It is made up of doxorubicin and FePt NPs loaded into hydrophobic NBs (Dox-FePt@NB) nanocomposite. This enables the drug to bypass the BBB and access the brain, enhancing the suppression of glioblastoma (Chan et al., 2021). The study employed a nanocomposite known as zeolitic imidazolate framework 8-coated Prussian blue (ZIF-8@PB), which underwent treatment with quercetin (QCT), a naturally occurring antioxidant. This nanocomposite demonstrated an excellent response to NIR radiation and was able to penetrate the BBB in order to access the location of mitochondrial dysfunction. This was achieved by utilizing the photothermal effect. The purpose of this approach was to treat Parkinsons disease (PD) (Liu et al., 2021).

14.2.4 VACCINE DELIVERY

Nanocomposites can function as vehicles for vaccination antigens, adjuvants, or immunomodulatory substances, augmenting antigen presentation, stimulating immune response, and enhancing vaccine effectiveness. The nanocomposite consisted of N-alkylated polyethyleneimine having molecular of 2 kDa, suprerparamagnetic iron oxide, and lactobionic acid, demonstrated effective siRNA transfection and MRI labeling abilities, enabling the observation of dendritic cell vaccines (DCV) transportation to lymph nodes (Wu et al., 2021). EUAC-Ins consists of a core complex made up of aminoclay-protein, which is coated with Ulex europaeus agglutinin 1 (UEA-1) and a pH-sensitive polymer called Eudragit® L100. This formulation effectively targets M cells, enhances the uptake of insulin in intestinal tissues, and significantly reduces blood glucose levels in diabetic mice (Song et al., 2021).

Nanocomposites are essential in vaccine delivery methods as they improve stability and effectiveness (Hussein et al., 2023; Zafar et al., 2023). These sophisticated carriers enclose vaccine components, safeguarding them from deterioration and facilitating precise administration. NPs possess distinctive characteristics that provide advantages such as enhanced immune responses, antigen stability, and regulated release (Foyez and Imran, 2022). Polymer-based nanoparticles (PBNs) have demonstrated potential as vaccine adjuvants and delivery systems, providing safety, targeting abilities, and excellent stability (Han, 2022). Nanotechnology in vaccine development tackles issues such as low immunogenicity and the requirement for several doses, hence facilitating the implementation of more potent and streamlined immunization approaches. In general, nanocomposites have a substantial potential to transform the way vaccines are delivered and improve the immune response of the host against different diseases (Hussein et al., 2023).

These findings demonstrate the potential of nanocomposites to show great potential in precise and directed medication administration (Kumari et al., 2023; Ashique et al., 2023). Systems like polymer nanocomposites offer benefits such as better biocompatibility, site-specific targeting, and improved drug release profiles (Desai and Shinde, 2022). Nanotechnology-based drug delivery technologies hold promise for overcoming the issue of crossing the BBB (Kumar et al., 2022; Wang et al., 2022). Researchers are developing hybrid NPs, such as polymeric NPs and liposomal systems, to precisely deliver medications to treat brain problems. Additionally, the use of nanocomposites in combination drug therapy allows for the incorporation of numerous therapeutic agents with minimal toxicity. In addition, researchers are investigating the use of nanocomposites for delivering vaccines. This is possible due to their tiny size and large surface area, which allow for effective loading of drug cargo. The achievements emphasize the critical role of nanocomposites in transforming drug delivery techniques for a variety of medical applications.

14.3 NANOCOMPOSITE IN TISSUE ENGINEERING

Tissue engineering is a multidisciplinary field that applies concepts from the biological sciences, chemistry, and technology to investigate the procedure of repairing injured tissues or organs (Zaffagnini et al., 2006). To provide temporary substrates for cell-sowing, tissue engineering combines molecular biology and material design. Scaffolds with the right amount of pores, pore size, biological compatibility, and degradation rate can treat different tissues and organs, as shown in Figure 14.2. These scaffolds must possess sufficient mechanical stability to withstand the stress of regeneration, whether it is in a laboratory setting (*in vitro*) or within a living organism (*in vivo*). Nanocomposites are intensively researched

TABLE 14.1 Examples of Few Trending Nanocomposites Used in Drug Delivery.

Nanocomposite Materials	Loaded Drug	Targets			Results	References
		In Vitro	***In Vivo***	**Antibacterial/ Antimicrobial Study**		
Mesoporous silica nanoparticle, chitosan, selenium	Metformin Cisplatin	MDA-MB-231 (Breast cancer cell)	5–6 weeks old breast tumor xenograft female athymic nude mice model	–	Increased drug absorption, regulated release, and inhibitory actions against cancer cells	Vimala et al. (2021)
Hydrogel-cellulose nanocrystals, poly-acrylic acid, calcium oxalate nanoparticles	Doxorubicin	*MDA-MB-231* (Breast cancer cell line)	–	*Escherichia coli, Staphylococcus aureus*	Bactericidal, antiarthritic activity, controlled, and targeted release of medication against cancer cells	Shahzadi et al. (2023)
Chitosan, Fe_3O_4, reduced graphene oxide	Doxorubicin	MCF-7 (Breast cancer cell line)	Zebrafish Danio rerio	–	Nanocomposite exhibited improved antibiofilm, antioxidant, and targeted chemotherapeutic characteristics	Karthika et al. (2020)
Chitosan, carbon quantum dots, Fe_2O_3	Curcumin	MCF-7 (Breast cancer cell line)	–	–	Cytotoxicity effects on MCF-7 cell lines and promising alternative for targeted drug delivery	Zoghi et al. (2023)
Graphene oxide, poly (epichlorohydrin), hyperbranched polyglycerol	Paclitaxel (PTX) Curcumin (Cur)	MCF-7 (breast cancer cell line)	–	–	Both PTX and Cur had a prolonged release and showed synergistic effects against MCF-7 breast cancer cells, inhibiting cell growth	Asgari et al. (2021)

TABLE 14.1 *(Continued)*

Nanocomposite Materials	Loaded Drug	Targets			Results	References
		In Vitro	***In Vivo***	**Antibacterial/ Antimicrobial Study**		
Hydrogel- Tragacanth gum, *Terminalia Bellirica* seeds, silver nanoparticles	5-fluorouracil (5-FU)	_	_	*Staphylococcus aureus, Pseudomonas aeruginosa, Escherichia coli, Klebsiella pneumonia*	The hydrogels swelling behavior at specific pH and temperature ranges enables controlled release of drugs, suitable for cancer treatment in biomedical areas and drug delivery application	Nagaraja et al. (2021)
Hydrogel-copper oxide nanoparticles, poly(vinyl alcohol), oxidized starch	Doxorubicin	*SW480* (Human colon cancer cell line)	_	*Staphylococcus aureus, Escherichia coli*	The CuO nanoparticle content in the oxidized starch/PVA hydrogels resulted in longer and more regulated drug releases	Gholamali et al. (2020)
Carrageenan, Chitosan (N,N,N-trimethyl chitosan chloride), 3% silver nanoparticle (NC-3% Ag)	Curcumin	Caco-2 (Human colon cancer cells)	_	*Streptococcus mutans, Staphylococcus aureus, Pseudomonas aeruginosa, Escherichia coli, Klebsiella pneumonia*	A very effective controlled drug delivery vehicle is being developed specifically for hydrophobic drugs, with a focus on targeting the intestines	El-Maadawy et al. (2022)
Hyaluronic acid, carboxyl terminated PAMAM dendrimer, fluorochrome indocyanine green	Doxorubicin hydrochloride	4T1(Metastatic breast cancer cell line) with expressing CD44 receptor	4T1 xenograft mice model	_	Created large levels of ROS, which reduced tumor growth and lung metastasis without systemic toxicity and caused cell apoptosis	Wu et al. (2019)

TABLE 14.1 *(Continued)*

Nanocomposite Materials	Loaded Drug	Targets			Results	References
		In Vitro	*In Vivo*	Antibacterial/ Antimicrobial Study		
Zinc oxide nanoparticles, g-C_3N_4 (Graphitic carbon nitride)	Curcumin	MDA-MB-231 (Human adenocarcinoma cells)	–	*Escherichia coli, Staphylococcus aureus*	This drug-loaded nanocomposite has the potential to be an excellent choice for targeting and treating breast cancer cells	Padmanabhan et al. (2023)
Chitosn, polyacrylic acid, g-C_3N_4 (graphitic carbon nitride)	Curcumin	MCF-7 (Breast cancer cell line)	–	–	The nanocomposite showed exceptional efficacy in removing cancerous cells and exhibited significant potential for pH-dependent controlled drug release	Abdouss et al. (2023)
Uio-66 MOF (zirconium-based metal organic framework), Ti_3C_2 Mxenes, Rosmarinus officinalis, clustered regularly interspaced short palindromic repeats (CRISPR)	Doxorubicin	HEK-293 (Human embryonic kidney cell line), HePG2 (liver cell line), HT-29 (colon cancer cell line), MCF-7 (breast cancer line), A549 (lung adenocarcinoma cell line)	–	–	This study found that green alteration of nanocomposites mostly affects DOX cellular internalizations, rather than CRISPR	Tavakolizadeh et al. (2023)

because of their capacity to improve mechanical characteristics, cellular attachment, and growth rate. People in both the academic and business worlds find polymer/layered nanocomposites, composed of hydroxyapatite (HAp), carbon nanotubes, and multilayer silicates, very interesting (Tsivintzelis et al., 2007; Sinha Ray et al., 2003). Incorporating a limited quantity of extended silicate NPs can greatly enhance the physical as well as mechanical characteristics of a polymer matrix.

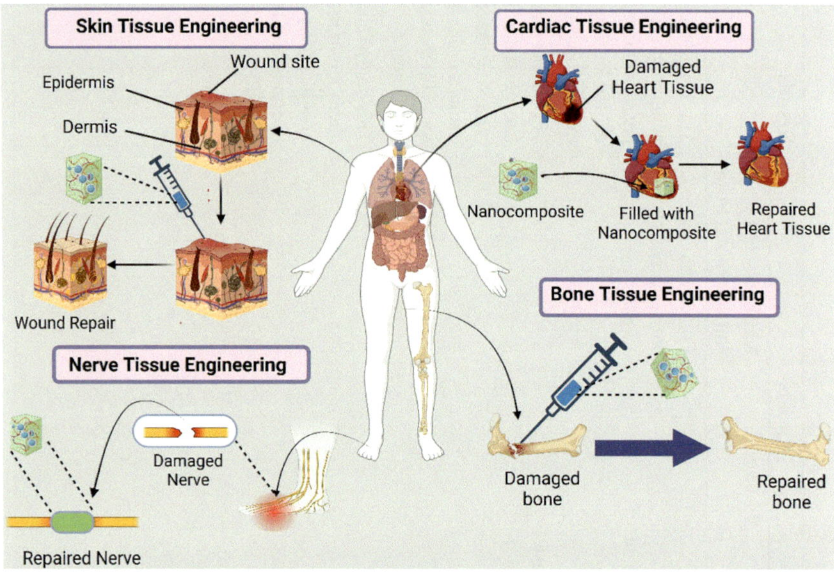

FIGURE 14.2 Application of nanocomposites in various forms of tissue engineering.

14.3.1 CARDIAC TISSUE ENGINEERING

In recent times, cardiovascular disorders have emerged as a significant concern impacting the well-being of many individuals. A coronary artery disease, myocardial infraction (MI), is the primary cause of illness and death on a global scale (Muniyandi et al., 2020). Myocardial infarction (MI) is a condition that arises when there is an obstruction in the circulation of blood to the heart, resulting in irreversible damage to the heart's pumping ability and eventually leading to persistent heart failure. Given the limited ability of cardiac tissue to regenerate, a heart transplant is the sole solution for those suffering from severe heart failure to regain normal cardiac function. Nevertheless, the scarcity of individuals willing to donate their hearts and the exorbitant expenses associated with surgical procedures render it unattainable for a significant number of patients.

Cardiac transplantation holds great potential for effectively replacing or repairing damaged myocardium (Nazari et al., 2019). The objective of cardiovascular tissue engineering is to fabricate three-dimensional scaffolds that replicate the extracellular matrix with the purpose of regenerating injured myocardium, as shown in Figure 14.2.

In order to replicate healthy heart tissue, cardiac tissue engineering must fulfil many crucial criteria, including biodegradability, porosity, and appropriate mechanical and electrical characteristics. Nanocomposite hydrogels that include conductive NPs are seen as appealing approaches for heart regeneration, according to their attributes (Marchesan et al., 2016). Multiple research investigations have been conducted to investigate the application of electrically charged nanocomposite hydrogels in cardiovascular tissue technology.

14.3.2 BONE TISSUE ENGINEERING

The cartilage and bones are intensively studied tissues within the realm of tissue engineering because of their exceptional regenerative capabilities, making them among the most explored tissues in the human body. Fractures and other defects have been repaired using bone graft materials, which possess osteoinductive and osteoconductive qualities (Moore et al., 2001). Nevertheless, there are potential hazards associated with illness spreading, infection, long-lasting discomfort, immunogenicity, and insufficient availability.

Bone tissues consist of connective tissues, mostly made of living cells and the surrounding matrix. These aforementioned tissues are durable and able to endure repetitive electrical stimulation without any damage or impairment to normal physiological functioning. The structure of bones is deeply connected to the physical motion of the body of an individual. Once impaired, it will significantly disrupt the patient's daily life. The primary objective of tissue engineering of bones is to create a resilient and authentic 3D milieu that has the necessary biological conditions for the formation of the desired tissue (Place et al., 2009). There have been a growing number of nanocomposite hydrogels created for bone tissue creation, specifically emphasizing the field of cartilage and bone-related tissue engineering.

The mineral composition of bone closely resembles that of HAp, but it also includes additional ions that enhance its suitability for biological applications. Yazaki et al. (2011) improved the DNA-fibronectin-apatite laminating layer to boost its functionality in tissue engineering applications by including carbonate or fluoride, which allowed for precise control over the layer's solubility. Adding carbonate makes the transmission of genes have a bigger effect on the effectiveness layer. Adding fluoride, on the other hand, lowers the effectiveness and increases the time of gene transmission in a way that depends on the dose. Furthermore, manganese (Mn), which is found in small amounts in teeth and bones, regulates the process of bone remodeling.

14.3.3 SKIN TISSUE ENGINEERING

The skin serves as the primary barrier against infection, and any damage to it, like burns, might result in substantial health consequences. Tissue engineering of skin has great potential as a strategy for regenerating skin. The scaffolding component used in the tissue engineering of skin must not only have the appropriate mechanical qualities but also exhibit antibacterial and anti-infective characteristics. Additionally, it needs to be able to facilitate medication administration to enhance the healing of wounds. Nanocomposite hydrogels are extensively used in skin tissue regeneration to accomplish this objective. Research has shown that zinc oxide (ZnO) NPs have significant antimicrobial properties and do not cause any harm to healthy tissue. In their 2019 study, Rakhshaei et al. added ZnO NPs to a hydrogel made of chitosan and gelatin (called CS-gel/ZnO) to make it more antibacterial and better at moving medicines around. ZnO nanomaterials greatly improve the adaptability of the scaffold by being present on its surface. The antimicrobial characteristics of these nanocomposite hydrogels are achieved by successfully destroying the cell walls of bacteria. The hydrogel contains reactive oxygen species or ROS and zinc ions (Zn^{2+}), which possess the ability to target the negatively charged cell walls of microorganisms, leading to their destruction.

14.3.4 NERVOUS TISSUE ENGINEERING

The primary role of the central nervous system is to collect sensory input from various cells throughout the body, interpret this resulting data, and transfer impulses to other tissues and organs in order to evoke appropriate reactions (Vijayavenkataraman et al., 2020). In today's society, ischemia, whether biochemical, mechanical, or thermal, frequently causes damage to the nervous system. Commonly used in cases of nervous system injuries, surgical techniques such as nerve reconstruction and allogenic nerve implantation are utilized to get the most favorable outcome in terms of recovery. Nevertheless, the prognosis is poor. Recent research indicates that a mere 40–50% of individuals get neurological recovery after allogenic nerve transplantation (Lee and Wolfe, 2000). Therefore, experts consider tissue engineering technology as the most effective method for restoring nerve injuries.

The objective of nervous tissue technology is to develop nerve grafting alternatives to alleviate the pain of nerve injuries and achieve sustained physiological restoration. Graphene-based composite hydrogels are an optimal option for augmenting neuron restoration. Researchers led by Qiao et al. (2018) mixed GO with polyacrylic acid (PAA) to make an electrically responsive nanocomposite hydrogel. They did this using a process called in situ polymerization, which they wrote about in their 2018 paper. Research has shown that electrical stimulation enhances both the overall proliferation of neurons and their average extent. Researchers used hydrogels made of carbon nanotube (CNT) nanocomposites and PEG to achieve this. From the above information and Figure 14.2, you can easily understand the utilization of nanocomposites in the realm of tissue engineering.

Nanocomposites are crucial to the progress of tissue engineering for cardiac, bone, skin, and brain applications. Metallic, carbon, and ceramic NPs enhance scaffolds to improve cell adhesion, regulate growth factor release, and monitor tissue regeneration (Vijayan et al., 2022; Züger et al., 2022). The combination of nanocomposite scaffolds and regenerative stem cells shows potential for the development of functional tissues (Pryjmaková et al., 2020). Tissue engineering tries to copy natural extracellular matrices by adding nanomaterials with special properties, like better electrical and mechanical performance, to boost cell activity and tissue formation (Singla et al., 2019). Also, scientists are looking into how to use nanocomposites in biofabrication methods to make tissues that are conductive and respond to stimuli. They are focusing on heart and brain tissue engineering in particular (Aruna et al., 2022). Nanocomposites have the potential to enhance tissue regeneration and create new possibilities for developing medical devices, making them a promising field in tissue engineering.

14.4 APPLICATIONS OF NANOCOMPOSITES IN WOUND HEALING

Nanomaterial usage in wound healing is growing, with clinical research investigating nanotherapy-based therapies (Ghobadi et al., 2018; Hamdan et al., 2017). Table 14.2 displays a few examples of these nanocomposites. The morphology and dimensions of NPs show great potential for enhancing the process of wound recovery because they affect cellular actions, permeation, and active drug transportation (Mihai et al., 2019). Figure 14.3 illustrates how wound therapy widely employs nanomaterials with inherent properties that enhance wound healing and those that transport therapeutic medicines.

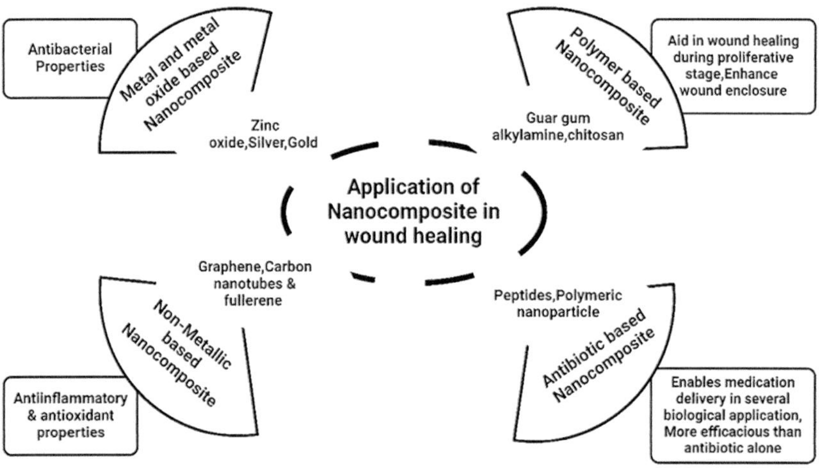

FIGURE 14.3 Applications of various types of nanocomposites in wound healing.

14.4.1 METAL AND METAL OXIDE-BASED NANOCOMPOSITES

Metal oxide and other metal NPs are more advantageous in wound-healing therapy than traditional materials. Furthermore, their intrinsic characteristics contribute to their comparable capacities in terms of antibacterial properties and wound-healing promotion. Researchers have extensively researched on different metallic NPs for their antibacterial properties and appealing physiochemical attributes (Naskar et al., 2020).

14.4.2 NONMETALLIC-BASED NANOCOMPOSITES

Nonmetallic NPs may heal wounds. Carbon-derived nanomaterials, including graphene, CNTs, and fullerenes, have demonstrated potential in nanomedicine for numerous biological applications (Zhou et al., 2010). The researchers discovered that fullerenes possess potent anti-inflammatory and antioxidant characteristics, making them highly useful in promoting wound healing. Because of its antioxidant properties, it has the ability to remove and neutralize nitrogen and ROS, promoting wound healing. Adding α-cyclodextrin, tris-dicarboxyl, and hexadicarboxyl to it may stop it from making the cell system worse (Gao et al., 2010). Khan et al. (2015) successfully employed GO nanosheets of material for the thermal treatment of fungi and microbial wounds. The combined use of NPs resulted in a synergistic impact that enhanced the speed of wound healing in comparison to employing the NPs individually.

Polymeric NPs can serve as delivery vehicles or wound dressings due to their wound healing and antibacterial capabilities (Korrapati et al., 2016). Chitosan, a natural polymer, has been intensively researched for its wound-healing properties among NPs. Chitosan, known for its anti-bacterial, biodegradable, and biocompatible properties, was initially the chosen element for wound recovery (Matica et al., 2019). Chitosan-related nanocomposites are frequently more effective than chitosan individually for wound repair. In their study, cellulose has been extensively researched by numerous scientists for its ability to heal wounds and its antibacterial capabilities. Nano-biocomposite made of AgNPs and bamboo cellulose nanocrystals were shown by Singla et al. (2017) to be able to treat diabetic wounds in living organisms. Black phosphorus (BP) is a potential nano-material for many biological applications, such as wound repair. Huang et al. (2018) proposed that water-based BP nanosheets have the potential to serve as a versatile substrate for promoting wound recovery.

14.4.3 POLYMER-BASED NANOCOMPOSITES

Several studies (Kalashnikova et al., 2015) have investigated the application of polymers incorporating NPs for the purpose of wound repair. Researchers conducted wound repair experiments on adult rats using nanocomposites that included zinc, copper, silver, and chitosan nanomaterials. Research suggests that copper–chitosan nanocomposites may aid in the proliferation phase of wound recovery (Gopal et al., 2014). Surface-modified magnetite NPs containing limonene and eugenol showed antiadherence and antibacterial properties, making them essential for the procedure of wound recovery (Anghel et al., 2012). Vasile et al. (2014) showed that a gel made of three parts—gentamicin-loaded ZnONPs, chitosan, and a gellan gum—worked together to kill bacteria more effectively. Researchers found that this gel was more effective in preventing the development of bacteria as compared to the use of gentamicin alone as a control. This implies that the nanocomposite has advantageous properties for facilitating wound healing when a pathogen is the source of the infection. Rigo et al. (2013) employed polyethylene fabric to distribute AgNPs in wound models, both *in vivo* and *in vitro*. This led to either cell death or the absence of cytotoxicity in repaired skin specimens. Furthermore, *in vivo*, the application of AgNPs permitted cutaneous layer to redevelop rapidly. Ghosh Auddy et al. (2013) also used guar gum alkylamine, a new cationic biopolymer, along with AgNPs to help wounds heal faster.

14.4.4 ANTIBIOTICS-BASED NANOCOMPOSITES

Researchers have used a variety of biodegradable as well as biologically compatible nanomaterials to deliver antibiotics to specific wounds (Hadinoto et al., 2013). Liposome NPs phospholipid bilayer structure enables medication delivery in several biological applications. Staphylococcal lactamase may prevent liposome-enclosed piperacillin and γ-lactam antibiotics from hydrolyzing (Mao et al., 2018). Piperacillin encapsulated in liposomes showed superior antibacterial action against *S. aureus* compared to piperacillin alone. In 2005, Mugabe et al. did a study that showed that liposome-encapsulated gentamicin antibiotics were better at fighting *P. aeruginosa* biofilms than gentamycin monotherapy.

Li et al. (2013) discovered that liposomes containing daptomycin effectively stopped the growth of *S. aureus* biofilm at the site of infection in mice with subcutaneous illness. Despite promising findings, further investigation is necessary to address the inadequate chemical and physical stability of antibiotic-encapsulated liposomes. Intensive research is underway to investigate the use of poly (lactic-co-glycolic acid) NPs for the purpose of wound recovery (Chereddy et al., 2016). This nanomaterial, like liposomes, is biocompatible and biodegradable, making it ideal for next-generation wound care and healing applications. A novel category of NPs with hybrid properties has been proposed in a recent study and has the potential to substitute polymers and liposomes (Hadinoto et al., 2013). These hybrid NPs combine the advantages of polymeric and liposome NPs. Functionalized antibiotics with NPs may assist in wound healing. Consider factors such as time-released drug delivery, international toxicology evaluation, less scarring, increased affordability, quicker recovery, and surface optimization when using nanomaterials with antibiotics (Gould, 2016). Further investigation is necessary to improve the efficiency of antibiotic of encapsulation and prevent the early release of the drug.

Nanocomposites have significant potential in wound healing due to their wide range of uses and advantages. NPs possess improved characteristics such as antibacterial, anti-inflammatory, and antioxidant qualities, which make them highly helpful during the course of therapy of diabetic wounds (Sari et al., 2023). Liposomes, micelles, and polymeric NPs are examples of NPs made of soft materials. They have better rheological properties and controlled drug release, which helps wounds heal (Qin et al., 2022). Nanomaterials, encompassing both organic and inorganic substances, are transforming the field of wound care by offering several antimicrobial properties. This has the potential to reduce the financial strain associated with therapy (Rajwade et al., 2023). Nanotechnology, specifically nanodrug delivery devices, has potential in the management of diabetic wounds by enabling controlled drug release and enhancing treatment efficacy (Akgöl et al., 2021). Researchers are investigating the use of chitosan-based nanocomposites in wound dressings because of their ability to promote effective wound closure. These nanocomposites offer biocompatibility, antibacterial effects, and antioxidant characteristics, making them beneficial for wound healing (Nie et al., 2023). These improvements suggest that nanocomposites have a great potential for the future in improving wound repair processes.

14.5 BIOSENSING TECHNOLOGIES UTILIZING NANOCOMPOSITES

Nanotechnology has enhanced the functionality of several devices and technologies, including sensor technology, particularly biosensors. Integrating nanomaterials into the design of biosensors has yielded significant benefits and successfully addressed several obstacles in biosensor technology (Jiang et al., 2021). Nanostructure inclusion has successfully

TABLE 14.2 Examples of Some Recently Developed Nanocomposite Used in Wound Healing.

Nanocomposite Materials	*In Vitro*	*In Vivo*	**Result**	**References**
Graphene oxide (GO)-cellulose nanocomposite	EA. hy 926 human endothelial cells	Rats	This nanocomposite enhanced wound healing through fibroblast proliferation.	Soliman et al. (2021)
Bacterial cellulose containing montmorillonite modified with silver nanocomposite	Mouse fibroblast L929 cells		Destroy biofilms of *S. aureus* and *P. aeruginosa*, preventing septicemia in chronic wounds	Horue et al. (2020)
Chitosan functionalized halloysite nano tubes and poly (caprolactone) (PCL) nanocomposite	N1H3T3 fibroblast cells		Nanocomposite film containing functionalized HNTs demonstrated superior proliferation adhesion and migration rates making them suitable for wound healing	Kouser et al. (2021)
Nanofibers composed of chitosan (CS) and polyethylene oxide (PEO) include silver and zinc nanoparticles	L929 fibroblast cells		This nanocomposite has antioxidant ability and antimicrobial activity, accelerating the movement and multiplication of fibroblast cells	Bagheri et al. (2021)
Nanocomposite containing pectin and keratin incorporating silver nanoparticles and ferulic acid	L929 fibroblast cells	Male Sprague Dawley rats	The nanocomposite positively influences the wound recovery, by promoting cell division, re-epithelialization, and fibroblast emigration	Kishore et al. (2023)
Nanocomposite formed by encapsulating a nano emulsion inside Alginate-pectin macroparticles	Immortal Keratinocyte (HaCaT) cell line		This nanocomposite showed controlled release if lipophilic drugs, directly into the wound	Amante et al. (2022)
Nanocomposite containing PVA and kaolin loaded with penicillin and streptomycin	NIH 3T3 fibroblast cell line		This nanocomposite showed potent homeostatic and antibacterial wound repairing	Tamer et al. (2021)
CuO and Ago comodified ZnO nanocomposite	L929 fibroblast cells	BALB/c mice	This nanocomposite showed negligible inflammatory, cellulose or pathological abnormalities, good biocompatibility	Ye et al. (2023)

addressed many issues, such as enhancing biorecognition binding efficiency, enhancing sensibility, and reducing the threshold for detection (Huang et al., 2021). Furthermore, it enhanced the specificity of sensors and reduced the reaction time required for a response. Electrochemical sensors improve the method of detection by increasing the speed at which electrons are transported to the detecting electrodes following reactions (Jiang et al., 2021).

14.5.1 IMPLICATIONS OF NANOCOMPOSITES IN BIOSENSING

Several nanocomposites have been studied in biosensors, exhibiting positive results in the field. This following section will explore several nanocomposite materials and their varieties, highlighting the benefits they provide to the field of biosensors. Nanocomposites may be categorized according to the filler type, which refers to the nanomaterial composition, or according to the matrix type, as previously indicated. Primarily, nanocomposites may consist of various nanostructures (Alghamdi, 2024), such as follows:

14.5.1.1 NANOCOMPOSITES BASED ON GRAPHENE AND GRAPHENE OXIDE

Nanocomposites incorporating graphene and GO provide a wide range of outstanding characteristics that have attracted considerable attention in diverse applications. Because graphene is so good at conducting heat and electricity, has an enormous ratio of surface area to volume, and has many other great physical properties, it is highly sought after for use in biological sensing and detection (Kazemi et al., 2020). When graphene is used as an additive in polymer matrices, it improves the characteristics of the finished nanocomposites. These nanocomposites are particularly useful in biosensors since they have enhanced sensing capacities (Kausar et al., 2023). The heightened electrical conductivity of nanocomposites based on graphene enables improved electron transfer, leading to enhanced sensing abilities. Biosensors that use field-effect transistors (FETs) or electrochemical-potentiometric sensors to detect Zika virus proteins and volatile gases like ammonia and sulfur compounds have proven this (Kausar et al., 2023).

GO derivatives have similar extraordinary features, which make them potential candidates for improving biosensors. Having functional groups on the surfaces of GO makes it easier for biomolecules to bind, which improves biorecognition in biosensors (Ramesh et al., 2022; Sharma et al., 2016). Biswas et al. (2019) and Kausar et al. (2023) say that adding GO to polymers like polyaniline (PANI) makes them better at finding hydrogen, methane, lead, and cadmium, among other gases and ions. Moreover, researchers have created composites using GO for the purpose of detecting hydrogen gas at room temperature. These composites have quick response times and a high level of sensitivity, as illustrated in a research investigation by Mohammadi et al. in 2020. Nanocomposites consisting of reduced graphene oxide (rGO) and polymers such as polypyrrole (PPY) are used in glucose sensors. These nanocomposites provide support for enzyme bioreceptors, which enhance electron transport and improve the performance of the sensors (Gosai et al., 2021). Adding rGO to biosensors made of ZnO nanorods also makes them more sensitive and reduces the threshold for detection (Gosai et al., 2021).

14.5.1.2 METAL AND METAL OXIDE NANOPARTICLES BASED NANOCOMPOSITES

According to Kausar et al. (2023), and other researchers, the exceptional properties of NPs made of metal and metal oxides have garnered considerable attention in biosensing applications. These NPs possess remarkable sensitivity, specificity, exceptional conductivity, and stability, rendering them indispensable in biosensing technologies. Gold nanoparticles (Au NPs), known for their chemical durability, are widely used in electrochemical and optical sensors due to their biocompatibility and catalytic activity (Ramesh et al., 2022). In addition, the performance of sensors has been improved by using nanocomposites that contain Au NPs and PANI. Studies by Korzeniewska et al. (2020) and Naseri et al. (2018) showed that this improvement was present in the detection of a particular antigen known as CEA. NPs of silver (Ag NPs) have been used in many types of biosensors, including a colorimetric biosensor on paper for finding the MERS coronavirus and strong, responsive sensors for finding hepatitis B (Huang et al., 2021).

Because they are more stable, conduct electricity better, and are cheaper than noble metals like gold, copper nanoparticles (Cu NPs) are being considered as possible candidates for biosensing applications (Naresh et al., 2021). Cu NPs have been effectively employed in composites to detect chloroform, leading to the development of highly efficient sensors (Liang et al., 2017). Platinum nanoparticles (Pt NPs) and palladium nanoparticles (Pd NPs) have demonstrated outstanding efficacy in biosensing, exhibiting similar features to other types of NPs. Their extensive accessibility, especially palladium, makes them attractive for sensing applications (Naresh et al., 2021). According to Su et al. (2020), tin oxide (SnO_2) composites have demonstrated significant sensitivity and selectivity in the identification of hydrocarbons and NOx gases. Additionally, Kannan et al.'s (2023) research showed that the use of nanocomposites consisting of mixed metal oxides like NiO, ZnO, and Fe_3O_4 has improved the selectivity and electrochemical responsiveness of dopamine detection. Adding metal and metal oxide NPs to nanocomposites makes it possible to make biosensors that are very good at finding a wide range of substances and last a long time.

14.5.1.3 CARBON NANOTUBE-BASED NANOCOMPOSITES

Extensive research has been conducted on CNTs with numerous applications, including biosensors, medication delivery, and diagnostics. CNTs, or carbon nanotubes, are hollow cylindrical tubes that may be created by rolling up one or more layers of graphene. Single-walled carbon nanotubes (SWCNTs) typically have a narrow diameter of 1–2 nm, but multiwalled carbon nanotubes (MWCNTs) can have much larger sizes up to 100 nm (Curulli, 2023; Ramesh et al., 2022). CNTs possess a multitude of distinctive features that make them very appealing in the realm of biosensors. These features include their strong thermos endurance, strong tensile strength, good electrical efficiency, good surface area, and impressive electrocatalytic activities. Additionally, they enhance the method of detection with their exceptional capacity to identify a broad spectrum of substances, including ionized particles and gaseous compounds. Also, they are very good at conducting electricity, which lets electrons and charges move quickly. This makes it easier for important biological molecules to respond to changes in electricity (Ferrier and Honeychurch, 2021; Naresh et al., 2021).

Adding CNTs to polymeric nanocomposites has a big effect on improving their ability to stretch, their porosity, and their ability to conduct electricity. These composites may be included in several categories of biological sensors, including electrolytic and optical biosensors (Ferrier and Honeychurch, 2021). Nanocomposites made up of MWCNTs and atomically printed polymers (MIPs) have been used successfully in a number of biosensors to identify HIV and other infectious diseases very accurately (Ferrier and Honeychurch, 2021).

14.5.2 APPLICATIONS OF BIOSENSORS COMPOSED FROM NANOCOMPOSITES

Currently, there is a pressing need for biosensors that exhibit dramatically enhanced performance, given the recent surge in the number of illnesses and outbreaks, as well as the growing global population. Consequently, there will be a heightened need for enhanced biosensors that can efficiently monitor vital aspects of human existence and the surroundings, including healthcare, water quality, and food safety. Some of the materials are hybrid nanocomposites, which contain more than one type of nanofiller (Ferrier and Honeychurch, 2021). Nanocomposites are important for biosensing applications because they aid in biorecognition, transduction, and signal generation *via* many channels. Some things, like the type and amount of nanofiller material, the polymer that is used, and the way the composite is put together, can change how the sensor works (Kausar et al., 2023).

In general, a variety of applications, including medication administration, environmental monitoring, illness detection, food and water quality monitoring, and health care, have made use of nanocomposites-based biosensors (Naresh et al., 2021). Below is a list of a few biosensing applications for nanocomposites-based sensors.

1. Biomedical mentoring biosensors are employed for the purpose of monitoring the well-being of patients and monitoring the development of their illness, assisting them in their recuperation (Ramesh et al., 2022).
2. Tumor-detecting biological sensors are designed to detect cancer biomarkers, which are specific analytes like DNA, proteins, RNA, and enzymes that are produced as a result of genetic alterations.

Identifying these biomarkers aids in determining the suitable treatment (Hasan et al., 2021).

3. Diagnostic sensors: The incorporation of NPs into biological sensors has the potential to enable accurate identification of specific biological markers in a short period of time. Glucose biosensors have a vital function in monitoring glucose levels to avoid any health issues caused by their fluctuations (Ramesh et al., 2022).

Nanocomposites have great potential for numerous applications in future biosensing technologies. Nanomaterial-based biosensors provide improved sensitivity, specificity, and fast identification of analytes. These biosensors have the capability to identify a diverse array of biomolecules, such as cancer biomarkers, poisons, infections, and different pollutants, even when present in very low amounts (Kumar et al., 2022). Nanomaterials are incorporated into biosensors to create very sensitive and small devices for diagnosing diseases, monitoring the environment, and ensuring food safety (Tripathi et al., 2023). Nanocomposite-based biosensors have the potential to significantly transform healthcare by facilitating immediate disease detection, point-of-care diagnostics, and individualized therapy (Singh and Singh, 2022). The ongoing progress in nanotechnology and biosensing is creating opportunities for the development of cutting-edge diagnostic instruments that are both novel and efficient, and have wide-ranging practical applications (Tripathi and Bonilla-Cruz, 2023).

14.6 REGULATORY CHALLENGES OF NANOCOMPOSITES BIOMEDICAL FIELD

14.6.1 BIOCOMPATIBILITY

Nanocomposites' biocompatibility issues in the biomedical field are a matter of great concern. Agrawal et al. (2022) and Ramezani and Ripin (2023) said that nanocomposites are better for medical implants and tissue engineering, but they need to be tested for biocompatibility, toxicity, and long-term stability first. Nanocomposites, such as polymer-based materials containing nanofillers, have demonstrated promise in enhancing the performance of implants (Meera et al., 2023). However, it is imperative to conduct comprehensive assessments of their biocompatibility and toxicity profiles. Biopolymers like polylactic acid can be made stronger and more biocompatible by adding nanofillers like ceramic nanobioglass (Castro et al., 2022). However, there are still difficulties to overcome, such as the need for standardization and quality control. Understanding the things that affect how well nanomaterials work with living things and how much damage they can do is important for making sure they can be used safely in biomedical settings (Darbasizadeh et al., 2022).

14.6.2 ENVIRONMENTAL IMPACTS

The production and disposal of nanocomposites used in biomedical applications have substantial environmental consequences. When released into soil and water, nanomaterials, including nanocomposites, have the potential to cause environmental pollution and negatively impact plants and animals (dos Santos et al., 2022; Ali, 2023). Moreover, the extensive utilization of nanotechnology across several sectors leads to the ongoing discharge of nanomaterials into the environment, which may have detrimental consequences for living species and their habitats (Shrivastava et al., 2022). Nanomaterials created to capture and store CO_2, including liquid, metal-based, and carbon-based nanomaterials, can have harmful effects on the environment, specifically on aquatic organisms and their food chains, causing disruptions in aquatic cycles (Mir et al., 2023). Engineered NPs can accumulate in soil from a variety of sources, including consumer products, resulting in soil pollution. This contamination can have negative impacts on agriculture and present difficulties in comprehending the way these nanomaterials behave and their impact on the environment (Kokilavani et al., 2023).

14.6.3 LONG-TERM EFFECTS

Comprehending the long-term impacts of nanocomposites on human health presents notable difficulties because of their potential toxicity (Sharma and Alam, 2023). NPs' physicochemical properties are very important for figuring out how they affect human health, such as their ability to harm cells, the immune system, and genes (Xia et al., 2023). In addition, the interplay between NPs and physiological systems might result in negative consequences, such as the development of disorders related to oxidative stress and pulmonary diseases (Costa et al., 2022). Variables such as size, shape, surface charge, and composition influence these intricate interactions, making it challenging to evaluate their long-term impact on human health (Williams and Bannon, 2022). Moreover, the scarcity of historical data regarding manufactured NPs and nanomaterials poses a significant obstacle to gaining a thorough comprehension of their environmental and health effects. We must develop safer-by-design methods to regulate the hazardous characteristics of nanomaterials to address these difficulties.

14.6.4 REGULATORY FRAMEWORK

To fix regulatory issues with nanocomposites used in biomedical settings, strategies must be put in place that involve stakeholders working together and making changes to the current regulatory framework. Stakeholders, such as regulatory authorities like the Environmental Protection Agency and the Food and Drug Administration, have a vital role in participating toward progress of efficient regulatory frameworks (Zee et al., 2022). The REFINE initiative highlights the significance of cooperation between regulatory and research communities in tackling industry-specific obstacles and improving regulatory decision-making by focusing on generating specific data and sharing knowledge (Gupta et al., 2022). The need to be able to quickly adapt to regulations is emphasized, particularly for nanomaterials such as the nTRACK nano-imaging agent. In these cases, traditional testing procedures may not be appropriate, making it necessary to have constant communication with authorities during the development of the product (Halamoda-Kenzaoui et al., 2022). International collaboration and standardization initiatives are crucial in order to build effective regulatory strategies for the safe utilization of nanomaterials (Chauhan et al., 2023).

14.7 FUTURE DIRECTIONS AND OPPORTUNITIES

Looking ahead, several exciting opportunities and directions emerge for nanocomposite biomedical applications:

1. **Advanced Functionalization:** Continuing research into surface modification techniques will provide meticulous regulation of the characteristics and capabilities of NPs. Modifying the surface

chemistry and enhancing the structure of a substance will enhance its capacity to interact with biological molecules and make it more compatible with biological systems (Kulkarni et al., 2022; Kumar et al., 2023).

2. **Multifunctional Platforms:** The integration of many functionalities into nanocomposites, including drug delivery, diagnostic, and therapeutic capacities, will lead to the development of adaptable biomedical platforms. These systems provide individualized treatment alternatives and enhanced patient results (Akgöl et al., 2021).
3. **Smart Nanocomposites**: Integrating stimuli-responsive materials into nanocomposites will allow for adaptive reactions to variations in the biological environment. Intelligent nanocomposites possessing the capacity to administer medications in a regulated way, detect changes in their environment, and repair themselves have great promise for targeted medical treatments and diagnostic applications.
4. **Translation and Commercialization:** To speed up the use of nanocomposite technologies in medical care, it matters to make the process of getting regulatory approval easier and to deal with safety issues more effectively. Collaborative efforts involving researchers, business partners, and regulatory bodies will facilitate the commercialization of novel biomedical products.
5. Nanocomposites provide promising prospects for developing biomedical applications in several areas, such as enhancing biocompatibility, facilitating medication delivery, and advancing tissue engineering. We can speed up the development of safe, effective, and adaptable biomedical devices based on nanocomposites by tackling regulatory issues and looking into what the future holds.

14.8 CONCLUSION

Nanocomposites have exceptional prospects for tackling urgent healthcare difficulties, encompassing drug delivery, tissue engineering, biosensing, and wound recovery. Through the utilization of the distinct characteristics of nanocomposites, scientists and medical professionals can develop new and innovative therapeutic approaches, increase the effectiveness of treatments, and boost the results for patients. Furthermore, the adaptability of nanocomposites allows for customized solutions to address unique biomedical requirements, thus facilitating the development of personalized medicine and precision healthcare. Nevertheless, it is essential to recognize and confront the regulatory obstacles associated with incorporating nanocomposites into biomedical applications. To promote responsible innovation and guarantee the safety and effectiveness of nanocomposite-based technologies, it is crucial to prioritize tasks such as ensuring biocompatibility, evaluating environmental implications, comprehending long-term repercussions, and developing strong regulatory frameworks. In order to completely utilize the potential of nanocomposites in healthcare, it is crucial to focus on future research, interdisciplinary collaborations, and breakthroughs in production procedures. By embracing these upcoming trends and possibilities, we may utilize the potential of nanocomposites to completely transform the way healthcare is provided, introducing a new era of medicine that focuses on the needs of the patient and enhances their overall well-being.

KEYWORDS

- **biosensor**
- **drug delivery**
- **nanocomposites**
- **regulatory challenges**
- **tissue engineering**

REFERENCES

Abdouss, H., Pourmadadi, M., Zahedi, P., Abdouss, M., Yazdian, F., Rahdar, A., & Díez-Pascual, A. M. (2023). Green synthesis of chitosan/polyacrylic acid/graphitic carbon nitride nanocarrier as a potential pH-sensitive system for curcumin delivery to MCF-7 breast cancer cells. *International Journal of Biological Macromolecules, 242*, 125134. https://doi.org/10.1016/j.ijbiomac.2023.125134.

Agrawal, A. A., Pawar, K. A., Ghegade, V. N., Kapse, A. A., & Patravale, V. B. (2022). Nanobiomaterials for medical devices and implants. *Nanotechnology in Medicine and Biology*, 235–272. https://doi.org/10.1016/b978-0-12-819469-0.00008-3

Ahmed, Z. A. G., Hussein-Al-Ali, S. H., Ibrahim, I. A. A., Haddad, M. Kh., Ali, D. K., Hussein, A. M., & Abu Sharar, A. A. (2022). Development and evaluation of amlodipine-polymer nanocomposites using response surface methodology. *International Journal of Polymer Science, 2022*(2), 1–15. https://doi.org/10.1155/2022/3427400.

Akgöl, S., Ulucan-Karnak, F., Kuru, C. İ., & Kuşat, K. (2021). The usage of composite nanomaterials in biomedical engineering applications. *Biotechnology and Bioengineering, 118*(8), 2906–2922. https://doi.org/10.1002/bit.27843.

Alghamdi, N. (2024). Nanocomposites for advanced biosensing applications. In *Smart and sustainable applications of nanocomposites* (pp. 222–244). IGI Global. https://doi.org/DOI:10.4018/979-8-3693-1094-6.ch008.

Ali, E. Y. (2023). Environmental impact of the use of nanomaterials for CO2 capture and conversion technologies. In *Nanomaterials for carbon dioxide capture and conversion technologies* (pp. 401–422). https://doi.org/10.1016/b978-0-323-89851-5.00012-3

Amante, C., Andretto, V., Rosso, A., Augusti, G., Marzocco, S., Lollo, G., & Del Gaudio, P. (2022). Alginate-pectin microparticles loaded with nanoemulsions as nanocomposites for wound healing. *Drug Delivery and Translational Research, 13*(5), 1343–1357. https://doi.org/10.1007/s13346-022-01257-9

Anghel, I., Holban, A. M., Grumezescu, A. M., Andronescu, E., Ficai, A., Anghel, A. G., Maganu, M., Lazăr, V., & Chifiriuc, M. C. (2012). Modified wound dressing with phyto-nanostructured coating to prevent staphylococcal and pseudomonal biofilm development. *Nanoscale Research Letters*, *7*(1), 690. https://doi.org/10.1186/1556-276x-7-690

Aruna, K., Manovasuki, J., Mohandoss, S., Balakrishnana, A., Prabu, D., & Sridhar, T. M. (2022). Challenges and future prospect of nanoparticles in tissue engineering. In *Application of nanoparticles in tissue engineering* (1st Ed., pp. 119–133). Springer. https://doi.org/10.1007/978-981-16-6198-3_7

Asgari, S., Pourjavadi, A., Setayeshmehr, M., Boisen, A., & Ajalloueian, F. (2021). Encapsulation of drug-loaded graphene oxide-based nanocarrier into electrospun pullulan nanofibers for potential local chemotherapy of breast cancer. *Macromolecular Chemistry and Physics, 222*(15). https://doi.org/10.1002/macp.202100096.

Ashique, S., Upadhyay, A., Gulati, M., Singh, D., Chawla, P. A., & Chawla, V. (2023). One-dimensional polymeric nanocomposites in drug delivery systems. *Current Nanoscience*, *19*(6), 825–839. https://doi.org/10.2174/1573413719666230110110706

Bagheri, M., Validi, M., Gholipour, A., Makvandi, P., & Sharifi, E. (2021). Chitosan nanofiber biocomposites for potential wound healing applications: Antioxidant activity with synergic antibacterial effect. *Bioengineering & Translational Medicine*, *7*(1), e10254. https://doi.org/10.1002/btm2.10254

Bardania, H., Jafari, F., Baneshi, M., Mahmoudi, R., Ardakani, M. T., Safari, F., & Barmak, M. J. (2023). Folic acid-functionalized albumin/graphene oxide nanocomposite to simultaneously deliver curcumin and 5-Fluorouracil into human colorectal cancer cells: An *in vitro* study. *BioMed Research International, 2023*, 1–11. https://doi.org/10.1155/2023/8334102.

Biswas, M. R. U. D., & Oh, W.-C. (2019). Comparative study on gas sensing by a schottky diode electrode prepared with graphene–semiconductor–polymer nanocomposites. *RSC Advances, 9*(20), 11484–11492. https://doi.org/10.1039/c9ra00007k.

Castro, J. I., Valencia Llano, C. H., Tenorio, D. L., Saavedra, M., Zapata, P., Navia-Porras, D. P., Delgado-Ospina, J., Chaur, M. N., Hernández, J. H. M., & Grande-Tovar, C. D. (2022). Biocompatibility assessment of polylactic acid (PLA) and nanobioglass (n-BG) nanocomposites for biomedical applications. *Molecules*, *27*(11), 3640. https://doi.org/10.3390/molecules27113640

Chan, M. H., Chen, W., Li, C. H., Fang, C. Y., Chang, Y. C., Wei, D. H., Liu, R. S., & Hsiao, M. (2021). An advanced *in situ* magnetic resonance imaging and ultrasonic theranostics nanocomposite platform: Crossing the blood–brain barrier and improving the suppression of glioblastoma using iron-platinum nanoparticles in nanobubbles. *ACS Applied Materials & Interfaces*, *13*(23), 26759–26769. https://doi.org/10.1021/acsami.1c04990

Chauhan, P., Sharma, P., Chaudhary, S., & Kumar, R. (2023). Risk governance policies for sustainable use of nanomaterials. In *Advanced functional nanoparticles "Boon or Bane" for environment remediation applications* (pp. 305–331). https://doi.org/10.1007/978-3-031-24416-2_11

Chereddy, K. K., Vandermeulen, G., & Préat, V. (2016). PLGA based drug delivery systems: Promising carriers for wound healing activity. *Wound Repair and Regeneration*, *24*(2), 223–236. https://doi.org/10.1111/wrr.12404

Costa, G. M. D., Machado, A. B., Osôrio, D. M. M., Berlese, D. B., & Hussain, C. M. (2022). Risks and effects on human health of nanomaterials. In *Environmental, ethical, and economical issues of nanotechnology* (pp. 61–72). https://doi.org/10.1201/9781003261858-3

Curulli, A. (2023). Functional nanomaterials enhancing electrochemical biosensors as smart tools for detecting infectious viral diseases. *Molecules, 28*(9), 3777. https://doi.org/10.3390/molecules28093777.

Darbasizadeh, B., Feyzi-barnaji, B., & Naderi, N. (2022). Toxicity and biocompatibility of nanomaterials: *In vivo* studies. In *Emerging nanomaterials and nano-based drug delivery approaches to combat antimicrobial resistance* (pp. 701–732). Elsevier. https://doi.org/10.1016/B978-0-323-90792-7.00014-2

Dong, Y., Li, S., Li, X., & Wang, X. (2021). Smart MXene/agarose hydrogel with photothermal property for controlled drug release. *International Journal of Biological Macromolecules*, *190*, 693–699. https://doi.org/10.1016/j.ijbiomac.2021.09.037

dos Santos, O. A. L., Pizzorno Backx, B., Abumousa, R. A., & Bououdina, M. (2022). Environmental implications associated with the development of nanotechnology: From synthesis to disposal. *Nanomaterials*, *12*(23), 4319. https://doi.org/10.3390/nano12234319

El-Maadawy, M. W., Mohamed, R. R., & Hanna, D. H. (2022). Preparation of carrageenan/chitosan-based (N,N,N-Trimeth(Yl Chitosan Chloride) silver nanocomposites as pH sensitive carrier for effective controlled curcumin delivery in cancer cells. *OpenNano, 7*, 100050. https://doi.org/10.1016/j.onano.2022.100050.

Ferrier, D. C., & Honeychurch, K. C. (2021). Carbon nanotube (CNT)-based biosensors. *Biosensors, 11*(12), 486. https://doi.org/10.3390/bios11120486.

Foyez, T., & Imran, A. B. (2022). Nanotechnology in vaccine development and constraints. In *Nanovaccinology as targeted therapeutics* (pp. 1–20). https://doi.org/10.1002/9781119858041.ch1

Gao, P., Liu, M., Tian, J., Deng, F., Wang, K., Xu, D., Liu, L., Zhang, X., & Wei, Y. (2016). Improving the drug delivery characteristics of graphene oxide based polymer nanocomposites through the "One-Pot" synthetic approach of single-electron-transfer living radical polymerization. *Applied Surface Science, 378*, 22–29. https://doi.org/10.1016/j.apsusc.2016.03.207.

Gao, J., Wang, H. L., & Iyer, R. (2010). Suppression of proinflammatory cytokines in functionalized fullerene-exposed dermal Keratinocytes. *Journal of Nanomaterials, 2010*(1), 1–9. https://doi.org/10.1155/2010/416408

Ghobadi, M., Gharabaghi, M., Abdollahi, H., Boroumand, Z., & Moradian, M. (2018). $MnFe_2O_4$-graphene oxide magnetic nanoparticles as a high-performance adsorbent for rare earth elements: Synthesis, isotherms, kinetics, thermodynamics and desorption. *Journal of Hazardous Materials*, *351*, 308–316. https://doi.org/10.1016/j.jhazmat.2018.03.011

Gholamali, I., Hosseini, S. N., & Alipour, E. (2020). Doxorubicin-loaded oxidized starch/poly (vinyl alcohol)/CuO bio-nanocomposite hydrogels as an anticancer drug carrier agent. *International Journal of Polymeric Materials and Polymeric Biomaterials, 70*(14), 967–980. https://doi.org/10.1080/00914037.2020.1767616.

Ghosh Auddy, R., Abdullah, M. F., Das, S., Roy, P., Datta, S., & Mukherjee, A. (2013). New guar biopolymer silver nanocomposites for wound healing applications. *BioMed Research International*, *2013*, 1–8. https://doi.org/10.1155/2013/912458

Gopal, A., Kant, V., Gopalakrishnan, A., Tandan, S. K., & Kumar, D. (2014). Chitosan-based copper nanocomposite accelerates healing in excision wound model in rats. *European Journal of Pharmacology*, *731*, 8–19. https://doi.org/10.1016/j.ejphar.2014.02.033

Gosai, A., Khondakar, K., Ma, X., & Ali, Md. (2021). Application of functionalized graphene oxide based biosensors for health monitoring: Simple graphene derivatives to 3D printed platforms. *Biosensors, 11*(10), 384. https://doi.org/10.3390/bios11100384.

Gould, L. J. (2016). Topical collagen-based biomaterials for chronic wounds: Rationale and clinical application. *Advances in Wound Care*, *5*(1), 19–31. https://doi.org/10.1089/wound.2014.0595

Gupta, A., Gupta, S., & Mukherjee, G. (2022). Nanotechnology and regulatory issues. In *Handbook of nanotechnology in nutraceuticals* (pp. 441–460). https://doi.org/10.1201/9781003244721-18

Gupta, M., Bala, R., Madaan, R., Chauhan, S., Chawla, R., Kaur, J., Sharifi-Rad, J., & Calina, D. (2023). Unfolding the potential of nanocomposites as drug carriers and their future scenarios. *Polymer Bulletin, 81*(7), 1–21. https://doi.org/10.1007/s00289-023-04987-z.

Hadinoto, K., Sundaresan, A., & Cheow, W. S. (2013). Lipid–polymer hybrid nanoparticles as a new generation therapeutic delivery platform: A review. *European Journal of Pharmaceutics and Biopharmaceutics*, *85*(3), 427–443. https://doi.org/10.1016/j.ejpb.2013.07.002

Halamoda-Kenzaoui, B., Geertsma, R., Pouw, J., Prina-Mello, A., Carrer, M., Roesslein, M., Sips, A., Weltring, K. M., Spring, K., & Bremer-Hoffmann, S. (2022). Future perspectives for advancing regulatory science of nanotechnology-enabled health products. *Drug Delivery and Translational Research*, *12*(9), 2145–2156. https://doi.org/10.1007/s13346-022-01165-y

Hamdan, S., Pastar, I., Drakulich, S., Dikici, E., Tomic-Canic, M., Deo, S., & Daunert, S. (2017). Nanotechnology-driven therapeutic interventions in wound healing: Potential uses and applications. *ACS Central Science*, *3*(3), 163–175. https://doi.org/10.1021/acscentsci.6b00371

Han, Y. (2022). Applications of polymer-based nanomaterials in vaccine, drug delivery, and antibacterial agents. In *International Conference on Biomedical and Intelligent Systems (IC-BIS 2022)*. https://doi.org/10.1117/12.2659848

Desai, H., & Shinde, U. (2022). Hybrid nanoparticles to cross the blood–brain barrier. In *Nanocarriers for drug-targeting brain tumors* (1st Ed., pp. 565–586). Elsevier. https://doi.org/10.1016/B978-0-323-90773-6.00002-6

Hasan, M. R., Ahommed, M. S., Daizy, M., Bacchu, M. S., Ali, M. R., Al-Mamun, M. R., Saad Aly, M. A., Khan, M. Z. H., & Hossain, S. I. (2021). Recent development in electrochemical biosensors for cancer biomarkers detection. *Biosensors and Bioelectronics: X, 8*, 100075. https://doi.org/10.1016/j.biosx.2021.100075.

Hassan, T., Salam, A., Khan, A., Khan, S. U., Khanzada, H., Wasim, M., Khan, M. Q., & Kim, I. S. (2021). Functional nanocomposites and their potential applications: A review. *Journal of Polymer Research*, *28*(2). https://doi.org/10.1007/s10965-021-02408-1

Horue, M., Cacicedo, M. L., Fernandez, M. A., Rodenak-Kladniew, B., Torres Sánchez, R. M., & Castro, G. R. (2020). Antimicrobial activities of bacterial cellulose – Silver montmorillonite nanocomposites for wound healing. *Materials Science and Engineering: C*, *116*, 111152. https://doi.org/10.1016/j.msec.2020.111152

Huang, X. W., Wei, J. J., Zhang, M. Y., Zhang, X. L., Yin, X. F., Lu, C. H., Song, J. B., Bai, S. M., & Yang, H. H. (2018). Water-based black phosphorus hybrid nanosheets as a moldable platform for wound healing applications. *ACS Applied Materials & Interfaces*, *10*(41), 35495–35502. https://doi.org/10.1021/acsami.8b12523

Huang, X., Zhu, Y., & Kianfar, E. (2021). Nano biosensors: Properties, applications and electrochemical techniques. *Journal of Materials Research and Technology, 12*, 1649–1672. https://doi.org/10.1016/j.jmrt.2021.03.048.

Hussein, M., Mumtaz, M., Nasir, I., & Abdullahi, A. (2023). Nanotechnology-based vaccines. *Biology, Medicine, & Natural Product Chemistry*, *12*(1), 343–361. https://doi.org/10.14421/biomedich.2023.121.343-361

Jayalekshmi, A. C. (2018). Nanocomposites used for drug delivery applications. In *Drug delivery nanosystems for biomedical applications* (pp 181–199). Elsevier. https://doi.org/https://doi.org/10.1016/B978-0-323-50922-0.00009-2.

Jiang, W., Huang, L., Zhang, D., Wang, Y., & Pan, G. (2021). Graphene-based composites for electrochemical sensor fabrication and their application for drug detection. *International Journal of Electrochemical Science, 16*(3), 21037. https://doi.org/10.20964/2021.03.69.

Kalashnikova, I., Das, S., & Seal, S. (2015). Nanomaterials for wound healing: Scope and advancement. *Nanomedicine*, *10*(16), 2593–2612. https://doi.org/10.2217/nnm.15.82

Kannan, P., & Maduraiveeran, G. (2023). Metal oxides nanomaterials and nanocomposite-based electrochemical sensors for healthcare applications. *Biosensors, 13*(5), 542. https://doi.org/10.3390/bios13050542.

Karthika, V., AlSalhi, M. S., Devanesan, S., Gopinath, K., Arumugam, A., & Govindarajan, M. (2020). Chitosan overlaid Fe3O4/rGO nanocomposite for targeted drug delivery, imaging, and biomedical applications. *Scientific Reports*, *10*(1). https://doi.org/10.1038/s41598-020-76015-3

Kausar, A., Ahmad, I., Zhao, T., Aldaghri, O., Ibnaouf, K. H., & Eisa, M. H. (2023). Multifunctional polymeric nanocomposites for sensing applications—design, features, and technical advancements. *Crystals, 13*(7), 1144. https://doi.org/10.3390/cryst13071144.

Kazemi, F., Naghib, S. M., Zare, Y., & Rhee, K. Y. (2020). Biosensing applications of polyaniline (PANI)-based nanocomposites: A review. *Polymer Reviews, 61*(3), 553–597. https://doi.org/10.1080/15583724.2020.1858871.

Khan, M. S., Abdelhamid, H. N., & Wu, H. F. (2015). Near infrared (NIR) laser mediated surface activation of graphene oxide nanoflakes for efficient antibacterial, antifungal and wound healing treatment. *Colloids and Surfaces B: Biointerfaces*, *127*, 281–291. https://doi.org/10.1016/j.colsurfb.2014.12.049

Khonina, T. G., Demin, A. M., Tishin, D. S., Germov, A. Y., Uimin, M. A., Mekhaev, A. V., Minin, A. S., Karabanalov, M. S., Mysik, A. A., Bogdanova, E. A., & Krasnov, V. P. (2023). Magnetic nanocomposite materials based on Fe_3O_4 nanoparticles with iron and silica glycerolates shell: Synthesis and characterization. *International Journal of Molecular Sciences, 24*(15), 12178. https://doi.org/10.3390/ijms241512178.

Kishore, K., Rajesh, S., Sivadas, S., Selvasudha, N., Barathidasan, R., & Vasanthi, H. R. (2023). Pectin encapsulated novel nanocomposite augments wound healing in Sprague Dawley rats. *Carbohydrate Polymer Technologies and Applications*, *6*, 100370. https://doi.org/10.1016/j.carpta.2023.100370

Kokilavani, S., Janani, B., Balasurya, S., & Khan, S. S. (2023). Accumulation of engineered nanomaterials in soil, water, and air. In *Agricultural and environmental nanotechnology* (pp. 551–582). https://doi.org/10.1007/978-981-19-5454-2_21

Korrapati, P. S., Karthikeyan, K., Satish, A., Krishnaswamy, V. R., Venugopal, J. R., & Ramakrishna, S. (2016). Recent advancements in nanotechnological strategies in selection, design and delivery of biomolecules for skin regeneration. *Materials Science and Engineering: C*, *67*, 747–765. https://doi.org/10.1016/j.msec.2016.05.074

Korzeniewska, E., Szczęsny, A., & Kasprowicz, D. (2020). IDEs structures created in the physical vacuum deposition process on textile substrates. *Journal of Physics: Conference Series*, *1534*(1), 012004. https://doi.org/10.1088/1742-6596/1534/1/012004

Kouser, S., Prabhu, A., Sheik, S., Prashantha, K., Nagaraja, G., Neetha Dsouza, J., Navada, K., & Manasa, D. (2021). Chitosan functionalized halloysite nanotube/poly (caprolactone) nanocomposites for wound healing application. *Applied Surface Science Advances*, *6*, 100158. https://doi.org/10.1016/j.apsadv.2021.100158

Kulkarni, D., Musale, S., Panzade, P., Paiva-Santos, A. C., Sonwane, P., Madibone, M., Choundhe, P., Giram, P., & Cavalu, S. (2022). Surface functionalization of nanofibers: The multifaceted approach for advanced biomedical applications. *Nanomaterials*, *12*(21), 3899. https://doi.org/10.3390/nano12213899

Kumar, S., Gogoi, A. S., Shukla, S., Trivedi, M., & Gulati, S. (2022). Conclusion and future prospects of chitosan-based nanocomposites. In *Chitosan-based nanocomposite materials* (1st Ed., pp. 305–341). Springer. https://doi.org/10.1007/978-981-19-5338-5_14

Kumar, Y., Sinha, A. S. K., Nigam, K. D. P., Dwivedi, D., & Sangwai, J. S. (2023). Functionalized nanoparticles: Tailoring properties through surface energetics and coordination chemistry for advanced biomedical applications. *Nanoscale*, *15*(13), 6075–6104. https://doi.org/10.1039/d2nr07163k

Kumari, S., Singh, J., Kharavtekar, S. S., Grewal, D., Ghosh Ray, S., Bhardwaj, A., Dubey, S. K., Kesharwani, P., Arora, T., & Taliyan, R. K. (2023). Futuristic aspect of nanocarriers on targeted delivery for dementia. In *Nanomedicine-based approaches for the treatment of dementia* (1st Ed., pp. 265–294). Elsevier. https://doi.org/10.1016/B978-0-12-824331-2.00005-4

Lee, S. K., & Wolfe, S. W. (2000). Peripheral nerve injury and repair. *Journal of the American Academy of Orthopaedic Surgeons*, *8*(4), 243–252. https://doi.org/10.5435/00124635-200007000-00005

Li, C., Zhang, X., Huang, X., Wang, X., Liao, G., & Chen, Z. (2013). Preparation and characterization of flexible nanoliposomes loaded with daptomycin, a novel antibiotic, for topical skin therapy. *International Journal of Nanomedicine*, *8*, 1285–1292. https://doi.org/10.2147/ijn.s41695

Liang, J., Wei, M., Wang, Q., Zhao, Z., Liu, A., Yu, Z., & Tian, Y. (2017). Sensitive electrochemical determination of hydrogen peroxide using copper nanoparticles in a polyaniline film on a glassy carbon electrode. *Analytical Letters*, *51*(4), 512–522. https://doi.org/10.1080/00032719.2017.1343832

Liu, Y., Hong, H., Xue, J., Luo, J., Liu, Q., Chen, X., Pan, Y., Zhou, J., Liu, Z., & Chen, T. (2021). Near-infrared radiation-assisted drug delivery nanoplatform to realize blood–brain barrier crossing and protection for Parkinsonian Therapy. *ACS Applied Materials & Interfaces*, *13*(31), 37746–37760. https://doi.org/10.1021/acsami.1c12675

Mao, C., Xiang, Y., Liu, X., Cui, Z., Yang, X., Li, Z., Zhu, S., Zheng, Y., Yeung, K. W. K., & Wu, S. (2018). Repeatable photodynamic therapy with triggered signaling pathways of fibroblast cell proliferation and differentiation to promote bacteria-accompanied wound healing. *ACS Nano*, *12*(2), 1747–1759. https://doi.org/10.1021/acsnano.7b08500

Marchesan, S., Bosi, S., Alshatwi, A., & Prato, M. (2016). Carbon nanotubes for organ regeneration: An electrifying performance. *Nano Today, 11* (4), 398–401. https://doi.org/10.1016/j.nantod.2015.11.007.

Matica, M. A., Aachmann, F. L., Tøndervik, A., Sletta, H., & Ostafe, V. (2019). Chitosan as a wound dressing starting material: Antimicrobial properties and mode of action. *International Journal of Molecular Sciences*, *20*(23), 5889. https://doi.org/10.3390/ijms20235889

Meera, A. P., Pillai, R. R., & Sreelekshmi, P. B. (2023). Novel polymer nanocomposites: Synthesis, designing and cost-effective biomedical applications. In *Bio-inspired nanotechnology* (1st Ed., pp. 56–72). Bentham Science. https://doi.org/10.2174/9789815080179123010006

Mihai, M. M., Dima, M. B., Dima, B., & Holban, A. M. (2019). Nanomaterials for wound healing and infection control. *Materials*, *12*(13), 2176. https://doi.org/10.3390/ma12132176

Mir, T. U. G., Katoch, V., Angurana, R., Wani, A. K., Shukla, S., El Messaoudi, N., Sher, F., Mulla, S. I., & Américo-Pinheiro, J. H. P. (2023). Environmental and toxicological concerns associated with nanomaterials used in the industries. In *Nanomaterials for bioreactors and bioprocessing applications* (pp, 141–193). https://doi.org/10.1016/b978-0-323-91782-7.00010-2

Moore, W. R., Graves, S. E., & Bain, G. I. (2001). Synthetic bone graft substitutes. *ANZ Journal of Surgery, 71*(6), 354–361. https://doi.org/10.1046/j.1440-1622.2001.02128.x.

Mugabe, C., Azghani, A. O., & Omri, A. (2005). Liposome-mediated gentamicin delivery: Development and activity against resistant strains of Pseudomonas aeruginosa isolated from cystic fibrosis patients. *Journal of Antimicrobial Chemotherapy*, *55*(2), 269–271. https://doi.org/10.1093/jac/dkh518

Muniyandi, P., Palaninathan, V., Veeranarayanan, S., Ukai, T., Maekawa, T., Hanajiri, T., & Mohamed, M. S. (2020). ECM mimetic electrospun porous poly (L-lactic acid) (PLLA) scaffolds as potential substrates for cardiac tissue engineering. *Polymers, 12*(2), 451. https://doi.org/10.3390/polym12020451.

Nagaraja, K., Rao, K. M., Reddy, G. V., & Rao, K. S. V. K. (2021). Tragacanth gum-based multifunctional hydrogels and green synthesis of their silver nanocomposites for drug delivery and inactivation of multidrug resistant bacteria. *International Journal of Biological Macromolecules, 174*, 502–511. https://doi.org/10.1016/j.ijbiomac.2021.01.203.

Naresh, V., & Lee, N. (2021). A review on biosensors and recent development of nanostructured materials-enabled biosensors. *Sensors, 21*(4), 1109. https://doi.org/10.3390/s21041109.

Naseri, M., Fotouhi, L., & Ehsani, A. (2018). Recent progress in the development of conducting polymer-based nanocomposites for electrochemical biosensors applications: A mini-review. *The Chemical Record*, *18*(6), 599–618. https://doi.org/10.1002/tcr.201700101

Naskar, A., Lee, S., & Kim, K. S. (2020). Easy one-pot low-temperature synthesized Ag-ZnO nanoparticles and their activity against clinical isolates of methicillin-resistant Staphylococcus aureus. *Frontiers in Bioengineering and Biotechnology*, *8*. https://doi.org/10.3389/fbioe.2020.00216

Nazari, H., Heirani-Tabasi, A., Hajiabbas, M., Khalili, M., Alavijeh, M. S., Hatamie, S., Gorabi, A. M., Esmaeili, E., & Ahmadi Tafti, S. H. (2019). Incorporation of two-dimensional nanomaterials into silk fibroin nanofibers for cardiac tissue engineering. *Polymers for Advanced Technologies*, *31*(2), 248–259. https://doi.org/10.1002/pat.4765.

Nie, L., Wei, Q., Zhao, C., Ding, P., Valentine Okoro, O., & Shavandi, A. (2023). Chitosan-based nanocomposites as efficient wound dressing materials. In *Chitosan nanocomposites* (1st Ed., pp. 181–199). Springer. https://doi.org/10.1007/978-981-19-9646-7_8

Padmanabhan, V. P., Sivashanmugam, P., S. M., M., Sagadevan, S., & Kulandaivelu, R. (2023). The development of ZnO nanoparticle-embedded graphitic-carbon nitride towards triple-negative breast cancer therapy. *RSC Advances, 13*(35), 24333–24342. https://doi.org/10.1039/d3ra02887a.

Place, E. S., Evans, N. D., & Stevens, M. M. (2009). Complexity in biomaterials for tissue engineering. *Nature Materials, 8*(6), 457–470. https://doi.org/10.1038/nmat2441.

Pryjmaková, J., Kaimlová, M., Hubáček, T., Švorčík, V., & Siegel, J. (2020). Nanostructured materials for artificial tissue replacements. *International Journal of Molecular Sciences*, *21*(7), 2521. https://doi.org/10.3390/ijms21072521

Qiao, K., Guo, S., Zheng, Y., Xu, X., Meng, H., Peng, J., Fang, Z., & Xie, Y. (2018). Effects of graphene on the structure, properties, electro-response behaviors of GO/PAA composite hydrogels and influence of electro-mechanical coupling on BMSC differentiation. *Materials Science and Engineering: C, 93*, 853–863. https://doi.org/10.1016/j.msec.2018.08.047.

Qin, W., Wu, Y., Liu, J., Yuan, X., & Gao, J. (2022). A comprehensive review of the application of nanoparticles in diabetic wound healing: Therapeutic potential and future perspectives. *International Journal of Nanomedicine*, *17*, 6007–6029. https://doi.org/10.2147/ijn.s386585

Rajwade, J. M., Kawle, K., Kulkarni, S., & Kowshik, M. (2023). Wound treatment using nanomaterials. In *Nanobiomaterials – Perspectives for medical applications in the diagnosis and treatment of diseases* (1st Ed., Vol. 145, pp. 207–235). Materials Research Foundations. https://doi.org/10.21741/9781644902370-8

Rakhshaei, R., Namazi, H., Hamishehkar, H., Kafil, H. S., & Salehi, R. (2019). *In Situ* synthesized chitosan–gelatin/ZnO nanocomposite scaffold with drug delivery properties: Higher antibacterial and lower cytotoxicity effects. *Journal of Applied Polymer Science*, *136*(22). https://doi.org/10.1002/app.47590.

Ramesh, M., Janani, R., Deepa, C., & Rajeshkumar, L. (2022). Nanotechnology-enabled biosensors: A review of fundamentals, design principles, materials, and applications. *Biosensors, 13*(1), 40. https://doi.org/10.3390/bios13010040.

Ramezani, M., & Ripin, Z. M. (2023). An overview of enhancing the performance of medical implants with nanocomposites. *Journal of Composites Science*, *7*(5), 199. https://doi.org/10.3390/jcs7050199

Rigo, C., Ferroni, L., Tocco, I., Roman, M., Munivrana, I., Gardin, C., Cairns, W., Vindigni, V., Azzena, B., Barbante, C., & Zavan, B. (2013). Active silver nanoparticles for wound healing. *International Journal of Molecular Sciences*, *14*(3), 4817–4840. https://doi.org/10.3390/ijms14034817

Rostami, M., Nasab, A. S., Fasihi-Ramandi, M., Badiei, A., Ganjali, M. R., Rahimi-Nasrabadi, M., & Ahmadi, F. (2021). Cur-loaded magnetic $ZnFe_2O_4$@mZnO-Ox-p-g-C3N4 composites as dual pH- and ultrasound responsive nano-carriers for controlled and targeted cancer chemotherapy. *Materials Chemistry and Physics*, *271*, 124863. https://doi.org/10.1016/j.matchemphys.2021.124863

Sari, M. H. M., Cobre, A. D. F., Pontarolo, R., & Ferreira, L. M. (2023). Status and future scope of soft nanoparticles-based hydrogel in wound healing. *Pharmaceutics*, *15*(3), 874. https://doi.org/10.3390/pharmaceutics15030874

Singh, P., & Singh, R. B. (2022). Beneficial utility and perspective of nanomaterials toward biosensing. In *Biotechnological advances for microbiology, molecular biology, and nanotechnology* (1st Ed., p. 34). Apple Academic Press.

Singh, R. P., Singh, P., & Singh, K. R. B. (2021). Introduction to composite materials: Nanocomposites and their potential applications. In *Composite materials* (pp. 1–28). CRC Press. https://doi.org/DOI:10.1201/9781003080633-1.

Shahzadi, I., Islam, M., Saeed, H., Shahzadi, A., Haider, J., Haider, A., Imran, M., Rathore, H. A., Ul-Hamid, A., Nabgan, W., & Ikram, M. (2023). Facile synthesis of copolymerized cellulose grafted hydrogel doped calcium oxide nanocomposites with improved antioxidant activity for anti-arthritic and controlled release of doxorubicin for anti-cancer evaluation. *International Journal of Biological Macromolecules*, *235*, 123874. https://doi.org/10.1016/j.ijbiomac.2023.123874.

Sharma, R., & Alam, M. N. (2023). A mini review on toxic effects of nanomaterials on human health. *International Journal of Applied Research*, *9*(4), 200–202. https://doi.org/10.22271/allresearch.2023.v9.i4c.10759

Sharma, D., Kanchi, S., Sabela, M. I., & Bisetty, K. (2016). Insight into the biosensing of graphene oxide: Present and future prospects. *Arabian Journal of Chemistry, 9*(2), 238–261. https://doi.org/10.1016/j.arabjc.2015.07.015

Shrivastava, N., Shrivastava, V., Singh Tomar, R., & Jyoti, A. (2022). Environmental and ecotoxicological impacts of engineered nanomaterials. In *Environmental toxicology and ecosystem* (pp. 11–33). https://doi.org/10.1201/9781003244349-2

Singla, R., Abidi, S. M. S., Dar, A. I., & Acharya, A. (2019). Nanomaterials as potential and versatile platform for next generation tissue engineering applications. *Journal of Biomedical Materials Research Part B: Applied Biomaterials, 107*(7), 2433–2449. https://doi.org/10.1002/jbm.b.34327

Singla, R., Soni, S., Patial, V., Kulurkar, P. M., Kumari, A., S., M., Padwad, Y. S., & Yadav, S. K. (2017). *In vivo* diabetic wound healing potential of nanobiocomposites containing bamboo cellulose nanocrystals impregnated with silver nanoparticles. *International Journal of Biological Macromolecules, 105*, 45–55. https://doi.org/10.1016/j.ijbiomac.2017.06.109

Sinha Ray, S., & Okamoto, M. (2003). Polymer/layered silicate nanocomposites: A review from preparation to processing. *Progress in Polymer Science, 28*(11), 1539–1641. https://doi.org/10.1016/j.progpolymsci.2003.08.002.

Soliman, M., Sadek, A. A., Abdelhamid, H. N., & Hussein, K. (2021). Graphene oxide-cellulose nanocomposite accelerates skin wound healing. *Research in Veterinary Science, 137*, 262–273. https://doi.org/10.1016/j.rvsc.2021.05.013

Song, J. G., Lee, S. H., & Han, H. K. (2021). Development of an M cell targeted nanocomposite system for effective oral protein delivery: Preparation, *in vitro* and *in vivo* characterization. *Journal of Nanobiotechnology, 19*(1). https://doi.org/10.1186/s12951-020-00750-y

Su, P.-G., & Lu, P.-H. (2020). Electrical and humidity-sensing properties of impedance-type humidity sensors that were made of Ag microwires/PPy/SnO2 ternary composites. *Chemosensors, 8*(4), 92. https://doi.org/10.3390/chemosensors8040092.

Tamer, T. M., Sabet, M. M., Omer, A. M., Abbas, E., Eid, A. I., Mohy-Eldin, M. S., & Hassan, M. A. (2021). Hemostatic and antibacterial PVA/Kaolin composite sponges loaded with penicillin–streptomycin for wound dressing applications. *Scientific Reports, 11*(1). https://doi.org/10.1038/s41598-021-82963-1

Tavakolizadeh, M., Atarod, M., Seyyed Tabaei, S. J., Sojdeh, S., Nazarzadeh Zare, E., Rabiee, M., & Rabiee, N. (2023). Green modified-UiO-66/MXene sandwich composites for gene-chemotherapy synergistic cancer suppression: Co-delivery of doxorubicin and pCRISPR. *Alexandria Engineering Journal, 80*, 144–154. https://doi.org/10.1016/j.aej.2023.08.046.

Tripathi, A., & Bonilla-Cruz, J. (2023). Review on healthcare biosensing nanomaterials. *ACS Applied Nano Materials, 6*(7), 5042–5074. https://doi.org/10.1021/acsanm.3c00941

Tripathi, D., Katiyar, S., Chaurasia, A. K., & Mishra, A. (2023). Nanomaterials for biosensing applications. In *Recent advances in biosensor technology* (1st Ed., pp. 1–29). Bentham Science. https://doi.org/10.2174/9789815123739123010004

Tsivintzelis, I., Marras, S. I., Zuburtikudis, I., & Panayiotou, C. (2007). Porous poly(l-lactic acid) nanocomposite scaffolds prepared by phase inversion using supercritical CO_2 as antisolvent. *Polymer, 48*(21), 6311–6318. https://doi.org/10.1016/j.polymer.2007.08.021.

Vasile, B. S., Oprea, O., Voicu, G., Ficai, A., Andronescu, E., Teodorescu, A., & Holban, A. (2014). Synthesis and characterization of a novel controlled release zinc oxide/gentamicin–chitosan composite with potential applications in wounds care. *International Journal of Pharmaceutics*, *463*(2), 161–169. https://doi.org/10.1016/j.ijpharm.2013.11.035

Vijayan, V. M., Hernandez-Moreno, G., & Thomas, V. (2022). Future of nanotechnology in tissue engineering. *Tissue Engineering*, 193–236. https://doi.org/10.1016/b978-0-12-824064-9.00003-4

Vijayavenkataraman, S. (2020). Nerve guide conduits for peripheral nerve injury repair: A review on design, materials and fabrication methods. *Acta Biomaterialia, 106*, 54–69. https://doi.org/10.1016/j.actbio.2020.02.003.

Vimala, K., Maruthupandian, A., Thangaraj, R., & Kannan, S. (2020). Selenium tethered mesoporous silica nanocomposite enhances drug delivering efficiency to target breast cancer. *Journal of Cluster Science, 32*(6), 1475–1489. https://doi.org/10.1007/s10876-020-01906-4.

Wang, L., Shi, Y., Jiang, J., Li, C., Zhang, H., Zhang, X., Jiang, T., Wang, L., Wang, Y., & Feng, L. (2022). Micro-nanocarriers based drug delivery technology for blood-brain barrier crossing and brain tumor targeting therapy. *Small*, *18*(45). https://doi.org/10.1002/smll.202203678

Wang, Z., Zhang, P., Ding, X., Wang, J., Sun, Y., Yin, C., Wang, W., Fan, C., & Sun, D. (2021). Co-delivery of ampicillin and β-lactamase inhibitor by selenium nanocomposite to achieve synergistic anti-infective efficiency through overcoming multidrug resistance. *Chemical Engineering Journal*, *414*, 128908. https://doi.org/10.1016/j.cej.2021.128908

Williams, M. A., & Bannon, D. I. (2022). Knowledge gaps, future directions, and the emergence of nanoplastics as an environmental threat pollutant. In *Toxicology of nanoparticles and nanomaterials in human, terrestrial and aquatic systems* (pp. 423–451). https://doi.org/10.1002/9781119316329.ch16

Wu, P., Sun, Y., Dong, W., Zhou, H., Guo, S., Zhang, L., Wang, X., Wan, M., & Zong, Y. (2019). Enhanced anti-tumor efficacy of hyaluronic acid modified nanocomposites combined with sonochemotherapy against subcutaneous and metastatic breast tumors. *Nanoscale, 11*(24), 11470–11483. https://doi.org/10.1039/c9nr01691k.

Wu, C., Zhu, W., Jin, R., Ai, H., & Xu, Y. (2021). The MRI-visible nanocomposite facilitates the delivery and tracking of siRNA loaded DC vaccine in the breast cancer model. *Frontiers in Oncology*, *10*. https://doi.org/10.3389/fonc.2020.621642

Xia, T., Zhang, H., Wang, S., Xin, W., & Liu, S. (2023). Environmental health and safety of engineered nanomaterials. *Nanomedicine*, 801–846. https://doi.org/10.1007/978-981-16-8984-0_23

Yazaki, Y., Oyane, A., Sogo, Y., Ito, A., Yamazaki, A., & Tsurushima, H. (2011). Control of gene transfer on a DNA–fibronectin–apatite composite layer by the incorporation of carbonate and fluoride ions. *Biomaterials, 32*(21), 4896–4902. https://doi.org/10.1016/j.biomaterials.2011.03.021.

Ye, L., He, X., Obeng, E., Wang, D., Zheng, D., Shen, T., Shen, J., Hu, R., & Deng, H. (2023). The CuO and AgO co-modified ZnO nanocomposites for promoting wound healing in Staphylococcus aureus infection. *Materials Today Bio*, *18*, 100552. https://doi.org/10.1016/j.mtbio.2023.100552

Zafar, A., Arshad, R., Ur.Rehman, A., Ahmed, N., & Akhtar, H. (2023). Recent developments in oral delivery of vaccines using nanocarriers. *Vaccines*, *11*(2), 490. https://doi.org/10.3390/vaccines11020490

Zaffagnini, S., Giordano, G., Vascellari, A., Bruni, D., Neri, M. P., Iacono, F., Kon, E., Presti, M. L., & Marcacci, M. (2006). Arthroscopic collagen meniscus implant results at 6 to 8 years follow up. *Knee Surgery, Sports Traumatology, Arthroscopy, 15*(2), 175–183. https://doi.org/10.1007/s00167-006-0144-4.

Zee, M. V. D., de Vries, C., Masa, M., Morales, M., Rayo, M., & Hegger, I. (2022). Regulatory aspects of a nanomaterial for imaging therapeutic cells. *Drug Delivery and Translational Research, 13*, 2693–2703. https://doi.org/10.21203/rs.3.rs-2212824/v1.

Zhang, Y., Hou, Z., Ge, Y., Deng, K., Liu, B., Li, X., Li, Q., Cheng, Z., Ma, P., Li, C., & Lin, J. (2015). DNA-hybrid-gated photothermal mesoporous silica nanoparticles for NIR-responsive and aptamer-targeted drug delivery. *ACS Applied Materials & Interfaces, 7*(37), 20696–20706. https://doi.org/10.1021/acsami.5b05522.

Zhang, G., Jiang, B., Wu, C., Liu, Y., He, Y., Huang, X., Chen, W., Xi, K., Guo, H., Zhao, X., & Jia, X. (2020). Thin platelet-like COF nanocomposites for blood brain barrier transport and inhibition of brain metastasis from renal cancer. *Journal of Materials Chemistry B*, *8*(20), 4475–4488. https://doi.org/10.1039/d0tb00724b

Zhou, Z., Joslin, S., Dellinger, A., Ehrich, M., Brooks, B., Ren, Q., Rodeck, U., Lenk, R., & Kepley, C. L. (2010). A novel class of compounds with cutaneous wound healing properties. *Journal of Biomedical Nanotechnology*, *6*(5), 605–611. https://doi.org/10.1166/jbn.2010.1157

Zoghi, M., Pourmadadi, M., Yazdian, F., Nigjeh, M. N., Rashedi, H., & Sahraeian, R. (2023). Synthesis and characterization of chitosan/carbon quantum dots/Fe2O3 nanocomposite comprising curcumin for targeted drug delivery in breast cancer therapy. *International Journal of Biological Macromolecules, 249*, 125788. https://doi.org/10.1016/j.ijbiomac.2023.125788.

Züger, F., Marsano, A., Poggio, M., & Gullo, M. R. (2022). Nanocomposites in 3D bioprinting for engineering conductive and stimuli-responsive constructs mimicking electrically sensitive tissue. *Advanced NanoBiomed Research*, *2*(2), 2270021. https://doi.org/10.1002/anbr.202270021

CHAPTER 15

India, US, and Other Foreign Regulatory Guidelines for Nanocomposites

MOHAMMAD KASHIF IQUBAL[1], SWAROOP J. PANSARE[1], RIZWAN SHAIKH[1], BHANU P. DONGALA[1], SUNIL K. THOTA[1], NASEEM A CHAROO[2], MANSOOR A. KHAN[1], and ZIYAUR RAHMAN[1,*]

[1]*Department of Pharmaceutical Sciences, Irma Lerma Rangel School of Pharmacy, Texas A&M Health Science Center, Texas A&M University, College Station, TX, USA*

[2]*Adcan Pharma LLC, Mussafah, Abu Dhabi, UAE*

*Corresponding author.

ABSTRACT

The sponsor has to submit NDA (new drug application) and ANDA (abbreviated NDA) for nanotechnology-based brand and generic versions of the drug product, respectively. The Food and Drug Administration's (FDA) basic requirements of safety and efficacy apply to nanomedicines as well. The FDA may request additional data from the sponsor regarding the safety and effectiveness of nanomedicines. For example, environmental and toxicological assessments may be included in the safety evaluation of nanotechnology-based therapeutic products. Particle size distribution has a major impact on product safety and efficacy. Nanomedicines should be characterized for quality especially particle size by appropriate methods. Other regulatory agencies also provided regulatory and technical guidance for nanotechnology-derived drug products. For example, the Indian government recently published guidelines on the evaluation of nanotechnology-derived drug products. Similarly, the European Medicines Agency

Nanocomposites and Nanomaterials in Biomedical Applications. Vivek Dave, Deependra Singh, and Pallerla Naveen Reddy (Eds.)

DOI: 10.1201/9781779641236-15

also published technical guidelines on specific nanomedicines such as ion-based colloidal products, liposomes, micelles, etc. This chapter provides nanotechnology-based drug products' science and regulatory aspects in the USA, Europe, Canada, Japan, China and India.

15.1 INTRODUCTION

Nanocomposites are nano-sized admixtures of organic and/or inorganic materials (Sen, 2020). Certain features such as controlled drug release and biocompatibility, can be engineered into the nanocomposite materials (Chabria et al., 2021). Utilizing nanocomposite materials as drug delivery systems to target specific organs or tissues can dramatically increase the therapeutic efficacy and lower the incidence of adverse events (Abd Elkodous et al., 2021). Drug delivery systems can use a variety of nanocomposite materials, such as micelles, lipid-based nanoparticles, and polymeric nanoparticles (Chabria et al., 2021). The biodistribution of nanocomposites can be broad or localized in specific tissues or organs depending on size, shape, charge, and surface chemistry. Thus, nanocomposites can improve the safety and efficacy of drugs; however, these attributes may be inconsistent or compromised if quality is not controlled and monitored. Therefore, to ensure the quality, efficacy, and safety of nanomedicines, various countries broadened the existing regulatory framework to cover nanomedicines.

To understand and control matter at the nanoscale and reap its potential benefits, the United States launched the National Nanotechnology Initiative (NNI) in 2001. Nanotechnology is "the understanding and control of matter at dimensions between approximately 1 and 100 nanometers (nm), where unique phenomena enable novel applications." The US Food and Drug Administration (FDA) also engages with NNI to uphold its mandate to safeguard public health and deliver vital human services (NNI, 2023b). In anticipation of submissions of drug products involving the application of nanotechnology, the agency launched the "Nanotechnology Task Force (NTF)" in 2006 to help the FDA evaluate the current state of nanotechnology. NTF provided a comprehensive recommendation to the agency in 2007 (FDA, 2014a). The use of nanotechnology-based medicines in illness prevention and therapy is known as nanomedicines. The European Science Foundation defines nanomedicines as "the science and technology of diagnosing, treating, and preventing disease and traumatic injury, of relieving pain, and of preserving and improving human health, using molecular tools and molecular knowledge of the human body" (Sorg, 2005). Whereas the National Institutes of Health (NIH) defines nanomedicine as "An offshoot of nanotechnology which refers to highly specific medical interventions at the molecular scale for curing disease or repairing damaged tissues, such as bone, muscle, or nerve." As far as the FDA is concerned, it has yet to establish a regulatory definition of "nanotechnology," "nanomaterial," "nanoscale," "nanomedicine" or other related terms (FDA, 2014b, 2022b).

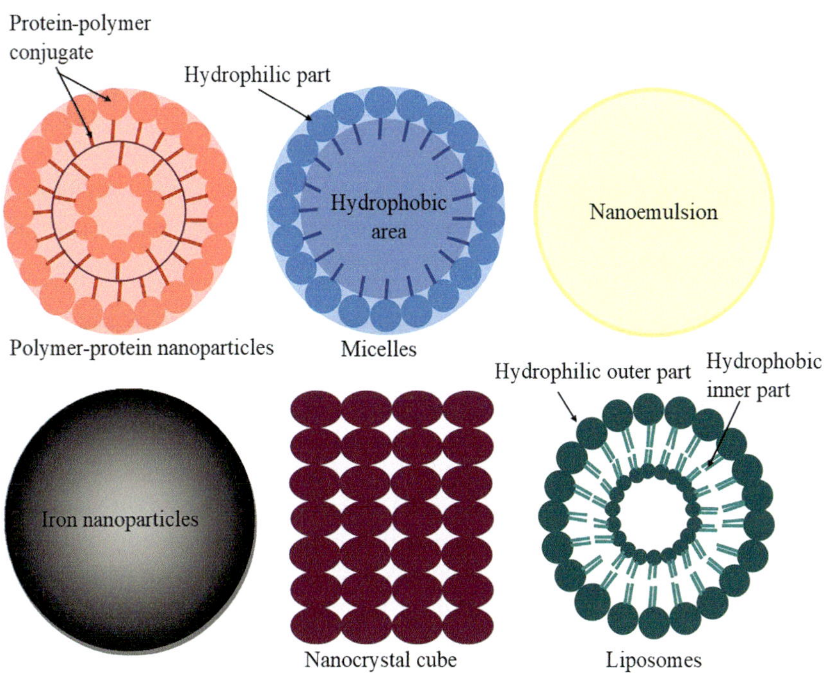

FIGURE 15.1 Various types of clinically available nanomedicines.

Due to the safety and efficacy concerns of nanomedicines, the FDA is allocating more resources to scientific research to identify techniques that can reliably characterize quality attributes and identify and forecast the possible impact of those modifications on the health of humans and/or animals. To obtain more insight into these products, the FDA has taken several actions, including starting internal research projects, collaborating with academia and industry, and compiling data on the efficacy, safety, and quality of nanomedicine products that have already received the agency approval (FDA, 2018b). The agency intends to use this information in the review process and regulatory decisions on submissions involving nanomedicines. According to the agency, the challenges pertaining to the regulation of nanomedicines can be adequately addressed by the current regulatory framework (Tinkle et al., 2014). The quality by design (QbD) paradigm can also be used to explore the relationship between critical product quality attributes (CQAs), and the safety and efficacy of nanomedicines. Additionally, the industry is required to follow numerous general and product-specific guidelines issued by the FDA regarding product development and data requirements for the filing of new drug applications (NDAs) and abbreviated NDAs (ANDAs). Several nanomedicines, including micelles, liposomes, nanocrystals, biologicals, polymeric nanoparticles, and nanoemulsions (Figure 15.1 and Table 15.1) have already been approved by the FDA and EMA (European Medicines Agency). These products are nanotechnology-derived medicines although do not meet the NNI size criteria for nanotechnology (NNI, 2023a).

15.2 QUALITY BY DESIGN

The FDA recommends the implementation of the QbD approach in the formulation and process development of a drug product. It is a proactive, scientific, and risk-based approach to new product development that is not limited to standard dosage forms like pills and capsules only; it can also be deployed to complex products such as nanomedicines. Product development using the QbD approach starts with the identification of quality, safety, and efficacy aspects of nanomedicine to propose a quality target product profile (QTPP). The QTPP profile for a nanomedicine can be determined by considering the mode of administration, type of nanomedicine (liposomes, nanocrystals, etc.), container closure system, strength, and characteristics that may impact pharmacokinetics and quality features. The next step in product development is to describe CQAs. These are "physical, chemical, biological, or microbiological attributes that should fall within a proven range or limit to ensure the desired product quality."

TABLE 15.1 Approved Nanomedicines in Various Regulated Markets.

Nanocarriers	Active Ingredients	Brand Name	Therapeutic Use	Company	Approving Agency and Approval Year	Application Number	References
Polymer-protein Conjugate nanoparticles	PEGylated interferonalpha-2A	Pegasys®	Hepatitis B and Hepatitis C	Zr Pharma and GmbH Pharma and GmbH	FDA and EMA—2002	BLA103964 EMEA/H/C/000395	Hui and Lau (2005), Pharmaand (2002), ZrPharma (2002)
	Pegvisomant (PEGylated recombinant human growth hormone receptor antagonist)	Somavert®	Acromegaly	Pfizer Europe MA EEIG; Pharmacia & Upjohn Company LLC	EMA—2002 FDA—2003	EMEA/H/C/000409 BLA021106	Pfizer (2002), Pharmacia-&-Upjohn-Company-LLC (2003)
	Epoetin beta	Mircera®	Renal anemia	Vifor Pharma	EMA and FDA—2007	EMEA/H/C/000739 BLA125164	Bartnicki et al. (2013), Roche (2007), ViforPharma (2023)
	Certolizumab pegol (PEGylated certolizumab pegol)	Cimzia®	Rheumatoid arthritis, Crohn's disease, psoriatric arthritis, ankylosing spondylitis	UCB, Inc.	FDA—2008 EMA—2009	BLA125160 EMEA/H/C/001037	Curtis et al. (2019), UCB-Inc (2009), UCB-Pharma (2009)
	Peginterferon beta-1a	Plegridy®	Relapsing remitting multiple sclerosis in adults	Biogen Inc.	FDA—2014	BLA125499	BiogenInc (2023), Gohil (2014), https://dailymed.nlm.nih.gov/dailymed/drugInfo.cfm?setid=08f0ea03-4e6d-195d-aef4-886e32befa95
	Pegaspargase (Modified L-asparaginase covalently conjugated with monomethoxypolyethylene glycol)	Oncaspar®	Acute lymphoblastic leukemia	Servier Pharmaceuticals LLC	FDA—1994	BLA103411	Ettinger (1995), Servier Pharmaceuticals-LLC (1994a, 1994b)

TABLE 15.1 *(Continued)*

Nanocarriers	Active Ingredients	Brand Name	Therapeutic Use	Company	Approving Agency and Approval Year	Application Number	References
Albumin nanoparticle	Paclitaxel	Abraxane®	Metastatic breast cancer	Bristol-Myers Squibb Co	FDA—2005 EMA—2008	NDA021660 EMEA/H/C/000778	Abraxis-Bioscience (2022), Bristol-Myers-Squibb-Pharma (2007), Yuan et al. (2020)
Micelle	Paclitaxel	Apealea®	Ovarian cancer, peritoneal cancer, fallopian tube cancer	Inceptua AB Gustavslundsv	EMA—2018	EMEA/H/C/004154	Borgå et al. (2019), Oasmia (2018)
Nanoemulsion	Propofol	Diprivan®	Anesthetic agent for induction and maintenance of anesthesia. For sedation of patient under critical care and those requiring mechanical ventilator.	Fresenius-Kabi USA LLC	FDA—1996	NDA019627	Fresenius Kabi (1996, 2020), Terblanche and Coetzee (2008)
	Cyclosporine	Restasis®	Chronic dry eye	Abbvie Inc.	FDA—2002	NDA050790	Allergan (2003), Lallemand et al. (2017)
	Estradiol hemihydrate	Estrasorb®	Treatment of moderate to severe vasomotor symptoms in postmenopausal women.	Exeltis USA, Inc	*FDA—2003	NDA021371	Ei (2015), Novavax (2003a, 2003b), Simon (2006)
Liposome	Doxorubicin hydrochloride	Doxil®	Metastatic ovarian cancer, HIV associated Kaposi's sarcoma, multiple myeloma	Baxter Healthcare Corp	FDA—1995	NDA050718	Barenholz (2012), Baxter (1995, 2019), James (1995)

TABLE 15.1 *(Continued)*

Nanocarriers	Active Ingredients	Brand Name	Therapeutic Use	Company	Approving Agency and Approval Year	Application Number	References
	Poractant alfa	Curosurf®	Respiratory distress syndrome (RDS)	Chiesi USA, Inc.	FDA—1999	BLA020744	Chiesi (1999), Mousseau et al. (2020), Ramanathan et al. (2004), Vrbanović Mijatović et al. (2017)
	Doxorubicin hydrochloride and an anthracycline cytotoxic agent	Myocet®	Metastatic breast cancer	Teva Pharmaceutical Industries Ltd.	EMA—2000	EMEA/H/C/000297	Brucker et al. (2016), Teva (2000)
	Verteporfin	Visudyne®	Severe eye conditions, such as macular degeneration, decreased vision, ocular, histoplasmosis, and pathologic myopia	Valeant Pharmaceuticals Luxembourg SARL	FDA and EMA—2000	NDA021119 EMEA/H/C/000305	Bressler and Bressler (2000), Cheplapharm (2000), QLT-Ophthalmics (2000), Valeant (2000)
	Morphine sulfate	DepoDur®	Postoperative analgesia	Pacira Pharmaceuticals Inc.	*FDA—2004	NDA021671	Pacira (2004, 2008), Pasero and McCaffery (2005)
	Mifamurtide	Mepact®	High grade non-metastatic Osteosarcoma and myosarcoma	Takeda France SAS	EMA—2009	EMEA/H/C/000802	Takeda (2009a, 2009b)
	Vincristine	Marqibo®	Philadelphia chromosome- Negative acute lymphoblastic leukemia in adult patients	Acrotech Biopharma LLC	*FDA—2012	NDA202497	Silverman and Deitcher (2013), Takeda (2009a)

TABLE 15.1 *(Continued)*

Nanocarriers	Active Ingredients	Brand Name	Therapeutic Use	Company	Approving Agency and Approval Year	Application Number	References
	Irinotecan	Onivyde®	Metastatic pancreatic cancer	Ipsen Biopharmaceuticals Inc	FDA—2015	NDA207793	Ipsen (2015), Milano et al. (2022)
	Paclitaxel	Lipusu®	Lung squamous cell carcinoma	Luye Pharma	#NRLD—2020	NA	Luye (2020); Zhang et al. (2022)
	Doxorubicin	Caelyx®	Metastatic breast cancer, ovarian cancer, AIDS-related Kaposi's sarcoma	Baxter Holding B.V.	EMA—1996	EMEA/H/C/000089	Baxter (1996), Ranson et al. (2001)
	mRNA vaccine	Pfizer-BioNTech vaccine	Prevention of COVID-19 infection	Pfizer Pharmaceuticals	FDA—2020	NA	Attia et al. (2021), Pfizer (2023)
	mRNA vaccine	Moderna Covid-19 Vaccine	Prevention of COVID-19 infection	Moderna US, Inc.	FDA—2020	NA	Attia et al. (2021), (ModernaTX (2023)
	Amphotericin B	AmBiosome®	Fungal infections in febrile neutropenic patients; Aspergillosis, candidiasis, and cryptococcosis infections refractory to amphotericin B	Astellas Pharma US, Inc.	FDA—1997	NDA050740	Astellas (2020), Boswell et al. (1998)
	Doxorubicin hydrochloride	Celdoxome pegylated liposomal	Breast Neoplasms, Ovarian Neoplasm, Multiple Myeloma, Sarcoma, Kaposi	YES Pharmaceutical Development Services GmbH	EMA—2022	EMEA/H/C/005330	YES Pharmaceutical (2023)

TABLE 15.1 *(Continued)*

Nanocarriers	Active Ingredients	Brand Name	Therapeutic Use	Company	Approving Agency and Approval Year	Application Number	References
Lipid complex	Amphotericin B, dimyristoylphosphatidylcholine, dl, dimyristoylphosphatidylglycerol, dl	Abelcet®	Aspergillosis in patients refractory to or intolerant of conventional amphotericin B and for invasive fungal infections	Leadiant Biosciences, Inc.	FDA—2010	NDA050724	Leadiant (2010), Rust and Jameson (1998)
	Patisiran sodium	Onpattro®	Polyneuropathy of hereditary transthyretin-mediated amyloidosis in adults	Alnylam Pharmaceuticals, Inc	FDA and EMA—2018	NDA210922 EMEA/H/C/004699	Akinc et al. (2019), Alnylam (2018a, 2018b)
Iron nanoparticles	Iron sucrose	Venofer®	Iron deficiency in chronic kidney disease	American Regent Inc.	FDA—2000	NDA021135	American-Regent (2007), Bhandari et al. (2015)
Nanocrystal	Aprepitant	Emend®	Antiemetic	Merck & Co. Inc	FDA—2003	NDA021549	Merck (2003), Prommer (2005)
	Fenofibrate	Tricor®	Antihyperlipidemia	Abbvie Inc.	FDA—2004	NDA021656	Abbvie (2016), Saurav et al. (2012)
	Megestrol acetate	Megace® ES	Anorexia, cachexia, and AIDS-related weight loss	Endo Pharmaceuticals	FDA—2005	NDA021778	Atlantic (2005), Endo (2005), Tuca et al. (2013)

*Discontinued.

#NRLD, National Reimbursement Drug List (The main pathway for pharmaceutical reimbursement in China, aimed at improving affordability of drug treatments).

A repetitive formal risk assessment is typically carried out early in the product development to identify the risks to CQAs arising from drug substances, excipients, and the process. Ishikawa or fishbone diagram is created for risk assessment to pinpoint possible material and process factors that may affect CQAs. Risk identification is followed by risk analysis. Failure mode analysis or comparable techniques can be used to rank order variables according to probability, severity, and detectability (ICH, 2008, 2009, 2015). For example, size of the bead, percentage of bead, motor speed and current, mill pressure, and product temperature and flow rate could be the critical process parameters (CPPs) of the media milling process used to produce nanocrystals. Similarly, the characteristics of excipients, such as polymers, lipids, stabilizers, and surfactants, and the properties of active pharmaceutical ingredients (APIs) including particle size distribution (PSD), density, crystalline content, stability, etc., are examples of critical material attributes (CMAs) (Halwani, 2022; Holland, 2010). CMAs and CPPs are defined as "a material or process whose variability has an impact on a critical quality attribute and therefore it should be monitored or controlled to ensure the desired drug product quality." Risk evaluation is performed to evaluate the impact of critical characteristics of drug substances, excipients (such as lipids and polymers), intermediates (in-process materials), and unit processes on the CQAs of nanomedicine. During evaluation, tools such as process analytical technology (PAT), multivariate analysis, the design of experiments (DOE), and historical data (prior knowledge) are frequently used to comprehensively understand and control products and processes. The DOE is a systematic and structured approach to ascertain the correlation among CMAs, process variables, and the process output, or response. When coupled with the process through a feedback system, PAT tools may monitor, control, and adjust the process parameters, providing real-time information regarding the impact of CMAs and CPPs on the CQAs of products. When producing nanocrystals, PAT tools can offer important insights into product behavior, including nucleation, crystal growth/dissolution, agglomeration, coalescence, breakage/attrition, and segregation.

Product and process understanding would culminate in establishing design space for nanomedicines. Design space refers to the "multidimensional combination and interaction of input variables (e.g., material attributes) and process parameters that have been demonstrated to assure quality." The design space allows for the modification of process parameters within the boundary space without requiring prior permission from a regulatory agency. A design space may be proposed for nanomedicine in an NDA or ANDA submission. Any changes in the manufacturing process outside the approved design space boundary would require prior regulatory approval. In essence, the design space gives the sponsor more freedom about process and product modifications while also lessening the post-approval submissions to the FDA and regulatory burden. Modest changes outside the design space in the manufacturing process could dramatically impact nanomedicine CQAs.

A control strategy is an integral part of the marketing authorization submissions. The control strategy ensures that processes remain under control, resulting in the creation of consistently high-quality products. "Parameters and attributes related to drug substance and drug product materials and components, facility and equipment operating conditions, in-process controls, design space, finished product specifications, and the associated methods and frequency of monitoring and control" are included in the control strategy (ICH, 2008, 2009, 2015).

15.3 PARTICLE SURFACE PROPERTIES AND PARTICLE SIZE DISTRIBUTION

The size, shape, crystal structure, surface area, and surface energy of a nanomaterial define its surface characteristics. These attributes affect the solubility, permeability, pharmacokinetics, and biodistribution of nanomedicines (FDA, 2007). A greater surface area is available for solubility and permeability due to the smaller size of the nanoparticles. In fact, particle shape has also been shown to influence absorption. For instance, rod-shaped nanoparticles exhibited better intestinal absorption than spherical and flaky nanoparticles (Guo et al., 2019). Their higher surface charge and surface area promote faster drug dissolution. However, the interactions between the nanoparticles and body fluids following administration may alter their PSD and surface properties (Wang et al., 2012). It is apparent that the PSD of nanomedicines not only defines them, but also impacts their distinct biological, chemical, and physical characteristics. In other words, PSD influences the clinical outcome of nanomedicines and hence their safety and efficacy.

A validated PSD method capable of identifying and quantifying the different size ranges contained in the product must be submitted to the agency. Various methods with differing degrees of precision and accuracy exist for determining PSD (Horiba, 2010; Measurlabs, 2022). In general, the FDA or other regulatory bodies do not recommend any specific technique or method for PSD characterization of nanomedicines. Instruments based on "ensemble" and "single-particle" techniques are available for PSD determination. A signal generated by a mixture or superposition of responses by a wide range of sizes and/or concentrations is detected by ensemble methods. A mathematical procedure is used to invert or deconvolute the signal, assuming a Gaussian distribution of particles to obtain PSD. Contrastingly, single-particle methods identify a signal produced by a single particle at any moment. Compared with ensemble methods, single-particle technologies yield data that is more accurate, resolved, and more closely matches the true value. This is demonstrated by the polydisperse system, which distinctly displays a peak of larger particles from the main population of particles. Light scattering, light extinction, or a combination of the two phenomena are used in single-particle optical sizing (SPOS) detection techniques (Owen et al., 2018; Satinover et al., 2014). AccuSizer 780 AD system (Sizing Systems: AccuSizer 780 AD) and nanoparticle tracking analysis (Malvern Instruments: Nanoparticle Tracking Analysis) are examples of commercially available instruments based on SPOS. Sufficient sample dilution is necessary for SPOS to function, allowing only one particle to travel through the optical area of an instrument at a time. This makes it challenging to apply SPOS methods for nanomedicines stabilized by polymers or surfactants, such as surfactant-stabilized nanocrystals, as this could cause aggregation and produce inaccurate results. In contrast, no dilution is required in an ensemble approach. Dynamic light scattering (DLS, sometimes called quasi-light scattering photon correlation spectroscopy or PCS) is the most used ensemble method. It is an effective method for describing nanomedicine in the size range of 1–1000 nm. By examining the variations in light intensity brought on by the Brownian motion of particles, one can ascertain the diffusion coefficient or diffusivity of suspended particles in a liquid. Another benefit of this method is that it is absolute, meaning that uniform-size particles' diffusivity depends only on their size and not on their makeup (light scattering Instrument – DLS). Before selecting a PSD measuring method, one must consider the method's capabilities, including its lower detection limit, traceability, noise, drift, bias, accuracy, precision, resolution, measurement range, and analysis time. A complete method description is required to be submitted by the sponsor to the regulatory agency, including the parameters chosen for the measurements, dilution, diluting fluid, and if applicable, the addition of salt, acid, or base during the measurements. It is important to present data in the correct format to avoid misinterpretation of otherwise very valuable data. For example, data might be provided in number, intensity, or volume formats, and each format produces a distinct set of data. Reporting the results in the principal format of the technique is often advised. The results of laser diffraction and acoustic spectroscopy are based on number by image analysis and DLS by intensity and volume analysis. The conversion of PSD data into a different format may substantially change the data and how it is interpreted. The one exception to this rule is image analysis data conversion from volume to number format. It is recommended to identify the PSD data format (number, intensity, or volume) when submitting data to the FDA due to these complexities. The agency has provided a clear recommendation for sizing technique and the data format for some of the nanomedicine products, such as z-average technique (based on intensity for ferric carboxymaltose colloidal dispersion and cyclosporine ophthalmic nanoemulsions) (FDA/CDER/ChoiStephnie, 2016; FDA, 2021). PSD varies from coarsest to finest particles in a sample as characterized in terms of D_{10}, D_{50}, and D_{90}. However, reporting data as D_{99}, or D_{100}, can distort measurement results because of air bubbles or temperature fluctuations. The width of PSD is measured by the polydispersity index (PDI), also known as span $(D_{10}–D_{90})/D_{50}$, is an essential parameter required to be submitted to regulatory bodies. According to Horiba (A Manual to Particle Size Analysis), DLS generates the PDI as a measure of the data's polydispersity. The sponsor must provide statistics from several batches to show that the method is repeatable. Reproducibility test guidelines are also available from the United States Pharmacopoeia (USP) and the International Organization for Standardization (ISO). Both documents recommend performing these experiments in triplicate. According to ISO criteria, the coefficient of variation (CV) for D_{50} should be <3%, and for D_{10} and D_{90}, it should be <5%. On the other hand, according to USP<429>, the CV limit is <15% for D_{10} and D_{90} and <10% for D_{50} (ICH, 2021; USP-NF, 2016).

15.4 AMORPHOUS AND CRYSTALLINE CONTENT

The stability of drug phases in nanomedicine is required during storage and in-use conditions for a consistent clinical outcome. Changes in the crystalline or amorphous content of a product may affect its clinical efficacy. Therefore, it is paramount to control and monitor the amorphous/crystalline ratio in nanomedicines. It is prudent to correlate a product's *in vitro* performance with its crystalline or amorphous content during the early stages of product development, identify the factors that lead to crystallization, and develop stabilizing methods. The crystalline–amorphous ratio should be monitored, preferably included in the product specifications as the ratio may impact the product's *in vitro* release, bioavailability, and hence effectiveness. The crystal content in nanomedicines frequently increases during storage or on exposure to high temperature and/or humidity, frequently accompanied by an increase in crystal size. The analytical method used should be able to quantify the lowest quantity of crystalline/amorphous expected in the product (ICH, 2021). Techniques based on vibration spectroscopy, solid-state nuclear magnetic resonance, thermal, and X-ray diffraction have been documented in the literature for amorphous/crystalline quantification. X-ray powder diffraction is still the most acceptable and accurate method for determining the amount of amorphous/crystalline material, although vibration spectroscopies have many advantages such as quick collection times and a lack of sample preparation requirements. Multivariate methodology can applied to spectral or other measurement techniques data if there are overlapping peaks, excipient interference, or poorly defined peaks (Kayalar et al., 2023; Mohamed et al., 2023).

15.5 NANOMEDICINE SPECIFIC GUIDANCE

The FDA has published general and product-specific guidelines for the industry on nanomedicine development and submission requirements (FDA, 2018i). The agency has also published a Manual of Policies and Procedures (MAPP) for reporting information about nanotechnology in CMC (chemistry, manufacturing, and control) reviews. Using the MAPP, CMC reviewers can report information about a drug product that uses nanomaterials. Reviewers take great care to ensure the accuracy of the information they cite regarding nanotechnology in their reviews. These data are intended to be used in the creation of nanotechnology-related policies and laws. According to MAPP, "drug products containing nanocarriers broadly refer to a material that has been deliberately manipulated to be in the nanoscale range (1–100 nm) or to exhibit properties or phenomena, including physical, chemical, or biological effects that are attributable to its dimensions, up to 1000 nm" (Cruz et al., 2013; Jordan Paradise, 2019).

The FDA released a guidance sheet in 2014 on nanotechnology to help the industry and reviewers determine whether drug products involve the application of nanotechnology. In assessing if nanotechnology has been used in the manufacturing of a drug product, two factors are considered: engineered substances or finished goods (1) have at least one external or internal dimension in the nanoscale range, roughly 1 nm to 100 nm, and (2) display physical or chemical characteristics, phenomena, or biological effects that are related to dimensions up to 1000 nm. The top limit of 1000 nm is used to further screen the materials for study and find out if they show characteristics and phenomena that are related to their size and connected to the use of nanotechnology. These factors not only need to be considered when developing new products but also when making modifications to products that have already been approved that could alter their dimensions, and consequently, their qualities. Additionally, both excipients and APIs are covered by the aforementioned principles. This guidance states that when assessing the safety, effectiveness, and quality of the drug product under the current regulatory framework, the FDA will consider both size and characteristics. Additionally, the FDA evaluates the safety of nanomedicines using the current regulatory framework. As a result, while evaluating a product's quality, its unique qualities, the impact of nanomedicines on the biological system, and its intended purpose are all considered. Additionally, the FDA has published several guidance documents specifically for nanomedicines, informing the industry about the types of studies and data needed for NDA/ANDA submissions, such as those involving liposomes, microemulsions, nanoemulsions, polymer nanoparticles, nanocrystals, and among others (FDA, 2014a).

15.6 IN VITRO RELEASE/DISSOLUTION

As one of the measures of quality, compliance with dissolution or *in vitro* release specifications ensures clinical performance throughout the lifecycle of the drug product. For liposomes, the *in vitro* release method may help in estimating the amount of free *versus* encapsulated drugs and assessing the integrity of the system. The discriminatory characteristics of dissolution or the *in vitro* release method are invaluable in detecting any changes in product quality occurring during shelf life or due to changes in formulation composition or manufacturing process. Needless to mention, the *in vitro* release/dissolution test is an important CQA of nanomedicines. Similarly, the discriminatory dissolution method will be able to differentiate between formulations having different PSD or amorphous/crystalline ratios. The FDA recommends using *in vitro* methods including *in vitro* release and dissolution testing to demonstrate the sameness between reference and test formulations when pharmacokinetic and/or clinical studies are impractical (FDA, 2014a). Generally, the FDA does not recommend a specific pharmacopeial method for *in vitro* release or dissolution tests for nanomedicines. The onus of proving the discriminatory power of the developed dissolution/ *in vitro* release method lies with the manufacturer of the product.

Some of the challenges in developing *in vitro* release and dissolution methods for nanomedicines include the inability to quickly and effectively separate dissolved drugs, the lack of knowledge of release mechanisms, and the complexity of nanomedicines including the fact that some products are designed not to release drugs until they reach the target tissue or organ. Discriminatory attributes can be imparted to the release/dissolution method by varying the pharmacopeial or non-pharmacopeial apparatus, medium (pH, buffer type and strength, volume), surfactant (type, concentration), flow rate (USP IV), speed (basket/paddle), and temperature, etc. Drug solubility, stability, and test sensitivity should be considered while selecting a release medium. The testing can be performed in both sink and/ or non-sink conditions with adequate agitation to prevent nanomedicine's aggregation. The *in vitro* release techniques for nanomedicines that have been documented in the literature include membrane diffusion, dynamic dissolution, USP apparatus IV, and microdialysis (Gupta et al., 2023; Rahman et al., 2014; Sangwai and Vavia, 2013; Shen and Burgess, 2013).

15.7 NEW DRUG APPLICATION

Under section 505 of the FDC Act, an NDA for a nanomedicine can be submitted to the FDA for evaluation via one of two regulatory tracts, namely, 505(b)(1) or 505(b)(2) (Figure 15.2). NDAs are submitted under the 505(b)(1) route for a new chemical entity (NCE) that has never been approved and requires comprehensive data on safety and efficacy. The first step is to submit an investigational new drug (IND) application in the review and approval process. Preclinical data in the IND application submitted to demonstrate drug is safe to use on humans. In a clinical trial, the sponsor must present substantial treatment benefits, safety and effectiveness goals, and outcomes. In Phase I, safety data are collected from a limited sample size of human subjects. Phase II generates data on dosage range and efficacy in a larger human sample size. A much higher number of individuals than in Phase II are enrolled in Phase III which provides crucial safety and efficacy data. The sponsor, upon completion of all the three clinical phases, files the NDA for a nanomedicine product via the 505(b)(1) regulatory route. Comprehensive data on nonclinical, preclinical, clinical, bioanalytical, pharmacological, and pharmacokinetics are included in the submission. The sponsor also needs to exhibit that the proposed labeling is suitable and that the manufacturing method(s) used are sufficient to ensure the drug's identity, strength, quality, and purity. Under 505(b)(1) NDA submission, the applicant either generates the data or receives the right to use or refer to it (FDA, 2022f; Gaffney, 2015). The exclusivity period is granted to the sponsor once the NDA is approved. During this period, no generic applications or NDAs under 505(b)(2) can be submitted for FDA review. Patents and exclusivity may run concurrently, or exclusivity may expire before the patents. Orphan drug (OD), NCE, and pediatric (PED) exclusivities are available to NDAs under 505(b)(1) (Lal, 2015). Conversely, OD exclusivity is awarded to drug products that are approved for the treatment of diseases or ailments that affect $<$200,000 people in the United States (or $>$200,000 people with no chance of cost recovery), and is granted for 7 years after the date of approval (FDA, 2023d, 2023c). NCE exclusivity is granted for 5 years after the date of approval (FDA, 2023b). When the sponsor completes and submits data from PED clinical trials on the active moiety in response to a formal request from the FDA, PED exclusivity is awarded for a period of 6 months and added to the existing exclusivity and patent period (FDA, 2018e).

The Drug Price Competition and Patent Term Restoration Act of 1984 (Hatch-Waxman Amendments) provides an avenue for NDA submission under the 505(b)(2) pathway (FDA, 2022f, 2022e; Gaffney, 2015). FDA may approve an NDA for a nanomedicine formulation under 505(b)(2) based on data from published literature or the agency's assessment of the efficacy and/or safety of a drug that has already received approval. The 505(b)(2) NDA process is typically used for major modifications to previously approved

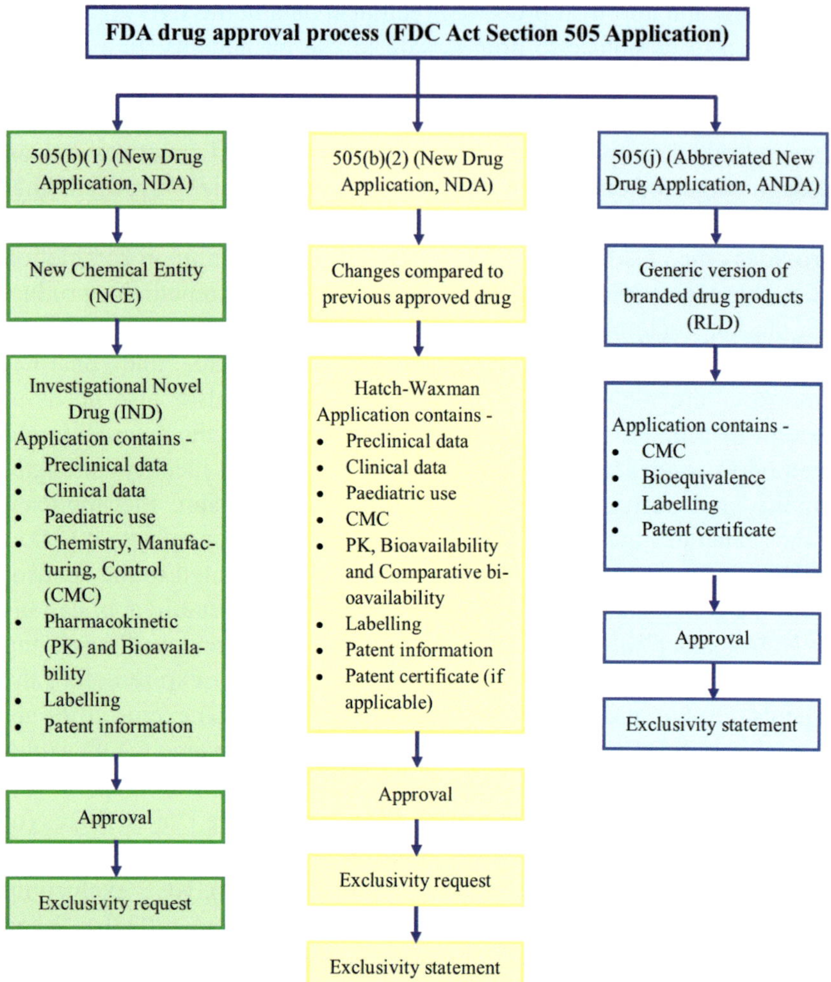

FIGURE 15.2 Drug product approval process in the United States.

products, such as the addition of new dosage forms, new nanomedicine formulations, strength (lower dose due to higher efficacy of nanomedicines), route of administration, dosing schedules, indications, active ingredients (different salts, esters, complexes, chelates, clathrates, racemates, or enantiomers), combination products of two previously approved products, monograph deviations, and active ingredient substitution in a combination formulation. NCE for nanomedicine may also be filed via the 505(b)(2) regulatory pathway if a portion of the data needed for approval is derived from studies that the applicant did not conduct or for which the applicant does not have a right of reference (FDA, 1999). The NDA under the 505(b)(2) route must include patent certifications as specified in 21 CFR 314.50(i) and provide notice of certain patent certifications to the NDA holder and the patent owner. Under Waxman-Hatch, NDA 505(b)(2) is eligible for 3 years of exclusivity (21 CFR 314.52, 2023; FDA, 2023a). NDA under 505(b)(2) is also eligible for NCE, OD, and PED exclusivities, similar to NDA under 505(b)(1) (Lal, 2015). Patent and exclusivity rights that may apply to an approved NDA may cause a delay in NDA approval or filing under 505(b)(2). Identification of those portions of the application that rely on information the applicant does not own or to which the applicant does not have a right of reference, "if the 505(b)(2) seeks to rely on the agency's previous finding of safety or efficacy for a listed drug or drugs, identification of any and all reference listed drugs (RLD) by established name, proprietary name (if any), dosage form, strength, route of administration, name of the listed drug's sponsor, and the application number" are required for the NDA submission under 505(b)(2) (FDA, 1999).

Under the authority granted by Congress, the FDA charges the sponsor a fee for reviewing an NDA under the Prescription Drug User Fee Act (PDUFA) 1992. Payment of the PDUFA fee is crucial for the approval of new drugs. PDUFA needs to be reauthorized every 5 years, and it was recently extended to 2028 (PDUFA VII). Table 15.2 shows the PDUFA fee structure for a review of NDAs submitted under 505(b)(1) and 505(b)(2). PDUFA offers standard and priority reviews, which are the two tiers of the NDA review process. NDA submissions are subject to the legal review process, even if they represent a slight improvement over the currently marketed therapy. And 10 months is the standard review period according to the PDUFA amendments of 2002. NDAs that offer significant therapeutic advancements or introduce a treatment where none previously existed are given priority review which should be completed in six months. There are fast-track, accelerated, and breakthrough therapy reviews available under NDA priority review. "Fast track is a procedure intended to speed up and simplify the development of medications intended to treat severe illnesses and address unmet medical needs." The FDA can review surrogate endpoints for extreme conditions that address unmet medical needs to expedite approval. The FDA's fast-track designation of a drug as a breakthrough involves accelerating the drug's development and review to treat extreme medical conditions (FDA, 2023g).

TABLE 15.2 Fees that FDA Charge for NDA and ANDA Reviews.

PDUFA Fee			GDUFA Fee			
Conditions	**2023**	**2024**	**Conditions**		**2023**	**2024**
NDA with clinical data	$3,242,026	$4,048,695	ANDA		$240,582	$252,453
NDA without clinical data	$1,621,013	$2,024,348	DMF		$78,293	$94,682
Program	$393,933	$416,734	Program	Large size	$1,620,556	$1,729,629
				Medium size	$648,222	$691,852
				Small size	$162,056	$172,963
			Facility	Domestic API	$37,544	$40,464
				Foreign API	$52,544	$55,464
				Domestic FDF	$213,134	$220,427
				Foreign FDF	$228,134	$235,427
				Domestic CMO	$51,152	$52,902
				Foreign CMO	$66,152	$67,902

API, active pharmaceutical ingredients; FDF, Finished dosage form; CMO, Contract manufacturing organization.

EMA facilitates marketing authorization of nanomedicines by publishing guidelines for the industry. The agency also categorizes nanomedicines based on their function, organicity, degradability, and approval status (EMA, 2024; Euractiv, 2020). A company or sponsor submits a dossier called a marketing authorization application (MAA), which is similar in content to the NDA and ANDA, to the EMA for marketing approval of nanomedicine. There are four pathways for MAA submission: centralized procedures (CP), decentralized procedures (DCA), mutual recognition procedures (MRP), and national procedures (NP) (Table 15.3). In the CP, one application and one assessment are required for the entire EU approval. The Committee for Medicinal Products for Human Use reviews the submitted application and sends its recommendation to the European Commission (EC). Medicines containing new active substances to treat HIV infections, cancer, diabetes, neurodegenerative diseases, autoimmune and other immune diseases, viral infections, biological products, advanced therapy (gene therapy, cell therapy, or tissue-engineered medicine), and OD are required to follow the CP procedure for medicine approval. If the sponsor intends to market only in one or more EU countries instead of the union, the MAA is submitted through the DCP or MRP procedure. Generally, generic medicines, over-the-counter (OTC), and older medicines are approved by these procedures, whereas NP is very specific to a particular national authority. The majority of medicines available in the EU are authorized by individual European nations (EMA, 2021; Eupati, 2024). More than 25 NDAs have received FDA and EMA approval for nanomedicines (Table 15.1) (Jia et al., 2023; Rahman et al., 2018a; Thapa and Kim, 2023).

In Canada, the review and assessment of safety, efficacy, and quality data for medication is the responsibility of Health Canada's Health Products and Food Branch (HPFB). To guarantee that all required components of the application are included in the New Drug Submission (NDS), a screening mechanism is part of the process. The HPFB, or more specifically, the Therapeutic Product Directorate (TPD) or the Biologic and Genetic Therapies Directorate (BGTD), respectively, reviews and authorizes medicines and biologics to be marketed or sold in Canada. There are particular offices and bureaus in each of these directorates. Only after successfully completing the relevant Bureau review process (in which safety, efficacy, and quality of drugs are evaluated), and receiving a positive decision are drugs allowed to enter that market (HealthCanada, 2002; MuscularDystrophyCanada, 2019; SPharm, 2023).

The National Medical Products Administration (NMPA), China oversees the drug product approval process which is similar to the USFDA review process. The following categories represent how the NMPA has organized the approval process: Fast-track approval, special approval procedure, breakthrough, and priority review. The approval process for drugs in China can take up to 3 years, but with the recent introduction of fast-track approvals, it can be completed in as little as 16 months. There are three steps in the procedure, and each one is crucial to guarantee the effectiveness and safety of drug products supplied in China. The NMPA review procedure consists of the following three stages (Chapman, 2021; NMPA, 2020; PacificBridgeMedical, 2023):

1. Technical Review: The quality, safety, and efficacy of the drug's technical data are all thoroughly examined at this stage.
2. On-site Inspection: During this phase, the drug's manufacturing facilities and quality control procedures are examined.
3. Expert Panel Review: In this phase, a group of specialists reviews the technical information on drug products and makes recommendations to the NMPA regarding whether or not it should approve them.

The NMPA mandates post-marketing monitoring of drug products' effectiveness and safety after they are authorized for sale in China. The sponsor is expected to monitor adverse events continuously and provide safety data on a regular basis. To verify adherence to GMP regulations, the NMPA may also carry out inspections of the manufacturing site. Pharmaceutical firms are required to follow stringent regulatory criteria during the drug development and post-approval processes (PacificBridgeMedical, 2023).

The Ministry of Health, Labor and Welfare (MHLW) and the Pharmaceuticals and Medical Devices Agency (PMDA) oversee the drug approval process for drug products in Japan under the "Law for Ensuring the Quality, Efficacy, and Safety of Drugs and Medical Devices" (Law No. 145, 1960; hereinafter referred to as the Law). Information regarding the most recent advancements in nanomedicine technology is being gathered and examined by regulators and reviewers. The process of approving drugs involves several steps, such as reviewing and verifying the effectiveness, safety, and quality of each product; PMDA inspections and conformance audits; manufacturing and distribution business license approval; MHLW approval of manufacture and distribution; and reporting any unfavorable effects resulting from product marketing (Figure 15.3). After submitting the application, the review period is 12 months for a conventional file and 9 months for an "OD Designation," which is comparable to that of the US and EU (Credevo, 2022; Hara, 2015; JPMA, 2020).

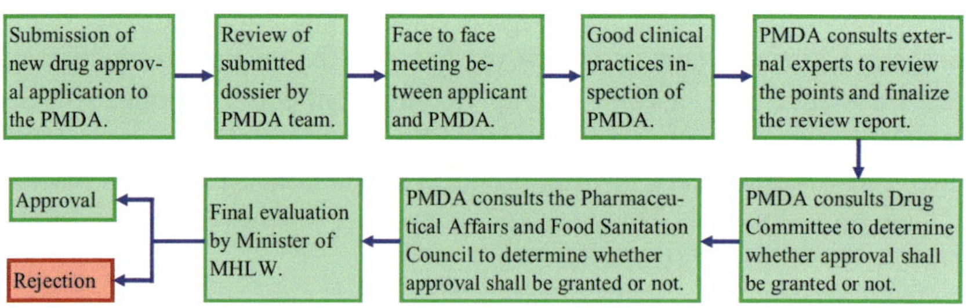

FIGURE 15.3 Drug product approval process in Japan.

Central Drugs Standard Control Organization (CDSCO) is the regulatory body in charge of manufacturing, selling, registering, importing, and exporting nanomedicines in India. The organization is commonly identified by the title of its chief officer, the Drugs Controller General of India (DCGI). Guidelines for the evaluation of nanomedicines in India have been made so that pharmaceutical preparations containing nanomaterials can be evaluated. In the case of nanomedicine, if the drug is a new molecular entity and not approved in any regulated country and the nanocarrier is new or approved in any regulated country, then that nanomedicine is considered an IND. The quality, safety, and efficacy of IND nanomedicines shall be evaluated as per Schedule Y of the Drug & Cosmetic Rules of 1945. Whereas the drug is approved in regulated countries and the nanocarrier is new, Schedule Y for IND is not required, but the dossier must have safety and efficacy data for nanomedicine. If the drug and nanocarrier are both approved in any regulated country, they shall be considered for abbreviated studies. Schedule Y of the Drug & Cosmetics Rules, 1945, and its Rules 122A, 122B, 122D, and 122E are followed in approving the new nanomedicine. A pharmaceutical firm must request permission from the licensing authority (DCGI) to manufacture or import a new nanomedicine by submitting Form 44 together with the information required under Schedule Y of the Drugs & Cosmetics Act 1940 and Rules 1945 (ISNM, 2022; Musyuni, 2016a, 2016b).

TABLE 15.3 EMA Drug Approval Processes.

Regulatory Pathways	Regulatory Body Involved	Timelines	Approval For
Centralized Procedures (CP)	CHMP and EC	210 days for CHMP including 67 days for EC decision	One approval for all EU and Iceland + Liechtenstein + Norway
National Procedures (NP)	National competent authorities	As per the specific guidelines of a particular country	One EU country or Iceland, Liechtenstein, Norway
Decentralized Procedures (DCP)	Reference Member State (RMS) and Concerned member state (CMS) are handled by the Coordination Group for Mutual Recognition and Decentralized Procedures-Humans (CMDh)	120 or 210 days including the National phase	Parallel approval in more than one country
Mutual Recognition Procedures (MRP)	Reference Member State (RMS) and Concern member state (CMS) are handled by the Coordination Group for Mutual Recognition and Decentralized Procedures-Humans (CMDh)	National approval including 90 days	First National approval then more countries added

TABLE 15.4 *(Continued)*

Brand Name	NDA Number	NDA Approval Year	Generic	ANDA Number	ANDA Approval Year	Recommended Studies	References
AmBisome®	50740	1997	Amphotericin B liposomal injection	212514	2021	The formulation should be Q1 and Q2, the same as the RLD Liposome size distribution, number of lamellar, electrical surface potential or charge, lipid bilayer phase transition, and *in vitro* leakage rates comparable to the RLD *In vivo* bioequivalence study with PK endpoints study (Submission of a Bio IND Application is required prior to conducting a bioequivalence study for a cytotoxic drug product such as amphotericin B) Analyze the free amphotericin B (liposome-unbound) and liposome-encapsulated amphotericin B in plasma Bioequivalence based on (90% CI): Liposome-encapsulated amphotericin B. Submit AUC and C_{max} of free amphotericin B as supportive data Equivalence based on (95% CI): PBE based on D_{50} and SPAN or alternatively on the harmonic intensity-weighted average particle diameter (z-average) and PDI derived from cumulant analysis of the size intensity distribution.	AmphotericinB (2020)
Ferrlecit®	20955	1999	Ferric Oxyhydroxide injection	78215	2011	The formulation should be Q1 and Q2, the same as the RLD Determination of stoichiometric ratios/composition and Fe(II) content Analysis of particle size distribution using a PBE statistical approach Determination of particle morphology Electrical surface potential or charge measurement Crystalline structure analysis Determination of magnetic properties Determination of Fe(III) to Fe(II) reduction potential and reduction kinetics Determination of labile iron under multiple conditions Comparative *in vivo* PK study (quantification of colloidal ferric oxyhydroxide in serum, or quantification of total iron and transferrin-bound iron in serum)	Kozak (2022)

TABLE 15.4 *(Continued)*

Brand Name	NDA Number	NDA Approval Year	Generic	ANDA Number	ANDA Approval Year	Recommended Studies	References
Restasis®	205894	2002	Cyclosporine ophthalmic emulsion	205894	2022	In vitro The formulation should be Q1 and Q2, the same as the RLD Analysis of globule size distribution using an appropriate histogram comparator and PBE statistical approach Electrical surface potential or charge measurement Determination of viscosity, pH, drug distribution, and surface tension *In vitro* drug release test In vivo Comparative clinical endpoint research in individuals whose keratoconjunctivitis sicca-related ocular inflammation is thought to be the cause of their tear production suppression	Kozak (2022)
Durezol	22212	2008	Difluprednate ophthalmic emulsion	211776 211526	2021	In vitro The formulation should be Q1 and Q2, the same as the RLD Determination of globule size distribution, viscosity profile as a function of applied shear, pH, zeta potential, osmolality, and surface tension. Sponsors should also submit information on the drug distribution in different phases within the formulation *In vitro* drug release test In vivo Bioequivalence based on (95% upper confidence bound). Comparative bioequivalence study with a PK endpoint in individuals undergoing indicated cataract surgery.	Amneal (2017)

TABLE 15.4 *(Continued)*

Brand Name	NDA Number	NDA Approval Year	Generic	ANDA Number	ANDA Approval Year	Recommended Studies	References
Feraheme™	22180	2009	Ferumoxytol intravenous	206604	2021	The formulation should be Q1 and Q2, the same as the RLD. Equivalence in the stoichiometric ratios of polyglucose sorbitol carboxymethyl ether, iron, and other relevant components needs to be established. Comparable physicochemical properties should be established with orthogonal analytical methods: Iron core characterizations include but are not limited to core size determination, iron oxide crystalline structure, and iron environment Analysis of the composition of carbohydrate shell Determination of magnetic properties Study of particle morphology Determination of labile iron determination under physiologically relevant conditions The test can be performed with ultra-filtration, *in vitro* hemodialysis system, and the catalytic bleomycin assay of spiked human serum samples The spectrophotometric measurement of Fe reduction, or other methods that are validated for accuracy and precision Bioequivalence based on (90% CI): Z-average size PDI or D_{50} and SPAN as appropriate as RLD Bioequivalence based on (95% upper confidence bound): Z-average and PDI or D_{50} and SPAN using the PBE statistical approach	Covis (2023)

15.9 COMMON TECHNICAL DOCUMENT

The majority of the regulatory agencies have reached an agreement to assemble quality, safety, and efficacy information in a common technical document (CTD) format for the review and approval of brand or generic drug products. The "CTD is a set of specifications for drug application dossiers for registration in the European Union, Japan, and the United States" (FDA, 2018g). It provides a common format and recommendations for the technical documentation of the dossier, with the aim of reducing the time and resources both for application submissions and their regulatory review. The CTD consists of five modules. Briefly, the contents of each module are as follows:

Module 1—Regional administrative and prescribing information.
Module 2—Overview of modules 3 and 5.
Module 3—Quality information (pharmaceutical).
Module 4—Preclinical data (animal pharmacology and toxicology).
Module 5—Clinical study data (clinical trials).

As per the regional regulatory requirements, the contents and subheadings in Module 1 may vary (FDA, 2018g). With online systems, the FDA and other regulatory authorities have transitioned toward electronic CTD (eCTD) submission of drug applications. It has streamlined the process of reviewing, approving, and monitoring the drugs after their approval by the FDA. Sponsors have immensely benefited from the simplification of the document preparation process. Since 2008, the FDA has used the eCTD format for all applications, modifications, supplements, and reports. Started in 2017, the FDA has made it "mandatory for the sponsors to submit applications in eCTD for NDAs, ANDAs, BLAs, DMFs (drug master files), combination products, and all subsequent submissions to these types of applications, including amendments, supplements, and reports, even if the original submission was paper filed before the requirements went into effect" (FDA, 2020). Since 2018, the FDA has even extended these changes to eCTD submissions for commercial INDs. However, non-commercial INDs are exempted (FDA, 2020, 2022c; Garg et al., 2017).

The eCTD format for the FDA includes an application form within the administrative part. In Modules 4 and 5, there are specific formats for study data required, so-called Study Tagging Files (STF). The EMA also requires electronic submissions in the eCTD format with minor differences from those in the US. For example, submissions in the EU must include specific country sections in Module 1 and detailed specific sections for the product information. Electronic submissions in the eCTD format are also required for submissions in Japan, with the PMDA imposing specific requirements for certain submission types. The China NMPA also accepts eCTD submissions for INDs and NDAs. NMPA seeks documentation in the Chinese language for M2–M5 as legally binding. These documents can also be provided in "English" as a second set of documents in the respective sections (Extedo, 2023; FDA, 2014a; GaBi, 2011; Jordan, 2014; Vaibhav Subhash Janjal et al., 2021).

15.10 PRE-APPROVAL INSPECTION

Pre-approval inspections (PAI) are part of the pharmaceutical cGMP (current good manufacturing practices) for the 21st-century initiative (FDA, 2015b) with the objective of ensuring that manufacturing facilities adhere to the FDA rules and regulations. PAI provides the FDA with an assurance that a manufacturing establishment named in a submission can manufacture a drug product and that the submitted data are accurate and complete. Due to this assurance and reliability, NDA and ANDA submissions of nanomedicine products also trigger the PAI of the manufacturing facilities. The PAI program lays down the guidelines for the Office of Regulatory Affairs district office and/or Office of Process and Facility on how to evaluate and conduct on-site inspections and establish file reviews when the firm is named in the CMC section of NDAs/BLAs (biologic license applications)/ANDAs, CMC amendments to pending original submissions, and CMC supplements to approved drug applications, including their analytical method evaluations by FDA laboratories. For ANDAs and NDAs, both domestic and international PAIs are conducted depending on the location of the proposed manufacturing site. This PAI covers "all facilities associated with a submission, including drug component manufacturing (such as APIs), finished drug product manufacturing, control testing laboratories, and primary packaging and labeling" (FDA, 2022d). Intermediate manufacturers are exempted from PAI by the FDA, however, they may be evaluated or inspected if the intermediate is considered critical to the quality of the drug product or if the review identifies concerns. Similarly, this exemption is extended to component manufacturers such as syringes, vials or stoppers, and component-only sterilization sites; however, if there is reason for concern, the decision is re-evaluated. Excipient manufacturers are also exempted from PAI unless the excipient is novel and/or the excipient manufacturing process is considered a critical step in the overall drug product manufacturing process. PAI exemptions also apply to labelers and/or secondary packagers.

Though these sites are exempted from PAI, the basic statutory requirements laid down by the FD&C Acts must always be met. A risk-based approach is utilized by the FDA to determine the need for a PAI, considering critical facility, product, and process-related factors. Various scenarios trigger PAI, such as if the product is a new molecular entity, the first ANDA of an approved drug, RLD has complaints about adverse events and stability issues, the product is for serious indication, breakthrough therapy, the facility has cGMP issues, field alert reports, recalls and numerous applications filed at once, the process involved manufacturing of narrow therapeutic index drug, API is derived from animal products, PAT, and QbD, development data are incomplete and the process is complex or complicated, complex drug product (nanomedicine), etc. The PAI team consists of a team leader, the investigator, and other specialists from relevant fields of chemistry, microbiology, process/facility, and/or formulation (FDA, 2015a). The recommendation is made by the lead investigator for approval or withholding approval based on the inspection if there are issues with cGMP compliance, information in CMC is not consistent with site records, or information submitted is not accurate and complete (FDA, 2019). The three primary objectives of a PAI are as follows:

1. Readiness for commercial manufacturing: To determine whether the establishment(s) has a robust quality system in place to control facility and commercial manufacturing operations.
2. Conformance to the application: To verify that the formulation, manufacturing/processing methods and analytical methods align with the descriptions provided by the NDAs/ANDAs for the bio/exhibit batch (or pivotal clinical batches when applicable), proposed commercial scale batches, and APIs.
3. Data integrity audit: To verify the CMC data submitted by the FDA by auditing the raw data, whether in hardcopy or electronic format. Also, to verify the completeness and accuracy of all relevant data (e.g., stability, bio batch data).

The FDA conducts the PAI of drug manufacturer premises located in foreign countries to ensure GMP compliance. Form 438 is issued to firm management if the investigator(s) observe conditions that may violate the FDA norms. Objectionable conditions and offenses are notified to the company's management by the FDA in Form 483. At the end of an inspection, the FDA Form 483 is presented and discussed with the company's senior management. Form 483 is not a final determination of violation but is considered, along with the responses made by the manufacturing company, while making the final determination. The FDA may also conduct specific 'for-cause inspections' in response to a specific incident or communication. These triggers may include recalls, outbreaks, and information from whistle-blowers.

In China, PAI are regulated by the NMPA. The Center for Drug Evaluation (CDE) acts as the initiator and the Center for Food and Drug Inspection (CFDI) acts as the inspection organizer and implementer. The NMPA has established specific requirements for PAI, which are overseen by competent authorities. These inspections serve several key purposes, including verifying the authenticity and consistency of data presented in drug registration applications, evaluating and qualifying drugs for commercial manufacture, evaluating the regulatory compliance of drug R&D (research and development), and auditing the data integrity at the R&D level. The sites that can potentially be inspected include R&D sites, manufacturing units, facilities that are involved in the production of APIs, excipients, packaging materials, manufacturers, suppliers, and contract organizations. The R&D sites include pharmacology and toxicology laboratories, clinical trial sites, and manufacturing sites. Similar to the USFDA, a risk-based approach is used in deciding the criteria for PAI in China. Risk is identified by considering various factors, such as product characteristics, R&D site compliance, product novelty, previous inspection records, manufacturing process, facilities, etc. Subsequently, a risk-based analysis is performed to assign these factors to high-or low-risk categories. For high-risk parameters, PAI is required and made mandatory while for low-risk factors PAI is waived. Once the high-risk elements have been identified, the NMPA proceeds through several well-defined inspection process steps. The PAI is initiated after the CDE issues a notification and posts PAI tasks on the CDE website. The PAI agenda is set in the following steps where the inspection schedule is finalized, and the readiness of the on-site conditions is ensured. The PAI team is composed of observers and officials. The NMPA conducts a PAI of the site and prepares a detailed report on the findings and observations made during the inspection. The report is forwarded to the CDE for the final release (FDA, 2004). In Japan, PAI is conducted by the PMDA, while it is performed by Health Canada in Canada (CFDI, 2021, 2022; Nautiyal, 2023; NMPA, 2021).

15.11 AMENDMENTS, SUPPLEMENTS, AND COMPARABILITY PROTOCOLS

Amendments to nanomedicine can be submitted to the FDA while the NDA/ANDA is still under review. It is mandatory for these amendments to be reviewed by the agency before final approval of the application. The amendments can be solicited, unsolicited, and administrative. Solicited amendments are the responses that an applicant submits in response to the FDA's letter. This can be either information requests or complete responses. These are further classified into major, minor, and easily correctable deficiencies. Major amendments involve significant data or information that is required but was not previously submitted by the applicant to the FDA. As a result, these amendments require a substantial review by the agency. Some examples of these types of amendments are changes in the starting material of API synthesis, a new manufacturing site, CQAs not identified or controlled, requests for repeating bioequivalence studies, etc. Examples of minor changes are additional or enhanced in-process controls needed for the manufacturing process, additional stability data, or additional or tightened specifications for release or stability, etc. Finally, easily correctable deficiencies are minor issues like missing data points, missing letters of authorization, or incorrect established names in the labeling. As opposed to solicited amendments, unsolicited amendments or gratuitous amendments are submitted to the FDA on the applicant's own initiative, and not in response to any FDA's letter or observation. Examples of unsolicited amendments are risk evaluation and mitigation strategies (REMS) and REMS modifications, changes to the RLD's labeling or updates to the USP monograph, etc. Administrative amendments are submitted in the annual report and do not require scientific review (FDA, 2018a).

An applicant wanting to make changes to already approved drug products including nanomedicine, must submit a supplement to the FDA. Essentially, the purpose of supplements is to modify already approved drug products. Different types of changes may be included in the supplement, some of which include: (1) components and composition, (2) manufacturing sites, (3) manufacturing process, (4) specifications, (5) container closure system, (6) labeling, (7) miscellaneous changes, and (8) multiple related changes (FDA, 2004). These changes may have a potential impact on product quality and therefore must undergo an FDA review before they can be implemented by the sponsor. Depending on the significance and severity of the changes, they can be categorized as major, moderate, or minor changes. Major changes can have a significant impact on the product's quality attributes, which may impact its safety and efficacy. These changes are filed in a prior approval supplement (PAS). Examples of PAS are manufacturing site changes, manufacturing methods, etc. PAS review can be expedited for public health reasons such as drug shortages. Moderate changes in CMC are filed as change being affected in 30 days (CBE-30) and change being affected immediately (CBE-0) supplements owing to their significant impact on product quality. The CBE-30 supplement should be submitted to the FDA at least 30 days before the drug product is manufactured and distributed using the proposed changes. These changes include relocating testing or primary packaging to a different site, etc. During the review of the CBE-30 by the FDA, if the agency determines that the changes are major, then the changes mentioned in the CBE-30 should not be implemented. Instead, these changes must be submitted via PAS. CBE-0 supplements are intended for specific moderate changes that can be implemented. Examples of CBE-0 changes include changes to the manufacturing site for the manufacture or processing of the final intermediate. The applicant must cease the distribution of drug products manufactured using the proposed changes if the FDA disapproves the changes in both types of supplements (CBE-0 and CBE-30). A minor change has a very minimal potential to adversely affect the drug product quality. These minor changes are reported by the applicant in the Annual Report, for example, changes are different manufacturing sites for secondary packaging, different manufacturing sites for labeling, etc. (FDA, 2018a).

Submission of a comparability protocol (CP) to the FDA provides an alternate regulatory pathway for post-approval modifications to the CMC section of nanomedicine drug products. Food and Drug Administration defines CP as "a well-defined, detailed, written plan for assessing the effect of specific CMC changes in the identity, strength, quality, purity, and potency of a specific drug product as these factors relate to the safety and effectiveness of the product." The CP contains proposed changes in the CMC section that have minimal or no impact on product quality. These changes include specific tests and studies that will be performed, analytical procedures, and acceptance criteria that have very minor or no adverse impact on product quality. CP has submitted either PAS or as part of the original market applications (NDA/ANDA). The CP allows the sponsor to implement CMC changes, and thus distribute the product in the market sooner than if this option was not used (FDA, 2022a). The FDA recently released new guidance for assisting the industry in submitting CP for post-approval manufacturing changes. A draft guidance was issued in April 2016, and it incorporates the principles of the International Council on Harmonization (ICH) Q12 principles for post-approval changes. With this new change, the manufacturers can now change the active API suppliers through a CP as opposed to the earlier PAS reporting category. Apart from that, the guidance now includes the DMFs and drug–device combination products within its scope (Eglovitch, 2022).

15.12 POST-APPROVAL STUDIES

The Food and Drug Administration Amendments Act of 2007 (FDAAA) grants authority to the agency regarding post-marketing studies requirements (FDA, 2018f). These studies include clinical trials that the sponsor must conduct after the approval of NDAs, ANDAs, or supplements to gather additional information about a nanomedicine's safety, efficacy, or optimal use. The primary objective of these studies is to evaluate or identify any known or potential long-term risks associated with approved products including nanomedicines. For nanomedicine products, additional information about the chemistry and pharmacology of the drug would be required for further understanding. In scenarios where the FDA becomes aware of new safety information, these studies become mandatory. Post-marketing requirements (PMRs) include studies and clinical trials that sponsors are required to conduct under the FDAAA and 21 CFR Part 601, the Pediatric Research Equity Act, and the Animal Efficacy Rule. Post-marketing commitments (PMCs) are studies or clinical trials that a sponsor has agreed to conduct but are not required as per the statutory requirement of 505(o)(3)(B). FDAAA mandates very specific post-marketing clinical trials such as (a) “assess a known serious risk related to the use of the drug”; (b) “assess signals of serious risk related to the use of the drug,” and (c) “identify an unexpected serious risk when available data indicate the potential for a serious risk” (CBER, 2023).

15.13 ENVIRONMENT ASSESSMENT

In the United States, the regulatory framework for drug approval is composed of two key laws: the Federal Food, Drug, and Cosmetic Act (FFDCA) and the National Environmental Policy Act (NEPA). The NEPA of 1969 was signed into law by President Richard Nixon on January 1, 1970. The NEPA mandates that all federal agencies, including the FDA, evaluate the environmental impacts of their actions. Within the context of drug approval, the FDA considers the environmental consequences under the NEPA. The NEPA process aims to inform public officials’ decisions by providing a thorough understanding of environmental effects, with the goal of protecting, restoring, and enhancing the environment. The NEPA’s implementation falls under 21 CFR Part 25, and NEPA regulations are detailed in 40 CFR Parts 1500–1508 (EPA, 2023; FDA, 2023e; Kümmerer, 2004). Unless the action is exempt from categorization, 21 CFR 25.20 mandates that environment assessments must be filed with NDAs, ANDAs, BLAs, supplements to such applications, INDs, and for various other actions (21CFR25.20, 2023). It consists of environmental assessments (EAs, which include environmental impact statements, or EISs) and categorical exclusions. According to the FDA regulations, the claim of categorical exclusion or an EA must be included with any NDA, ANDA, BLA, or supplement submission package. The agency may refuse to accept the submissions if not providing an EA or making a claim of categorical exclusion (21CFR25.15, 2023; 21CFR314.101, 2023; 21CFR601.2, 2023).

The agency determines categorical exclusion based on prior experiences that taken separately or together, do not significantly impact the human environment. EAs or EISs are not needed for these actions. EAs are not necessary for the following submissions: “(a) approval of submission does not result in increased use of active moiety, (b) active moiety at the point of entry into an aquatic environment is less than 1 ppb (part per billion); (c) when naturally occurring substances are approved, and their approval does not significantly change the concentration, distribution, metabolite, or degradation product in the environment; (d) INDs and (e) BLAs for transfusable human blood or blood components and plasma" (21CFR25.31, 2023). It is plausible that nanomedicines do not meet the criteria for categorical exclusion because of their distinct characteristics and consequential impact on the environment and biological system. Moreover, in 2007, the FDA NTF recommended a case-by-case regulatory decision regarding the categorical exclusion of nanomedicines from the EA process (NTF, 2018). However, an EA is required in the following cases: (a) submissions containing naturally occurring substances whose approval would significantly alter the substance's concentration, distribution, metabolite, or degradation product in the environment; (b) submissions whose approval would increase the use of active moiety and whose estimated concentration of active moiety at the point of entry into aquatic environments is greater than 1 ppb; and (c) unusual circumstances indicating that FDA approval might significantly affect the quality of the human environment (21CFR25.15, 2023; 21CFR25.31, 2023). Because of the mobility, transportation, reactivity, durability, and bioaccumulation of nanomaterials, EA assessment is essential for nanomedicines. Nanomaterials can get into spaces that are inaccessible to larger particles, travel quickly through soils and aquifers before settling down, act as transporters, absorb smaller contaminants because of their larger and more reactive surface, and combine with other substances to create hazardous compounds (Kurwadkar et al., 2015). Adequate information must be included in the EA to address pertinent nanomaterial environmental issues.

In the European Union, guidelines for medical product environmental risk assessments (ERAs) have been made available by the EMA. An ERA is carried out prior to a new medication's approval including nanomedicines. Safe drug concentrations in an ERA are determined by environmental fate and toxicity data in compliance with global guidelines. Formal ERA reports are submitted with the resultant data to regulatory bodies (EMA, 2019). Assessment of medicinal products' possible environmental risks is outlined in the EMA guideline. Protecting terrestrial and aquatic ecosystems, such as soil, groundwater, surface water, secondary poisoning, and the microbial community in sewage treatment plants, is the goal of the ERA. The guideline explains how to conduct the ERA and assess possible environmental risks associated with the use of the medicinal product (EMA, 2006).

In Canada, to comply with the requirements of the Canadian Environmental Protection Act of 1999, the New Substances Notification Regulations (Chemicals and Polymers), and the New Substances Notification Regulations (Organisms), Health Canada is proposing changes to the Food and Drug Regulations in order to modernize the risk assessment and management of direct and indirect human health risks associated with ingredients used in drug products. Under the proposed amendments, the Food and Drugs Act would have a regulatory framework for ERA and management that is comparable to the specifications found in the Canadian Environmental Protection Act, 1999 (HealthCanada, 2023a). According to the Canadian Environmental Protection Act of 1999, before being produced or imported into Canada, all new substances, including nanomaterials must undergo an ERA (HealthCanada, 2023b). To improve the quality of Health Canada's regulatory assessments, the organization has embraced and incorporated the use of risk management plans (RMPs) and the ICH E2E guideline into the regulatory review process for drugs in Canada (HealthCanada, 2015). In Japan, the guidelines for ERA on new drug products have been made available by the Japanese Ministry of Health, Labor and Welfare (MHLW). The guidelines clarify the need for testing environmental toxicity and fate, as well as the computation of predicted environmental concentrations (PECs) (Inohana et al., 2019).

15.14 REGULATORY RESEARCH

Nanotechnology has opened up new possibilities in drug delivery, diagnostics, and therapeutics, but it also brings unique challenges and risks that must be thoroughly understood and managed. Understanding how nanomedicines interact with biological systems and evaluating the efficacy, safety, and quality of these products are the main goals of regulatory research. The rules, guidelines, and procedures established by regulatory bodies (e.g., FDA, EMA, Health Canada Authority, China Health Authority, Indian Health Authority, and Japan Health Authority) for the development, production, and use of nanoparticles are the subject of regulatory research on nanomedicines (EuropeanCommission, 2023; FDA, 2014a, 2018h, 2018d).

15.14.1 USFDA

Materials at the nanoscale have unique chemical, biological, physical, and other characteristics. For these reasons, the FDA launched a Regulatory Science research program (RSRP) in nanotechnology to investigate how materials at the nanoscale affect the safety and effectiveness of products that contain them. A component of the FDA's 2011 strategy plan is the RSRP on nanotechnology. The agency has prioritized eight areas, according to the strategic plan. FDA preparedness to assess novel emerging technologies, such as nanotechnology, is one of the key areas (FDA, 2011). Establishing a framework, carrying out a plan, and offering leadership on regulatory science activities and issues pertaining to products containing nanomaterials or otherwise involving the application of nanotechnology are the goals of RSRP. Through these steps, the FDA has been able to address the major gaps in knowledge, approaches, and/or tools, which are crucial for regulatory assessments. Additionally, by using the best available science to support regulatory actions and decisions on premarket review and post-market surveillance of NDAs and ANDAs, it assists the FDA in carrying out its mission of safeguarding public health. Keeping up with science and technology, making the most of innovations, and supporting the creation of novel products that help patients and consumers is another goal of RSRP. In addition, the FDA formed the NTF in 2007 to promote collaboration and communication regarding RSRP throughout the agency. RSRP approaches include core laboratory facilities, staff training, and professional development (to strengthen the agency's scientific expertise in nanotechnology), coordination, and collaborative opportunities for the research Excellence in Science (CORES) program. The goal of CORES is to improve cooperation both inside and outside the organization (academic institutions, industry, and other government agencies). Additionally, it encourages and makes domestic and international cooperation easier. Within the FDA, there are nanotechnology core centers that offer the technical staff and equipment necessary to conduct quality and safety assessments of products containing nanomaterials and to support regulatory decisions on the submissions. These centers evaluate the safety of nanomaterials, create and validate methods for evaluating quality and efficacy, and if necessary, identify and create standard methods (NTF, 2018).

15.14.2 EMA

The EMA proposes the following actions: (i) "raise awareness of new nanomedicines and materials via the EU-Innovation Network and foster collaboration with Director-General Joint Research Center (DG JRC) and other international partners (e.g., IPRP), to share knowledge and harmonize regulatory practices: Generate guidance addressing pharmacokinetics/pharmacodynamics (including modeling) requirements and long-term efficacy and safety, (ii) develop and standardize new testing methods related to the quality and safety assessment of nanomedicines, and (iii) understand the CQA of a given product and the relationship between CQA and the biological activity and *in vivo* behavior of the product" to develop necessary understanding of the nanotechnology (EMA, 2020). Furthermore, the EU regulates new chemicals through REACH (registration, evaluation, authorization, and restriction of chemical substances). REACH imposes uniform regulatory requirements on all chemicals manufactured or imported within the EU. Companies are mandated to furnish chemical data to the European Chemical Agency (ECA) as part of this regulation. This data encompasses guidelines for safe handling to mitigate the potential risks associated with chemical substances and information about the environmental, health, and safety attributes of these chemicals (Jarvis and Richmond, 2011).

15.14.3 HEALTH CANADA AUTHORITY

Health Canada has highlighted nanotechnology as a top priority for the Canadian government. Canada aims to establish itself as a global leader in advancing and applying 21st century technologies, including biotechnology, environmental technology, information and communication technology, health technology, and nanotechnology. To achieve these objectives, the Canadian government allocates an annual budget of approximately $45–50 million specifically for nanotechnology research and development. Moreover, the Canadian government has recognized the importance of breaking down traditional barriers between various research disciplines and promoting a multidisciplinary approach. Instead of funding isolated laboratories working on individual projects, Canada now favors a collaborative approach involving experts from diverse backgrounds, such as physicists, biologists, and chemists, who can collaborate on nanotechnology-related challenges. This approach is fostering significant progress, allowing Canada to make substantial advancements in the field of nanotechnology. During the establishment of regulations for nanotechnology-based drug products, the Health Canada faces the following challenges:—(i) "limited experience with novel nanotechnology-based medicines, (ii) based on current information, concerns regarding potential toxicity and lack of efficacy, (iii) limitations for post-market surveillance within the current regulatory framework, and (iv) limitations for post-market surveillance being addressed through product lifecycle approach" (HealthCanada, 2011; Vu and Gallivan, 2010).

15.14.4 MHLW/ PMDA, JAPAN

The goal of MHLW/PDMA is to elevate scientific and technological benchmarks, leading to substantial contributions to economic and societal advancement, both within Japan and globally. The MHLW has been actively involved in supporting research and development activities related to nanomedicines over the years (Garg et al., 2017; Inohana et al., 2019). Efforts involve initiating discussions among MHLW regulators, PMDA reviewers, and NIH researchers regarding nanomedicine regulation. Emphasis is placed on promoting research in the analysis of nanoparticles and the evaluation of nanomedicine biodistribution within the human body, with corresponding regulatory discussions. An open dialogue between industry, academia, and regulatory authorities to establish appropriate regulations for nanomedicines while advancing their medical applications. International cooperation with relevant organizations is sought to address global challenges and opportunities in the field of nanomedicines.

15.14.5 NMPA, CHINA

The United States has been at the forefront of global investment in research and development in nanotechnology; China has rapidly ascended as a significant global contender. By 2005, China had positioned itself as the second-largest investor in nanotechnology, followed by the United States. Furthermore, it had secured the second position in the number of peer-reviewed research publications related to nanotechnology, contributing to 15% of the world's total output in this field. Additionally, China has solidified its leadership in carbon nanotube technology and production on a global scale. In response to the increasing complexity of nano-based research, both in pure and applied contexts, driven by augmented government funding and the expanding range of industrial applications for nanomaterials, China has taken proactive steps to establish comprehensive standards related to measurement, handling, exposure, toxicity, and safety. These nanotechnology standards are subject to review by the National Nanotechnology Standardization Technical Committee (NSTC) and its subsidiary Technical Committee 279, which focuses specifically on nanomaterials. This sub-committee operates under the Standardization Administration of China (SAC) and is situated within the National Center for Nanoscience and Technology (NCNST). The SAC/TC279 plays a pivotal role in coordinating the development of crucial nanotechnology standards, encompassing terminology, methodologies, and safety guidelines across various domains, including nanoscale measurements, materials, and nanoscale biomedicine. The regulatory framework for overseeing nanotechnology chemicals, which is still in the process of being implemented, is expected to address risks similar to those outlined in the European Union's REACH regulation.

15.14.6 CDSCO, INDIA

The Indian government has been at the forefront of advancing nanotechnology R&D in the country. It has undertaken numerous initiatives to stimulate and facilitate R&D activities, spanning various government departments. Several schemes and programs have been introduced to support infrastructure development and the growth of skilled human resources in the field of nanotechnology. Furthermore, the government has engaged in collaborative efforts with numerous countries, both bilaterally and multilaterally, to drive nanotechnology research, development, and innovation. In addition, the government has actively promoted public–private partnerships as a means to foster the development of nano-based products. Department of Science and Technology and the Department of Biotechnology are the key government departments that have been actively involved in nanotechnology in India (Kumar, 2014; Samal and Manohara, 2019).

15.15 SUMMARY

Nanoscale materials possess unique physical, chemical, biological, and other properties. To comprehensively understand, mitigate and control the quality, safety, and efficacy of nanomedicines, regulatory authorities have launched various research programs. FDA established the NTF in 2006, and RSRP in 2011. EMA, Health Canada Authority, NMPA (China), CDSCO (India), and MHLW/PMDA (Japan) also established similar programs to regulate nanomedicines. The existing regulatory framework can be used to file, review, and approve nanomedicines.

KEYWORDS

- **abbreviated new drug application**
- **common technical document**
- **food and drug administration**
- **new drug application**
- **quality by design**

REFERENCES

21 CFR 314.52. Code of federal Regulations Title 21, Part 314, Sec. 314.52: Notice of certification of invalidity or noninfringement of a patent. https://www.accessdata.fda.gov/scripts/cdrh/cfdocs/cfcfr/CFRSearch.cfm?fr=314.52 (accessed Dec 03, 2023).

21CFR25.15. Code of federal Regulations Title 21, Part 25, Subpart B, Section 25.15: General procedures. https://www.accessdata.fda.gov/scripts/cdrh/cfdocs/cfcfr/cfrsearch.cfm?fr=25.15 (accessed Dec 04, 2023).

21CFR25.20. Code of federal Regulations Title 21, Part 25, Subpart B, Section 25.20: Action requiring preparation of an environment assessment. https://www.accessdata.fda.gov/scripts/cdrh/cfdocs/cfCFR/CFRSearch.cfm?fr=25.20 (accessed Dec 04, 2023).

21CFR25.31. Code of federal Regulations Title 21, Part 25, Subpart C, Section 25.31: Human drugs and biologics. https://www.accessdata.fda.gov/scripts/cdrh/cfdocs/cfcfr/CFRSearch.cfm?fr=25.31 (accessed Dec 04, 2023).

21CFR314.101. Code of federal Regulations Title 21, Part 314, Subpart D, Section 314.101: Filing an application and receiving an abbreviated new drug application. https://www.accessdata.fda.gov/scripts/cdrh/cfdocs/cfcfr/CFRSearch.cfm?fr=314.101 (accessed Dec 04, 2023).

21CFR601.2. Code of federal Regulations Title 21, Part 601, Subpart A, Section 601.2: Application for biologic licenses; procedures for filing. https://www.accessdata.fda.gov/scripts/cdrh/cfdocs/cfCFR/CFRSearch.cfm?fr=601.2 (accessed Dec 04, 2023).

Abbvie. Tricor®. https://dailymed.nlm.nih.gov/dailymed/drugInfo.cfm?setid=b693e68d-f812-4993-54b6-852e3517c344 (accessed Jan 09, 2024).

Abd Elkodous, M., Olojede, S. O., Morsi, M., & El-Sayyad, G. S. (2021). Nanomaterial-based drug delivery systems as promising carriers for patients with COVID-19. *RSC Advances, 11* (43), 26463–26480. https://doi.org/10.1039/D1RA04835J.

Abraxis-Bioscience. Abraxane®. https://dailymed.nlm.nih.gov/dailymed/drugInfo.cfm?setid=24d10449-2936-4cd3-b7db-a7683db721e4 (accessed Jan 09, 2024).

Akinc, A., Maier, M. A., Manoharan, M., Fitzgerald, K., Jayaraman, M., Barros, S., Ansell, S., Du, X., Hope, M. J., Madden, T. D., Mui, B. L., Semple, S. C., Tam, Y. K., Ciufolini, M., Witzigmann, D., Kulkarni, J. A., van der Meel, R., & Cullis, P. R. (2019). The onpattro story and the clinical translation of nanomedicines containing nucleic acid-based drugs. *Nature Nanotechnology,* 14 (12), 1084–1087. https://doi.org/10.1038/s41565-019-0591-y.

Allergan. Restasis®. https://dailymed.nlm.nih.gov/dailymed/drugInfo.cfm?setid=8e24af2b-bc1c-4849-94f2-6df950cdca89 (accessed Jan 09, 2024).

Alnylam. Onpattro®. https://dailymed.nlm.nih.gov/dailymed/drugInfo.cfm?setid=e87ec36f-b4b4-49d4-aea4-d4ffb09b0970 (accessed Jan 04, 2024).

Alnylam. Onpattro®. https://www.ema.europa.eu/en/medicines/human/EPAR/onpattro#ema-inpage-item-authorisation-details (accessed Jan 10, 2024).

Ambardekar, V. V., & Stern, S. T. (2015). NBCD pharmacokinetics and bioanalytical methods to measure drug release. In *Non-biological complex drugs* (pp. 261–287). Springer. https://doi.org/10.1007/978-3-319-16241-6_8.

American-Regent. Venofer®. https://dailymed.nlm.nih.gov/dailymed/drugInfo.cfm?setid=626dc9e5-c6b4-4f9c-9bf4-774fd3ae619a (accessed Jan 09, 2024).

Amneal. Draft Guidance on Difluprednate. https://www.accessdata.fda.gov/drugsatfda_docs/psg/Difluprednate_ophthalmic emulsion_RLD 022212_RV02-17.pdf (accessed Jan 04, 2024).

Amphotericin-B. Draft Guidance on Amphotericin B. https://www.accessdata.fda.gov/drugsatfda_docs/psg/PSG_050740.pdf (accessed 2024-01-04).

Astellas. AmBiosome®. https://dailymed.nlm.nih.gov/dailymed/drugInfo.cfm?setid=f7be6506-4d20-401e-a0ff-02ad7c33158a (accessed Jan 09, 2024).

Atlantic. Megace® ES. https://dailymed.nlm.nih.gov/dailymed/drugInfo.cfm?setid=654854bc-f847-4e3b-a0a1-8716553460ac (accessed Jan 09, 2024).

Attia, M. A., Essa, E. A., Elebyary, T. T., Faheem, A. M., & Elkordy, A. A. (2021). Brief on recent application of liposomal vaccines for lower respiratory tract viral infections: From influenza to COVID-19 vaccines. *Pharmaceuticals, 14* (11), 1173. https://doi.org/10.3390/ph14111173.

Barenholz, Y. (2012). (Chezy). Doxil®—The first FDA-approved nano-drug: Lessons learned. *Journal of Controlled Release, 160* (2), 117–134. https://doi.org/10.1016/j.jconrel.2012.03.020.

Bartnicki, P., Fijałkowski, P., Majczyk, M., Błaszczyk, J., Banach, M., & Rysz, J. (2013). Effect of methoxy polyethylene glycol-epoetin beta on oxidative stress in predialysis patients with chronic kidney disease. *Medical Science Monitor, 19*, 954–959. https://doi.org/10.12659/MSM.884024.

Baxter. Caelyx®. https://www.ema.europa.eu/en/documents/product-information/caelyx-pegylated-liposomal-epar-product-information_en.pdf (accessed Jan 09, 2024).

Baxter. Doxil®. https://dailymed.nlm.nih.gov/dailymed/drugInfo.cfm?setid=1c153e9e-4cf2-4ac7-9cf9-16f9b48d7dce (accessed Jan 09, 2024).

Baxter. Doxil®. https://www.accessdata.fda.gov/scripts/cder/ob/results_product.cfm?Appl_Type=N&Appl_No=050718#24015 (accessed Jan 09, 2024).

Bhandari, S., Kalra, P. A., Kothari, J., Ambühl, P. M., Christensen, J. H., Essaian, A. M., Thomsen, L. L., Macdougall, I. C., & Coyne, D. W. (2015). A randomized, open-label trial of iron isomaltoside 1000 (Monofer®) compared with iron sucrose (Venofer®) as maintenance therapy in haemodialysis patients. *Nephrology Dialysis Transplantation, 30* (9), 1577–1589. https://doi.org/10.1093/ndt/gfv096.

Biogen-Inc. Plegridy®. https://dailymed.nlm.nih.gov/dailymed/drugInfo.cfm?setid=08f0ea03-4e6d-195d-aef4-886e32befa95 (accessed Jan 09, 2024).

Borgå, O., Lilienberg, E., Bjermo, H., Hansson, F., Heldring, N., & Dediu, R. (2019). Pharmacokinetics of total and unbound paclitaxel after administration of paclitaxel micellar or nab-paclitaxel: An open, randomized, cross-over, explorative study in breast cancer patients. *Advances in Therapy, 36* (10), 2825–2837. https://doi.org/10.1007/s12325-019-01058-6.

Boswell, G. W., Buell, D., & Bekersky, I. (1998). AmBisome (liposomal amphotericin B): A comparative review. *The Journal of Clinical Pharmacology, 38* (7), 583–592. https://doi.org/10.1002/j.1552-4604.1998.tb04464.x.

Bressler, N. M., & Bressler, S. B. (2000). Photodynamic therapy with verteporfin (visudyne): Impact on ophthalmology and visual sciences. *Investigative Ophthalmology & Visual Science 41* (3), 624–628.

Bristol-Myers-Squibb-Pharma. Abraxane®. https://www.ema.europa.eu/en/medicines/human/EPAR/abraxane#ema-inpage-item-authorisation-details (accessed Jan 10, 2024).

Brucker, J., Mayer, C., Gebauer, G., Mallmann, P., Belau, A. K., Schneeweiss, A., Sohn, C., & Eichbaum, M. (2016). Non-pegylated liposomal doxorubicin for patients with recurrent ovarian cancer: A multicentric phase II trial. *Oncology Letters, 12* (2), 1211–1215. https://doi.org/10.3892/ol.2016.4740.

CBER. Center for Biologics Evaluation and Research (CBER). https://www.fda.gov/about-fda/center-biologics-evaluation-and-research-cber/center-biologics-evaluation-and-research (accessed Dec 02, 2023).

CFDI. Center for Food and Drug Inspection of NMPA 2021annual report on information disclosure. https://www.cfdi.org.cn/resource/news/14360.html (accessed Dec 03, 2023).

CFDI. Notice of the Food and Drug Review and Inspection Center of the State Drug Administration on the release of five documents including the "Drug Registration and Verification Work Procedures (Trial)." https://www.cfdi.org.cn/resource/news/14200.html (accessed Dec 03, 2023).

Chabria, Y., Duffy, G., Lowery, A., & Dwyer, R. (2021). Hydrogels: 3D drug delivery systems for nanoparticles and extracellular vesicles. *Biomedicines, 9* (11), 1694. https://doi.org/10.3390/biomedicines9111694.

Chapman, J. An inside look at China's regulatory and drug approval processes. https://redica.com/pharma-an-inside-look-at-chinas-regulatory-and-drug-approval-processes/ (accessed Jan 04, 2024).

Cheplapharm. Visudyne®. https://www.ema.europa.eu/en/medicines/human/EPAR/visudyne#ema-inpage-item-authorisation-details (accessed Jan 10, 2024).
Chiesi. Curosurf®. https://dailymed.nlm.nih.gov/dailymed/drugInfo.cfm?setid=99bef307-00bd-4379-9f45-ede1380e74d4 (accessed Jan 09, 2024).
Covis. Draft guidance on ferumoxytol. https://www.accessdata.fda.gov/drugsatfda_docs/psg/PSG_022180.pdf (accessed 2024-01-04).
Credevo. The drug approval process in Japan. https://credevo.com/articles/2020/04/15/the-drug-approval-process-in-japan/ (accessed Jan 06, 2024).
Cruz, C. N., Tyner, K. M., Velazquez, L., Hyams, K. C., Jacobs, A., Shaw, A. B., Jiang, W., Lionberger, R., Hinderling, P., Kong, Y., Brown, P. C., Ghosh, T., Strasinger, C., Suarez-Sharp, S., Henry, D., Van Uitert, M., Sadrieh, N., & Morefield, E. (2013). CDER risk assessment exercise to evaluate potential risks from the use of nanomaterials in drug products. *AAPS Journal, 15* (3), 623–628. https://doi.org/10.1208/s12248-013-9466-6.
Curtis, J. R., Mariette, X., Gaujoux-Viala, C., Blauvelt, A., Kvien, T. K., Sandborn, W. J., Winthrop, K., de Longueville, M., Huybrechts, I., & Bykerk, V. P. (2019). Long-term safety of certolizumab pegol in rheumatoid arthritis, axial spondyloarthritis, psoriatic arthritis, psoriasis and Crohn's disease: A pooled analysis of 11 317 patients across clinical trials. *RMD Open, 5* (1), e000942. https://doi.org/10.1136/rmdopen-2019-000942.
Eglovitch, J. S. Comparability protocols: Final guidance incorporates ICH Q12, allows for API supplier changes. https://www.raps.org/News-and-Articles/News-Articles/2022/10/Comparability-protocols-Final-guidance-incorporate (accessed Dec 02, 2023).
Ei. Estrasorb®. https://www.accessdata.fda.gov/drugsatfda_docs/label/2015/021371s007lbl.pdf (accessed 2024-01-09).
EMA. EMA regulatory science to 2025—Strategic reflection. https://www.ema.europa.eu/en/documents/regulatory-procedural-guideline/ema-regulatory-science-2025-strategic-reflection_en.pdf (accessed Dec 03, 2023).
EMA. Guideline on the environmental risk assessment of medicinal products for human use. https://www.ema.europa.eu/en/documents/scientific-guideline/guideline-environmental-risk-assessment-medicinal-products-human-use-first-version_en.pdf (accessed Dec 03, 2023).
EMA. Guideline on the environmental risk assessment of medicinal products for human use. www.ema.europa.eu/en/documents/scientific-guideline/draft-guideline-environmental-risk-assessment-medicinal-products-human-use-revision-1_en.pdf (accessed Dec 03, 2023).
EMA. Multidisciplinary: Nanomedicines. https://www.ema.europa.eu/en/human-regulatory-overview/research-and-development/scientific-guidelines/multidisciplinary guidelines/multidisciplinary-nanomedicines (accessed Jan 06, 2024).
EMA. Obtaining an EU marketing authorisation, step-by-step. https://www.ema.europa.eu/en/human-regulatory-overview/marketing-authorisation/obtaining-eu-marketing-authorisation-step-step (accessed Jan 06, 2024).
Endo. Megace® ES. https://www.accessdata.fda.gov/scripts/cder/ob/results_product.cfm?Appl_Type=N&Appl_No=021778#22979 (accessed Dec 10, 2023).
EPA. What is the National Environmental Policy Act? https://www.epa.gov/nepa/what-national-environmental-policy-act#:~:text=In 1978%2CCEQissuedregulations preparationofenvironmentalimpactstatements (accessed Dec 02, 2023).

Ettinger, A. R. (1995). Pegaspargase (Oncaspar). *Journal of Pediatric Oncology Nursing, 12* (1), 46–48. https://doi.org/10.1177/104345429501200110.

Eupati. EU regulatory procedures for a marketing authorisation (MA). https://learning.eupati.eu/mod/book/tool/print/index.php?id=893 (accessed Jan 05, 2024).

Euractiv. The future of nanomedicines in Europe. https://en.euractiv.eu/wp-content/uploads/sites/2/special-report/The-future-of-nanomedicines-in-Europe-Event-Report-EAASM.pdf (accessed Jan 06, 2024).

European Commission. European Commission—Nanotechnology. https://joint-research-centre.ec.europa.eu/scientific-activities-z/nanotechnology_en (accessed 2023-12-03).

Extedo. EXTEDO—What are the regional differences of eCTD? https://www.extedo.com/blog/what-are-the-regional-differences-of-ectd (accessed Dec 02, 2023).

FDA. Advancing regulatory science at FDA: A strategic plan. https://www.fda.gov/science-research/advancing-regulatory-science/strategic-plan-regulatory-science (accessed Dec 03, 2023).

FDA. ANDA submissions—Amendments to abbreviated new drug applications under GDUFA. https://www.fda.gov/regulatory-information/search-fda-guidance-documents/anda-submissions-amendments-abbreviated-new-drug-applications-under-gdufa (accessed Dec 02, 2023).

FDA. ANDA Submissions—Content and format of abbreviated new drug applications. https://www.fda.gov/regulatory-information/search-fda-guidance-documents/anda-submissions-content-and-format-abbreviated-new-drug-applications (accessed Dec 02, 2023).

FDA. Applications covered by section 505(b)(2). https://www.fda.gov/regulatory-information/search-fda-guidance-documents/applications-covered-section-505b2 (accessed Jan 06, 2024).

FDA. Center for drug evaluation and research nanotechnology programs. https://www.fda.gov/science-research/nanotechnology-programs-fda/center-drug-evaluation-and-research-nanotechnology-programs (accessed Dec 03, 2023).

FDA. Changes to an approved NDA or ANDA. https://www.fda.gov/regulatory-information/search-fda-guidance-documents/changes-approved-nda-or-anda (accessed Dec 02, 2023).

FDA. Code of federal regulations Title 21, Part 314, Sec. 314.50: Content and format of an application. https://www.accessdata.fda.gov/scripts/cdrh/cfdocs/cfcfr/CFRSearch.cfm?fr=314.50 (accessed Jan 03, 2024).

FDA. Code of federal regulations Title 21, Part 314, Subpart D, Section 314.108: New drug product exclusivity. https://www.accessdata.fda.gov/scripts/cdrh/cfdocs/cfcfr/CFRSearch.cfm?fr=314.108 (accessed Jan 04, 2024).

FDA. Code of federal regulations Title 21, Part 316, Subpart D, Sec. 316.31: Scope of orphan drug exclusive approval. https://www.accessdata.fda.gov/scripts/cdrh/cfdocs/cfcfr/CFRSearch.cfm?fr=316.31 (accessed Jan 04, 2024).

FDA. Code of federal Regulations Title 21, Part 316: Orphan drugs. http://www.accessdata.fda.gov/scripts/cdrh/cfdocs/cfcfr/CFRSearch.cfm?CFRPart=316 (accessed Jan 04, 2024).

FDA. Comparability protocols for postapproval changes to the chemistry, manufacturing, and controls information in an NDA, ANDA, or BLA. https://www.fda.gov/regulatory-information/search-fda-guidance-documents/comparability-protocols-postapproval-changes-chemistry-manufacturing-and-controls-information-nda (accessed Dec 02, 2023).

FDA. Cosmetics nanotechnology. https://www.fda.gov/cosmetics/cosmetics-science-research/cosmetics-nanotechnology (accessed Jan 02, 2024).
FDA. Drug products, including biological products, that contain nanomaterials—Guidance for industry. https://www.fda.gov/regulatory-information/search-fda-guidance-documents/drug-products-including-biological-products-contain-nanomaterials-guidance-industry (accessed Dec 02, 2023).
FDA. Environmental impact considerations. https://www.fda.gov/animal-veterinary/development-approval-process/environmental-impact-considerations (accessed Dec 02, 2023).
FDA. FD&C Act Chapter V: Drugs and devices, Part A, Section 355: New drugs. https://www.fda.gov/regulatory-information/federal-food-drug-and-cosmetic-act-fdc-act/fdc-act-chapter-v-drugs-and-devices#Part_A (accessed Dec 02, 2023).
FDA. FDA compliance program 7346.832 Chapter 46-ew drug evaluation. https://www.fda.gov/media/121512/download (accessed Dec 02, 2023).
FDA. FDA guidance for industry considering whether an FDA-regulated product involves the application of nanotechnology. https://www.fda.gov/regulatory-information/search-fda-guidance-documents/considering-whether-fda-regulated-product-involves-application-nanotechnology (accessed Dec 03, 2023).
FDA. FDA guidance for industry: Statistical approaches to establishing bioequivalence. https://www.fda.gov/regulatory-information/search-fda-guidance-documents/statistical-approaches-establishing-bioequivalence-0 (accessed Dec 02, 2023).
FDA. FDA Modernization Act (FDAMA) of 1997. https://www.fda.gov/regulatory-information/selected-amendments-fdc-act/food-and-drug-administration-modernization-act-fdama-1997 (accessed 2024-01-04).
FDA. FDA Prescription Drug User Fee Amendments (PDUFA). https://www.fda.gov/industry/fda-user-fee-programs/prescription-drug-user-fee-amendments (accessed Dec 06, 2023).
FDA. FDA's approach to regulation of nanotechnology products. https://www.fda.gov/science-research/nanotechnology-programs-fda/fdas-approach-regulation-nanotechnology-products (accessed Dec 03, 2023).
FDA. FDA's Pre-Approval Inspection (PAI) program and how to prepare for a successful outcome. https://www.fda.gov/files/drugs/published/FDA's-Pre-Approval-Inspection-(PAI) -Program-and-How-to-prepare-for-a-successful-outcome.pdf (accessed Dec 03, 2023).
FDA. Food and Drug Administration Amendments Act (FDAAA) of 2007. https://www.fda.gov/regulatory-information/selected-amendments-fdc-act/food-and-drug-administration-amendments-act-fdaaa-2007 (accessed Dec 02, 2023).
FDA. Guidance for industry: Safety of nanomaterials in cosmetic products. https://www.fda.gov/regulatory-information/search-fda-guidance-documents/guidance-industry-safety-nanomaterials-cosmetic-products (accessed Jan 02, 2024).
FDA. Hatch–Waxman letters. https://www.fda.gov/drugs/abbreviated-new-drug-application-anda/hatch-waxman-letters#:~:text=The %22Drug Price Competition and,Drug%2C and Cosmetic Act (FD%26C (accessed Dec 03, 2023).
FDA. M4 organization of the common technical document for the registration of pharmaceuticals for human use guidance for industry. https://www.fda.gov/regulatory-information/search-fda-guidance-documents/

m4-organization-common-technical-document-registration-pharmaceuticals-human-use-guidance-industry (accessed Dec 03, 2023).
FDA. (2007). Nanotechnology a report of the U.S. Food and Drug Administration. https://www.fda.gov/media/74257/download.
FDA. Nanotechnology fact sheet. https://www.fda.gov/science-research/nanotechnology-programs-fda/nanotechnology-fact-sheet (accessed Dec 03, 2023).
FDA. Nanotechnology guidance documents. https://www.fda.gov/science-research/nanotechnology-programs-fda/nanotechnology-guidance-documents (accessed Jan 04, 2024).
FDA. New drug application (NDA). https://www.fda.gov/drugs/types-applications/new-drug-application-nda (accessed Jan 04, 2024).
FDA. Pharmaceutical quality for the 21st century a risk-based approach progress report. https://www.fda.gov/about-fda/center-drug-evaluation-and-research-cder/pharmaceutical-quality-21st-century-risk-based-approach-progress-report (accessed Dec 02, 2023).
FDA. Product-specific guidances for ferric oxyhydroxide; Revised draft guidances for industry; availability. https://www.federalregister.gov/documents/2021/09/17/2021-20064/product-specific-guidances-for-ferric-oxyhydroxide-revised-draft-guidances-for-industry-availability (accessed Jan 04, 2024).
FDA. Providing regulatory submissions in electronic format—Certain human pharmaceutical product applications and related submissions using the eCTD specifications guidance for industry. https://www.fda.gov/regulatory-information/search-fda-guidance-documents/providing-regulatory-submissions-electronic-format-certain-human-pharmaceutical-product-applications (accessed Dec 02, 2023).
FDA/CDER/ChoiStephnie. Draft guidance on cyclosporine. https://www.accessdata.fda.gov/drugsatfda_docs/psg/Cyclosporine_ophthalmic emulsion_RLD 050790_RV09-16.pdf (accessed Jan 04, 2024).
Fresenius-Kabi. Diprivan®. https://dailymed.nlm.nih.gov/dailymed/drugInfo.cfm?setid=5cec3b97-5183-4a9d-accd-21e76c99d3dd (accessed Jan 01, 2024).
Fresenius-Kabi. Diprivan®. https://www.accessdata.fda.gov/scripts/cder/ob/results_product.cfm?Appl_Type=N&Appl_No=019627#3118 (accessed Jan 09, 2024).
GaBi. China's regulatory authority adopt CTD format for NDAs. https://gabionline.net/guidelines/China-s-regulatory-authority-adopt-CTD-format-for-NDAs (accessed Dec 02, 2023).
Gaffney, A. An increasing number of companies are using a once-obscure FDA drug approval pathway. Regulatory Affairs Professionals Society. https://www.raps.org/news-and-articles/news-articles/2015/4/an-increasing-number-of-companies-are-using-a-once (accessed Jan 04, 2024).
Garg, V., Chopra, S., Singh, S. K., Gulati, M., Kumar, B., & Mittal, N. (2017). Diabetic wound healing view project a comparative study of common technical document in different regulated market. *Journal of Pharmacy Research, 11*, 1015–1024.
Gohil, K. (2014). Pharmaceutical approval update. *P T, 39* (12), 823–845.
Guo, M., Wei, M., Li, W., Guo, M., Guo, C., Ma, M., Wang, Y., Yang, Z., Li, M., Fu, Q., Yang, L., & He, Z. (2019). Impacts of particle shapes on the oral delivery of drug nanocrystals: Mucus permeation, transepithelial transport and bioavailability. *Journal of Controlled Release, 307*, 64–75. https://doi.org/10.1016/j.jconrel.2019.06.015.

Gupta, M., Bala, R., Madaan, R., Chauhan, S., Chawla, R., Kaur, J., Sharifi-Rad, J., & Calina, D. (2023). Unfolding the potential of nanocomposites as drug carriers and their future scenarios. *Polymer Bulletin,* No. 0123456789. https://doi.org/10.1007/s00289-023-04987-z.

Halwani, A. A. Development of pharmaceutical nanomedicines: From the bench to the market. *Pharmaceutics, 14* (1), 1–21. https://doi.org/10.3390/pharmaceutics14010106.

Hara, R. Japanese submission/approval processes from programming perspective. https://www.pharmasug.org/proceedings/2015/SS/PharmaSUG-2015-SS02.pdf (accessed Jan 08, 2024).

HealthCanada. Drug and health product review and approval. https://www.canada.ca/en/health-canada/services/drug-health-product-review-approval.html (accessed Jan 06, 2024).

HealthCanada. Environmental risk assessment and management of active ingredients in drugs. https://www.canada.ca/en/health-canada/corporate/about-health-canada/legislation-guidelines/acts-regulations/forward-regulatory-plan/plan/environmental-risk-assessment-active-ingredients-drugs.html (accessed Dec 03, 2023).

HealthCanada. Guidance document—Submission of risk management plans and follow-up commitments. https://www.canada.ca/en/health-canada/services/drugs-health-products/reports-publications/medeffect-canada/guidance-document-submission-risk-management-plans-follow-commitments.html (accessed Dec 03, 2023).

HealthCanada. Nanotechnology-based health products and food. https://www.canada.ca/en/health-canada/services/drugs-health-products/nanotechnology-based-health-products-food.html (accessed Dec 03, 2023).

HealthCanada. Risk assessment of chemical substances. https://www.canada.ca/en/health-canada/services/chemical-substances/canada-approach-chemicals/risk-assessment.html (accessed Dec 03, 2023).

Holland, S. Special aspects of nanomedicines—Development, manufacturing and characterization. www.ema.europa.eu/en/documents/presentation/presentation-special-aspects-nanomedicines-development-manufacturing-characterisation-simon-holland_en.pdf (accessed 2023-12-03).

Horiba. A guidebook to particle size analysis. https//ats-scientific.com/uploads/PSA_Guidebook.pdf (accessed Jan 04, 2024).

Hui, C., & Lau, G. K. (2005). Peginterferon–A2a (40 KDa) (Pegasys®) for Hepatitis B. *Expert Review of Anti-infective Therapy, 3* (4), 495–504. https://doi.org/10.1586/14787210.3.4.495.

ICH. ICH guideline Q9 on quality risk management. www.ema.europa.eu/en/documents/scientific-guideline/international-conference-harmonisation-technical-requirements-registration-pharmaceuticals-human-use_en-3.pdf (accessed Jan 06, 2024).

ICH. ICH Q10 PQS guideline. https//database.ich.org/sites/default/files/Q10_Presentation.pdf (accessed Dec 03, 2023).

ICH. ICH Q2(R1) validation of analytical procedures: Text and methodology guidance for industry. https://www.fda.gov/regulatory-information/search-fda-guidance-documents/q2r1-validation-analytical-procedures-text-and-methodology-guidance-industry (accessed Jan 04, 2024).

ICH. International Conference on Harmonization Q8(R2) guideline for pharmaceutical development. https://database.ich.org/sites/default/files/Q8_R2_Guideline.pdf (accessed Jan 02, 2024).

Inohana, S., Kaminishi, M., de Roode, D. F., & Minami, H. (2019). Comparison of environmental risk assessment scheme on pharmaceuticals between EU and Japan. *JSOT Annual Meeting 2019 Poster P-174, 99*, 2019.

Ipsen. ONIVYDE- irinotecan hydrochloride injection, powder, for solution. https://dailymed.nlm.nih.gov/dailymed/drugInfo.cfm?setid=c8b58efa-1820-48a4-b70d-62918fc4abfc (accessed Jan 04, 2024).

ISNM. Guidelines for evaluation of nanopharmaceuticals in India. https://birac.nic.in/webcontent/1550639649_guidelines_for_evaluation_of_Nanopharmaceuticals_in_India_20_02_2019.pdf (accessed 2024-01-06).

James, J. S. (1995). DOXIL approved for KS. *AIDS Treatment News, 7*, 6.

Janjal, V. S., Dhamodkar, S. R., Jadhao, Y. P., Manmode, S. B., Pawar, A. K., Khandelwal, H. R. (2021). Recent drug regulatory affair and CTD module progress review for submission of pharmaceuticals product. *GSC Biological and Pharmaceutical Sciences, 16* (3), 200–221. https://doi.org/10.30574/gscbps.2021.16.3.0279.

Jarvis, D. S. L., & Richmond, N. (2011). Regulation and governance of nanotechnology in China: Regulatory challenges and effectiveness. *European Journal of Law and Technology*. https://www.ejlt.org/index.php/ejlt/article/download/94/155?inline=1 (accessed Dec 03, 2023).

Jia, Y., Jiang, Y., He, Y., Zhang, W., Zou, J., Magar, K. T., Boucetta, H., Teng, C., & He, W. (2023). Approved nanomedicine against diseases. *Pharmaceutics, 15* (3), 1–29. https://doi.org/10.3390/pharmaceutics15030774.

Jordan Paradise. (2019). Regulating nanomedicine at the Food and Drug Administration. *AMA Journal of Ethics, 21* (4), E347–E355.

Jordan, D. (2014). An overview of the common technical document (CTD) regulatory dossier. *Medical Writing, 23* (2), 101–105. https://doi.org/10.1179/2047480614Z.000000000207.

JPMA. Pharmaceutical regulations in Japan. https://www.jpma.or.jp/english/about/parj/eki4g6000000784o-att/2020e_ch03.pdf (accessed Jan 08, 2024).

Kayalar, C., Rahman, Z., Mohamed, E. M., Dharani, S., Khuroo, T., Helal, N., Kuttolamadom, M. A., & Khan, M. A. (2023). Preparation and characterization of 3d-printed dose-flexible printlets of tenofovir disoproxil fumarate. *AAPS PharmSciTech, 24* (6), 171. https://doi.org/10.1208/s12249-023-02623-7.

Kozak, D. (2016). Development and characterization of generic drug products containing nanomaterials. https://www.fda.gov/media/166985/download (accessed Jan 04, 2024).

Kumar, A. (2014). Nanotechnology development in India—An overview. *Research and Information System for Developing Countries*. www.ris.org.in/sites/default/files/Publication/DP 193 Amit Kumar.pdf (accessed Dec 03, 2023).

Kurwadkar, S., Pugh, K., Gupta, A., & Ingole, S. (2015). Nanoparticles in the environment: Occurrence, distribution, and risks. *Journal of Hazardous, Toxic & Radioactive Waste, 19* (3). https://doi.org/10.1061/(ASCE)HZ.2153-5515.0000258.

Lal, R. (2015). Patents and exclusivity. *FDA/CDER SBIA Chronicles, 505*, 1–3.

Lallemand, F., Schmitt, M., Bourges, J.-L., Gurny, R., Benita, S., & Garrigue, J.-S. (2017). Cyclosporine a delivery to the eye: A comprehensive review of academic and industrial efforts. *European Journal of Pharmaceutics and Biopharmaceutics, 117*, 14–28. https://doi.org/10.1016/j.ejpb.2017.03.006.

Leadiant. Abelcet®. https://dailymed.nlm.nih.gov/dailymed/drugInfo.cfm?setid=5587db37-f21a-4a39-a319-e1077032ced9 (accessed Jan 09, 2024).

Luye. (2020). Luye Pharma's innovative formulation Lipusu® included in the 2020 Chinese National Reimbursement drug list. https://www.luye.cn/lvye_en/view.php?id=1883 (accessed Jan 4, 2024).

Measurlabs. Particle size distribution analysis—guide to method selection. https://measurlabs.com/blog/particle-size-distribution-analysis-method-selection/ (accessed Jan 02, 2024).

Merck. Emend®. https://dailymed.nlm.nih.gov/dailymed/drugInfo.cfm?setid=696f9e80-9cae-403b-de9e-078343ce4713 (accessed Jan 09, 2024).

Milano, G., Innocenti, F., & Minami, H. (2022). Liposomal irinotecan (Onivyde): Exemplifying the benefits of nanotherapeutic drugs. *Cancer Science, 113* (7), 2224–2231. https://doi.org/10.1111/cas.15377.

ModernaTX. MODERNA COVID-19 VACCINE injection, suspension. https://dailymed.nlm.nih.gov/dailymed/drugInfo.cfm?setid=530786f0-58d9-4370-88f0-6e98e4b85183 (accessed Jan 04, 2024).

Mohamed, E. M., Dharani, S., Nutan, M. T. H., Cook, P., Arunagiri, R., Khan, M. A., & Rahman, Z. (2023). Application of sucrose acetate isobutyrate in development of co-amorphous formulations of tacrolimus for bioavailability enhancement. *Pharmaceutics, 15* (5), 1442. https://doi.org/10.3390/pharmaceutics15051442.

Mousseau, F., Oikonomou, E. K., Vacher, A., Airiau, M., Mornet, S., & Berret, J.-F. (2020). Revealing the pulmonary surfactant corona on silica nanoparticles by cryo-transmission electron microscopy. *Nanoscale Advances, 2* (2), 642–647. https://doi.org/10.1039/c9na00779b.

Muscular-Dystrophy-Canada. How new medicines are reviewed, approved and funded in Canada. https://muscle.ca/wp-content/uploads/2019/10/medicines.pdf (accessed Jan 04, 2024).

Musyuni, P. India: New drug approval: Regulatory scenario in India. https://www.appliedclinicaltrialsonline.com/view/regulatory-approval-india-updated-review (accessed Jan 06, 2024).

Musyuni, P. India: New drug approval: Regulatory scenario in India. https://www.mondaq.com/india/life-sciences-biotechnology-nanotechnology/477290/new-drug-approval-regulatory-scenario-in-india (accessed Jan 06, 2024).

Nautiyal, S. Indian pharma industry wants to end multiple inspections by US FDA that results in long timelines for market approval. http://www.pharmabiz.com/NewsDetails.aspx?aid=160439&sid=1 (accessed Dec 02, 2023).

NMPA. National Medical Product Agency (NMPA). https://globalregulatorypartners.com/chinas-nmpa-introduces-new-revised-regulation-for-drug-approval-by-foreign-companies/ (accessed Jan 04, 2024).

NMPA. Notice of the State Food and Drug Administration on issuing the "Measures for Drug Inspection and Management (Trial)." https://www.nmpa.gov.cn/xxgk/fgwj/gzwj/gzwjyp/20210528171603115.html (accessed Dec 03, 2023).

NNI. About nanotechnology. https://www.nano.gov/about-nanotechnology/what-is-so-special-about-nano (accessed Jan 06, 2024).

NNI. National nanotechnology initiative. https://www.nano.gov/national-nanotechnology-initiative (accessed Dec 03, 2023).

Novavax. Estrasorb®. https://www.accessdata.fda.gov/drugsatfda_docs/label/2003/21371_estrasorb_lbl.pdf (accessed Jan 09, 2024).

Novavax. Estrasorb®. https://www.accessdata.fda.gov/drugsatfda_docs/nda/2003/21-371_Estrasorb.cfm (accessed Jan 09, 2024).

NTF. (2007). Nanotechnology task force report. https://www.fda.gov/science-research/nanotechnology-programs-fda/nanotechnology-task-force-report-2007 (accessed Jan 04, 2024).

Oasmia. Summary of Apealea. https://www.ema.europa.eu/en/documents/product-information/apealea-epar-product-information_en.pdf (accessed Jan 04, 2024).

Owen, J., Crake, C., Lee, J. Y., Carugo, D., Beguin, E., Khrapitchev, A. A., Browning, R. J., Sibson, N., & Stride, E. (2018). A versatile method for the preparation of particle-loaded microbubbles for multimodality imaging and targeted drug delivery. *Drug Delivery and Translational Research, 8* (2), 342–356. https://doi.org/10.1007/s13346-017-0366-7.

PacificBridgeMedical. Navigating China's drug approval process: Key steps for success. https://www.pacificbridgemedical.com/resources/navigating-chinas-drug-approval-process-key-steps-for-success/ (accessed Jan 04, 2024).

Pacira. DepoDur®. https://dailymed.nlm.nih.gov/dailymed/drugInfo.cfm?setid=cc8f105c-c8ee-4c57-86ee-ee6bf917cf78 (accessed Jan 09, 2024).

Pacira. DepoDur®. https://www.accessdata.fda.gov/scripts/cder/ob/results_product.cfm?Appl_Type=N&Appl_No=021671#4366 (accessed Jan 10, 2024).

Pasero, C., & McCaffery, M. (2005). Extended-release epidural morphine (DepoDur™). *Journal of PeriAnesthesia Nursing, 20* (5), 345–350. https://doi.org/10.1016/j.jopan.2005.07.004.

Pfizer. PFIZER-BIONTECH COVID-19 VACCINE- covid-19 vaccine, mRNA injection, suspension. https://dailymed.nlm.nih.gov/dailymed/drugInfo.cfm?setid=d19d59cb-f1cd-479d-ab0c-a36971e65544 (accessed Jan 04, 2024).

Pfizer. Somavert®. https://www.ema.europa.eu/en/medicines/human/EPAR/somavert#ema-inpage-item-authorisation-details (accessed Jan 09, 2024).

Pharmaand. Pegasys. https://www.ema.europa.eu/en/medicines/human/EPAR/pegasys#ema-inpage-item-authorisation-details (accessed Jan 10, 2024).

Kümmerer, K., Ed. (2004). *Pharmaceuticals in the Environment*. Springer Berlin Heidelberg. https://doi.org/10.1007/978-3-662-09259-0.

Pharmacia-&-Upjohn-Company-LLC. Somavert®. https://purplebooksearch.fda.gov/productdetails?query=021106 (accessed 2024-01-09).

Prommer, E. (2005). Aprepitant (EMEND): The role of substance P in nausea and vomiting. *Journal of Pain and Palliative Care Pharmacotherapy, 19* (3), 31–39.

Propofol. Draft guidance on Propofol. https://www.accessdata.fda.gov/drugsatfda_docs/psg/Propofol injectable injection RLD 19627_RC06-16.pdf (accessed Jan 04, 2024).

QLT-Ophthalmics. Visudyne®. https://dailymed.nlm.nih.gov/dailymed/drugInfo.cfm?setid=31512723-9ff0-4e18-aa3a-55ab833038c6 (accessed Jan 09, 2024).

Rahman, Z., Charoo, N. A., Akhter, S., Beg, S., Reddy, I. K., & Khan, M. A. (2018). Nanotechnology-based drug products. In A. M. Grumezescu (Ed.), *Nanoscale fabrication, optimization, scale-up and biological aspects of pharmaceutical nanotechnology* (pp. 619–655). Elsevier. https://doi.org/10.1016/B978-0-12-813629-4.00016-4.

Rahman, Z., Siddiqui, A., Gupta, A., & Khan, M. (2014). Regulatory considerations in development of amorphous solid dispersions. In *Amorphous solid dispersions theory and practice* (pp. 545–563). https://doi.org/10.1007/978-1-4939-1598-9_17.

Ramanathan, R., Rasmussen, M. R., Gerstmann, D. R., Finer, N., & Sekar, K. (2004). A randomized, multicenter masked comparison trial of poractant alfa (curosurf) versus beractant (survanta) in the treatment of respiratory distress syndrome in preterm infants. *American Journal of Perinatology, 21* (3), 109–119. https://doi.org/10.1055/s-2004-823779.

Ranson, M. R., Cheeseman, S., White, S., & Margison, J. (2001). Caelyx (stealth liposomal doxorubicin) in the treatment of advanced breast cancer. *Critical Reviews in Oncology/Hematology, 37* (2), 115–120. https://doi.org/10.1016/S1040-8428(00)00107-4.

Roche. Mircera®. https://www.ema.europa.eu/en/medicines/human/EPAR/mircera#ema-inpage-item-authorisation-details (accessed Jan 10, 2024).

Rust, D. M., & Jameson, G. (1998). The novel lipid delivery system of amphotericin B: Drug profile and relevance to clinical practice. *Oncology Nursing Forum, 25* (1), 35–48.

Samal, S. S., & Manohara, S. R. (2019). Nanoscience and nanotechnology in India: A broad perspective. *Material Today: Proceedings, 10*, 151–158. https://doi.org/10.1016/j.matpr.2019.02.200.

Sangwai, M., & Vavia, P. (2013). Amorphous ternary cyclodextrin nanocomposites of telmisartan for oral drug delivery: Improved solubility and reduced pharmacokinetic variability. *International Journal of Pharmaceutics, 453* (2), 423–432. https://doi.org/10.1016/j.ijpharm.2012.08.034.

Satinover, S. J., Dove, J. D., & Borden, M. A. (2014). Single-particle optical sizing of microbubbles. *Ultrasound in Medicine and Biology, 40* (1), 138–147. https://doi.org/10.1016/j.ultrasmedbio.2013.08.018.

Saurav, A., Kaushik, M., & Mohiuddin, S. M. (2012). Fenofibric acid for hyperlipidemia. *Expert Opinion on Pharmacotherapy, 13* (5), 717–722. https://doi.org/10.1517/14656566.2012.658774.

Sen, M. (2020). Nanocomposite materials. In *Nanotechnology and the environment*. IntechOpen. https://doi.org/10.5772/intechopen.93047.

Servier-Pharmaceuticals-LLC. Oncaspar®. https://dailymed.nlm.nih.gov/dailymed/drugInfo.cfm?setid=d046ec54-9ac8-4c9f-87a7-69e9900d8016 (accessed Jan 09, 2024).

Servier-Pharmaceuticals-LLC. Oncaspar®. https://purplebooksearch.fda.gov/productdetails?query=103411 (accessed Jan 09, 2024).

Shen, J., & Burgess, D. J. (2013). *In vitro* dissolution testing strategies for nanoparticulate drug delivery systems: Recent developments and challenges. *Drug Delivery and Translational Research, 3* (5), 409–415. https://doi.org/10.1007/s13346-013-0129-z.

Silverman, J. A., & Deitcher, S. R. (2013). Marqibo® (vincristine sulfate liposome injection) improves the pharmacokinetics and pharmacodynamics of vincristine. *Cancer Chemotherapy and Pharmacology, 71* (3), 555–564. https://doi.org/10.1007/s00280-012-2042-4.

Simon, J. A. (2006). Estradiol in Micellar Nanoparticles. *Menopause, 13* (2), 222–231. https://doi.org/10.1097/01.gme.0000174096.56652.4f.

Sorg, C. (2005). Scientific forward look on nanomedicine. *European Science Foundation Policy Brief, 23*, 1–6.

SPharm. The drug review and approval process in Canada—An eGuide. https://spharm-inc.com/the-drug-review-and-approval-process-in-canada-an-eguide/?cn-reloaded=1 (accessed Jan 06, 2024).

Takeda. Mepact. https://www.ema.europa.eu/en/medicines/human/EPAR/mepact#ema-inpage-item-overview (accessed Jan 10, 2024).

Takeda. Mepact®. https://www.ema.europa.eu/en/documents/assessment-report/mepact-epar-public-assessment-report_en.pdf (accessed Jan 10, 2024).

Terblanche, N., & Coetzee, J. (2008). A comparison of induction of anaesthesia using two different propofol preparations. *Southern African Journal of Anaesthesia and Analgesia, 14* (6), 25–29. https://doi.org/10.1080/22201173.2008.10872573.

Teva. Myocet®. https://www.ema.europa.eu/en/documents/product-information/myocet-liposomal-previously-myocet-epar-product-information_en.pdf (accessed Jan 09, 2024).

Thapa, R. K., & Kim, J. O. (2023). Nanomedicine-based commercial formulations: Current developments and future prospects. *Journal of Pharmaceutical Investigation, 53* (1), 19–33. https://doi.org/10.1007/s40005-022-00607-6.

Tinkle, S., McNeil, S. E., Mühlebach, S., Bawa, R., Borchard, G., Barenholz, Y. C., Tamarkin, L., & Desai, N. (2014). Nanomedicines: Addressing the scientific and regulatory gap. *Annals of the New York Academy of Sciences, 1313*, 35–56. https://doi.org/10.1111/nyas.12403.

Tuca, A., Jimenez-Fonseca, P., & Gascón, P. (2013). Clinical evaluation and optimal management of cancer cachexia. *Critical Reviews in Oncology/Hematology, 88* (3), 625–636. https://doi.org/10.1016/j.critrevonc.2013.07.015.

UCB-Inc. Cimzia®. https://dailymed.nlm.nih.gov/dailymed/drugInfo.cfm?setid=b4c2c9dc-a0bb-4d64-a667-a67ebe88392d (accessed Jan 09, 2024).

UCB-Pharma. Cimzia®. https://www.ema.europa.eu/en/medicines/human/EPAR/cimzia-0#ema-inpage-item-authorisation-details (accessed Jan 10, 2024).

USP-NF. (429) Light diffraction measurement of particle size. www.usp.org/sites/default/files/usp/document/harmonization/gen-chapter/g13_pf_35_3_2009.pdf (accessed Jan 04, 2024).

Valeant. Visudyne®. https://www.accessdata.fda.gov/scripts/cder/ob/results_product.cfm?Appl_Type=N&Appl_No=021119#22516 (accessed Jan 10, 2024).

ViforPharma. Mircera®. https://dailymed.nlm.nih.gov/dailymed/drugInfo.cfm?setid=22c56f2a-f73c-60e7-e054-00144ff88e88 (accessed Jan 09, 2024).

Vrbanović Mijatović, V., Šerman, L., Gamulin, O. Analysis of Pulmonary Surfactant by Fourier Transform Infrared Spectroscopy after Exposure to Sevoflurane and Isoflurane. Bosn. J. Basic Med. Sci. 17 (1), 38–46. https://doi.org/10.17305/bjbms.2016.1680.

Vu, D., & Gallivan, J. (2010). Current issues with nanomedicines in Canada. In *1st International Workshop on Nanomedicines* (pp. 1–16).

Wang, G. D., Mallet, F. P., Ricard, F., & Heng, J. Y. (2012). Pharmaceutical nanocrystals. *Current Opinion in Chemical Engineering, 1* (2), 102–107. https://doi.org/10.1016/j.coche.2011.12.001.

YES-Pharmaceutical. Celdoxome pegylated liposomal. https://www.ema.europa.eu/en/medicines/human/EPAR/celdoxome-pegylated-liposomal#ema-inpage-item-authorisation-details (accessed Jan 10, 2024).

Yuan, H., Guo, H., Luan, X., He, M., Li, F., Burnett, J., Truchan, N., & Sun, D. (2020). Albumin nanoparticle of paclitaxel (abraxane) decreases while taxol increases breast cancer stem cells in treatment of triple negative breast cancer. *Molecular Pharmaceutics, 17* (7), 2275–2286. https://doi.org/10.1021/acs.molpharmaceut.9b01221.

Zhang, J., Pan, Y., Shi, Q., Zhang, G., Jiang, L., Dong, X., Gu, K., Wang, H., Zhang, X., Yang, N., Li, Y., Xiong, J., Yi, T., Peng, M., Song, Y., Fan, Y., Cui, J., Chen, G., Tan, W., Zang, A., Guo, Q., Zhao, G., Wang, Z., He, J., Yao, W., Wu, X., Chen, K., Hu, X., Hu, C., Yue, L., Jiang, D., Wang, G., Liu, J., Yu, G., Li, J., Bai, J., Xie, W., Zhao, W., Wu, L., & Zhou, C. (2022). Paclitaxel liposome for injection (lipusu) plus cisplatin versus gemcitabine plus cisplatin in the first-line treatment of locally advanced or metastatic lung squamous cell carcinoma: A multicenter, randomized, open-label, parallel controlled clinical study. *Cancer Communications, 42* (1), 3–16. https://doi.org/10.1002/cac2.12225.

ZrPharma. Pegasys®. https://purplebooksearch.fda.gov/productdetails?query=103964 (accessed Jan 01, 2024).

Index

45S5 Bioglass, 241

A

Acalypha indica, 381–382
Aceclofenac controlled release, 232–233
Achyranthes aspera, 380
Acinetobacter baumannii, 402
AcrAB-TolC efflux pump system, 434
Acute myeloid leukemia (AML), 318
Adipose-derived stem cells (ADSCs), 319
Adsorption and adsorbent
  polymer nanocomposites as dye removal, 254
  water purification, mechanism of action, 256
Aedes aegypti, 409
Aerobic bacteria, used in bioremediation, 273
$Ag/Fe_3O_4/ZrO_2$ Nanocomposite, 387
AGN/RSV Nanoparticles, 381
Air pollution
  nanocomposites in remediation, 270
Alanine aminotransferase (ALT), 315–316
Albumin, 378
Aldehyde oxidase, 314–315
Alginate, 377–378
Alginate nanocomposite hydrogels (ANH), 59
  antidot delivery, 60
  bone-related diseases treatment, 59
  effective delivery of vaccine, 59
  nasal delivery of drugs, 59–60
  ocular delivery of antibiotics, 60
Aloe vera, 381–382
Alzheimer's disease (AD)
  application of NCs for, 296–297
Aminophenylboronic acid (APBA), 386
3-Aminopropyl triethoxysilane (APTES), 434–435
Amlodipine, 433–434
Amorphous nanocomposites, 113–114
Amphiphilic folate-PEG-hydrophobically modified dextran (F-PEG-HMD) micelles, 320
Amphiphilic iridium complex-loading SOR, 316
Amp-Zn/Al-LDH NCs, 407
Anaerobic amphibious bacteria
  used in bioremediation, 273
Angelica gigas Nakai (AGN), 380
Animal Efficacy Rule, 503
Anopheles stephensi, 409
Antimicrobial activity
  recycled waste-derived NCs, used in, 156–157
Arabinoxylan, 411
Azadirachta indica, 381–382

B

Bacillus, 273
Ball milling, 8
B-Cell lymphoma 2, 326
Bedsores/pressure ulcers
  application of NCs, 299–300
Berberine (BER), 386
Bimetallic iron-doped ZIF-8 NCs, 319
Bioactive glass, 237, 411–412
Bioactive glass-ceramics, 237
Bioactive scaffold AZG-4, 411
Bioceramics
  applications in soft tissue repair, 241
Biodegradable nanocomposites, 26–27
Biodegradable NCs, 167
Biodegradation
  kinetics of, 179–183
Bio-encapsulated chitosan-Ag nanocomplex (BNC), 409
Biofilms
  application of NCs, 298–299
Bioglasses (BGs), 241

pulmonary tissue regeneration, 242–243
regeneration of nervous tissue, 242
Biological sensors, nanocomposites applications in
3D printed, 357
customized biosensors for disease detection, 357–358
personalized medicine and detection of disease at early stages, 358
potential for implantable, 358
Biologic and Genetic Therapies Directorate (BGTD), 489
Bio-nanomaterial CS coating iron oxide (Fe_3O_4/Cs) NC, 315
Biopiles, 276
Biopolymers, 156
Bioreactors, 276–277
Bioremediation, 271–272
microorganisms used in, 272
aerobic circumstances, 273
anaerobic amphibious bacteria, 273
types
ex situ, 276–277
in situ, 274–276
Biosensing technologies utilizing nanocomposites, 447, 449
implications
applications, 452–453
carbon nanotube-based, 450–451
graphene and graphene oxide, 449–450
metal and metal oxide, 450–451
Biosensors, 414–417
Biotechnological products, 221
Black phosphorus (BP), 445
Blood-brain barrier (BBB), 434
Bone marrow stromal (MS-5) cells, 318
Bone regeneration
CNH, 61–63
CsNH and, 60–61
Bone tissue engineering, 441–442
Bone tissue regeneration
application of NCs for diseases, 296
Brain cancer, 325–326
Brain-targeted liposomes (na-ATS/TMP@ lipoBX), 411
Breast cancer, 318–319
Burkholderia stabilis, 385

C

Calcium phosphates (CaPs), 237–238
Canadian Environmental Protection Act, 1999, 506
Canadian Environmental Protection Act of 1999, 506
Cancer, 311–312
environmental contaminants, 264–265
therapy
brain, 325–326
breast, 318–319
colorectal, 326–327
esophageal, 314–315
leukemia, 317–318
liver, 315–316
lung, 319–320
melanoma, 316–317
ovarian, 328–330
pancreatic, 327–328
prostate, 320–321, 325
treatment, nanocomposite based
chemotherapy, 330–331
photodynamic therapy, 331–332
sonodynamic therapy, 332–333
Candida albicans, 409
Carbapenem-resistant *A. baumannii* (CRAB), 404–405
Carbonaceous nanomaterials, 319
Carbonaceous polypyrrole (C/Ppy) concentrations, 317
Carbon-based materials, 178–179
Carbon dots (CDs), 407–408
Carbon nanocomposite, 268–269
Carbon nanotubes (CNTs), 360–361
Carboxymethyl cellulose (CCM), 434–435
Carcinogenesis, 264–265
Cardiac tissue engineering, 440–441
Cardiovascular diseases (CVDs)
application of NCs for diseases, 302
CCH-Based synergistic therapy, 327
Cellular toxicity in MOLT4 (T lymphoblast leukemia) cells, 318
Cellulose, 172–173
Cellulose nanocomposite hydrogels (CNH), 61–62
bone tissue engineering, 62

internal bleeding and wound healing reduction, 62–63
packaging industry, 62
Centaurea cyanus, 387
Center for Drug Evaluation (CDE), 500
Center for Food and Drug Inspection (CFDI), 500
Central Drugs Standard Control Organization (CDSCO), 490
Ceramic-ceramic-based nanocomposites, 376
Ceramic matrix composites (CMCs), 288
Ceramic matrix nanocomposite (CMNCs), 376
Ceramic matrix nanocomposites (CMNCs), 86, 252–253
basic methods of preparation, 94–96
polymer precursor method, 97
powder method, 96–97
vapor techniques, 97–99
Ceramic nanocomposite, 69
dental prosthesis, 69
promote bone/cartilage regeneration, 70
tissue engineering, 69
Cerium-titanium oxide, 331
Change being affected in 30 days (CBE-30), 502
Chemoresistant PANC-1 pancreatic cancer cells, 328
Chitin and polybutylene succinate, 155
Chitosan (CA), 171–172, 377
Chitosan/glycerophosphate (CS/GP), 387
Chitosan nanocomposite hydrogels (CsNH), 60
antibacterial activity, 61
bone regeneration, 60–61
cancer chemotherapy, 61
Chitosan-silver nanocomplex (Ch-AgNPs), 409
Chronic obstructive pulmonary disease (COPD)
multifunctional NCs, 291–292
particle shape, 291–292
particle size, 291
target, 292
nano-theranostic for, 293–294
NCs
multifunctional, 291–292
role and application, 289–290
simportance, 290–291
Chronic wounds
application of NCs, 298–299
CMCS-glutathione-glycyl sarcosine (CMCG-GS), 130
Collaborative Opportunities for Research Excellence in Science (CORES) program, 507
Collagen, 174–175
Colloidal powder/films, 7
Colorectal cancer, 326–327
Common technical document (CTD), 497–498
Comparability protocol (CP), 502–503
Composite materials, 86
Composites, 40
Compound curcumin (CUR), 314
Conjugated bio-nanomaterials, 329
Contaminants
removable, 254–255
Copper-cysteamine (Cu-Cy), 331
Copper oxide nanoparticles (CuONPs), 402, 403
Co-precipitation, 7
Coprecipitation approach, 155
Covalent organic framework (COF) nanocomposites, 434
cRGD-Labeled brain tumor cell membrane, 326
Critical micellar temperature (CMT), 24
Crosslinking networks (c-PLMC), 360–361
CS Oligosaccharide (CSO)-modified paclitaxel (PTX)-loaded liposomes, 319–320
CS-Sodium alginate-polyethylene glycol-crocin NCs (CSP-Cr-NCs) formulation, 314
Curcumin nanocomposites, 122
CUR-loaded PAOx NPs, 330–331
Cur-loaded ZnFe2O4@mZnO nanocarriers, 433

D

Decursin (D), 380
Decursinol angelate (DA), 380

Degradation process, 180
Dendritic cell vaccines (DCV), 435
DEN-Induced liver cancer, 315
Density functional theory (DFT) approach, 413
Dentistry
  dental prosthetics, 3D printed, nanocomposites use
    creating strong and lightweight dental crowns and bridges, 351–352
    customized orthodontic aligners using, 353–354
    implants with enhanced functionality, 353
  dental restorative materials, NC use in, 417
Dexamethasone, 229
Dextran nanocomposite hydrogels (DNHs), 58
  bone regeneration, 59
  cardiac tissue repair and regeneration, 58
  glucose sensor, 58
Diabetes
  application of NCs for diseases, 302–303
  ulcers, application of NCs for diseases, 300–301
Diethylnitrosamine (DEN)-induced hepatic carcinoma (HCC), 315
Direct ink writing, 348
Director-General Joint Research Center (DG JRC), 508
Distinctive materials used in biodegradable and nondegradable NCs, 167
  biomedical applications
    antimicrobial activity, 191–193
    bioimaging and biosensors, 190–191
    drug delivery, 185–188
    tissue engineering and regeneration, 188–190
    wound-healing application, 193–195
  carbon-based materials, 178–179
  cellulose, 172–173
  chitosan (CA), 171–172
  collagen, 174–175
  emerging trends and future directions, 205–206
  features and applications over last five years, 196–203
  gold (Au) nanoparticles, 178
  hybrid approaches, 195, 204–205
  hydroxyapatite (HA), 175–176
  nanoclay, 176–177
  polycaprolactone (PCL), 173–174
  polyhydroxyalkanoates (PHA), 168–169
  polylactic acid (PLA), 168
  silicon dioxide (SiO2) nanoparticles, 178
  starch, 169–171
  strategic combination, 195, 204–205
  titanium dioxide (TiO2), 177–178
  zirconia (ZrO2) nanoparticles, 178
Dopamine-coated Fe_3O_4 nanoparticles, 327
Doxorubicin (DOX), 317
Doxorubicin in glioma targeted drug delivery, 234
Doxorubicin-targeted drug delivery, 233–234
Drug delivery systems, 111
  biomedical applications, 185–188
  inhalation nanocomposite, 120–121
    dry powder inhaler (DPI), 121–122
  intranasal, 123–125
  ocular nanocomposite
    contact lens, 130–131
    eye depot, 131
    eye drop, 128, 130
    intraocular injection, 132
  oral nanocomposite, 113
    amorphous nanocomposites, 113–114
    exosome nanocomposites, 114
    films, 114–115
    hydrogel, 115–116
    sponges, 115
  parenteral nanocomposite, 117
    intracortical nanocomposite-based, 118
    intramuscular nanocomposite-based, 116, 118
    intravenous nanocomposite-based, 118–119
    subcutaneous nanocomposite-based, 119–120
  topical nanocomposite, 125
    hydrogel, 126
    microneedle patch, 127
    nanocomposites hydrogel film, 126

nanofiber, 127–128, 129
use of nanocomposites in, 437–439
combination therapy, 434
controlled and targeted drug release, 432–434
controlled porosity for sustained and targeted drug release, 354–355
crossing biological barriers, 434–435
design of nanocomposite filaments for fused deposition modeling printing, 355–356
potential for 4D printed nanocomposite, 356
vaccine delivery, 435–436
vaginal, 132–133
Drug Price Competition and Patent Term Restoration Act of 1984, 485
Hatch-Waxman Amendments, 485–486
Drugs Controller General of India (DCGI), 490
Dry powder inhaler (DPI), 121–122
Dual/multiple-responsive nanocomposite, 17–24

E

EATR technique, 321
Elaeagnus umbellata, 384
Electronic CTD (eCTD), 497–498
Engineered bioremediation, 274
bioslurping, 275
biosparing, 275
bioventing, 275
Enhanced drug delivery systems
nanocomposite system for treating cancer, 230–231
polymeric nanocomposites, 228
siRNA, nanocomposite-mediated, 229–230
using nanocomposites, 227, 229
Enterobacter species, 402
Enterococcus faecium, 402
Epoxy/Ag-TiO2 polymer NCs, 405–406
Escherichia coli, 380
ESKAPE pathogens, 402–406
Esophageal cancer (EC), 314–315
Etoposide (ETP), 386
Eudragit® L100, 435
European Chemical Agency (ECA), 508
European Union's REACH regulation, 510
Exopolysaccharides (EPS), 409
Ex situ bioremediation, 276–277
biopiles, 276
bioreactors, 276–277
land farming, 276
Eye depot, 131
Eye drop, 128, 130

F

Ferroptosis, 319
Field-effect transistors (FETs), 449
Flavobacterium, 273
Fluorouracil-loaded graphene oxide (GO), 317
Folic acid (FA or vitamin B9), 314
Folic acid receptor (FAR), 314
Food and Drug Administration Amendments Act of 2007 (FDAAA), 503–504
Free 6-carboxyfluorescein (FAM) labeled DNA (H1-FAM/H2-FAM), 327
Fungal-based CuONPs, 404
Fused deposition modeling, 347

G

Gadolinium oxide (GdO), 321
Gallium-doped carbon dots (Ga@CDs), 408
Gamma-glutamyl transferase (GGT), 315–316
Gel, 91
Gelatin-curcumin-coated NCs (GC-NCs), 419–420
Gelatine, 378
Gene therapy
recycled waste-derived nanocomposites, used in, 152–153
Gentamicin, 16
Glioblastoma. *see* Brain cancer
Glioblastoma multiforme (GBM), 325
Glucose transporter 1 (GLUT1), 410
Glutathione (GSH), 23
Gold clustoluminogens (AIE-Au) with x-ray radiation, 331

Gold nanoparticles, 361
Gold (Au) nanoparticles, 178
Gold/silver coupled carbon quantum dot NCs (Au/Ag-CQD-NC), 326
G-Protein chemokine receptor CXCR4, 318
Graphene, 255, 416
Graphene NCs, 316
Graphene oxide (GO)-chitosan (CHI) NCs, 417
Green platinum nanoparticles (PtNPs) using GO (GO-PtNPs), 321
Green synthesis, 385

H

Health Canada's Health Products and Food Branch (HPFB), 489
Heavy metal pollution, 260–261
Hench glass, 237
Hepatic carcinoma, 385
Herbal nanocomposites
  biomedical applications, 373–374, 376
    advantages and disadvantages, 389–392
    anti-bacterial activity, 384–385
    anti-microbial activity, 383–384
  biopolymers used, in preparation
    albumin, 378
    alginate, 377–378
    chitosan, 377
    gelatine, 378
  catalytic effect, 387
  hepatic carcinoma, 385
  lung cancer therapy, 386–387
  medical textile
    antibacterial activity, 380–381
    anti-oxidant, anti-UV, and anti-bacterial, 381–382
    carrier for plant extract, 382–383
  orthopedic implants, 386
  ovarian cancer therapy, 379–380
  packaging to increase shelf life, 387, 389
  tissue engineering, 387
High-energy ball milling, 102
Human serum albumin (HAS) nanocomposites, 386
Hybrid bio-NCs, 412
Hybridization chain reaction (HCR) amplification, 327
Hybrid nanocomposites, 66–67
  cancer diagnosis and therapy, 67
  detection of toxin, biosensor for, 67
  gene delivery, 68
  nerve substitute, 68
  promote bone and cartilage regeneration, 67–68
  therapeutic enzyme immobilization, 68
Hydrogel nanocomposites, 115–116
Hydrogels, 316
Hydrophobic NBs (Dox-FePt@NB) nanocomposite, 435
Hydroxyapatite (HA), 175–176
Hydroxyapatite nanocomposites (HA), 55
  orthopaedic and dentistry settings, 55–56
  tissue engineering and regenerative medicine, 55

I

Ibuprofen controlled release, 232
Implantable biosensors utilizing 3D printed nanocomposites, 358
Implantable devices by 3D printing, 348
  cardiac valves and stents using nanocomposites, 351
  improved biocompatibility, 349–351
  osseointegration in patient-specific implants, 349–351
  prospects, nanocomposite scaffolds in tissue engineering and neural implants, 352
Infiltration, 101
Inhalable APBA-HSA nanocomposites, 386–387
Inhalation nanocomposite drug delivery systems, 120–121
  dry powder inhaler (DPI), 121–122
  for nebulization, 122–123
Inhibited SQ aggregation, 328
Inorganic nanoparticles (NPs), 41
In situ bioremediation, 274–276
  engineered bioremediation
    bioslurping, 275
    biosparing, 275
    bioventing, 275
  intrinsic attenuation/natural attenuation, 274

In situ intercalative polymerization, 376
Interleukins (IL-1 beta and IL-6), 326
International Council on Harmonization (ICH) Q12 principles, 503
Interpenetrating polymer network (IPN), 16
Intranasal drug delivery, 123–125
Intraocular injection, 132
Intra-tumoral administration of GEM-POSS-SQ NPs, 328
Intrinsic attenuation/natural attenuation, 274
Investigational new drug (IND) application, 484–485
IPM@AgNPs-PEG-NOTA-NCs, 404–405
Iridium-based photosensitizer (IP), 316
Iron oxide (IO), 321
  and copper (Fe_3O_4@CS-Starch/Cu), 329
  NC (IO/GdO), 321
Iron oxide nanoparticles (IONPs), 328

K

Kinetics of stability, 183–185
Klebsiella pneumonia, 402

L

Land farming, 276
Large mesoporous silicon nanoparticles (LMSNs), 114
Law for Ensuring the Quality, Efficacy, and Safety of Drugs and Medical Devices, 490
Layered double hydroxides (LDHs), 407
Leishmaniasis, 407–408
Leukemia cancer, 317–318
Ligustrazine hydrochloride, 410–411
Liver cancer, 315–316
Localized drug delivery, 355
Lung cancer, 319–320
  therapy, 386–387

M

Magnesium phosphate ceramics (MPCs), 238–239
Magnetic nanocomposite (MNC) materials, 328
Magnetic nanocomposites, 289
Magnetic-sensitive nanocomposites, 23
Malaria, 408–411
Maleate Gellan Gum-Silk-Sericin-Chitosan (MA-GG-SS-CS) NCs, 401–402
Marketing authorization application (MAA), 488
MBG Microspheres (MBGMs), 241
Melanoma cancer, 316–317
Membranes nanocomposite, 269–270
Mesenchymal stem cells (MSCs), 412–413
Mesoporous bioglasses (MBGs), 241
Mesoporous silica nanoparticles (MSNs), 25
Mesoporous titanium dioxide (MTN) NP, 331
Metallic nanocomposites (Metal/PNCs), 63
  anti-adhesive membrane and wound dressing, 63
  antimicrobial activity, 65
  bone grafting, 65
  cancer chemotherapy, 64
  drug delivery system, 64
  gene delivery, 64
  trigger autophagic cell death, 63–64
Metal matrix nanocomposites (MMNC), 86, 288
  basic methods of preparation, 99
    colloidal method, 103
    high-energy ball milling, 102
    infiltration, 101
    RSP, 101–102
    spray pyrolysis, 100
  modern methods for the preparation
    biological methods, 105–106
    layer-by-layer (LPL) technique, 103–104
    plant-mediated biosynthesis, 106
    template-assisted methods, 104–105
Metal matrix nanocomposites (MMNCs), 252, 375–376
Metal-organic framework (MOF), 332
Metal-organic nanocomposite materials, 267–268
Metal oxides, 416
Metformin (MET)
  controlled release, 233
Methicillin-resistant *Staphylococcus aureus* (MRSA), 401
Methotrexate
  targeted drug delivery, 234–235

Methoxy poly (ethylene glycol), 24
Microorganisms
role in maintaining biological equilibrium, 272
aerobic circumstances, 273
anaerobic amphibious bacteria, 273
Microwaves, 7
Minimum bactericidal concentration (MBC), 403–404
Minimum inhibitory concentration (MIC), 401
Ministry of Health, Labor and Welfare (MHLW), 490, 506
Mitochondrial membrane potential (MMP), 314
Molybdenum and sulfur (MoS_2) nanoparticles (NPs), 315
Molybdenum-containing enzymes, 314–315
MUC4/CEACAM6/CD44v6-expressing PDAC cells, 328
Multiwalled carbon nanotubes (MWCNTs), 451
MXene@Hydrogel nanocomposite, 433
Mycobacterium, 273
Mycobacterium tuberculosis, 400

N

Nanoclay, 176–177
Nanocomposites (NCs), 1–2, 40–41, 86, 250–251, 311–312, 342, 397–398, 430–432
additive manufacturing, challenges for utilization, 362–363
advantages, 29
drug delivery, 31
drug targeting, 31
gas barriers, 30
mechanical strength, 29–30
thermal/electric conductivity, 30
thermal stability, 30
amorphous and semicrystalline thermoplastic polymer, 51–52
medical devices, 52
packaging materials, 52
tissue engineering, 52
ANH, 59–60
antimicrobial agents, as
ESKAPE pathogens, 402–406
leishmaniasis, 407–408
malaria, 408–411
tuberculosis, 400–402
applicability, 313, 388, 398–400
various human cancers, in, 313
application for diseases, 285–288
bedsores/pressure ulcers, 299–300
bone tissue regeneration, 296
cancer therapy, 294–295
cardiovascular diseases (CVDs), 302
chronic wounds and biofilms, 298–299
combinational therapy for AD, 296–297
COPD, 289–290
diabetes, 302–303
diabetic ulcers, 300–301
ocular illnesses, 301–302
Parkinson's disease (PD), 303–304
treatment of rheumatoid arthritis, 297–298
applications, 27–28
biomedical, 27–28
application through the pulmonary route, 124
biodegradable nanostructured
dental implants, 53
food packaging, 53
gene delivery, 53
tissue engineering and regenerative medicines, 52–53
biomedical applications, 43, 73–74
cell death, 329
ceramic, 69–70
chemotherapy based, 330–331
classification
CMNCs, 252–253, 376
compound involved, 44
material used for its preparation, 44
micro-structural makeup basis, 43–44
MMNCs, 252, 375–376
PMNC, 376
polymer matrix nanocomposites, 253
presence/absence of polymeric material in, 44–46

cleanup and wastewater treatment, 251
CNH, 61–62
CsNH, 60–61
disadvantages
 particle aggregation, 31–32
 sedimentation, 32
DNHs, 58–59
drug delivery systems by using, 111
 oral (*See* Oral nanocomposite drug delivery systems)
enzyme-responsive polymer based, 53–54
 biomedical implants, 54
 biosensors, 54
 drug delivery systems, 54
 environmental remediation, 55
 HA (*See* Hydroxyapatite nanocomposites (HA))
fundamentals
 biocompatibility, 344
 biological properties, 345
 mechanical strength, 344
 porosity, 344–345
 thermal and electrical conductivity, 345
future prospects, 304–305
history, 3–4
hybrid, 66–68
hydrogel, 56–57
 ANH, 59–60
 CsNH, 60–61
 DNHs, 58–59
hydrogel film, 126
importance of, 42
limitations, 42–43
mechanism
 biodegradable, 26–27
 dual pH/thermal-response composite, 17–24
 pH-responsive, 10–15
 stimuli-responsive, 9–10
 temperature-responsive, 15–17
 ultrasound-responsive, 25–26
Metal/PNCs, 63–65
method of preparation
 ball milling, 8
 co-precipitation, 7
 hydrothermal, 7
 melt blending, 6
 microwave synthesis, 7–8
 in situ method, 6
 sol-gel, 6–7
 solution dispersion, 9
 sonochemical method, 8
 vapor-phase reaction techniques, 9
methods for preparation for biomedical applications, 85
microneedle patch, 127
nanofiber, 127–128
natural polymer-based
 agriculture and analytical research, 49
 bioimaging, 48
 biomedical sensor and devices, 49
 bone tissue engineering, 48
 dental applications, 47
 drug delivery, 46–47
 food packaging industry, 49
 gene delivery, 48
 ophthalmology, 47
 regenerative medicine, 48–49
 skin regeneration, 47
 surgeries, 47–48
 tissue engineering, 48–49
 wound healing, 47
photodynamic therapy based, 331–332
PNH, 71–72
polymer act as adsorbents for contaminants, 253
 dye removal, 254
QD-based, 70–71
recycled waste materials for, 144–145
regulatory challenges
 biocompatibility, 453–454
 environmental impacts, 454
 long-term effects, 454–455
regulatory framework, 455
remediation of air pollution, 270
remediation of water pollution, 266
 carbon nanocomposite, 268–269
 membranes, 269–270
 metal-based, 266–267
 metal oxide, 267–268
 polymer nanocomposites (PNCs), 269
SNC, 65–66
sonodynamic therapy based, 332–333

Poly (amidoamine) dendrimer (PAMAM(G4)), 416–417
Polyether-ether-ketone/bioglass (PEEK/BG), 419
Polyether-ether-ketone/forsterite (PEEK/FT), 419
Polyethylene glycol (PEG), 41, 294
Polyethylene glycol-modified IONPs (IONPs-PEG) with triple scAbs, 328
Polyethyleneimine (PEI), 294
Polyethylene oxide (PEO), 294
Polyglycolic acid (PGA), 313
Polyhedral oligomeric silsesquioxane (POSS)-based nano-platform (GEM-POSS-SQ), 327
Polyhedral oligomeric silsesquioxane-polycaprolactone (POSS-PCL), 412–413
Polyhydroxyalkanoates (PHA), 168–169
Polylactic acid (PLA), 168
Poly lactic acid (PLA) nano-hydroxyapatite (n-HAp), (PLA-nHAP NCs), 419
Poly lactic-co-glycolic acid (PLGA), 41, 313
Poly-L-lysine functionalized rGOs, 319
Polymer-based nanoparticles (PBNs), 435–436
Polymeric nanocomposites, 41
Polymeric NCs, 320
Polymer matrix nanocomposites (PMNCs), 86, 253, 288–289, 376
  basic methods of preparation, 87–88
    electrospinning, 92–94
    melt method, 89–90
    in situ polymerization, 88–89
    sol-gel method, 91–92
    solution mixing, 90–91
  structure and advantages, 87
Polymer nanocomposites (PNCs), 4–5, 45, 269
  dye removal, as adsorbent, 254
Polypyrrole, 415
Polypyrrole-based adsorbents, 254
Polythiophene, 415
Polyvinyl alcohol (PVA), 294
Porosity-structured silk fibroin networks, 155
Post-marketing commitments (PMCs), 503
Post-marketing requirements (PMRs), 503
Prescription Drug User Fee Act (PDUFA) 1992, 487
4DPrinting, nanocomposites in, 359
  materials in biomedical research, 360–362
3D and 4D Printing technology, 342–343
  biological applications, 343
  materials
    ceramic-based, 346
    polymer-based, 345–346
  nanocomposites, applications in
    direct ink writing, 348
    fused deposition modeling, 347
    implantable devices, in, 348–352
    SLA, 346–347
Prior approval supplement (PAS), 502
Propranolol control release, 231–232
Prostate cancer, 320–321, 325
Prostate-specific membrane antigen (PSMA) protein, 321
Pseudomonas, 273
Pseudomonas aeruginos, 401
Pseudomonas aeruginosa, 402
Pseudomonas aeruginosa (5994 NLF), 384
Pullulan nanocomposite hydrogels (PNH), 71
  antibacterial agent, 72
  dressing membranes in wounds, 72
  preservation of biological samples, 72

Q

Quality target product profile (QTPP), 472
Quantum dot nanocomposite, 70
  drug carriers, 71
  photodynamic cancer therapy, 70–71
  in vivo imaging, 70
Quantum dots (QD), 230
Quercetin (QC), 13
Quercetin (QCT), 435

R

Raf family of kinases, 316
Rapid solidification process (RSP), 101–102
Rapid urbanization, 251
  advantages over other materials, 251–252

REACH (registration, evaluation, authorization, and restriction of chemical substances), 508
Reactive oxygen species (ROS), 314
Recycled waste-derived nanocomposites, 141–142
biomedical applications, 149–150
antimicrobial activity, 156–157
drug delivery, 150–152
gene therapy, 152–153
tissue engineering, 153–155
materials for, 144–145
need for, 143–144
preparation methods of, 145–146
extrusion, 147
hand lay-up/spray-mixing process, 148
melt mixing, 146
other methods, 149
in situ polymerization, 146–147
sol-gel process, 147
solution mixing, 146
Reduced GO-silver NCs (rGO-Ag), 329
Reference listed drugs (RLD), 487
Regeneration, 188–190
Regulatory Science Research Program (RSRP), 507
Resazurin-based studies, 402
Re-sponsive oncotherapy, 327
Resveratrol (Rv), 326
Resveratrol (RSV)-based NPs, 380
Rheumatoid arthritis
application of NCs for, 297–298
Rhodococcus, 273
Richard Feynman's theory, 417
Risk evaluation and mitigation strategies (REMS), 501
Risk management plans (RMPs), 506
R-Phycoerythrin (RPE), 329
Rubus spp., 384
Rv-Soy Protein (Rv-Sp) wet granules, 326

S

Scaffolds, 239
cartilage tissue engineering, used for, 239–240
Schedule Y of the Drug & Cosmetics Rules, 1945, 491
Scintillating nanoparticles (ScNP), 331
Selenium/CS/PEG/allyl isothiocyanate NCs (SCPg-AI-NCs), 315
SELEX technique, 321
Senna auriculata, 380
Sericin-chitosan-doped maleate gellan gum (MA-GG), 401
Silica nanocomposites (SNC), 65–66
cancer chemotherapy, 65–66
packaging material, 66
photodynamic therapy for cancer, 66
Silicon dioxide (SiO2) nanoparticles, 178
Silver nanoparticles (AgNPs), 116
Single-walled carbon nanotubes (SWCNTs), 451
Single-walled CNTs (SWCNTs), 181
Skin tissue engineering, 442
Soft tissue repair
applications of bioceramics in
injuries with bioactive glasses, 241
pulmonary tissue regeneration, bioglasses for, 242–243
regeneration of nervous tissue, bioglasses for, 242
Sol, 91
Solvothermal method, 7
Sonodynamic therapy (SDT), 332–333
Sorafenib (SOR), 316
Sphingomonas, 273
Spray pyrolysis, 100, 376
Standardization Administration of China (SAC), 510
Staphylococcus aureus, 380, 402
Starch, 169–171
Starch/ZnO coating, 314
Stereolithography (SLA), 346–347
Stimuli-responsive nanocomposites, 9–10
Study Tagging Files (STF), 497–498
Sulfite oxidase, 314–315
Superparamagnetic iron oxide nanoparticles, 350
Synthetic organic chemicals/dyes, 260

T

TAMs-Activated Cu2O@CaCO3@HA (CCH) NCs, 327
Targeted drug delivery
nanocomposites in, 233

doxorubicin in glioma, 234
doxorubicin-targeted, 233–234
methotrexate, 234–235
Technical Committee 279, 510
Telmisartan (TEL), 113–114
Temperature-responsive nanocomposite, 15–17
Therapeutic agents
future perspectives and challenges in, 243–244
nanocomposites for controlled release and targeted delivery, 231
aceclofenac controlled release, 232–233
controlled release of metformin, 233
ibuprofen controlled release, 232
propranolol control release, 231–232
Therapeutic Product Directorate (TPD), 489
Thermoplastic filaments, 347
Thermo-responsive nanocomposites (TNC), 349–350
Three-component HA/CH-gelatin network composites, 154
Tissue engineering, 188–190, 235
biomaterials employed in scaffold construction
bioactive glass, 237
bioactive glass-ceramics, 237
calcium phosphates (CaPs), 237–238
magnesium phosphate ceramics (MPCs), 238–239
biomimetic approaches utilizing, 239
characteristics and criteria for optimal nanocomposite scaffolds
biocompatibility, 235
degradability, 236
mechanical properties, 236
nutrient transport, 236
pore structure, 236
surface characteristics, 236
nanocomposite hydrogels, 240–241
recycled waste-derived nanocomposites, used in, 153–155
scaffolds for cartilage, 239–240
Tissue engineering, in NC, 411–414, 436, 440
bone, 441–442
cardiac, 440–441
nervous, 442–443
skin, 442
Tissue Engineering and Regenerative Medicine (TERM)
nanocomposites for, 235–236
Titanium dioxide (TiO2), 177–178
Titanium implantation, 420
Topical nanocomposites hydrogel, 126
Traditional drug administration methods, 354–355
Tridax procumbens, 381–382
Triple single-chain antibodies (scAbs) and IONPs, 328
Tuberculosis (TB), 400–402
Tumor-associated microenvironment (TAMs), 327
Tumor necrosis factor-alpha (TNF-alpha) in HCT-116 cells, 326
Tumor-targeting ligand (PDGFB-PEG), 321
TUNEL- and Ki-67-positive cells, 328
Two-nozzle electrospinning, 386
Tyrosine kinase inhibitor, 316

U

Ulex europaeus agglutinin 1 (UEA-1), 435
Ultrathin platinum (Pt) nanozyme, 318
Ulva/Ag/cellulose NCs, 418
US-FDA-Approved Fe_3O_4 (iron oxide) NPs, 325

V

Vaginal nanocomposites drug delivery systems, 132–133
Vancomycin-resistant *Enterococcus* (VRE), 401
Vanillin, 321
Vapor-phase reaction techniques, 9
Vascular endothelial growth factor (VEGF) receptors, 316
Viscos sintering, 92

W

Water-dispersible ferrimagnetic iron oxide nanocubes (wFIONs), 325

Water pollution
nanocomposites remediation, 266
carbon nanocomposite, 268–269
membranes, 269–270
metal-based, 266–267
metal oxide, 267–268
polymer nanocomposites (PNCs), 269
Water purification
mechanism of action
adsorption, 255–256
photocatalysis (*See* Photocatalysis)
Wound healing, applications of
nanocomposites in, 444
antibiotics-based, 446–447
metal and metal oxide-based, 444
nonmetallic-based, 445
polymer-based, 445–446
recently developed, 448

X

Xanthine oxidase, 314–315
Xanthomonas spp., 404

Z

Zeolitic imidazolate framework 8-coated Prussian blue (ZIF-8@PB), 435
Zinc-layered double hydroxides (ZLDH), 401
Zinc oxide NCs (ZnO-NCs), 384, 409
Zinc phthalocyanine, 316
Zirconia (ZrO2) nanoparticles, 178
ZnFe2O4@mZnO-Ox-p-g-C3N4 nanocarriers, 433
ZnO-TiO_2-CS-Amygdalin NCs, 318

For Product Safety Concerns and Information please contact our
EU representative GPSR@taylorandfrancis.com Taylor & Francis
Verlag GmbH, Kaufingerstraße 24, 80331 München, Germany